ADVANCES IN X-RAY ANALYSIS

Volume 6

ADVANCES IN X-RAY ANALYSIS

Volume 6

Edited by

William M. Mueller and Marie Fay

Proceedings of the Eleventh Annual Conference on Application of X-Ray Analysis Held August 8-10, 1962

Sponsored by
University of Denver
Denver Research Institute

ℚℙ

Distributed by
Plenum Press, New York

ACKNOWLEDGMENT

The University of Denver and Plenum Press are grateful to the following publishers for permission to reprint the illustrations indicated:

To Academic Press for Fig. 1 on p. 277, which originally appeared in *Advances in Electronics and Electron Physics, XIII* (1960), and Fig. 3 on p. 278, Figs. 11 and 12 on p. 285, and Fig. 15 on p. 288, which originally appeared in *Proceedings of the Third International Symposium on X-Ray Optics and X-Ray Microanalysis* (1963).

To the American Institute of Physics for Fig. 10 on p. 284, which originally appeared in the *Journal of Applied Physics* **32** (1961).

To the American Society for Metals for Fig. 3 on p. 48, Fig. 4 on p. 50, Fig. 5 on p. 51, Fig. 6 on p. 52, Fig. 7 on p. 53, and Fig. 8 on p. 54, which originally appeared in *Theory of Alloy Phases* (1956).

To the Czechoslovakian Academy for Figs. 1 and 2 on p. 3, which originally appeared in the *Czechoslovakian Journal of Physics* **4** (1954).

To the Elsevier Publishing Company for Fig. 7 on p. 281 and Fig. 8 on p. 282, which originally appeared in *X-Ray Microscopy and X-Ray Microanalysis* (1960).

To The Institute of Physics and the Physical Society for Figs. 4 and 5 on p. 280, which originally appeared in the *British Journal of Applied Physics* **11** (1960).

To the International Union of Crystallography for Fig. 3 on p. 4, Fig. 4 on p. 5, and Fig. 10 on p. 58, which originally appeared in *Acta Crystallographica* **2** (1949), **10** (1957) and **11** (1958), respectively.

To La Revue de Metallurgie for Fig. 9 on p. 283, which originally appeared in *La Revue de Metallurgie* **57** (1960).

To Metallurgia for Fig. 6 on p. 281, which originally appeared in *Metallurgia* **61** (1960).

To De Nederlandse Vereeniging voor Electronen Microscopie for Fig. 14 on p. 288, which originally appeared in *Proceedings of the European Regional Conference on Electron Microscopy, Delft*, 1960 (1961).

ISBN 978-1-4684-8785-5   ISBN 978-1-4684-8783-1 (eBook)
DOI 10.1007/978-1-4684-8783-1

FOREWORD

The torrential flow of technical information appearing in the world sources of literature is creating concern and apprehension among scientific people at all levels. It is extremely difficult to keep abreast of information flowing into a specific field. It is nearly impossible to transcend traditional confines of individual disciplines and put to effective use all pertinent information which stems from continuously increased trans-disciplinary research. At the same time the researcher is faced with problems of increasing complexity, with the requirement for new knowledge and new techniques, and must frequently, with little time, bridge the gap between his own sphere of experience and a sometimes apparently unrelated new interest. This is readily observed with X-ray analysis, where the chemist, physicist, metallurgist, and engineer are each faced with the solution of problems peculiar to specific disciplines but where solutions frequently correlate with the particular needs of the others.

The Annual Conference on Applications of X-Ray Analysis and the subsequent *Advances in X-Ray Analysis* contribute to better understanding of multidisciplinary accomplishments; they are a ready source of information for the researcher who must undertake an abrupt change in emphasis for new objectives. The scope of this conference is broad—concerning itself, as it does, with latest developments in high-temperature and cryogenic techniques, phase equilibria, crystal structures, polymers, microprobes, and new developments in instrumentation. Forty-four carefully screened papers delivered before the conference are presented, not as summary or abstract, but in full, with ample time for formal and informal discussion. The proceedings which follow consolidate the accomplishments of the conference, and with pertinent references incorporated, provide an appropriate and useful source book for important developments and new knowledge of the past year.

The most fruitful and productive function of the conference is the free interchange of information and knowledge between researchers, often stimulated by formal sessions, and reduced to a common denominator through informal discussions. Unfortunately much of this valuable acquisition of knowledge is difficult to reduce to print, but is carried away in the minds of people for immediate application in their own laboratories.

A satisfactory solution to the problems of information communication and retrieval is still remote. Certainly, knowledge obtained through informal discussions, unless put to use and ultimately reported, is available only to a few. But informal acquisition of knowledge, subtly blended with the wealth of information contained in the proceedings and with the new ideas discussed during the conference, engenders continued progress of X-ray analysis.

James P. Blackledge

PREFACE

The Eleventh Annual Conference on Applications of X-Ray Analysis sponsored by the University of Denver was held August 8, 9, and 10, 1962, at the Albany Hotel in Denver, Colorado. There were approximately 250 participants, several of whom came from foreign countries. Forty-four papers were presented and all of these are included in these Proceedings.

Financial assistance provided by the United States Office of Naval Research permitted the participation of four distinguished scientists from abroad, namely, Professor F. Laves, Institut für Kristallographie and Petrographie, Zurich, Switzerland; Professor P. M. deWolff, Technisch Physische Dienst T.N.O. en T.H. Delft, Netherlands; Dr. J. V. P. Long, University of Cambridge, Cambridge, England, and Dr. Jean Thomson, Imperial College of Science and Technology, London, England.

Much of the success of the conference was due to the following individuals who chaired the various technical sessions:

E. W. Filer, General Electric, Cincinnati, Ohio

Richard Marburger, General Motors, Warren, Michigan

Robert E. Michaelis, National Bureau of Standards, Washington, D.C.

Marion Semchyshen, Climax Molybdenum of Michigan, Detroit, Michigan

H. F. Quinn, IBM Federal Systems Division, Owego, New York

C. Manning Davis, International Nickel Company, Bayonne, New Jersey

George L. Clark, University of Illinois, Urbana, Illinois

Our most sincere thanks to the following persons who contributed generously to the planning and directing of a very successful conference: Mr. Robert A. McCune, Mrs. Esther Marie Capps, Mr. James P. Blackledge, and the many DRI staff members who served as meeting aids. Also, we extend our gratitude to Mrs. Mildred Cain for her invaluable help in the preparation of the manuscript and Mrs. Nedra Jenkins for assistance in the preparation of the subject index.

William M. Mueller
Marie J. Fay

CONTENTS

Indexing of Powder Diffraction Patterns.................................... 1
    P. M. deWolff
On the Use of a Modified Radial Distribution Analysis for Indexing Powder
    Patterns.. 18
    A. F. Berndt
The X-Ray Investigation of Preprecipitation in Supersaturated Solid Solutions .. 25
    A. Lutts
Factors Governing the Structure of Intermetallic Phases..................... 43
    F. Laves
The Zirconium–Iron System.. 62
    F. N. Rhines and R. W. Gould
X-Ray Diffraction Studies on the Titanium–Nickel System................... 74
    John V. Gilfrich
Helium Path Diffractometry and its Application to Determination of Retained
    Austenite and Macrostress in Steel..................................... 85
    R. A. McCune
The Crystal Structure of $ThPd_4$.. 91
    J. R. Thomson
The Effect of Cold-Work on the X-Ray Diffraction Pattern of a Copper–Silicon–
    Manganese Alloy... 96
    D. O. Welch and H. M. Otte
Lattice Spacings in Some Transition Metal Terminal Solid Solutions........... 121
    Henry Chessin, Sigurds Arajs, and Donald S. Miller
X-Ray Measurement of the Static Lattice Distortion in the Solid Solution of
    Oxygen in Titanium.. 136
    F. R. L. Schoening and F. Witt
Precision X-Ray Diffractometry Using Powder Specimens.................... 142
    L. F. Vassamillet and H. W. King
The Characterization of Large Single Crystals by High-Voltage X-Ray Laue
    Photographs... 158
    H. S. Peiser and E. P. Levine
Diffraction Effects from Irradiated Aluminum Single Crystals................. 164
    H. E. Kissinger
Oriented Single Crystals of Aluminum for X-Ray Analysis................... 172
    Joseph M. Dhosi, Charles P. Gazzara, and Raymond M. Middleton
An X-Ray Determination of Debye–Waller Factors for Cu_2O and UO_2 and the
    Atomic Scattering Factor for Cu in Cu_2O............................. 177
    C. J. Sparks, Jr., and B. S. Borie
An X-Ray Study of the Structure of the Alkaline Earth Oxide Cathode........ 185
    Paul Lublin
Applicability of Routine Methods of Crystallite Size Analysis 191
    Robert C. Rau
Low-Temperature Transitions of Some Ammonium Salts..................... 202
    M. Stammler, D. Orcutt, and P. C. Colodny

X-Ray Diffraction Analysis of Aerosols from Exploding Wires................ 210
    A. G. Barkow, F. G. Karioris, and J. J. Stoffels

The Preparation of Pole Figures for Polymers by Computer Techniques........ 223
    J. W. Jones

Recent Developments in the Measurement of Orientation in Polymers by X-Ray
Diffraction... 231
    Zigmond W. Wilchinsky

Measurement of the Lattice Constants of Neon Isotopes in the Temperature
Range 4 –24°K... 242
    L. H. Bolz and F. A. Mauer

A High-Temperature X-Ray Diffractometer Furnace Utilizing High-Frequency
Heating.. 250
    E. W. Franklin and S. M. Lang

Some X-Ray Generator Characteristics to Consider in Order to Realize the
Optimum Stability and Reproducibility of Intensity Measurements........... 262
    R. Torkildsen

Specifications and Performance Data for the ARL Electron Microprobe X-Ray
Analyzer... 268
    L. P. O'Brien

Recent Advances in Electron-Probe Analysis............................. 276
    J. V. P. Long

Oscilloscope Readout of Electron Microprobe Data........................ 291
    Kurt F. J. Heinrich

Practical Applications of Filters in X-Ray Spectrography................... 301
    Merlyn L. Salmon

The Daily Use of a Basic Norelco X-Ray Spectrograph in an Aluminum Reduction
Laboratory... 313
    W. B. Eastman

Experiences of X-Ray Analyses in Steel and Ferro-Alloy Production........... 328
    J. Baecklund

Determination of Aluminum in Iron–Aluminum Alloys by Vacuum X-Ray
Fluorescence... 339
    J. C. Wagner and F. R. Bryan

Continuous Determination of Zinc Coating Weights on Steel by X-Ray Fluorescence.. 345
    James A. Dunne

Some Aspects of Nondestructive X-Ray Spectrochemical Analysis of Alloys..... 352
    J. R. Rickenbach, Jr.

Sodium and Magnesium Fluorescence Analysis—Part I: Method.............. 361
    Burton L. Henke

Sodium and Magnesium Fluorescence Analysis—Part II: Application to Silicates 377
    A. K. Baird, D. B. McIntyre, and E. E. Welday

X-Ray Spectrographic Analysis of Rare Earths in Yttrium–Iron Garnet Powders. 389
    J. C. Lloyd and J. D. Kuptsis

The Use of X-Ray Emission Spectrography for Petroleum Product Quality and
Process Control... 396
    J. L. Caley

Absorption Effects in X-Ray Fluorescence Measurement of Elements in Oil..... 403
    E. L. Gunn

Determination of Catalyst Residues in Polyolefins by X-Ray Emission Spectroscopy 417
    G. D. Smith and R. L. Maute
Iron Oxide Determination by X-Ray Fluorescence for In-Process Control of Solid
    Propellant and Premixes.. 422
    Reuel E. Lamborn and Foster J. Sorenson
Design Considerations for On-Stream X-Ray Analysis....................... 429
    W. R. Kiley and R. W. Deichert
Particle Size and Mineralogical Effects in Mining Applications................ 436
    F. Bernstein
X-Ray Analysis of Mining and Mineral Processing Material.................. 447
    H. T. Dryer
Author Index... 459
Subject Index.. 465

INDEXING OF POWDER DIFFRACTION PATTERNS

P. M. de Wolff

Technische Hogeschool
Delft, Netherlands

ABSTRACT

Despite the advances of the electron microdiffraction technique, the determination of unit cells of microcrystalline compounds still depends in many cases on the interpretation of the X-ray powder pattern. The feasibility of this interpretation depends strongly on the precision of the data, the present-day level of which is far from its physical limit. It could be improved by application of monochromators in diffractometry. Some possible methods are discussed. With regard to interpretation, a survey of existing methods is given (Lipson, Ito, de Wolff, Szoldos). Some of these techniques are suitable for computers, at least in the initial phase of the procedure. This is illustrated by application of a zone-finding program for the complete solution of a given pattern. Finally, the influence of impurities and other disturbing factors is discussed.

INTRODUCTION

The powder diffraction method has been used for structure analysis almost since the time of its discovery. It is well known that many simple structures of metals, alloys and salts were established by powder methods some 40 years ago. Perhaps less generally known is the fact that, even then, very thorough attempts were made to find a method for indexing the complex patterns of low-symmetry compounds. This was done by the German mathematician C. Runge,[1] whose 1917 paper is still completely up to date. His method was applied to orthorhombic cementite by F. Wever[2] in 1923. The success of this first application is a tribute to Runge's method as well as to Wever's perseverance in handling data of the poor quality of that period.

It seems that few people, after Wever, possessed the same perseverance because Runge's method was almost forgotten until 1948 when Ito[3] used it as the basis of a method for unit cell determination. Since then, various other papers have appeared in this field, but Runge's principle is still the main foundation of the treatment of powder data.

We shall now proceed to describe the Runge principle and its variants as they are used in different analytical and graphical methods. An example of a complete analysis by a computer method will follow. Finally, a discussion will be given of the influence of some disturbing effects, especially errors of measurement, and some means of reducing the latter are described.

SURVEY OF INDEXING METHODS

For indexing purposes, the interplanar spacing d is a particularly unsuitable quantity.[4] However, because linear relations exist between values of $1/d^2$ this function is quite suitable. We shall mostly use the variable $10^4/d^2 = Q$, which has the additional advantage

---

[1] Superscripts pertain to references at the end of the paper.

1

that the error in Q, under normal conditions, is roughly a constant in a large interval of Q values, whereas the error in d strongly depends on the particular value of d.

Runge's principle consists of making use of linear relations between values of Q, in particular, those relations which are valid for Q pertaining to reflections from lattice planes belonging to the same zone. All such relations can be derived from one general equation expressing the Q for an arbitrary plane in this zone in the parameters Q' and Q'', belonging to two fixed planes of the zone:

$$Q = m^2 Q' + n^2 Q'' - 2mnR \tag{1}$$

The significance of m and n will be clear if we consider the geometry in reciprocal space. Suppose $\mathbf{h}'$ and $\mathbf{h}''$ are the vectors corresponding to the lattice planes for which $1/d^2$ is Q' and Q'', respectively, then the vector $\mathbf{h}$ belonging in the same way to Q can be expressed as

$$\mathbf{h} = m\mathbf{h}' + n\mathbf{h}'' \tag{2}$$

because of the assumed zonal relation between the three lattice planes. Equation (1) is nothing but the cosine rule applied to the triangle formed by the three vector terms occurring in (2), since Q is proportional to $|h|^2$. Accordingly we find

$$R = (Q' Q'')^{1/2} \cos \phi \tag{3}$$

where ϕ is the angle between $\mathbf{h}'$ and $\mathbf{h}''$.

Obviously a vector $\mathbf{h}$ of the zone is always obtained if m and n in (2) are integers. (The reverse is not generally true.) Consequently, equation (1) yields a possible value of Q (that is, a value commensurate with the crystal unit cell) if Q' and Q'' are two arbitrary observed values, and m and n are integers—provided R has the appropriate value given by (3) and depends only on the angle ϕ.

A zone detection method is obtained by reversing the last statement. A zone is first defined by taking for Q' and Q'' the values of Q of two observed lines, preferably among the lowest observed Qs. From equation (1) for two given integer values of m and n, setting $Q = Q_i$ equal to the value of Q for the ith observed line, we have

$$R = (m^2 Q' + n^2 Q'' - Q_i)/2mn \tag{4}$$

as a "possible" value of R in the sense that of all Rs calculated from (4) for fixed Q' and Q'' but varying i, m and n, some will be the correct value for this Q', Q'' pair.

This correctness can of course not be recognized as such, but it will be apparent if the correct value comes out more than once. Theoretically speaking, even finding the same R twice would be enough to conclude that it is the correct R. That is precisely Ito's prescription provided one takes $m = 1$ and $n = \pm 1$. If Q_j and Q_k are the two observed lines giving the same R, the following simple relations obtain:

$$Q_j + Q_k = 2(Q' + Q'') \tag{5}$$

$$Q_j - Q_k = 4R \tag{6}$$

which make it very easy to apply Ito's procedure. In practice the uniqueness of R is not safely established because the observed Qs are subject to errors of measurement. Therefore it cannot be expected that (5) will be fulfilled exactly, and it may accidentally be fulfilled by arbitrary Q_is within the limits imposed by error.

More certainty is obtained by enlarging the "index field," meaning the area covered by pairs of integers (m, n) in a hypothetical plane with coordinates m and n. This was first done graphically by Novak.[5] He constructed a set of concentric circles with radii $1/d_i = \sqrt{Q_i}$, corresponding to all observed lines. If the ϕ for a given pair Q', Q'' were

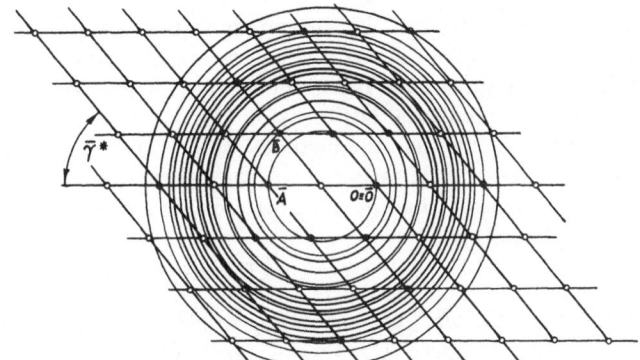

Figure 1. Coincidence of some nodal points with circles of radii $\sqrt{Q_i} = 1/d_i$, after Novak.[5]

known, the reciprocal net belonging to the zone defined by this pair could be constructed and the circles for which the reflecting lattice planes belong to this zone would pass through at least one point of the net (Figure 1).

Again, this argument has to be reversed for exploration purposes. Novak assumes integer values for m and n, and then by a straightforward construction finds a "possible" value of ϕ for each Q_i, which exactly corresponds to the "possible" value of R given by (4). Within predetermined limits for m and for n, all (m, n) pairs and all Q_i circles are combined and a frequency plot is made of the resulting ϕ values. In the example illustrated in Figure 2, the correct angle (30°) is seen to occur with a convincingly large frequency indeed. By using a computer technique, Novak's rather laborious procedure is easily carried out through computation of equation (4). This has been successfully applied by the present author,[6] and a detailed illustration will be given in due course. At present we continue our survey by mentioning two other important graphical methods.

The first is that of Zsoldos,[7] which is essentially identical with the method of Vand[8] for indexing long-spacing compounds. It is based on another form of equation (1), viz.,

$$(Q - m^2 Q') = n^2 Q'' - 2nRm = A - Bm \qquad (7)$$

implying that for a given Q', the left-hand side is a linear function of m if Q assumes in turn all the values corresponding to different m, but with fixed n and Q''. In reciprocal space this is equivalent to letting the point which corresponds to Q occupy lattice points on a row. The direction and period of that row is given by the vector $\mathbf{h}'$ corresponding to Q'.

For the solution of a given pattern, Q' is identified with the smallest observed Q. Now a graph is made by plotting $Q_i - m^2 Q'$ against m, so that each point corresponds to one pair of integer values (i, m). The value of m is limited by a predetermined range, say $-4 < m < 4$. According to (7), lines Q_i belonging to reciprocal points which are situated on a row with period $\mathbf{h}'$ yield collinear points in this graph and are thus, in

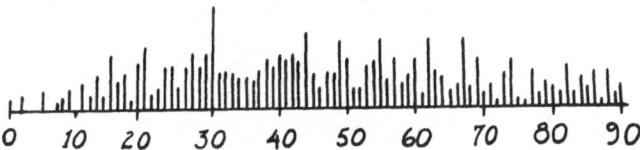

Figure 2. Frequency of occurrence of coincidences as in Figure 1, as a function of the angle γ^*, after Novak.[5]

principle, recognizable. The graph must be viewed very obliquely while the azimuth angle of view is continuously changing—until a substantial number of points are seen in line (see Figure 3). When some of these rows have been found, it is not difficult to construct the reciprocal lattice using (7).

Lastly, we mention the methods which can be used for nontriclinic crystals, notably Lipson's method.[9] It is based on the assumption that the reciprocal lattice contains at least one axis of rotation so that if $\mathbf{h}'$ is the smallest vector in the direction of that axis, we have from (1):

$$Q = m^2 Q' + n^2 Q'' \tag{8}$$

where the vector $n\mathbf{h}''$, corresponding to n^2Q'', is perpendicular to the axis. Accordingly, among the differences $\Delta Q = Q_i - Q_j$ between observed Qs, the values $(m_i^2 - m_j^2)Q'$, where m_i and m_j are integers can be expected to occur frequently, indeed, as often as the

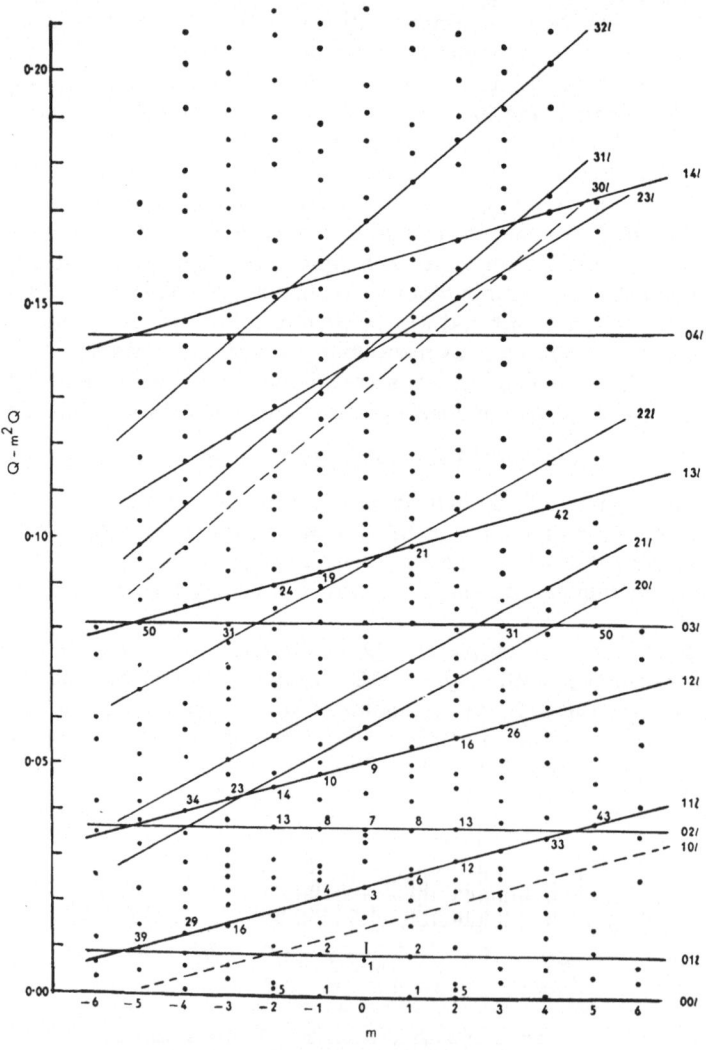

Figure 3. Plot of $Q_i - m^2Q'$ vs. m, after Zsoldos.[7]

Q values belonging to $nh'' + m_i h'$ and $nh'' + m_j h'$ are both observed for all possible values of nh''.

This somewhat generalized description of Lipson's principle is usually applied by plotting ΔQs on a line or on parallel lines (one for each Q_j) and inspecting the result for coinciding values of ΔQ (Figure 4). Alternatively, all ΔQs can be computed and co-incidences detected by a simple computer program.

This brings us to the question, which, if any, of these methods should be recommended with exclusive preference.[2] In one sense they are equivalent: they all serve primarily to establish zones, so they solve only the first part of the indexing problem. The second phase, consisting of building a reciprocal lattice, is essentially the same for all. It will be considered in detail in the example to be discussed. For zone detection, Lipson's procedure might seem to have the smallest scope. However, the fact that it is not applicable to triclinic crystals is not a very severe limitation, because these do not occur very often. As a matter of fact, the present author has used this procedure very often—even by way of a trial-and-error method without using graphical or computer techniques— in cases where the application of Ito's methods failed. A fully automatic computer program is now being developed in his laboratory, based on Lipson's principle in a modified version which, we hope, will be much more selective against spurious co-incidences caused by errors in the Q_i.

Among the more universal methods, the choice is, in the first place, between the Vand or Zsoldos method on one hand, and the Runge–Ito principle on the other. Here we find that Vand's method has the peculiarity that two unknown parameters of the zone are found simultaneously, namely, A and B in (7). Thus, success with his method depends largely on the remarkable capacity of the human eye to detect a collinear set among a cloud of points. For the same performance, even a fast computer needs a considerable amount of time. This is also the main drawback of the method: it cannot easily be programmed for a computer. With the Runge–Ito principle, on the other hand, only one parameter is determined at a time (as is also the case in Lipson's procedure). This can very well be done by a computer, so we think that this principle is to be preferred as a general method, yielding the obvious advantages of computer application. Lipson's procedure will probably be equally useful in a generalized computer version for non-triclinic crystals.

Which of these two (Lipson or Runge) should be applied first will depend largely on additional information, if available, and on the operator's judgment and experience. Finally, the use of the Runge principle with an extended index field—as developed by

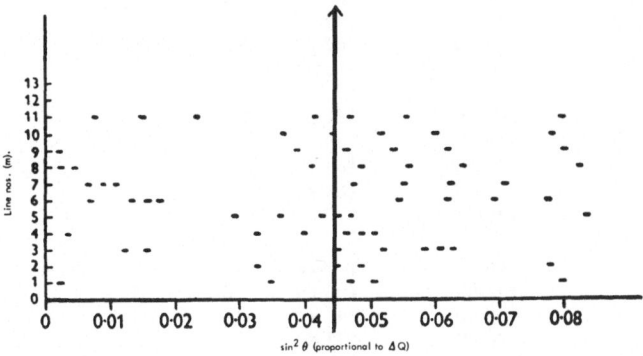

Figure 4. Plot of differences of Q: $\Delta Q = Q_{1m} - Q_i$ for $m = 1, 2, 3 \ldots$ (after Lipson[9]).

Novak and by the present author—is clearly more powerful than Ito's procedure in its original form.

COMPUTER PROGRAM FOR EXTENDED RUNGE–ITO METHOD

In its original form, the program was designed to carry out the calculation of equation (4) literally, for all values of i, m, and n within prefixed limits. These results were then sorted so as to form a list of increasing numbers R, and Q_i, m, and n were printed together with R on the same line (Table I).

However, such a list is not very convenient to find concentrations of R values (such as $R = 187$ in Table I). Accordingly, a new program was designed, which is still used with only minor modifications. It also produces a list of R values; this time, however, predetermined as a completely regular sequence of numbers increasing from zero up to a limit R_{max}, with equal intervals of, e.g., 0.1 unit. Such a list, of course, carries no information at all. The significant part of the machine's output is the second column. This shows a small integer p for each R of the first column. The meaning of p for, say, R_0 in the first column is that there are p different observed Q_i values satisfying equation (4) for this R_0, within a range $2t$; in other words, p values Q_i for which equation (4) yields a

Table I. Part of a List of R Values for the First Two Lines of $Na_4P_2O_7$

$Q_i(10^4/d_i^2)$	R	m	n
1107.4	184.07	4	2
1420.0	184.80	3	1
1242.2	184.84	4	3
1446.0	185.58	3	4
916.8	185.95	3	3
1217.9	187.42	4	1
884.3	187.54	1	3
794.8	187.61	2	3
1087.0	187.74	4	2
1217.9	187.75	4	3
271.7	187.78	2	1
100.5	187.92	1	1
1446.0	187.92	2	2
1446.0	187.92	2	4
642.9	188.06	3	2
642.9	188.21	3	1
884.3	188.50	1	2
1420.0	188.70	3	4
794.8	188.86	2	1
1406.3	190.34	3	4
884.3	191.15	3	3
1420.0	192.60	2	4
1380.6	193.43	3	4
769.0	193.80	2	3
1406.3	195.07	2	4

result between $R_0 - t$ and $R_0 + t$. For these p values, m and n will generally be different integers, but they are always smaller than the prefixed limits m_{max} and n_{max}.

This second version therefore directly "counts" the concentrations, because a concentration at R_0 will show up as a large p value. Table II illustrates this for a zone of triclinic $CuSO_4 \cdot 5H_2O$, where a concentration at $R = 7.4$ is immediately apparent. This table is only a small part of the actual list, in which R ran from 0 to 120. (The upper limit for R is $(Q'Q'')^{1/2} \cos 0° = (Q'Q'')^{1/2}$; in practice $\frac{1}{2}(Q' + Q'')$ is more convenient for our machine.) The remaining part, not shown in Table II, did not show any concentration nearly as large as the one at 7.4. Indeed, the peak number $p = 16$, being half the maximum possible value of combinations of $m = 1, 2, 3, 4$ and $n = -4, -3, -2, -1, 1, 2, 3, 4$, is unusually large.

The choice of the various program parameters is somewhat a matter of experience, though it is not overcritical. The limits m_{max} and n_{max} are chosen so that $m_{max}^2 Q'$ and $n_{max}^2 Q''$ are roughly $1\frac{1}{2}$ times larger than the largest Q_i value. Higher values would prolong the calculation and would produce an increase of the "background" of p at large R values. With smaller limits, on the other hand, coincidences can be missed. The most critical parameter is t, the error range. It should obviously be of the order of the average error in Q, but that error is not generally known very well. In the cases to be discussed here, we estimate it at about 0.3 units, whereas the best peaks in p were obtained with t slightly larger (0.5 for Table II). There are reasons to suppose that t should depend upon m and n [cf. equation (4)]. We tried $t = \text{constant} \times [(m^2 + n^2)/mn]$ but did not find any clear improvement.

As a technical improvement, a prefixed lower limit for p can be given to the computer, which then will print only those Rs and ps for which p exceeds that limit. Alternatively, only those entries for which p has the largest value of the whole list or is either 1 or 2 less can be printed. In both ways, the concentrations can be made extremely clear, as shown in the next section.

Table II. Part of a List of R and P Values for $CuSO_4 \cdot 5H_2O$

R	p	R	p	R	p	R	p	R	p
2.0	5	4.0	5	6.0	6	8.0	6	10.0	6
2.1	3	4.1	4	6.1	6	8.1	4	10.1	5
2.2	2	4.2	4	6.2	6	8.2	3	10.2	5
2.3	2	4.3	4	6.3	5	8.3	5	10.3	5
2.4	1	4.4	4	6.4	5	8.4	5	10.4	5
2.5	3	4.5	4	6.5	5	8.5	5	10.5	5
2.6	3	4.6	4	6.6	6	8.6	6	10.6	2
2.7	3	4.7	3	6.7	7	8.7	6	10.7	3
2.8	3	4.8	4	6.8	8	8.8	4	10.8	4
2.9	2	4.9	3	6.9	9	8.9	4	10.9	4
3.0	3	5.0	4	7.0	11	9.0	5	11.0	4
3.1	4	5.1	4	7.1	14	9.1	6	11.1	4
3.2	4	5.2	5	7.2	16	9.2	6	11.2	5
3.3	4	5.3	6	7.3	16	9.3	4	11.3	5
3.4	4	5.4	6	7.4	16	9.4	4	11.4	6
3.5	4	5.5	5	7.5	16	9.5	4	11.5	6
3.6	4	5.6	6	7.6	14	9.6	5	11.6	6
3.7	5	5.7	6	7.7	13	9.7	5	11.7	4
3.8	5	5.8	6	7.8	11	9.8	5	11.8	3
3.9	6	5.9	8	7.9	8	9.9	5	11.9	3

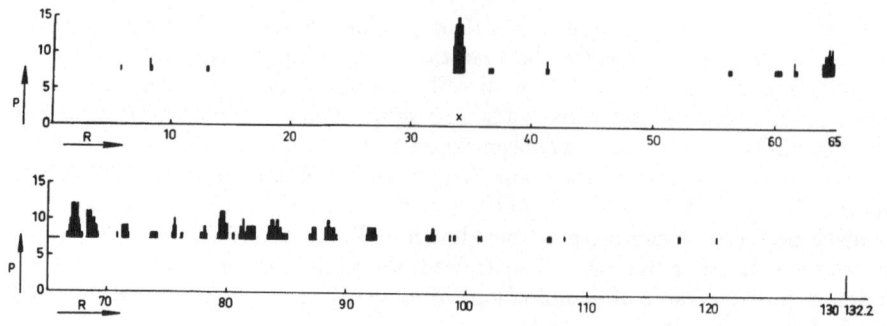

Figure 5. Graph of p $vs.$ R for a zone of $Na_3(PO_3)_3$.

The basic zone-defining Q' and Q'' are usually chosen from among the very first lines. They can also be identified with the same line, in which case a concentration will arise if this line occurs in an orthogonal zone.[11]

A still more obvious way of showing the concentrations is to make a graph of p $vs.$ R. This is essentially the same thing as Figure 2 taken from Novak's paper. Such a graph is reproduced in Figure 5 for one of our results [a zone of $Na_3(PO_3)_3$] for the complete range of that list, showing the prominence of the correct concentration as compared to spurious peaks in p.

EXAMPLE

A recent bulletin[10] of NBS contributions to the ASTM file contains the powder data of $LaNbTiO_6$, unindexed. We take this as an example since the NBS contributions (which usually are indexed) are typical of carefully measured diffractometer results.

They are obtained with a technique which can be applied as a routine by any reasonably well-equipped laboratory.

The values of d and of $Q = 10^4/d^2$ are given in the first and fourth columns of Table III.

The program described above was used to find the Rs for zones defined by combining the first line as Q' with each of the first five lines as Q'' (with the exception of $Q = 901$ since that is an obvious second order of the first: it was used to obtain a more accurate value of Q', namely, 225.0). The error range was set at $t = 1.0$; further we set $m_{max} = n_{max} = 4$, and R was this time increased in steps of 1 unit, whereas entries were printed only if p was at least 5. The results are shown in Figure 6. They have been plotted for clarity's sake but the original list in this case is almost equally clear: there are very prominent concentrations in several of the combinations. Analysis of these, once the unit cell had been found, showed that in each list except the first the largest concentration was true. In some lists, two true concentrations occur because the same pair of Q', Q'' may define more than one zone. For instance, if in a monoclinic crystal Q' corresponds to 110, it equally corresponds to $1\bar{1}0$, so there will be two zones if Q'' corresponds to any reflection except 001 reflections.

In order to find the complete unit cell, one may proceed in different ways:

a. If three zones have been detected which do not have a common line of intersection, they define the reciprocal lattice completely apart from a possible choice between two cases (because for two zones either the sharp angle ϕ or its supplement can be taken arbitrarily, but then the sharp and obtuse values for the third angle will yield different lattices). To verify which of these two possibilities is correct

Table III. Values of d and Q for LaNbTiO$_6$ Indexed from NBS Data

C—centered; $a = 11.20$; $b = 8.85$; $c = 5.27$; $\beta = 115.30°$

d	hkl	$10^4/d^2$ (calc.)	$10^4/d^2$ (obs.)
6.68	110	225	224.0
5.063	200	390	390.1
4.523	111	489	488.8
3.444	11$\bar{1}$	844	843.3
3.331	220	901	901.3
3.306	311	914	914.6
3.241	021	952	952.2
3.183	221	987	987.2
3.153	310	1005	1005.9
3.093	—	—	1045.3
2.833	130	1247	1245.7
2.632	202	1444	1443.3
2.573	131	1511	1510.7
2.531	400	1560	1561.4
2.421	312	1705	1705.9
2.355	421	1802	1802.9
2.273	331	1936	1935.9
2.249	31$\bar{1}$	1979	1977.4
2.223	330	2027	2024.0
2.170	511	2119	2123.5
2.096	022	2275	2276.1
2.065	11$\bar{2}$	2345	2344.9
2.033	422	2415	2419.7
2.006	041	2485	2485.1
1.993	241	2520	2517.5
1.9406	132	2657	2655.3
1.9155	332	2727	2725.3
1.8684	20$\bar{2}$	2864	2864.3
1.8249	33$\bar{1}$	3001	3002.6
1.7832	602	3144	3145.0
1.76606	24$\bar{1}$, 42$\bar{1}$	3230, 3222	3226.3
1.7440	150	3289	3287.9
1.7314	441	3335	3335.8
1.7214	22$\bar{2}$	3375, 3378	3374.6
1.7154	621	3397	3398.0
1.6777	151	3555	3552.9
1.6689	530	3587	3590.1
1.6513	114	3663	3667.5
1.6197	042	3808	3811.7
1.6034	51$\bar{1}$	3894	3889.6
1.5628	711	4104	4094.7
1.5072	334	4400	4402.6
1.4588	152	4701	4698.9
1.4269	641, 53$\bar{1}$	4930, 4916	4924.7

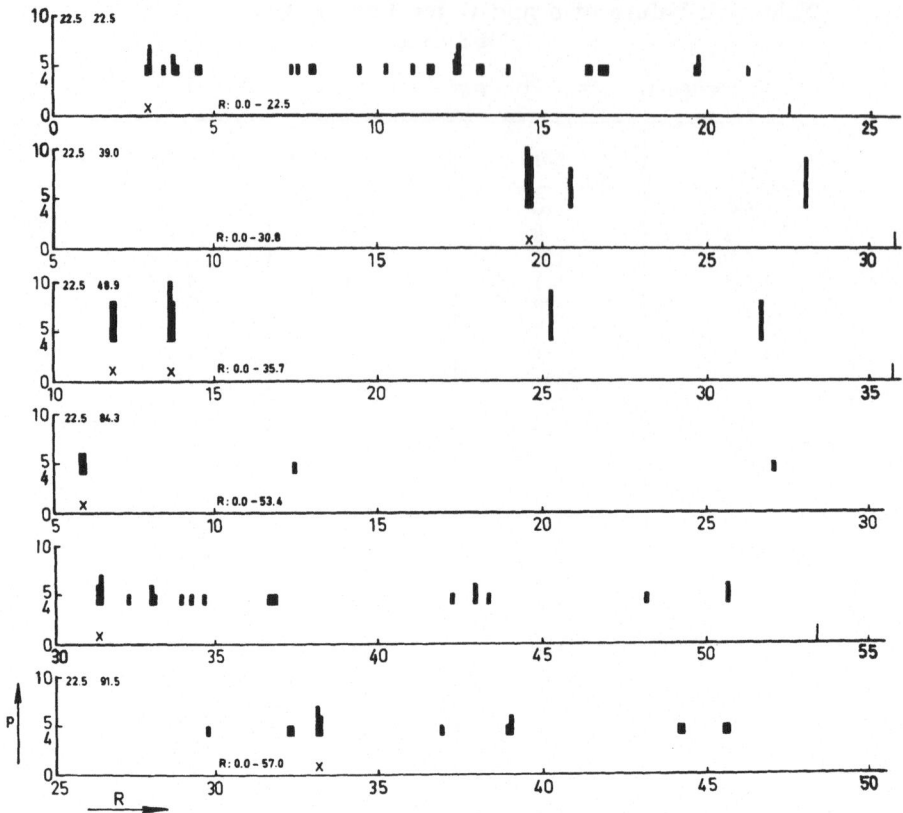

Figure 6. Graphs of p vs. R for 5 zones of LaNbTiO$_6$.

is a matter of straightforward calculation: an example is given in an earlier paper.[11]

b. Only two zones have led to a satisfactory determination of R. Since these zones have, of necessity, a line of intersection, the situation is as shown in Figure 7. Two net planes are known with all parameters of their nets, and the remaining unknown is the angle γ^* between these planes.

Again, therefore, we are confronted with the problem of determining a single parameter, just as R was first found for a zone. The fact that the problem now has one more dimension might seem to make it much more complex, but this is not the case. On the contrary, this last stage is by far the easier of the two. The reason is that now practically *all* observed lines will fit the correct value of F or γ^* whereas formerly only a small fraction of all Q_is fitted the zone for the correct R value.

We shall illustrate the procedure for the present example. Suppose we start with the two zones defined by $A: Q' = 225$, $Q'' = 390$, $R = 195$, $B: Q' = 225$, $Q'' = 489$, $R = 136.5$. These two zones can be schematized as shown in Table IV.

Each is drawn as an array of numbers which is a distortion of the actual net in reciprocal space. The common intersection is the line 0–225–900 . . . and is drawn horizontally both in S and T. The first row of reciprocal points parallel to this direction is represented by a corresponding row of Q values, calculated from equation (1) and placed so that the smallest number (e.g., 441 in T) is put above 0.

Table IV

	510	h $\uparrow$ 225	390	1005
(zone S)		0	225	$900 \to 1l$
	843	k $\uparrow$ 441	489	987
(zone T)		0	225	$900 \to 1l$

In order to explain the procedure, let us attach provisional indices to these Qs as shown in Table IV. The value of Q for an arbitrary lattice point hkl is then given by a generalized cosine rule:

$$Q = Ah^2 + Bk^2 + Cl^2 + Dkl + Ehl + Fhk = Q_0 + Fhk \qquad (9)$$

where

$$Q_0 = Ah^2 + Bk^2 + Cl^2 + Dkl + Ehl$$

By separating Q_0 from the term Fhk, we stress the fact that Q_0 is known and F is the only unknown. Indeed, in the present case $A = 225$, $B = 441$, $C = 225$, $D = 177.5$, $E = 60$ are completely defined by the zones S and T. The unknown F is, in general, equal to

$$F = 2(AB)^{1/2} \cos \gamma^*$$

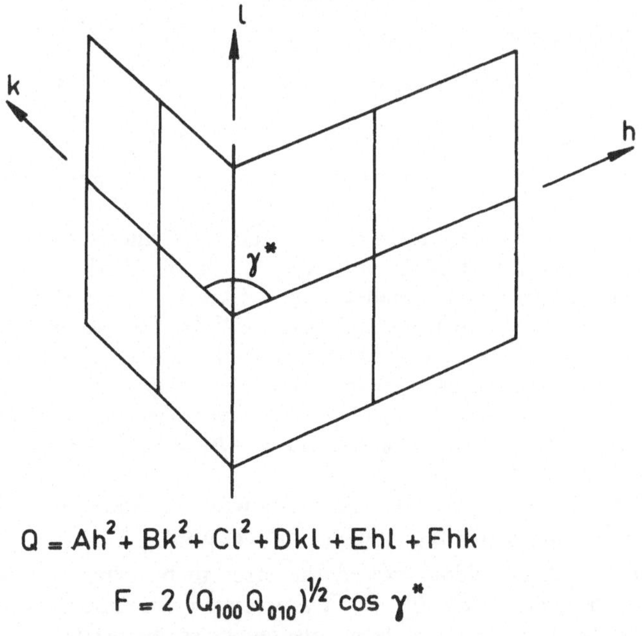

$$Q = Ah^2 + Bk^2 + Cl^2 + Dkl + Ehl + Fhk$$

$$F = 2 (Q_{100} Q_{010})^{1/2} \cos \gamma^*$$

Figure 7. Two intersecting nets in reciprocal space.

Therefore, F is directly related to the angle γ^* between two lines lying in planes of S and T, just as R in the first stage was related to the angle ϕ. The value of F can therefore be found exactly as before by using a similar computer program. It is so easy to find, however, that a computer is hardly needed. Instead, we calculate Q_0 for $k = 1$ and various small values of h and l:

Table V

		h ↑		
Q_0 for	1128	666	654	1092
$k = 1$	843	441	489	987 → $1l$
	1008	666	774	1332

The observed Q_i values not accounted for by zones S and T are 915, 952, 1045, etc. We now try to find a value of F which will explain them by equation (9), using the Q_0 scheme in Table V. It should be kept in mind that the first remaining Q_is have to correspond to the smallest Q_0s, otherwise an improbably large number of calculated Qs will not be observed. Try, for instance, $Q_{110} = 915$, which yields $F = 666 - 915 = -249$. This does not provide an explanation for $Q = 951$. After some further trials, we find $F = -177$ as the correct value, and the layer corresponding to Table V becomes shown in Table VI.

Table VI

Q for $k = 1$	951	489	477	915
	843	441	489	987
	1185	843	951	1509

Upon calculating Q for further values of h, k, and l, we find that the whole pattern is now satisfactorily explained. It is also apparent from Table VI that there is symmetry as shown by the dotted mirror line, parallel to one of those in zone S. The lattice must be schematized differently to show this clearly (Table VII). This schematization shows the lattice to be monoclinic C-centered (with respect to the new $h'k'l'$ indices of Table VII).

The agreement between observed and calculated Qs is seen to be very good (Table III), although the parameters $A' \dots F'$ have not yet been adjusted by least squares. (Only the constant B' is slightly increased so that $4B' = 511$ instead of 510 as in Table VII.)

One line ($Q = 1045$) is not explained. It is a weak one and must be caused by some impurity. The reliability of the indexing is good because:

  a. The discrepancies seldom exceed the error to be expected for measurements accurate to about 0.02° (2θ). Such a precision can indeed be obtained by the NBS technique. Somewhat large discrepancies do occur, but only at higher angles, where improved lattice parameters will doubtless reduce them.

Table VII

		k'			
		↑			
		510		900	
zone S $(l' = 0)$			225		1005
		0		390		1560 → h'

		k'			
		↑			
$k = 1$ $(l' = 1)$		951		987	
	843		489		915
	1186	441		476		1291 → k'

b. The discrepancies are also very much smaller than the average interval between successive calculated Q values. This can be verified by using formulas given elsewhere.[12]

c. The number of observed lines (43) is a satisfactorily large fraction of the total of 95 theoretical Q values in the same range of Q. The latter value becomes even smaller (86) if a systematic extinction for l' odd and $k' = 0$ is assumed (glide plane C), which is very probably present in view of the complete accordance with the absent lines.

COMMENTS

The example discussed above shows that a clear, well-measured pattern like this one can be solved easily by the present computer method. The selectivity could still have been improved (that is, the peaks be made more prominent) if the error range t had been reduced to correspond to the actually very small error, of which we became aware only after the computation. It is almost certain that other methods would have been equally successful. As a matter of fact, the lattice was first found without a computer, using Ito's criterion in the way described in an earlier paper.[11]

This illustrates the power of algebraic indexing methods. Compared to graphical methods, they have the advantage that a computer can be used, and, most important, that full justice is done to the precision of the measurements. The reliability of a result like the present one is practically complete.[12]

It is true that a number of unfavorable conditions can be summed up, which would make algebraic indexing difficult or impossible. In the author's experience, such circumstances occur very seldom. Accordingly, there is little reason for the still prevailing neglect of these methods in applied diffraction research.

It is of course possible, and often desirable, to use physical methods as an aid to indexing as well. Among these we mention especially, efforts to obtain preferred orientation; study of temperature effect; examination of powder particles by microdiffraction or by selected area electron diffraction. Generally speaking, however, the computer method is by no means more difficult or less reliable than these.

Another remark can be made regarding impurities. In computer applications, the author has encountered several cases where the correct solution was found despite severe contamination. Noncomputer procedures, on the other hand, are severely hampered by impurities, especially if impurity lines occurred among the first lines of the pattern. The systematic exploration of the pattern by a computer is the only means to overcome this difficulty, and the results are just as clear as they are for pure compounds. When the zones have been combined to form a lattice, 5–10 lines may be left over on a total of 30–50 lines without influencing the reliability of the unit cell, provided the above-mentioned criteria of reliability are fulfilled.

RELATED COMPUTER TECHNIQUES

Further extension of computer application can be made in two directions. First, it is possible to write a program for the combination of two zones (determination of F or γ^*), and also to connect this to the existing program in such a way that a fully automatic complete indexing program is obtained. The difficulty lies merely in taking account of various special lattices, e.g., lattices with orthogonal and centered nets. Apart from that writing such a program does not offer any essential problem.

The other extension is the computation of the indices for all observed lines and the determination of lattice parameters, once the unit cell is known approximately. We have developed a program for this purpose and are using it frequently. It is based on least-squares solution of equation (9) for the quantities $A \ldots F$. The peculiar aspect of the problem is the fact that hkl are not known beforehand (as is the case for the coefficients in a common least-squares problem); the only thing known is that they are integers. Therefore, the program has to be an iterating one. It starts by calculating $A \ldots F$ from a limited number of unambiguously indexed Q_is. Then it uses these provisional results to find the most probable indices for all lines. After that, it computes $A \ldots F$ once more by least squares, now using all observed lines except those which are still too far off the mark or which are doubly indexed. With this second set of results, new (and often different) indices are found, and a third cycle is started, etc.

Here again, the error range [meaning the permissible difference between both sides of equation (9)] is of utmost importance. We usually begin by putting it at 0.3% of Q, with a minimum of one unit. In three cycles it is then diminished down to 0.1%, same minimum. This yields a satisfactory convergence and a fourth cycle is seldom needed.

PRECISION OF MEASUREMENTS

For indexing purposes, the precision of diffraction angles is important for two reasons. In the first place, indexing is made much easier if the precision is high. Expressed in terms of our computer method: the higher the precision, the narrower the error limit t can be set, thereby suppressing spurious concentrations while leaving the true concentrations intact.

In this respect, the accuracy of the zone-defining lines Q' and Q'' is of primary importance. Wherever possible, it should be increased by using obvious higher orders, as was done in the example.

The effect of an error in Q' or Q'' is illustrated in Figure 8. Here, the concentrations in two zones both containing a line $Q' = 54.5$ are shown on plots of p vs. R, for several (hypothetic) values of Q' ranging from 54.0 to 55.2. It is clear that both concentrations, though very pronounced at the correct value of Q', almost disappear into the background even for as small an error as $\pm$ 0.5 in Q' (corresponding to $\pm$ 0.03°2θ). The accuracy of lines other than Q' and Q'' is less critical but still very essential.

The second aspect of precision is its effect on the reliability of the resulting unit cell. It can be shown that the reliability is an exponential function of the error.[12] So, if

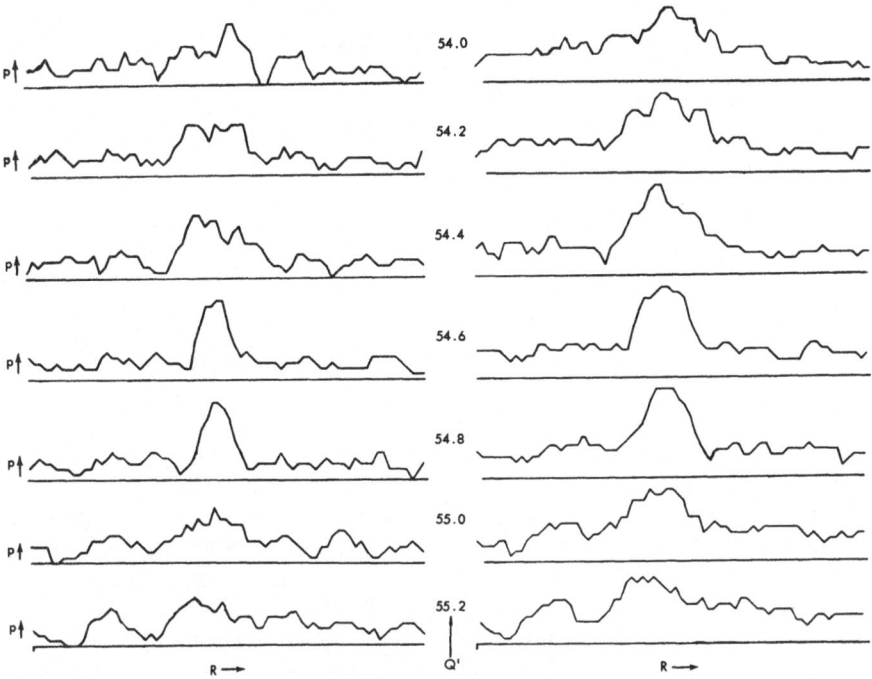

Figure 8. Influence of error in Q' on the shape of a concentration. Two zones from the pattern of $CuSO_4 \cdot 5H_2O$ are shown by plotting p vs. R in the neighborhood of a concentration.

the probability for a certain indexing to be wrong is, say, 10^{-4}, then for doubled precision it drops to 10^{-8}. Therefore even a moderate gain in precision can widen the scope of indexing enormously. It may, for instance, allow the complete analysis of patterns of mixtures or the analysis of extremely dense patterns.

The level of precision attainable by careful routine measurement is around $0.01°\, 2\theta$, both for cameras and diffractometers. The line width Δ is generally of the order of $0.15°$ in normal diffractometer work.

Now, the statistical error for a diffractometer measurement of 2θ is, roughly speaking $\Delta/\sqrt{N}$, where N is the total number of counts used for the measurement. Therefore, the precision can, in principle, be increased indefinitely by making N large; but by far the most effective manner is to reduce the line width Δ. Thereby we gain not only in accuracy, but in resolving power and peak-to-background ratio as well. This is equally true for photographic methods.

For that reason, it is to be expected that the use of focusing monochromators will become increasingly important. The Guinier-type cameras have already found wide application as a means of reducing background, geometrical broadening, and spectral line width. The latter effect is due to a mutual compensation of the spectral dispersion caused by monochromator crystal and specimen, respectively. In diffractometers, focusing crystals can be used equally well. The most useful design is perhaps the one in which the specimen is of the transmission type.[13] An instrument based on this "reversed Guinier" principle is shown in Figure 9. It is a modification of the standard Norelco diffractometer, in which the focusing quartz crystal is placed on the counter arm.

With Guinier-type cameras and with the modified diffractometer just mentioned,

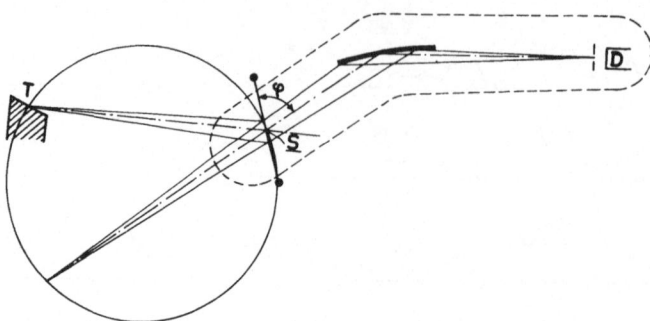

Figure 9. Diffractometer with transmission-type specimen.

the line width (expressed as the width at half height) can be reduced to less than 0.10° in a large range of angles.

Still further reduction is possible if the monochromator is made to reflect only the α_1-component of the K_α doublet. This has been realized in Guinier-type cameras.[14] The use of such devices is becoming more and more frequent now. Line widths of about 0.05° have thus been obtained in almost the whole forward reflection range, corresponding to the physical diffraction broadening caused by a crystal dimension of $0.1\,\mu$.

It looks as if the limit of practical and useful improvement in monochromator technique has been reached, and we may expect a similar development to take place in diffractometry. In the latter, the resulting accuracy can be realized more effectively than in photographic methods, because the nonlinear characteristic of a film causes systematic errors in the measurement of asymmetric line profiles, whereas a diffractometer record is completely free of such effects.

SUMMARY

Three main principles exist for direct indexing of powder patterns. In all three the primary result is a zone in the reciprocal lattice. The Runge–Ito principle, which had been generalized graphically by Novak, has now been programmed for a computer in such a way that the unknown zone parameter shows up very clearly. Computer programs for Lipson's method for complete indexing are being developed. A general iterating program now used routinely for refinement of lattice parameters is described.

Recently published unindexed diffractometer data are analyzed using the generalized Runge–Ito program. The relatively easy second stage, in which two zones are combined to form the entire reciprocal lattice, is worked out in detail. Finally, it is pointed out that the scope of direct indexing could be much increased by improving the accuracy of measurement. This is most effectively done by reducing line widths. The limit set by spectral line breadth can be overcome by using focusing crystal monochromators. Even with the present level of precision, around $0.01°2\theta$, direct indexing is, in general, no less reliable and no more difficult than physical methods.

ACKNOWLEDGMENT

The author wishes to express his appreciation for the active cooperation of Mr. A. A. Koene (Applied Mathematics Institute T.N.O., The Hague), who wrote the programs discussed here for the ZEBRA computer.

REFERENCES

1. C. Runge, *Phys. Z.* **18**: 509, 1917.
2. F. Wever, *Mitt. Kaiser-Wilhelm-Inst. Eisenforsch. Düsseldorf* **4**: 67, 1923.
3. T. Ito, *Nature* **164**: 755, 1949.
4. "Tables of Q as a Function of 2θ," *Acta Cryst.* **12**: 421, 1959.
5. C. Novak, *Czechoslov. J. Phys.* **4**: 496, 1954.
6. P. M. de Wolff, *Acta Cryst.* **11**: 664, 1958.
7. L. Zsoldos, *Acta Cryst.* **11**: 835, 1958.
8. V. Vand, *Acta Cryst.* **1**: 290, 1948.
9. H. Lipson, *Acta Cryst.* **2**: 43, 1949.
10. H. E. Swanson, Nat. Bur. Standards Rept. 7521 on Standard X-ray Diffraction Patterns, June 1962.
11. P. M. de Wolff, *Acta Cryst.* **10**: 590, 1957.
12. P. M. de Wolff, *Acta Cryst.* **14**: 579, 1961.
13. P. M. de Wolff, *Acta Cryst.* **13**: 835, 1960.
14. E. Hofmann and H. Jagodzinski, *Z. Metallk.* **46**: 601, 1955.

DISCUSSION

Robert Kelsey (Pratt and Whitney): Have you ever had occasion to work with mixtures of compounds and have this method break down completely?

P. M. de Wolff: To a certain extent, yes, I have had several cases where of a total of 30 to 50 lines there were 5 to 10 lines caused by the second component and that is not too bad if these extra lines do not occur among the first lines of the pattern. If they do, it's almost impossible to index the pattern by the noncomputive method. But in that case the computer method is not hindered very much and you can still use it.

ON THE USE OF A MODIFIED RADIAL DISTRIBUTION ANALYSIS FOR INDEXING POWDER PATTERNS

A. F. Berndt

Argonne National Laboratory
Argonne, Illinois

ABSTRACT

A modification of the theory of X-ray radial distribution analysis is presented. This modification can serve as a guide to the values of the unit cell dimensions and may be useful in indexing powder patterns of unknown structures, although the use of trial and error methods is not eliminated. This technique is shown to give consistent results with known structures and is applied to the indexing of powder patterns of Pu_3Ru and Pu_5Ru_3. The powder pattern of Pu_3Ru can be indexed on the basis of an orthorhombic unit cell with $a_0 = 6.216$ Å, $b_0 = 6.924$ Å, and $c_0 = 8.093$ Å, and Pu_5Ru_3 on the basis of a tetragonal unit cell with $a_0 = 8.092$ Å and $c_0 = 10.023$ Å.

INTRODUCTION

The problem of indexing the powder pattern of an unknown structure, which is trivial in the case of cubic symmetry, becomes increasingly difficult as the symmetry is lowered. Many methods, both graphical and analytical, for indexing powder patterns have been presented.[1,2] These are all basically trial and error methods. Therefore, any device which can give any information as to distances in the real or reciprocal lattice would be a useful tool to assist in the indexing of a powder pattern.

RADIAL DISTRIBUTION ANALYSIS

The theory for X-ray radial distribution analysis of diffraction data consisting of discrete peaks has been presented by Glusker and Miller.[3] From their theory:

$$4\pi r^2 \sum_M K_M[g'_M(r) - g_0] = (2r/\pi)(\sum_M Z_M{}^2)\{\sum_j F_jA_j[\exp(-B's_j{}^2)]\sin s_j r$$

$$-(\tfrac{1}{4})r\sqrt{\pi}(B + B')^{-3/2}\exp[-(\tfrac{1}{4})r^2/(B + B')]\} \qquad (1)$$

In equation (1), K_M is the effective weighting factor for the Mth atom and is approximately equal to Z_M, the atomic number; $g'_M(r)$ is the electron density function; g_0 is the average electron density; A_j is the observed intensity for the jth Bragg peak, F_j is the factor necessary to place the A_j on an absolute basis; $s_j = (4\pi \sin \theta_j)/\lambda = 2\pi/d_j$; and B is the temperature factor coefficient. The factor $\exp(-B's_j{}^2)$ is a convergence factor introduced[4] in order to decrease spurious detail in the calculated radial distribution function, which would be caused by the fact that data exist only to a finite value of s. The value of B' is usually chosen so that $\exp(-B's_j{}^2{}_{max}) = \tfrac{1}{10}$. The presence of a peak in the radial

---

[1] Superscripts pertain to references at the end of the paper.

distribution function at $r = r_1$ indicates that greater than average electron density is associated with atoms that have separations at this value.

The radial distribution function calculated according to equation (1) from the powder pattern of a given structure will have peaks at values of r equal to the unit cell dimensions. However, these will generally be unidentifiable in the maze of peaks from all other interatomic distances.

Consider a hypothetical structure which has the identical unit cell as the unknown structure whose powder pattern is to be indexed, but has only one atom per unit cell. The powder pattern of this hypothetical structure will have lines, all with the same structure factor, at the identical positions as the unknown structure (including lines corresponding to unobserved reflections from the unknown structure). The radial distribution function calculated for this hypothetical structure would therefore have peaks at values of r equal only to distances in the real lattice of the unknown structure, and should be of value for indexing the powder pattern of the unknown structure.

If it is assumed that $F_j A_j = 1$ for all observed reflections, then the radial distribution function for this hypothetical structure can be calculated from the powder pattern of the unknown structure by equation (2).

$$r \sum_M K_M [g'_M(r) - g_0] \propto rD(r) \propto \sum_j [\exp(-B's_j{}^2)] \sin 2\pi(r/d_j) \qquad (2)$$

This $rD(r)$ vs. r curve will be an approximation and will not be proportional to the exact radial distribution function for the hypothetical structure because multiplicity has not been taken into account and because reflections which are unobserved for the unknown structure cannot be included. The latter factor may cause several peaks, due to interatomic distances not corresponding to lattice translations, to persist. Neglect of the term,

$$\tfrac{1}{4}r\sqrt{\pi}(B + B')^{-3/2} \exp[-\tfrac{1}{4}r^2/(B + B')],$$

will result in a strong peak near the origin, where there can be no interatomic distances, and will not appreciably affect the curve for larger values of r. Equation (2), in addition to suppressing the peaks not corresponding to lattice translations, has the additional advantage over equation (1) that careful, laborious determination of the factors F_j and the intensities A_j are not required.

APPLICATION TO KNOWN STRUCTURES

As a preliminary check of the applicability of this method, it has been applied to known structures. The $rD(r)$ vs. r curves calculated from the powder patterns of Sb and α-U, for the hypothetical unit cell, are shown in Figures 1 and 2, respectively. The unit cell of Sb is rhombohedral[5] with $a_0 = 4.4976$ kX (4.507 Å) and $\alpha = 57°6'27''$. The unit cell of α-U is orthorhombic[6] with $a_0 = 2.854$ Å, $b_0 = 5.869$ Å, and $c_0 = 4.955$ Å. Vertical lines are drawn under the curves to indicate the values of the known distances in the real lattice as well as several other interatomic distances. In both cases the agreement between the theoretical predictions and experimental data is pleasing.

APPLICATION TO UNKNOWN STRUCTURES

Pu₃Ru

As a final test of the applicability of this method, it has been applied to the indexing of the powder patterns of unknown structures. A melt of Pu_3Ru containing 12.34 wt % Ru (calculated for Pu_3Ru, 12.36 wt % Ru) was prepared, annealed at 540°C for 35 days,

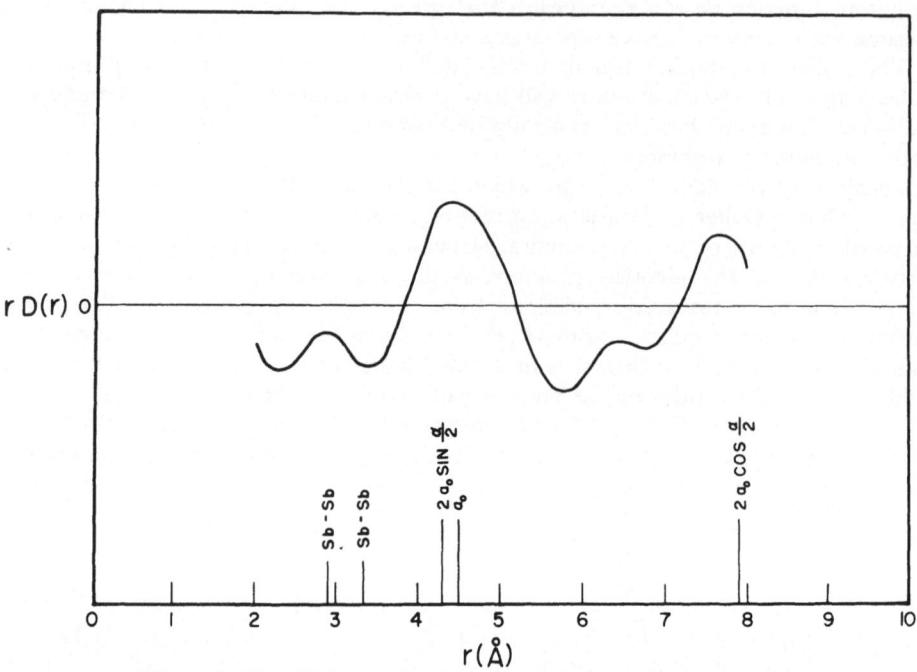

Figure 1. The *rD(r) vs. r* curve for the hypothetical unit cell of Sb.

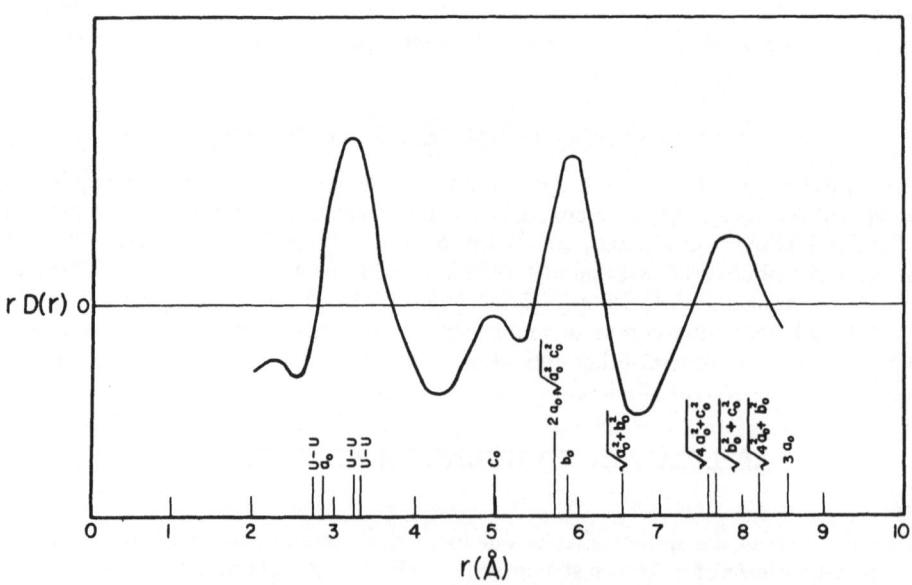

Figure 2. The *rD(r) vs. r* curve for the hypothetical unit cell of α-U.

and powdered. Debye–Scherrer patterns were made with Cu K_α radiation ($\lambda = 1.5418$ Å). Attempts at indexing this pattern by the standard methods[1,2] were unsuccessful.

The $rD(r)$ *vs.* r curve for the hypothetical unit cell is shown in Figure 3 and has strong peaks at 3.3 and 6.6 Å, a minor peak at 5.3 Å, and a broad peak between 8.4 and 9.2 Å. The assumption of the existence of distances in the real lattice near 3.3 or 5.3 Å did not lead to successful indexing. However, when the existence of distances near 6.6 and 8.4 Å was assumed, it was possible to index the powder pattern on the basis of an orthorhombic unit cell. The lattice constants, refined by least squares, are $a_0 = 6.216$ Å, $b_0 = 6.924$ Å, $c_0 = 8.093$ Å, all $\sigma = 0.001$ Å. (These values are illustrated by the vertical lines on Figure 3 to show the agreement obtained.) The density, calculated for $Z = 4$, is 15.60 g/cm³ (observed, 15.50 g/cm³). The agreement between observed and calculated values of Q ($= 4 \sin^2\theta/\lambda^2$) is shown in Table I. The lack of any systematic absences suggests that the space group, if centrosymmetric, is *Pmmm*. The strong peak at 3.3 Å probably corresponds to Pu–Pu distances. This value is approximately twice the metallic radius of Pu, 1.58 Å.[7]

Pu_5Ru_3

A melt of Pu_5Ru_3 containing 20.21 wt % Ru (calculated for Pu_3Ru_5, 20.24 wt % Ru) was prepared and annealed at 860°C for 11 days. The $rD(t)$ *vs.* r curve for the hypothetical

Table I. Observed and Calculated Values of Q for Pu_3Ru*

Q_{obs}	Intensity	hkl	Q_{calc}	Q_{obs}	Intensity	hkl	Q_{calc}
0.0341	vvw	011	0.0362	0.3778	w	322	0.3774
0.0863	vvw	102	0.0870	0.3837	ms	005	0.3817
0.1045	w	200	0.1035	0.3966	m	042	0.3949
0.1113	vw	120	0.1093	0.4097	m	105	0.4076
0.1178	vw	201	0.1188	0.4145	s	400	0.4141
0.1230	w	210	0.1244	0.4228	vvw	330	0.4206
0.1259	vw	121	0.1246			142	0.4208
0.1359	s	003	0.1374	0.4351	vw	410	0.4350
0.1415	s	211	0.1397			331	0.4359
0.1446	ms	022	0.1445	0.4497	m	411	0.4503
0.1572	w	013	0.1583	0.4590	w	134	0.4579
0.1607	m	103	0.1633	0.4776	vvw	304	0.4772
0.1678	vs	122	0.1703			412	0.4961
0.1896	s	030	0.1877			143	0.4971
0.206†	m	221	0.2022	0.4973	m	420	0.4975
		031	0.2030			314	0.4981
		130	0.2136			242	0.4984
0.2309	vw	131	0.2289	0.5205	ms	050	0.5215
		300	0.2329	0.5378	vvw	051	0.5368
0.2511	vw	032	0.2488	0.5466	mw	150	0.5473
		310	0.2538	0.5717	m	016	0.5705
0.2668	vw	014	0.2652			413	0.5724
0.3262	s	033	0.3251	0.6097	s	152	0.6085
		024	0.3277	0.6302	mw	342	0.6278
0.3332	vw	040	0.3338			026	0.6330
0.352†	vvw	133	0.3510	0.6382	m	251	0.6403
		232	0.3523	0.6540	mw	206	0.6531
		124	0.3536				

* A total of 137 reflections with $Q < 0.658$ are possible.
† Extremely broad feature.

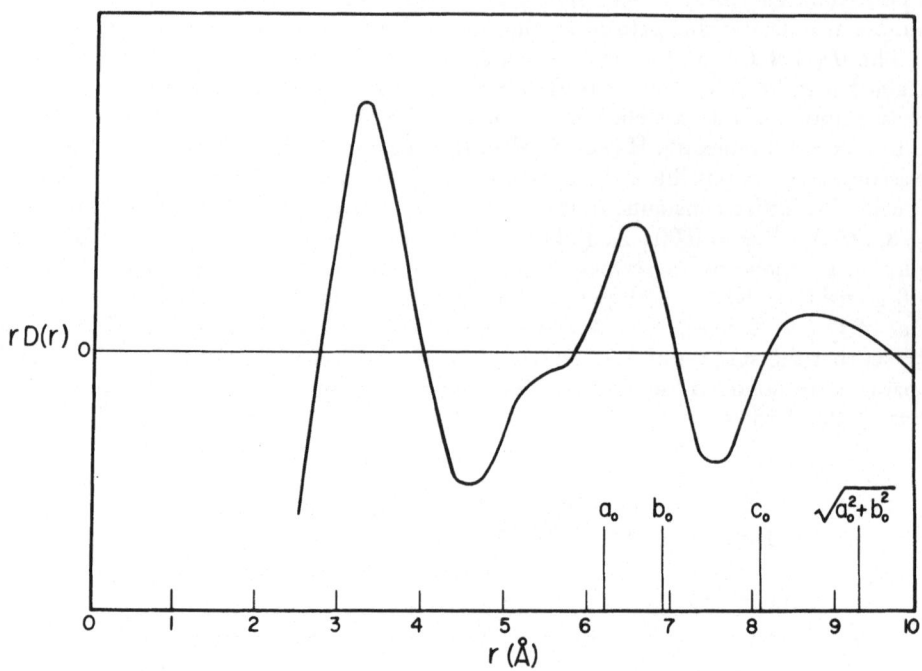

Figure 3. The $rD(r)$ *vs.* r curve for the hypothetical unit cell of Pu$_3$Ru.

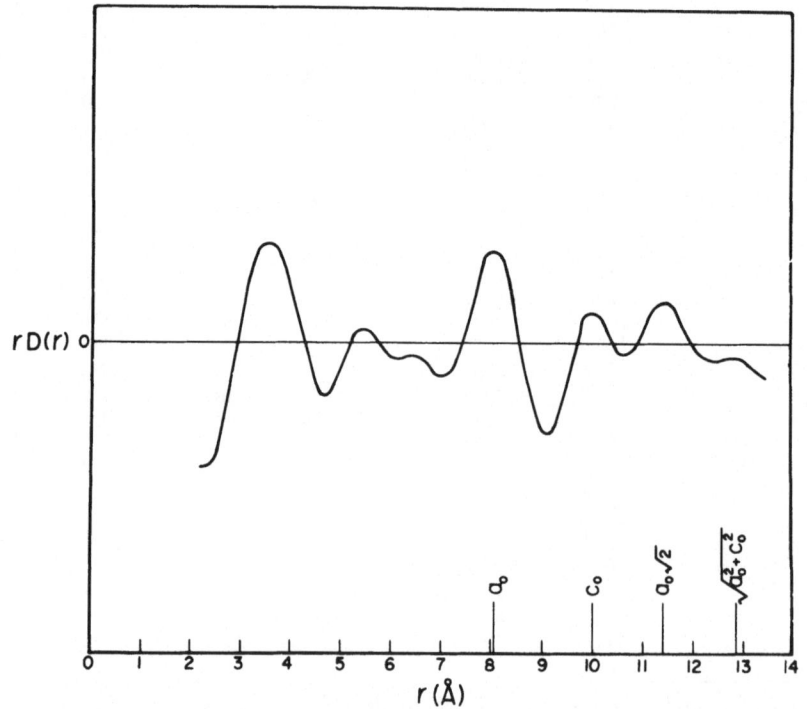

Figure 4. The $rD(r)$ *vs.* r curve for the hypothetical unit cell of Pu$_5$Ru$_3$.

Table II. Observed and Calculated Values of Q for Pu_5Ru_3*

Q_{obs}	Intensity	hkl	Q_{calc}
0.0248	vvw	101	0.0253
0.0299	vvw	110	0.0305
0.0746	vvw	210	0.0764
0.0877	w	211	0.0864
		003	0.0896
0.1029	vw	202	0.1009
		103	0.1047
0.1203	m	113	0.1201
0.1242	w	220	0.1222
0.1374	vw	300	0.1374
0.1436	s	301	0.1474
0.1551	ms	310	0.1527
0.1583	vw	004	0.1593
0.1648	w	222	0.1620
		311	0.1627
		213	0.1660
0.1741	s	104	0.1746
0.1784	vs	302	0.1772
0.1894	m	114	0.1898
0.2592	w	410	0.2596
0.2818	mw	224	0.2815
0.3277	vw	215	0.3253
0.3464	ms	422	0.3452
0.3806	ms	430, 500	0.3818
0.3855	w	305	0.3863
0.3971	w	510	0.3971
0.4143 | †	ms	414	0.4189
0.4200 /		206	0.4195
		432, 502	0.4216
0.4329	vs	334	0.4342
		216	0.4348

* A total of 81 reflections with $Q < 0.435$ are possible.
† Broad line, not completely resolvable.

unit cell is shown in Figure 4. Peaks are observed at 3.55, 5.4, 6.5, 8.05, 10.0, 11.4, and 12.8 Å. The powder pattern could not be indexed with the assumption of the existence of distances in the unit cell equal to any of the first three values. Since the first two peaks correspond to similar peaks for Pu_3Ru, it does not seem unreasonable to assume that these peaks also represent Pu–Pu or Pu–Ru interactions not corresponding to lattice translations. The observation that $11.4 \approx 8.05 \sqrt{2}$ and $12.8^2 \approx 10.0^2 + 8.05^2$ enabled the first 25 lines of the pattern to be indexed* on the basis of a tetragonal unit cell with $a_0 = 8.092$ Å $abd\ c_0 = 10.023$ Å. (These values are illustrated by the vertical lines on Figure 4.) The density, calculated for $Z = 4$, is 15.17 g/cm³ (observed, 14.82 g/cm³). A plot of reciprocal density $vs.$ composition for the phases Pu_3Ru, $PuRu$, $PuRu_2$,[8] and Ru suggests that the value 14.82 g/cm³ may be too low. A value of 15.20 g/cm³ lies on the straight line determined by the other four densities. The space group, if centrosymmetric, is probably $P4/mmm$. The agreement between the observed and calculated values of Q is shown in Table II.

* The remainder of the pattern cannot be unambiguously indexed because of the large size of the unit cell.

CONCLUSIONS

A modification of the theory of X-ray radial distribution analysis has been presented in order to make this technique applicable to the problem of indexing the powder pattern of an unknown structure. Although the use of trial and error methods for indexing powder patterns is not eliminated, the modified radial distribution curve calculated from the powder pattern of an unknown structure can serve as a valuable guide to the values of the unit cell dimensions. The modified theory has given consistent results when applied to known structures and has been used successfully to index powder patterns of Pu_3Ru and Pu_5Ru_3 after standard methods had failed.

ACKNOWLEDGMENTS

The author wishes to thank Dr. A. Miller of RCA Laboratories for several helpful discussions. This work was carried out under the auspices of the U.S. Atomic Energy Commission.

REFERENCES

1. H. P. Klug and L. E. Alexander, *X-ray Diffraction Procedures*, John Wiley & Sons, Inc., New York, 1954, Chap. 6.
2. S. L. Nudelman, *Kristallografiya* **5**: 819, 1960.
3. D. L. Glusker and A. Miller, *J. Chem. Phys.* **26**: 331, 1957.
4. J. Waser and V. Schomaker, *Rev. Mod. Phys.* **25**: 671, 1953.
5. E. R. Jette and F. Foote, *J. Chem. Phys.* **3**: 605, 1935.
6. E. F. Sturken and B. Post, *Acta Cryst.* **13**: 852, 1960.
7. W. H. Zachariasen and F. Ellinger, *J. Chem. Phys.* **27**: 811, 1957.
8. A. S. Coffinberry and W. N. Miner, ed., *The Metal Plutonium*, The University of Chicago Press, Chicago, p. 300, 1961.

DISCUSSION

D. Rodier (Grumman Aircraft Co.): You said this method would work with equal facility for all crystal symmetries.

A. F. Berndt: Since this method gives distances in the real lattice, and since in order to index a powder pattern you are working with distances in the reciprocal lattice it should work for the monoclinic or triclinic systems where you have non-90° angles, but not with equal facility.

P. Lublin (General Telephone and Electronic Labs): Is it possible that the low density might be due to the fact that you don't have a pure phase, that you have something else in there which would lower the density but not show up in the diffraction pattern?

A. F. Berndt: There is certainly a possibility that there could be a second phase. If so, it would have to be a ruthenium-rich phase because the plutonium is more dense. The only thing I can say is that it did not show up on the X-ray pattern. Low density is not unusual—voids and cracks do exist.

Z. Wilchinsky (Esso Research and Engineering Labs): In the case of Pu_3Ru how did you place the A and B parameters under the peak near 6 Å?

A. F. Berndt: I used the existence of this peak as a guide that there are distances in this range. The final choice of the parameters, of course, is agreement with the observed Q values. In the final result the average of these two chosen parameters does lie under the peak.

THE X-RAY INVESTIGATION OF PREPRECIPITATION IN SUPERSATURATED SOLID SOLUTIONS

A. Lutts

Centre National de Recherches
Métallurgiques, Liège, Belgium

ABSTRACT

This paper briefly traces the study of the preprecipitation or cold-aging stage of the decomposition of supersaturated solutions by means of diffuse scattering of X-rays.

This part of the overall precipitation process is of considerable practical as well as theoretical interest because it is during preprecipitation that changes in many physical and mechanical properties take place without the formation of a precipitate phase.

Examples will be given to show the types of zones encountered during this stage in various age-hardening alloys. These will attempt to illustrate the contribution which the study of diffuse scattering of X-rays, guided by present-day concepts of crystal imperfections, has made toward a better understanding of the precipitation process in alloys.

INTRODUCTION

Precipitation in a solid solution is a phenomenon of great practical as well as theoretical interest. From the technological point of view it is the basis of the many so-called age-hardening alloys so extensively employed in our modern civilization.

It is also a scientific curiosity of the first order, whose study during the last 50 years has given us much valuable information concerning the nature of the solid state.

It is now generally admitted that the process called "precipitation" actually can take place in at least two more-or-less-distinct stages. The first, which is usually detected at relatively low aging temperatures, has come to be known as the preprecipitation or cold-aging stage to distinguish it from the warm-aging or precipitation stage occurring at higher temperatures.

Such a classification has become necessary in order to explain the changes in physical and mechanical properties as well as the difference in certain X-ray effects detected during the decomposition process.

The general nature of the preprecipitation stage as well as a brief description of several X-ray techniques used in its study will be presented. This will be followed by a short résumé of preprecipitation phenomena detected in several alloys. These examples will illustrate not only the diverse types of preprecipitation encountered, but also indicate the role which is believed to be played by certain crystal imperfections during this first stage of aging.

GENERAL NATURE OF PREPRECIPITATION

There exists considerable experimental evidence that the preprecipitation stage does not involve the formation of precipitate particles characterized by a three-dimensional crystalline lattice.

X-ray diffraction studies during the 1920's revealed, for example, that the diffraction spectra of the precipitate phase could not be detected until after the hardness had passed through its maximum value. In addition, no precipitate phase could be observed under the optical microscope during the hardness increase. Such observations were not, in themselves, surprising since the particle size corresponding to the maximum hardness could well be submicroscopic. Thus, in addition to being invisible, they might also be too small to produce sharp diffraction effects.

The careful measurements of the matrix lattice parameter by Schmidt and Wasserman[1] in 1926 gave results which were quite disturbing. These authors observed that the parameter of the matrix in some aluminum-base alloys remained unchanged during the increase in hardness. Such a finding was most difficult to explain since the lattice parameter of the matrix was known to be a sensitive measure of its solute-atom content. Even if precipitate particles, otherwise too small to be detected, were being formed, it was only natural to expect a depletion of solute atoms in the matrix and a corresponding change in its lattice parameter.

In addition, Fraenkel and Seng[2] had shown in 1920 that the electrical resistance of duralumin initially increased slightly during the early part of low-temperature aging. According to theory, the electrical resistance should decrease continuously as solute atoms are rejected from the solid solution.

These as well as several other anomalies could not be satisfactorily explained by the classical precipitation theory as first proposed in 1919 by Merica, Waltenberg, and Scott.[3] Merica[4] was thus induced in 1932 to propose a modified theory. In the case of the alloy Al–Cu, which had been the most extensively studied, it was proposed that structural changes other than the formation of true precipitate particles took place during low-temperature aging. It was postulated that copper atoms, believed to be distributed at random in the alloy by the heat-treatment before quenching, diffused through the matrix at low temperatures to form copper-rich regions or "knots." These heterogeneities were supposed to be responsible for the changes in various physical and mechanical properties during low-temperature aging, where no precipitate phase could be detected. Heating at higher temperatures was supposed to result in the transformation of these knots in some unknown manner into nuclei of the precipitate phase. These eventually grew to form particles sufficiently large to produce diffraction effects and to be resolved with the optical microscope. The various mechanical properties such as hardness, which had previously attained their maximum values at a time when only the knots were present, were now much lower. The specimen is now, as the heat-treat metallurgist would say, "overaged" or "softened."

In 1938, working independently, Guinier[5,6] and Preston[7] simultaneously published results of great importance concerning the low-temperature aging of Al–Cu. Both these authors observed diffuse X-ray scattering from aged single crystals which they attributed to segregates of solute atoms in the supersaturated matrix. These segregates are now called Guinier–Preston or, more simply, G.P. zones. Extensive study of other age-hardening alloys has revealed the presence of various types of segregates, which are usually called zones.

Let us at this point consider some of the principal characteristics of zones. Following Guinier[8] zones can be divided into two classes: the ideal zone without distortion and the zone with distortion. The first, shown schematically in Figure 1(a), is formed when solute atoms occupy the lattice points of the matrix. Such zones are, of course, most easily formed when solute and solvent atoms have very nearly the same "size" as is the case in

---

[1] Superscripts pertain to references at the end of the paper.

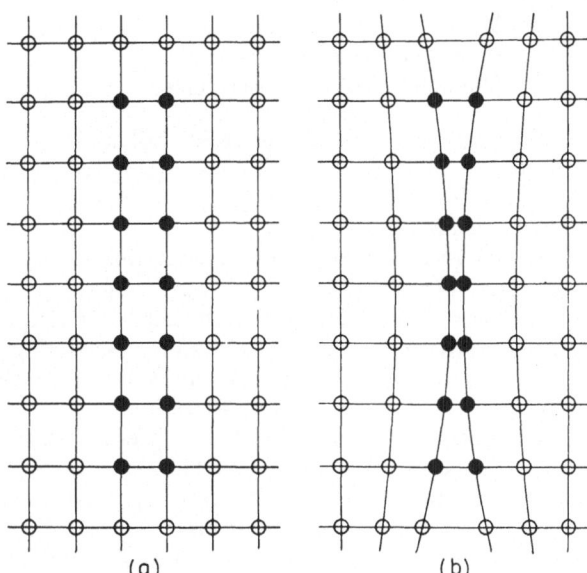

Figure 1. Schematic representation of zones:[8] (a) without distortion; (b) with distortion. Open circles: matrix atoms. Filled circles: zone atoms.

(a)   (b)

the systems Al–Ag and Al–Zn. As one can imagine, these are special cases as it is seldom that both solvent and solute atoms are of the same dimension. The more general case, illustrated schematically in Figure 1(b), involves the formation of zones in which the solute atoms do not occupy the lattice points of the solvent matrix. The displacement of the former results in a distortion of the matrix region near the zone. The solute atoms, although displaced from the matrix atomic sites, do not attain a periodicity sufficiently different from that of the matrix to be characteristic of a new phase. Another characteristic of zones is that they are coherent at their interface with the matrix. It is now generally held that coherency strains are responsible for the hardening effect.

Even though the interface between zone and matrix is probably difficult to define, it is possible from the shape and extent of the diffusely scattered X-ray effects to assign a shape to the zone, and to calculate its average dimensions.

Another characteristic of zones is their size, as shall be evident below. Such limited dimensions seem to be best explained as being due to stability. In order that the zone remain stable, so that its constituent atoms do not assume a structure entirely different from that of the matrix, all the zone atoms must be kept under the influence of the atoms of the matrix. This is, of course, the case when the atoms at the interior of the zone are never far from the zone–matrix interface. Experience up to now has shown that the largest dimension of a zone rarely exeeds 100 Å.

X-RAY STUDY OF PREPRECIPITATION

The non-periodic distribution of atoms and, for this reason, that of electronic density, in alloys during the preprecipitation stage results in the diffuse scattering of a small amount of the incident X-ray energy. The presence of scattering from this origin is evidenced by the appearance in diffraction patterns of streaks, halos, etc. in the region outside the diffraction maxima.

The electronic density in a crystal ceases to be periodic for one or more of the following reasons:

1. The distance between atoms is not periodic.

2. Atoms of differing scattering factors, i.e., atoms of differing chemical nature, are not arranged in a periodic sequence.
3. The number of otherwise correctly arranged atoms is too small.

The first and second have been called displacement and substitutional disorder, respectively, by Guinier,[9] who has attempted to explain X-ray scattering observed during low-temperature aging by a preprecipitation stage characterized by the formation of solute-rich regions or zones.

It can be shown that the intensity due to displacement disorder, which vanishes at the trace of the direct beam (zero scattering angle) increases with this angle. In contrast the intensity of the scattering produced by substitutional disorder increases with decreasing scattering angle since the atomic scattering factor attains its maximum value when the scattering angle becomes zero. It is thus possible, at least in theory, to differentiate between these two causes of scattering.

Examples of the third cause of scattering are very small particles or the presence of short-range order. Geisler and Hill,[10] as we shall note below, have attempted to explain scattering effects during early aging by employing the particle-size concept. That is to say, they consider that discrete precipitate particles, which may not in every case be structurally identical to the final or equilibrium phase, exist from the earliest moments of the decomposition process. These particles are so small in one or several directions that they do not satisfy the Laue conditions for the production of sharp diffraction spots.

Changes in the form of the scattered effects is, according to these authors, the result of growth during which the particles finally become large enough in all directions to diffract sharply.

The nature of the zones must be determined from a study of the shape, extent, and intensity of the scattered X-rays. As has been pointed out by Jagodzinski[11] the problem is very difficult and a unique solution does not seem possible. Up until now the problem has been attacked by comparing the observed scattering to that calculated from a proposed model of zone structure.

Experimentally, the study is rendered difficult for two reasons. First of all, the intensity scattered by zones is nearly always very low. In addition, even in the absence of zones, all crystals scatter X-rays. Also, the experimental equipment and even the air contribute their share of parasitic scattering. As we shall see, in order to most effectively study scattering due to zones, one must operate under two conditions: employment of intense sources of X-rays and either elimination or correction for all perturbing causes of scattering which might mask the effect being studied.

In the present study, parasitic scattering has the following origins:

1. Compton effect
2. Thermal agitation of the atoms
3. Fluorescent radiation
4. Accidental scattering from objects other than the specimen
5. Air

In accurate intensity measurements, it is necessary to know the components due to 1 and 2.

It is possible to calculate an approximate correction for Compton scattering following the method proposed by Compton and Allison[12]. The intensity of this scattering decreases with the scattering angle and, fortunately, becomes important only in the case of light elements. The component due to thermal scattering can be corrected for if one assumes,

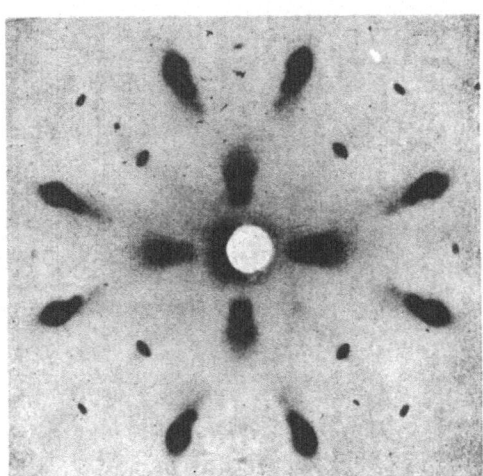

Figure 2. Laue transmission pattern obtained during aging of an Al–20%Ag alloy. Single-crystal specimen aged 2 years at 30°C plus 3 hr at 200°C. Crystal irradiated in cube orientation.[10]

according to Lambert,[13] that it is the same in an identical crystal free from structural defects. In the case of dilute alloys, it is possible to use the value of the pure solvent metal. As is the case with Compton scattering, the thermal component has a minimum intensity at low scattering angles. Guinier[14] has profited from this behavior by employing the low-angle scattering technique.

Fluorescence can be eliminated by the proper choice of the characteristic radiation. The intensity of the accidental scattering from pinhole systems, slits, specimen holders, etc., can usually be reduced to an acceptable level by proper design and construction and careful alignment. Scattering due to air can be effectively eliminated by either substituting a less absorbing atmosphere (hydrogen or helium) or, more simply, by placing the X-ray path in a vacuum chamber.

The X-ray study of preprecipitation involves the determination of the shape, extent, and intensity of the scattered effects. Since, as remarked above, the scattered intensity is nearly always very low, it is usually necessary to employ single-crystal specimens. In this manner, the intensity scattered at any given angle attains a maximum value.

The simplest technique is to obtain a diffraction pattern of an immobile crystal whose crystallographic orientation with respect to both the incident X-ray beam and the film is known.

The scattered effects are most easily interpreted by using the concept of the reciprocal lattice.[15] Such a lattice is constructed from the crystal lattice in the following manner.* One erects a normal to each set of planes in the crystal. Each of the latter is represented by a point whose distance from the origin of the reciprocal lattice is proportional to the reciprocal of the interplanar spacing in the crystal.

The reciprocal lattice is of considerable aid since both diffraction and scattered effects can be more easily interpreted with this "artificial" than with the "real" crystal lattice. The diffraction pattern of a fixed single crystal is, in effect, a distorted representation of a part of the reciprocal lattice. This pattern is composed of the matrix diffraction

---

* An excellent résumé is to be found in the review article of Hardy and Heal,[16] and a more complete description appears in the Guinier[9] text.

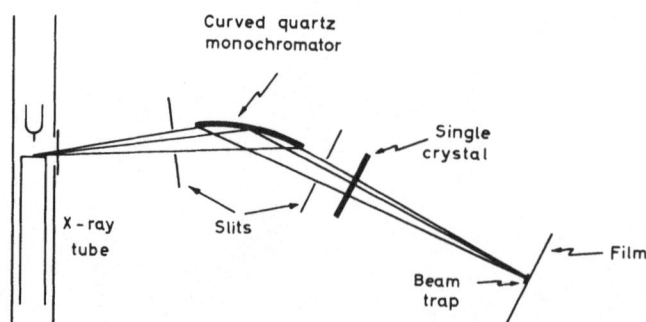

Figure 3. Schematic representation of curved quartz monochromatic technique.

maxima and the regions of scattered intensity situated in reciprocal space between the former.

If the crystal is rotated through a known angle and a second pattern obtained in the new fixed position, another region of reciprocal space is recorded. The distribution of scattering in reciprocal space is obtained from a series of such patterns. This time-consuming operation can be shortened by oscillating the crystal, thus obtaining the same information on one film.

The simplest experimental arrangement is the classical Laue transmission method employing polychromatic radiation. This method has been extensively employed by Preston[7,17,18] and by Barrett and Geisler et al.[10,19-22] An example of the diffraction pattern obtained with this technique is shown in Figure 2. The specimen employed was an aged single crystal of an Al–20%Ag alloy.

Although this technique has the advantage of being extremely simple, the presence of white radiation complicates the interpretation of the patterns. As has been suggested by Silcock, Heal, and Hardy,[23] this method can perhaps be most profitably employed after a preliminary study by more precise techniques has revealed the nature of the scattered regions.

A more elegant but complicated technique employing crystal monochromated radiation has been extensively employed by Guinier. As shown in Figure 3, the white radiation emitted by the X-ray tube irradiates the cylindrically curved quartz mono-chromator, placed in such a position that the K_α doublet is diffracted and brought to a short-line focus. The crystal to be examined is oriented in the convergent direct beam and the diffraction pattern obtained on a flat film placed normal to and at the focus of the incident beam.

An example of the diffraction patterns obtained with the aid of this technique will be illustrated below. This method possesses several advantages worth mentioning. The incident monochromatic beam is much more intense than that produced by a plane crystal monochromator. Used in conjunction with a fine-focus high-intensity X-ray tube the convergent beam is not only intense but is also geometrically well defined.

These characteristics of the curved monochromator have been utilized in the con-struction of cameras for the study of low-angle scattering. In order to eliminate as much of the parasitic scattering as possible, it has been found useful to employ two curved monochromators placed in series, as shown in Figure 4. A detailed study of the geometry by Fournet[24] has shown that the most effective arrangement is the antiparallel position as indicated in the above figure. The focus of the first monochromator serves as the sources for the second. The parasitic scattering of the first monochromator provoked by the intense beam from the X-ray tube as well as from the metallic supports of the crystal

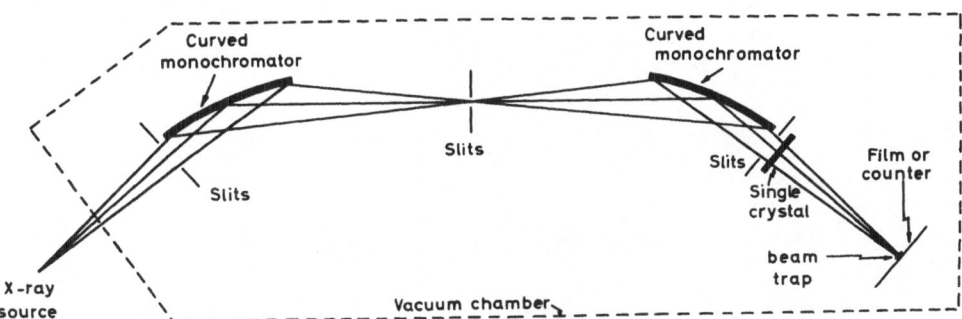

Figure 4. Schematic representation of double monochromator technique employed for low-angle scattering studies.

is eliminated by the second monochromator. According to Guinier and Fournet[25]* scattering is practically eliminated beyond an angle of about 10' from the direct beam. The film can be replaced by a counter tube to obtain accurate intensity measurements. The loss of intensity suffered at the two crystals renders necessary the use of a powerful source of X-rays, such as a rotating anode tube.

It must be remembered, however, that crystal monochromators also diffract the second and third harmonics of the fundamental component λ. These harmonics can be discriminated by the use of the proper filters. A second possibility is to decrease the high voltage of the X-ray tube so that the harmonics are not excited. This also greatly reduces the intensity of the fundamental component. The problem becomes much simpler in the case of proportional and scintillation counters since it is possible to eliminate the harmonics by an electronic discrimination.[26]

Several moving-film techniques, which produce patterns representing an essentially undeformed section of reciprocal space, have also been employed. Glocker *et al.*,[27] for example, have used the Schiebold–Sauter type of camera during the study of Al–Ag age-hardening alloys.

As a general rule the classical Debye–Scherrer and Seeman–Bohlin powder methods are incapable of giving information concerning structural changes taking place during preprecipitation. It has been possible, however, to increase the sensitivity of these techniques by substituting a crystal monochromated for the filtered radiation.

RESULTS OF X-RAY STUDIES DURING PREPRECIPITATION

The original papers published by Guinier[5,6] and by Preston[7] in 1938 announced the first experimental verification of segregate formation during aging in the alloy Al–Cu. During the following 24 years this system (especially the alloy Al–4%Cu) has been the most extensively studied.

In his two original papers, Guinier reported that the segregates or zones form on the {100} planes of the matrix. A good example of the scattering effects obtained with the monochromatic technique described above is shown in Figure 5. This pattern, due to Graf[28] was taken with a single crystal of Al–4%Cu aged 18 hr at 100°C oscillated about a matrix cube axis. A schematic representation of the intensity distribution in reciprocal space is shown in Figure 6. The streaks, passing through the modes of the matrix, are oriented along the matrix cube directions. This indicates that the zones are platelets.

* This work contains an extensive description of low-angle scattering techniques as well as a complete bibliography of the subject.

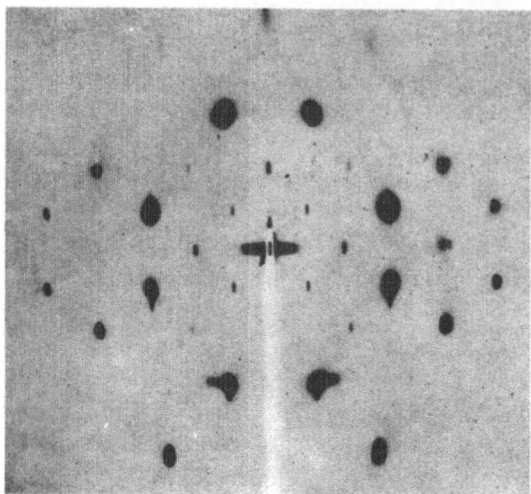

Figure 5. Scattering produced by an Al–Cu alloy aged 18 hr at 100°C. Single-crystal specimen oscillated about a vertical matrix cube axis. Crystal monochromatized Mo K_α radiation.[28]

The presence of a relatively intense scattering at low angles, as shown in Figure 7, reveals the existence of a substitutional disorder. This is to be expected in view of the considerable difference in the atomic scattering factors of $Al(Z = 13)$ and $Cu(Z = 29)$. The presence of scattered intensity on the high-angle side of the matrix reciprocal lattice points (Figure 6) makes it necessary to assume the existence of a displacement disorder. That is to say, the copper atoms in the zones are displaced from the lattice sites of the matrix atoms. The zones in Al–Cu are thus a good example of the type as shown schematically in Figure 1(b).

Measurement of the length of the streak passing through the origin of reciprocal space (the trace of the direct beam in Figure 7) gives an approximate idea of the average thickness of the zones. This turns out to be one or two atomic diameters. The width of the streak produced by an incident beam of limited length is a measure of the average diameter of the zones. The zones producing low-angle scattering, such as illustrated in

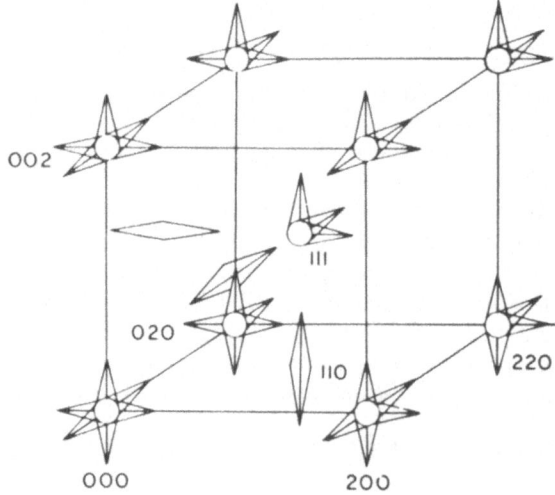

Figure 6. Schematic representation of diffuse X-ray scattering in the reciprocal space of the alloy Al–Cu.[28]

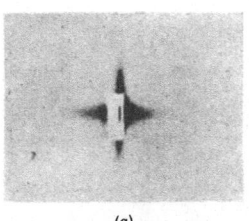 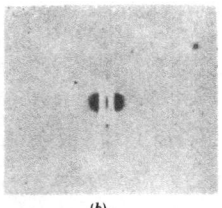

Figure 7. Low-angle scattering from Al–Cu alloys:[23] (a) Al–4%Cu, aged 6 months at 30°C, G.P. [1] zones, diameter 46 Å; (b) Al–4.5%Cu, aged 15 days at 130°C, G.P. [2] zones, diameter 400 Å.

(a)   (b)

Figure 7, are 46 Å in diameter. Graf[28] has reported that the diameters of the zones in Al–4% Cu aged at 20 and 100°C are 30 and 80 Å, respectively.

During aging at temperatures above 100°C, the streaks of the original zones begin to gradually disappear. At temperatures below 200°C, a series of diffuse spots appear. These are attributed to a structure which Guinier has called θ'' and Hardy[29] G.P. [2]. Although the structural nature of θ'' is still unknown, Guinier has maintained that G.P. [1] and θ'' are two different entities. Silcock, Heal, and Hardy[23] have reported two indications which substantiate the above assumption. A comparison of hardness curves and the diameter of the original zones G.P. [1] suggests that these are not thin platelets of θ''. In addition, these workers have shown that the change in the c parameter of θ'' suggests an ordering process resulting in the removal of Al atoms from the structure, thus decreasing coherency between it and the matrix.

According to Guinier,[30] θ'' possesses characteristics of both zones and precipitates. It appears to be different from an ordinary precipitate in that these entities never seem to grow large enough to produce sharp diffraction spots. On the other hand, one may think of θ'' as being a sort of superlattice of Al characterized by a tetragonal structure. Both G.P. [1] and θ'' produce a considerable hardening effect.

The structure θ'' transforms in some still unknown manner into the transition precipitate phase θ', first reported by Wassermann and Weerts[31] in 1936. With subsequent aging θ' is replaced by the equilibrium precipitate phase θ (CuAl₂).

Zones similar to those in Al–Cu are also detected in the alloy Cu–Be. Guinier and Jacquet[32,33] investigated a 2.3%Be alloy and obtained low-angle scattering results indicating the formation of Be-rich platelets on the matrix cube planes. It was observed that their thickness increased much more rapidly with respect to their diameter than did the zones in Al-Cu. As is the case in Al–Cu, the difference in size between Cu and Be atoms produces considerable distortion. The distribution of scattered intensity along the cube axes is different than in Al-Cu. With aging, the intensity along this direction, which was initially nearly uniform, suddenly exhibits the formation of a rather broad maximum corresponding to a lattice spacing of 2.7 Å. This value corresponds to the lattice parameter of the γ-phase (Be–Cu). The width of this Bragg spot cannot, however, be explained by the low-angle scattering results, which indicate a particle size of about 30 Å. It seems, according to Guinier,[30] that the cube planes of the platelets do not attain the regular periodicity characteristic of a well-defined crystalline structure.

In a more recent study, Geisler, Mallery, and Steigert[34] investigated an alloy containing 1.73% Be. These authors have attempted to explain, partly on the basis of metallographic evidence, that a sequence of structures are formed during aging. Following these authors the structures γ'', γ', and γ have the matrix habit planes {100}, {112}, and {111}, respectively.

The exact nature of the preprecipitation stage in this alloy is not well known at the present time. There seems to be some evidence[16] that there might exist at least one transition precipitate phase of unknown structure.

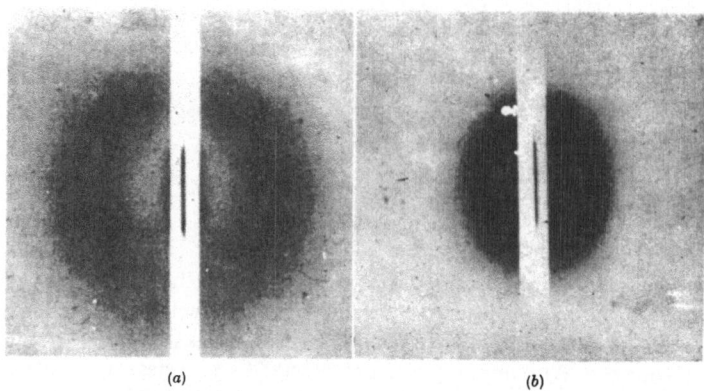

(a)   (b)

Figure 8. Low-angle scattering from Al–20%Ag. Radiation Cu K_α.[37] (a) specimen in as-quenched condition. (b) specimen aged 4 hr at 140°C.

In contrast to the large amount of distortion produced by zones in Al–Cu and Cu–Be, there exist at least two alloys in which zone formation is accompanied by little or no distortion. The most extensively studied example has been the alloy Al–Ag. The pure constituents both exhibit a face-centered cubic structure, and have very nearly identical lattice parameters. The resulting high solid solubility of Ag in Al, together with the large difference in atomic scattering factors (Al, $Z = 13$; Ag, $Z = 47$), as well as the absence of an appreciable difference in atomic size, makes it an interesting alloy from a theoretical standpoint.

This alloy has thus been extensively studied by three groups of workers, employing three different X-ray techniques.

Barrett and Geisler[20] and Barrett, Geisler, and Mehl[19,35] have investigated single crystals of an alloy containing 20% Ag by means of the Laue transmission method. They interpreted the experimentally observed scattering as being due to the presence of extremely small particles of an intermediate hexagonal precipitate phase, γ', coherent with the {111} matrix planes. This transition phase was replaced during the aging process by the equilibrium phase γ (Ag_2Al). A similar alloy has been studied by Guinier,[36,37] who employed monochromatic radiation with oscillating and immobile single crystals as well as the low-angle scatter technique. The latter is especially interesting because the rather large difference in atomic scattering factors of Al and Ag produces comparatively intense scattering in this region.

The type of low-angle scattering encountered during preprecipitation in Al–Ag is shown in Figure 8. In the as-quenched condition a halo is observed about the direct beam trace as shown in Figure 8(a). With continued aging, the diameter of this halo becomes smaller, as can be seen in Figure 8(b). This decrease is an indication that the average zone size increases with aging. It was also noted that the form of the halo is not influenced either by the orientation of a single crystal or the use of a fine-grain foil. The zones producing this scattering were for this reason considered to be roughly spherical in form.

Walker and Guinier[38] have developed a method of interpreting these X-ray results. They advance the idea that each zone is composed of a spherical silver-rich central region surrounded by a silver-poor spherical shell. Since identically shaped regions were observed about the matrix diffraction spots, these authors assumed that the silver atoms

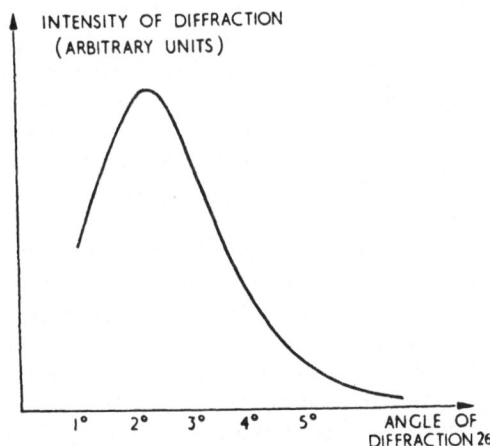

Figure 9. Intensity distribution near trace of direct beam of an Al–20%Ag alloy quenched from 500°C.[38]

are not displaced from the matrix lattice sites. They also consider that the zones are distributed at random throughout the crystal.

Precise intensity measurements as a function of the scattering angle obtained from a specimen in the as-quenched condition give the curve shown in Figure 9. A Fourier transform of this intensity distribution permitted Walker and Guinier to calculate the probability, P_{Ag-Ag}, of finding a silver atom at a distance r from a given solute atom. This probability curve for a quenched alloy containing 20%Ag is shown in Figure 10. It can be seen that the curve decreases to a value equal to that characteristic of a random distribution of 0.05 (indicated by the horizontal dotted line) at a distance of about 16 Å. The outer, silver-poor region has a diameter of about 80 Å. The excess of silver in the central region is supposed to be equal to the deficiency in the outer shell.

Precise intensity measurements obtained by Walker, Blin, and Guinier[39] later showed that extremely small silver-rich clusters also exist at temperatures corresponding to those of a solution heat-treatment. The average size of these clusters increases with decreasing temperature above that corresponding to the limit of solid solubility. Rudman, Flinn, and Averbach[40] have also obtained similar results from several Al–Ag alloys.

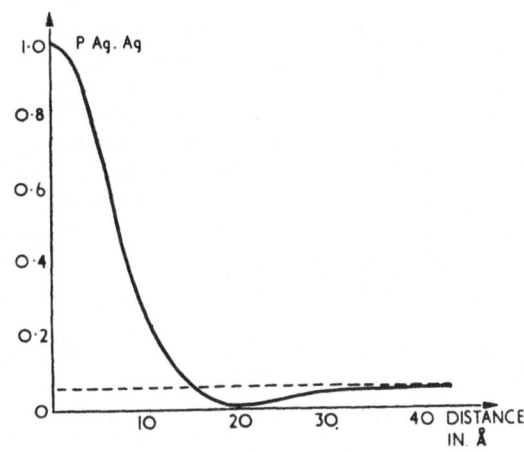

Figure 10. Probability, P_{Ag-Ag} of finding a Ag atom at a distance r from a given Ag atom. Horizontal broken line gives value of P_{Ag-Ag} for a random distribution of Ag atoms.[38]

A third group of investigators has studied an Al–38%Ag alloy using the Laue and Schiebold–Sauter techniques.

Glocker, Koster, Scherb, and Ziegler[27] and Ziegler[41] interpreted the streaks on their patterns as being produced mainly by staking faults in successive layers of the close-packed atoms of the matrix. Ziegler[41] has explained the intensity distribution by the interposition of hexagonal plates coherent with the matrix. The plates were estimated to be about six atom layers in thickness and are supposed to be irregularly spaced in the matrix.

The first group of workers has investigated the behavior of the 38% Ag alloy during reversion. They report the surprising observation that the streaks, present after an aging treatment of 6 hr at 150°C, were still visible after a reversion treatment of 2 min at 220°C. Both hardness[42] and electrical resistivity[43] measurements, however, have shown that the value of these two parameters attain those characteristic of the quenched and unaged alloy after an identical treatment. It thus seems that hardening is not produced by the structural changes producing the X-ray streaks. Hardy and Heal[16] have discussed this point in some detail and conclude that the spherical silver-rich segregates are probably the most likely cause of the first change in hardness during low-temperature aging.

Guinier,[37] however, has indicated that when an alloy aged at room temperature is heated to higher temperatures, the zones seem to grow instead of dissolve. Hardy and Heal have attempted to explain this apparent contradiction in the following manner. Many small spherical zones are present in the matrix after treatment at 150°C. During the short-time heating to 220°C, a large number of these zones probably decompose or dissolve. A few favored zones could, however, grow to a larger size during reversion. Resistance to slip, being greater in the presence of a large number of small zones, decreases, as does the hardness, to a value near that of the as-quenched specimen. There is, however, as yet no satisfactory explanation of the Laue streaks during reversion.

The preprecipitation stage in the alloy Al–Zn is similar in many respects to that of Al–Ag. Using the low-angle scattering technique, Guinier[44] obtained scattered effects in a quenched Al–34%Zn alloy which were attributed to the formation of spherical zinc-rich zones. These were also present after an aging treatment of up to 44 hr at 78°C. Geisler, Barrett, and Mehl[19] have investigated an Al–25%Zn alloy with the transmission Laue method and found streaks on their films after a treatment of 8 days at 100°C.

As was the case in Al–Ag, small spherical zinc-rich clusters seem to exist at equilibrium in the high-temperature single-phase region.[40]

An interesting needle-shaped zone appears to exist during preprecipitation in the commercial alloy Al–Mg–Si. When the Mg and Si contents are correctly proportioned this alloy is actually a pseudobinary system (Al–Mg_2Si) in which the stable precipitate is the intermetallic compound Mg_2Si.

Preprecipitation in this alloy has been extensively investigated by Guinier and Lambot,[45] Lambot,[46–48] and Geisler and Hill.[10]

In general, the results reported by these two groups of workers are not in good agreement. Such lack of accord can perhaps be attributed to the experimental techniques employed as well as to the difference in chemical composition and heat-treatment of the alloys.

During the early aging period both groups of workers reported the presence of X-ray scattering along planes of reciprocal space parallel to the matrix cube faces. Guinier and Lambot[45] have investigated an alloy containing 0.79% Mg_2Si with the monochromatic single-crystal technique. They observed that the intensity along these vertical streaks became non-uniform producing what Lambot[47] has called the "checker-board" pattern, as shown schematically in Figure 11. This effect was interpreted as being provoked by an ordering in groups of originally randomly distributed zones. Measurement of the length and width of the lines showed that the zones, when fully developed, are several hundred Angstroms in length but only a few atomic diameters in width and thickness. In addition, it was possible to show that the long axis of the zone is parallel to a matrix cube direction along which it is coherent. It is interesting to note at this point that there is a marked similarity between the shape as well as the orientation with respect to the matrix of both zones and Mg_2Si particles, as revealed by the electron microscope results of Castaing and Guinier.[49] A similar resemblance also appears to exist in the alloy Al–Cu.[50] Geisler and Hill[10] have employed the particle-size concept in the interpretation of their Laue transmission patterns obtained during the aging of an Al–1.4% Mg_2Si alloy. The changes in X-ray scattering, according to these authors, can be explained by the growth of extremely thin needles of a precipitate phase into three-dimensional particles sufficiently large to produce sharp Bragg diffraction spots. This growth is supposed to take place first by a thickening of the initial "stringlets" into "platelets." The latter then grow in the third direction until all three of the Laue conditions are satisfied. These workers never observed the "checkerboard" effect reported by Lambot. Hardy and Heal[16] are of the opinion that this could be due to the lack of sensitivity of the Laue technique.

Another type of zone seems to be present in the ternary alloys Al–Cu–Mg and Al–Mg–Zn. The X-ray patterns produced during the preprecipitation stage of these alloys are quite complex, and at the present time, the form and nature of the zones is not at all well known. We will briefly discuss the effects observed in Al–Mg–Zn, since this alloy has been the most extensively studied.

Graf[51] has investigated an alloy containing 7% Zn and 3% Mg. The general nature of the early diffuse scattering stage is shown in Figure 12. It can be seen that the scattered regions on the film are nearly circular. In addition, they appear to be situated at well-defined sites of the matrix. For example, the most intense regions are at the 100 matrix points, of which three are visible near the center of the pattern: at the left, at the right, and just above the trace of the direct beam.

Guinier[30] has discussed this type of diffuse scattering in some detail. Such an effect is observed when the zones are believed to contain more than one type of solute atom. In addition, there also seems to be a similarity between the present patterns and those obtained from single crystals of incompletely ordered alloys. Schmalzried and Gerold[52] have advanced the idea that there might exist an ordering of alternate (100) planes of

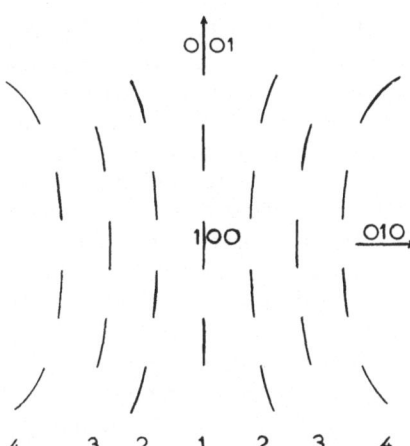

Figure 11. Schematic representation of "checker board" pattern produced during aging of a Al–Mg–Si alloy.[46]

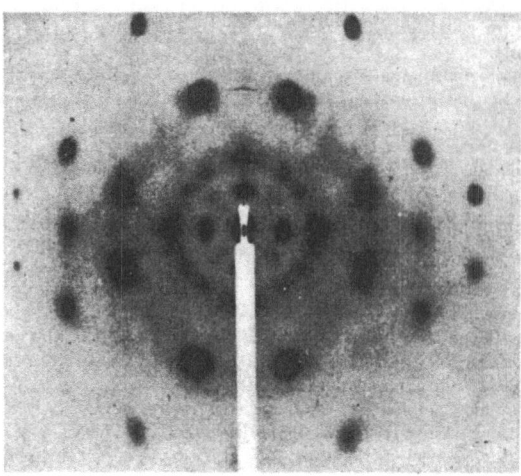

Figure 12. Scattering produced by a single crystal of an Al–7%Mg–3%Zn alloy aged 240 hr at 20°C. Specimen in cube orientation.[51]

Mg and Zn. A third interesting point mentioned by Guinier is that there may actually be two types of zones simultaneously present in such ternary alloys.

The low-angle scattering observed by Graf[51] could be due to zones rich in zinc, such as those which exist in the binary alloy Al–Zn. The second type of zone, which produces scattering at higher angles, could be composed of an ordered array of Mg and Zn atoms. This possibility seems to be supported by the observations reported by Lambot[47] in the case of the alloy Al–Mg–Cu. Low-angle scattering diagrams revealed the presence of streaks characteristic of zones in Al–Cu whereas zones containing the two species of solute atoms produced scattering only at higher angles.

In the alloy Al–Mg–Zn, the preprecipitation stage seems to be further complicated by the change of the initial scattered effects to other sites in reciprocal space as the result of aging at temperatures above 50°C.

It is quite evident that the complex scattering taking place in these ternary alloys has to be the subject of considerable study before the nature of zone formation can be known with a degree of certainty.

ROLE OF CRYSTAL IMPERFECTIONS IN ZONE FORMATION

There can be but little doubt that zones are formed and grow as a result of atomic diffusion of solute atoms through the matrix. That is, solute atoms which, in a first approximation, can be considered to be more or less randomly distributed through the matrix at elevated temperatures, must, at lower temperatures, find their way through the matrix to certain regions, where they collect to form zones. As we have seen, extremely small solute-rich regions can exist even at the elevated temperatures corresponding to the single-phase region of the phase diagram. Although these segregates might perhaps be potential nuclei for zones, they never grow beyond a certain size at a given temperature.

Zones are, however, frequently detected at room temperature in some alloys not long after the quench from elevated temperatures. Such a rapid zone formation indicates that atomic diffusion is very high.

It is possible to obtain an idea of the diffusion coefficient at low temperatures and compare it to that which one would expect from the extrapolation of the high-temperature value. The observed value is nearly always larger than that obtained by extrapolation by several powers of ten. The observed value is, of course, only a first approximation since one is obliged to assume values for the chemical composition of the zones as well as the variation of concentration in the matrix immediately surrounding the zone, etc.— parameters which are not at all well known. The difference between the two values is, however, much too great to be explained by either the above considerations or by the method of calculation employed. Hardy,[53] for example, has calculated the coefficient of diffusion of Cu during the formation of G.P. [1] zones in an Al–5.2%Cu alloy. The calculated value was 2.8×10^{-13} cm^2/day while that obtained by the extrapolation of the value measured by Mehl and co-workers[54] was 4.6×10^{-20} cm^2/day.

In order to explain the high rate at which zones are formed at low temperatures, solid state physicists have considered the possibility that the diffusion of solute atoms is aided by crystal imperfections. Among these imperfections, dislocations and unoccupied lattice sites or vacancies have been and continue to be the most extensively studied.

Turnbull[55] has assumed that dislocations form convenient and easy paths for the diffusion of copper atoms in Al–Cu. Since only about 1 in 10^6 atoms will be near a dislocation in an annealed crystal, he assumed that some of the dislocations had to sweep through the matrix during low-temperature aging. Although such a model could perhaps explain rapid zone formation in alloys where lattice distortions are involved (Al–Cu for example) such a mechanism is difficult to imagine in the case of alloys such as Al–Ag and Al–Zn.

Attention has thus turned toward the vacancy. This highly mobile point defect is generated in metals during heating. The number of vacancies present in a metal at equilibrium at a given temperature is proportional to the Boltzmann factor $\exp(-Qv/RT)$, where Qv is the activation energy for vacancy formation, R is the gas constant, and T the absolute temperature. Near the melting point of pure metals the vacancy content is believed to be about 10^{-4}. In quenched pure metals, the high vacancy content rapidly decreases or "anneals out" to a value characteristic of the lower temperature. This rapid recovery has been demonstrated by the electrical resistivity measurements of pure Au by Kauffman and Koehler.[56]

The situation seems to be quite different in the case of alloys. Here, there is considerable evidence which indicates that nonequilibrium vacancy concentrations can be retained by quenching. Instead of rapidly diffusing toward other imperfections (dislocations, subgrain and grain boundaries, free surfaces, etc.) at least some of the vacancies can be thought of as being "trapped" by solute atoms. These retained vacancies are thus available to aid in zone formation.

Perhaps the first mention of the role played by vacancies during preprecipitation is that due to Röhner.[57] This author tried to explain zone formation in Al–Cu by the interstitial diffusion of Cu atoms and the simultaneous formation of vacancies. The latter were supposed to be a partial cause of the hardening observed. Seitz,[58] however, is of the opinion that such vacancy complexes should rapidly condense and, for this reason, impart little or no hardness. He considers, on the other hand, the possibility that vacancies are bound to copper atoms and aid in the diffusion of the latter through the matrix at low temperatures.

Recent results[59–61] obtained during the study of preprecipitation in high-purity Al–Mg–Si and Al–Mg–Ge appear to indicate that a high vacancy content is associated with primitive zones in the former and probably in the latter alloy as well. As mentioned above, the preprecipitation stage in the alloy Al–Mg–Si is characterized by the formation of needle-shaped zones, which are linearly coherent with the matrix along its cube axes.

The alloy Al–Mg–Ge is crystallographically similar to Al–Mg–Si in that the intermetallic compound Mg$_2$Ge isomorphous to Mg$_2$Si is the equilibrium precipitate phase. In the alloy Al–Mg–Si we have the reverse situation to that which exists in Al–Ag. The three elements Al, Mg, and Si have very nearly the same X-ray scattering factors. In addition, while the diameter of the Si atom is only slightly smaller than that of Al, the size of Mg is nearly 10% greater than the solute atom. One would thus think that during zone formation, X-ray scattering at high angles, characteristic of a displacement disorder, only would be observed. In the alloy Al–Mg–Ge both low and high-angle scattering indicating the existence of substitutional and displacement disorders might be expected since both the large solute atom (Mg) and a heavier element (Ge) are both present.

It now appears that zones in these two alloys are formed in at least two steps. The primitive zones do not exhibit any internal periodicity whatever. With continued aging a periodicity equal to that of the matrix along its cube axis (4.04 Å) is established along the axis of the acicular zone.

During the early part of preprecipitation, the X-ray patterns of the two alloys are identical. As soon as the establishment of ordering in the zones appears, however, the scattered effects become different. The scattering from the Al–Mg–Si alloy which, in the first stage, was continuous near the center of the diagram suddenly decreased to a very low value. The same effect in the case of the alloy Al–Mg–Ge remained essentially constant during the second stage of zone development.

This difference can be explained in the following manner. The primitive zones in both alloys probably contain a high vacancy content, which prevents the establishment of an internal order. The low-angle scattering in Al–Mg–Si could be due to the difference in X-ray scattering factors between a vacancy or a group of vacancies and Al, Mg, and Si. The primitive zones grow by the diffusion of solute atoms, which gradually replace the vacancies. The number of vacancies present decreases thus resulting in a decrease of the intensity of X-rays scattered at low angles. Simultaneously, a linear periodicity can be established in the zone. In the alloy Al–Mg–Ge, the low-angle scattering does not change since some of the vacancies are being replaced by the Ge atoms. The elimination of the originally high vacancy content should permit the formation of a linear periodicity in the zone in Al–Mg–Si as well.

Using this model one can describe zone formation and growth in Al–Mg–Ge and Al–Mg–Si in the following manner. Many of the vacancies produced by the high-temperature treatment are attracted to and trapped by large solute (Mg) atoms. In the as-quenched condition some of these solute atoms must have vacancies associated with them. Although the nature of this "association" is not known, it is possible to suppose, for the moment at least, that the vacancy is situated relatively close to its solute atom. Such a complex probably diffuses relatively easily through the matrix. The primitive zone can be thought of as a collection of such complexes as well as other solute atoms (Si or Ge depending upon the alloy) and perhaps also some Al atoms. Growth of the zones takes place by an enrichment in solute atoms which upon reaching the zone, find it easier to simply replace a vacancy than eject a nearby atom. The result is a simultaneous enrichment of the zone in solute atoms and a decrease in its vacancy content. Vacancies thus "liberated" from the zone can diffuse away from it to a crystal imperfection or become reassociated with large solute atoms.

The above model of zone formation and growth in these two alloys seems to be reasonable in light of the results reported by Panseri, Gatto, and Federighi.[62] These authors have shown that there is no significant recovery in the electrical resistivity at room temperature when high-purity Al–Mg alloys were quenched from 550°C. It must be stressed here that the Mg content of these alloys never exceeded the solid solubility in Al at room temperature. Furthermore, it was found that nearly all of the increase in electrical resistivity, with respect to pure Al, could be recovered by an annealing treatment of 80 to 120°C. This recovery was attributed either to the thermal dissociation of the Mg atom–vacancy couples or to their diffusion through the matrix to dislocations or other crystal imperfections.

Vacancies can be produced by deformation and by particle irradiation as well as by heating. These two techniques are of special interest because they can be employed at any time during the preprecipitation stage. It has also been observed that the kinetics of zone formation and growth can be greatly modified by the addition of small quantities of an "impurity," which in many respects seems to behave as a catalyst. An excellent review of these last three topics as well as the properties of zones can be found in the paper of Guinier.[30]

CONCLUDING REMARKS

Despite the experimental and theoretical difficulties encountered, the study of diffuse X-ray scattering has greatly enriched our knowledge of structural changes taking place during preprecipitation. These X-ray techniques have made it possible to detect the small structural and chemical inhomogeneities called zones, to determine their shape and to obtain an idea of their average dimensions.

Such studies have shown that the nature as well as the shape of zones can be quite different from one alloy to another.

In addition to providing a better understanding of the structural changes taking place during preprecipitation, these studies are beginning to give us more direct information concerning the role played by crystal imperfections.

As is fitting here we have stressed the contribution of X-ray techniques. However, one must not neglect the fact that a simultaneous study of preprecipitation by one or several additional techniques frequently provides valuable information without which the X-ray investigation might be inconclusive.

ACKNOWLEDGMENTS

The author wishes to express his appreciation to Dr. L. Habraken, Chief Research Engineer and Dr. J. Gouzou, Head of the Fundamental Research Department of the C.N.R.M. for reading and commenting on this manuscript.

REFERENCES

1. E. Schmidt and E. Wasserman, *Naturwissenschaften* **14**: 980, 1926.
2. W. Fraenkel and R. Seng, *Z. Metallkunde* **12**: 225, 1920.
3. P. D. Merica, R. G. Waltenberg, and H. Scott, U.S. Bureau of Standards Scientific Paper 347, 1919; *AIME Trans.* **64**: 41, 1921.
4. P. D. Merica, *AIME Trans.* **99**: 13, 1932.
5. A. Guinier, *Compt. rend.* **206**: 1641, 1938.
6. A. Guinier, *Nature* **142**: 569, 1938.
7. G. D. Preston, *Nature* **142**: 570, 1938.
8. A. Guinier, *J. Metals* **8**: 673, 1956.
9. A. Guinier, *Théorie et technique de la radiocristallographie*, Dunod, Paris, 1956.
10. A. H. Geisler and J. K. Hill, *Acta Cryst.* **1**: 238, 1948.
11. H. Jagodzinski, *Z. Metallkunde* **46**: 491, 1955.
12. A. H. Compton and S. K. Allison, *X-rays in Theory and Experiment*, D. Van Nostrand Co., Inc., Princeton, N. J., 1935.
13. M. Lambert, Thesis, University of Paris, 1958.
14. A. Guinier, *Ann. phys.* **12**: 161, 1939.
15. P. P. Ewald, *Proc. Phys. Soc. (London)* **44**: 257, 1914.
16. H. K. Hardy and T. J. Heal, "Report on Precipitation," *Prog. Metal Phys.* **5**, 1954.
17. G. D. Preston, *Proc. Roy. Soc. (London)* A **167**: 526, 1938.
18. G. D. Preston, *Phil. Mag.* **26**: 855, 1938.
19. A. H. Geisler, C. S. Barrett, and R. F. Mehl, *AIME Trans.* **152**: 182, 1943.
20. C. S. Barrett and A. H. Geisler, *J. Appl. Phys.* **11**: 733, 1940.

21. A. G. Guy, C. S. Barrett, and R. F. Mehl, *AIME Trans.* **175**: 216, 1948.
22. A. H. Geisler and R. F. Mehl, *AIME Trans.* **143**: 134, 1941.
23. J. M. Silcock, T. J. Heal, and H. K. Hardy, *J. Inst. Metals* **82,** 1953–54, p. 239.
24. G. Fournet, *Bull. soc. franç. minéral. et crist.* **74**: 39, 1951.
25. A. Guinier and G. Fournet, *Small-Angle Scattering of X-rays,* John Wiley & Sons, Inc., New York, 1955.
26. P. H. Dowling, C. F. Hendee, T. R. Kuhlin, and W. Parrish, *Phillips Tech. Rev.* **18**: 262, 1956.
27. R. Glocker, W. Köster, J. Scherb, and G. Ziegler, *Z. Metallkunde* **43**: 208, 1952.
28. R. Graf, Publs. sci. et tech. ministère air, No. 315, 1956.
29. H. K. Hardy, *J. Inst. Metals* **75**: 707, 1948–49.
30. A. Guinier, "Heterogeneities in Solid Solutions," in *Solid State Physics, Vol. 9,* Academic Press, Inc., New York, p. 345.
31. G. Wassermann and J. Weerts, *Metallwirtschaft* **14**: 605, 1935.
32. A. Guinier and P. Jacquet, *Compt. rend.* **217**: 22, 1943.
33. A. Guinier and P. Jacquet, *Rev. mét.* **41**: 1, 1944.
34. A. H. Geisler, J. H. Mallery, and F. E. Steigert, *J. Metals* **4**: 307, 1952.
35. C. S. Barrett, A. H. Geisler, and R. F. Mehl, *AIME Trans.* **143**: 134, 1941.
36. A. Guinier, *J. phys. rad.* **8**(3): 124, 1942.
37. A. Guinier, *Z. Metallkunde* **43**: 217, 1952.
38. C. B. Walker and A. Guinier, *Compt. rend.* **234**: 2379, 1952.
39. C. B. Walker, J. Blin, and A. Guinier, *Compt. rend.* **235**: 254, 1952.
40. P. S. Rudman, P. A. Flinn, and B. L. Averbach, *J. Appl. Phys.* **24**: 365, 1953.
41. G. Ziegler, *Z. Metallkunde* **43**: 213, 1952.
42. W. Köster and F. Braumann, *Z. Metallkunde* **43**: 193, 1952.
43. W. Köster, H. Steinert, and J. Scherb, *Z. Metallkunde* **43**: 202, 1952.
44. A. Guinier, *Métaux et corrosion* **18**: 209, 1943.
45. A. Guinier and H. Lambot, *Compt. rend.* **227**: 74, 1948.
46. H. Lambot, Doctorate Thesis, University of Paris, 1950.
47. H. Lambot, *Rev. mét.* **47**: 709, 1950.
48. H. Lambot, *Bull. acad. roy. soc. Belg.* **26**: 1609, 1950.
49. R. Castaing and A. Guinier, *Compt. rend.* **229**: 1146, 1949.
50. R. Castaing and A. Guinier, *Compt. rend.* **228**: 2033, 1949.
51. R. Graf, *Compt. rend.* **242**: 1311, 1956; **244**: 337, 1957.
52. H. Schmalzried and V. Gerold, *Z. Metallkunde* **49**: 291, 1958.
53. H. K. Hardy, *J. Inst. Metals* **79**: 321, 1951.
54. R. F. Mehl, F. W. Rhines, and K. A. van der Steinen, *Metals & Alloys* **13**(1): 41, 1941.
55. D. Turnbull, *Report Bristol Conference on Defects in Crystalline Solids, 1954,* 1955, p. 203.
56. J. W. Kauffman and J. S. Koehler, *Phys. Rev.* **97**: 555, 1955.
57. F. Röhner, *J. Inst. Metals* **73**: 285, 1947.
58. F. Seitz, *L'état solide,* R. Stoops, Brussels, 1952, p. 405.
59. A. Lutts and H. Lambot, *Rev. mét.* **54**: 775, 1957.
60. A. Lutts, Doctorate Thesis, University of Liège, 1960.
61. A. Lutts, *Acta Met.* **9**: 577, 1961.
62. C. Panseri, F. Gatto, and T. Federighi, *Acta Met.* **6**: 198, 1958.

DISCUSSION

M. Semchyshen (Climax Molybdenum Company of Michigan): In the copper-beryllium system you have several transitions. Do you observe the same phenomenon in each of these?

A. Lutts: Yes, one observes an effect which is somewhat similar to the zone in aluminum copper, although there are some minor differences.

Z. Wilchinsky (Esso Research): Is it necessary to have single crystals to observe the diffuse scattering?

A. Lutts: The interpretation is much easier if one uses single crystals because the scattering effects are much more intense with single crystals than with powdered specimens or polycrystalline specimens. And the interpretation is much easier because one can use the reciprocal lattice, which is uniquely determined if you have one monocrystal.

FACTORS GOVERNING THE STRUCTURE OF INTERMETALLIC PHASES

F. Laves

*Institut für Kristallographie und
Petrographie der Eidg. Techn. Hochschule, Zürich, Switzerland*

ABSTRACT

A review of the crystal structures of the elements shows some prevailing tendencies of atomic arrangement. These are discussed as space, symmetry, and connection principles. Counteracting temperature and bond factors can be recognized.

The same principles and factors are responsible for the formation of alloy structures, taking into account additional factors due to the component's similarities and dissimilarities in size and electronegativity.

Similarity favors solid solution and dissimilarity favors compound formation. A q compound is here defined as a phase in a q-component system not connected with any other phase of the system by continuous solid solution. Similarly a q structure is defined as a structure type which needs only q components to be formed (considering present-day knowledge). For example, the binary compound $Mg_{17}Al_{12}$ has the elementary $(1-)$ structure of α-manganese. As a rule, q compounds tend to form p structures with $q > p$.

A discussion of q structures with $q = 1$, 2, and 3 is given in some detail on the basis of known representatives to show: (1) the competition of geometrical principles and physicochemical factors in determining atomic arrangements of alloys; and (2) the value of rules for making guesses on the probable occurrence of compounds and their chemical composition in polycomponent systems yet unknown.

INTRODUCTION

As is well known, the chemical compositions of intermetallic compounds usually do not follow the normal valency concepts of chemistry. Take for example the binary system K–Na. Why should the compound KNa_2 exist? Why is there no compound KNa, as both elements are univalent, or even K_7Na?

One of the main reasons for the difficulty in answering such questions is that the attracting forces between the atoms in metallic phases are of a rather general character. Therefore, *the geometrical properties of three-dimensional space* become more significant than in other classes of compounds and they, to a great extent, determine the chemical composition of intermetallic compounds.

GEOMETRICAL PRINCIPLES

When metals crystallize, their individual bonding tendencies must adjust themselves to the properties of space. They must meet, as well as possible, the requirements of certain geometrical principles which are recognized, on the basis of many structure determinations, as being primarily responsible for the formation of the structure of metals and their compounds.

A discussion of the structure of the elements will make clear the main principles involved. Let us divide the periodic table by a vertical line (the *Zintl line*) which runs between the rows C, Si, Ge, Sn, and Pb on the right side, and B, Al, Ga, In, and Tl on the left, and let us call the elements to the left of this line "metals." Then, looking at the crystal structures of these metals, we note that most structures are cubic or hexagonal close-packed with the coordination number (CN) equal to 12. (This is actually true for 52 out of 91 structures.) If we admit some tetragonal or rhombohedral deformation, and if we consider distances differing by not more than 10% as equivalent, we even have 58 close-packed structures. (A survey is given elsewhere.[1])

One may well conceive that some—if not most—metals would prefer arrangements with higher CN's, e.g., 18 or another number greater than 12. This is, however, geometrically impossible if the atoms of a metal are to show indistinguishable behavior. Under this condition, 12 is the highest CN possible.

If we consider the atoms as spheres in contact, space will be filled in the best possible way when the atoms are arranged in cubic or hexagonal close-packed structures with the "ideal" CN = 12. As most of the metals crystallize in these two types, the tendency to good space filling may be called the "space principle."

The next most frequent type of structure—23 of the remaining 33—is the body-centered cubic structure with CN = 8. Thus, there must be factors (e.g., temperature and bond factor, to be discussed later) which counteract the space principle. In this connection, however, it appears remarkable that the CN drops from 12 to 8, whereas sphere packings are available with a CN less than 12 but greater than 8, which would lead to a better space filling than that reached by the body-centred cubic arrangement. However, sphere packings with CN equal to 11, 10, or 9 would not be as symmetrical as the body-centered cubic structure. Thus a tendency to form arrangements of high symmetry becomes apparent. This tendency may be called the "symmetry principle."

A third geometrical tendency which plays an important role in the formation of compound structures may be called the "connection principle." Let us first define this concept of "connection." Imagine a structural arrangement where each atom is connected with all the others. There will be a shortest link. If all the links except the shortest are dropped, those atoms that are still connected with each other form a "connection." If the connection consists of structurally equivalent atoms, it is called *homogeneous*; if it consists of structurally nonequivalent atoms, it is called *heterogeneous*. The connections can be of finite or infinite extent and be one, two, or three-dimensional. Accordingly, we name them islands, chains, and nets or lattices. They are symbolized by the letters *I*, *C*, *N*, or *L* (capital letters in the case of homogeneous, small letters in the case of heterogeneous connections). By adding the CN's to these letters, the main geometrical features of structures that are not too complex can be represented by a short symbol. We may now define the *connection principle*: It is the tendency of the atoms to form connections of "high" dimension.

Whereas the importance of the connection principle can best be seen in a discussion of compound structures, it is already observable in the structures of the elements. In these, three-dimensional lattice connections are certainly the most frequent.

A discussion of the hexagonal close-packed metals may be of special interest. As is known, the CN = 12 is only exactly realized if the ratio $c/a = 1.63$. If c/a deviates from this value, and if we retain the exact definition of the CN, the latter drops from 12 to 6. Any deviation of c/a to higher values leads to *net* connections with a CN of 6, whereas any deviation of c/a to smaller values leads to *lattice* connections with a CN also 6.

---

[1] Superscripts pertain to references at the end of the paper.

Table I. Axial Ratios c/a of Metals with Hexagonal Close-Packed Structure

c/a	Elements*
1.56	Li (in the low-temperature state)
1.57	Be, Y, Tm
1.58	Tb, Dy, Er, Ru, Os
1.59	Sc, Gd, Ti, Zr, Hf, Lu
1.60	Tc, Tl
1.61	La, Pr, Nd, Sm, Ho, Re, Am
1.62	Mg, Ce, Co
1.63	Ca, Sr
1.86	Zn
1.89	Cd

* Some of the elements listed are polymorphous.

Table I shows the actual data. Note that only zinc and cadmium have c/a values above 1.63; the majority has values lower than 1.63. Thus we see that, whereas the space principle is somewhat violated by the deviation from the ideal c/a value, the connection principle is not (except in the cases of zinc and cadmium to be discussed later).

If all these three principles, space principle, symmetry principle, and connection principle, are exactly followed, the cubic close-packed structure results. However, many metals or some of their modifications do not crystallize in the cubic close-packed structure and do not exactly follow the main principles. Thus, there are counteracting factors, and these will now be discussed.

FACTORS THAT MAY COUNTERACT THE GEOMETRICAL PRINCIPLES

I. Formation of Special Bonds (Bond-Factor)

   a. Completion of electron shells where s and p levels are involved.

   (1) *Hume-Rothery rule*, which holds for the elements to the right of the Zintl line and leads to e.g., to the CN 4 in the cases of germanium and gray tin.

   (2) *Hume-Rothery tendency*, which causes white tin to have the CN 6 and influences the structures just to the left of the Zintl line; e.g., gallium has a CN 1 or 7 (depending on the extent to which distances may differ and still be treated as equal); indium has a CN 4, and zinc and cadmium form $6N$ connections, if only the very shortest distances are considered.

   b. Completion of electron shells where d or f levels are involved. Examples are the α and β modifications of manganese, uranium, and neptunium and the $8L$ connections of the modifications of iron and chromium stable at room temperature.

II. Effect of Temperature (Temperature-Factor)

   a. *Goldschmidt's Rule:* Rising temperature favors lower CN; e.g., titanium, zirconium, hafnium, thorium, thallium have $12L$ connections at low, but $8L$ connections at high temperatures.

   b. Sometimes the effect of temperature is reversed: The thermal movement counteracts the bond factor in such a way that the CN increases with rising temperatures; e.g., gray tin and white tin; α and β chromium; α and γ iron; α and γ manganese; γ and δ plutonium.

The same principles and factors can also be recognized in discussing the structure of metallic compounds. However, additional factors play a role due to differences between the components. Before discussing these factors, the space principle must be reconsidered in the light of some very important discoveries published by Kasper.[2]

KASPER'S POINT OF VIEW

Kasper observed that several complex structural types in which transition metals play a role (σ-phase, e.g., $Co_{14}Cr_{16}$; μ-phase, e.g., Fe_7Mo_6, and others) have some very interesting characteristics in common which (1) make it relatively easy to describe their geometrical features and (2) open up a very important new line of insight into the problems of CN and space filling.

The main point of the discovery is as follows: if one disregards the chemical difference between the components and considers only the number and positions of the atoms X which surround an atom Y, one finds that compounds of transition elements very frequently show four YX_n groupings with $n = 12, 14, 15$, or 16, and having the following characteristics: (1) The X atoms form the corners of convex polyhedra bounded by triangular faces only; (2) at least five but not more than six triangles meet in each corner; (3) each atom of the structure can be considered as a Y atom in the above sense. We shall call these four polyhedra "*Kasper Polyhedra.*"

As it is not easy to grasp the geometry of such arrangements in three dimensions, Figure 1 indicates its essence in two. One sees an AB arrangement with small A dots and larger B dots. Looking at this as an AB compound, one would "conventionally" say that it is a structure with a CN 4 as each A has four B neighbors and vice versa. However, such a description would miss essential points. Following Kasper, one should say: Let us forget about the A–B difference, but let us look simply at the coordination of the dots, regardless of whether they are A or B dots. Then, we see that A has the CN 5 and that B has the CN 7, and on the average the CN is 6. Using the X and Y letters in the above sense, we notice that there are YX_5 groupings and YX_7 groupings, and that each dot can be considered as a Y center. Thus, the arrangement can be considered to be the mutual interpenetration of polygons with 5 and 7 edges. Figure 2 shows two of them marked more clearly. Such two dimensional polygons correspond to the three-dimensional Kasper polyhedra, and Figure 3 shows these four polyhedra as originally drawn by Kasper.[2]

Two years later Frank and Kasper[3] showed that these four polyhedra are the only geometrically possible ones which are (1) convex, (2) bounded only by triangles, and (3) have corners in which at least five triangles meet. Table II lists their main topological features; e.g., the polyhedron having the CN 16 with respect to its center has 12

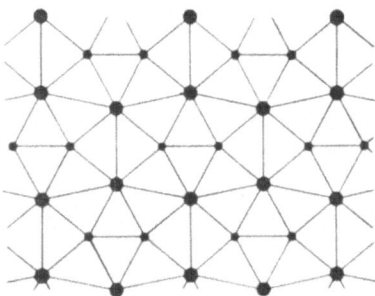

Figure 1. Two-dimensional AB arrangement for demonstrating Kasper's[2] concept of the coordination number. If the small dots represent A atoms and the larger ones B atoms, this arrangement corresponds to an AB compound with the "classical" CN 4, as each A has four B neighbors and vice versa. If, however, only the number of neighbors be considered but not their chemical difference, and if some variation in distance be admitted, the A dots have five neighbors and the B dots seven. See in addition Figure 2.

Table II. Geometrical Features of the Kasper Polyhedra

Co-ordination number (CN)	Corners	Number of triangular faces	Edges
12	12 C_5	20	30
14	12 C_5, 2 C_6	24	36
15	12 C_5, 3 C_6	26	39
16	12 C_5, 4 C_6	28	42

corners in which 5 triangles (or 5 edges) meet, and 4 corners in which 6 triangles (or 6 edges) meet, and it is bounded by 28 triangles.

The fact that the Kasper polyhedra are bounded by triangles has a very important consequence for the concept of space filling. We consider a grouping YX_n in which the X are at the corners of a Kasper polyhedron, bounded by triangles. If we connect these triangles formed by the X atoms with the Y center, we divide the space of the polyhedron into irregular tetrahedra. Therefore, if a structure is made up of atoms, all of which can be considered as centers of Kasper polyhedra, the whole structure can be divided into irregular tetrahedra. Of course, these tetrahedra *must* be irregular, for space cannot be divided up entirely into regular tetrahedra. However, the important point is that structures in which each atom can be considered as the center of a Kasper polyhedron are made up of a compact packing of somewhat deformed tetrahedra.

If for reasons of merit and convenience we call such structure types "Kasper types," it is interesting to compare these with the "close-packed structures." As can easily be realized, the close-packed structures have tetrahedral and octahedral voids, or, expressed differently, they are packings of tetrahedra and octahedra. In the close-packed structures, the corners of these tetrahedra and octahedra are occupied by the centers of the spheres which touch each other at the middle of the edges. Thus, there are relatively large fluctuations in density within a close-packed structure, a high density in the tetrahedral regions and a low density in the octahedral regions. On the other hand, in the Kasper types, the density distribution is more even, as they are entirely divided into tetrahedra, albeit irregular ones. Thus, the space principle can be satisfied in two ways:

1. Ideal packing of spheres with one CN, namely, 12, but with several types of voids (tetrahedra and octahedra).
2. Ideal packing of tetrahedra with several CN's, namely, two or more of 12, 14, 15, and 16, but with a single type of void (tetrahedra only).

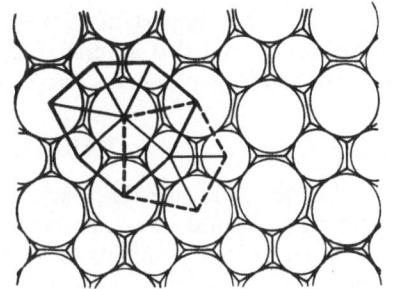

Figure 2. Two-dimensional analog of Kasper polyhedra. The small and large circles represent A and B atoms in the positions of Figure 1. Two different Kasper polyhedra, here sketched as polygons, are marked by dotted and solid lines. Note that the plane is entirely divided into irregular triangles. They correspond to the irregular tetrahedra into which three-dimensional space can be divided completely and uniquely if each atom of a structure can be considered to be the center of a Kasper polyhedron.

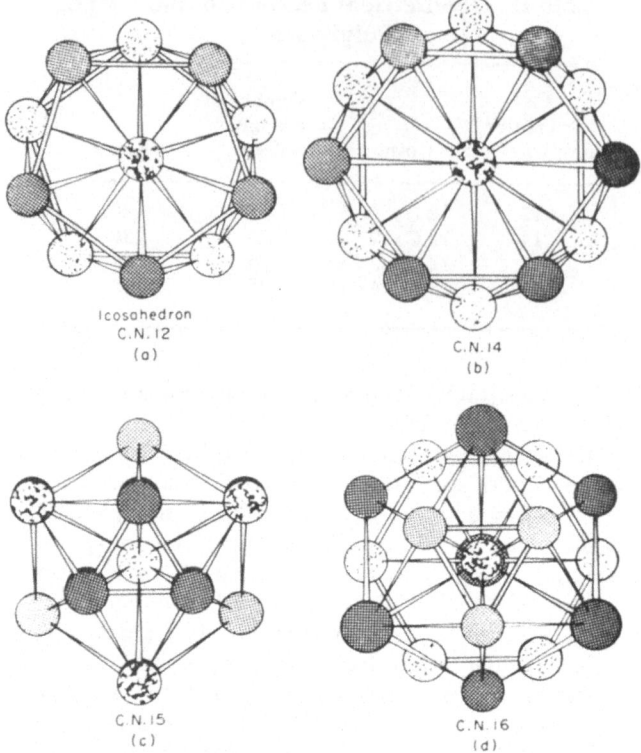

Figure 3. The four Kasper polyhedra, as first drawn by Kasper.[2] (Reprinted by permission from *Theory of Alloy Phases*, American Society for Metals, 1956.)

These two subcases of the space principle can be classified as realizations of a long-range space principle and short-range space principle.

Whereas the metal elements prefer following the space principle by the formation of sphere packings, the metallic compounds follow it rather frequently by the formation of tetrahedral packings, probably for two reasons:

1. The irregularity of the tetrahedra, occasioned by the varying length of their edges, allows for differences in the size of the components.
2. The irregularity of the tetrahedra, occasioned by the varying size of the edge angles, provides some flexibility in approaching special bond angles which may contribute to the stability of a compound, especially in cases where transition elements or elements near to, or to the right, of the Zintl line are involved.

It is rather important to note that both these subdivisions of the space principle are extremes and that several structure types (e.g., α-manganese) are known in which the two subprinciples enter into competition with one another.

DEFINITION OF A COMPOUND

For the following discussion of alloy structures, it may be useful to define what I mean when I speak of compounds.

A phase in a q-component system may exhibit either of two forms: (1) The phase may have a range of homogeneity; i.e., it may form solid solutions by changing its composition without changing its structure type. If the phase has no range of homogeneity or if its range of homogeneity does not extend continuously to a phase in a p-component system which is part of the q-component system (p smaller than q), it is called a compound or more specifically a q-component compound. (The words binary, ternary, and so on are usually used if q is small.) (2) If the range of homogeneity does extend continuously to a phase of the same structure type occurring in a p-component system which is part of the q-component system, it is not a q-component compound, but a q-component solid solution. It may be a solid solution of a p-component compound if the approached p-component phase satisfies the compound definition just given in (1) above.

For some purposes it may be convenient to classify structure types in a similar fashion. A structure type may be called a q-component type if (considering present-day knowledge) q components are needed to form such a type. By the use of this definition we can formulate a rule which is frequently followed by compounds: q compounds tend to crystallize in p-structure types, where p is smaller than q.

As examples of compounds that crystallize in the structure type of elements I mention the binary Hume-Rothery compounds crystallizing in the magnesium, tungsten, and β-manganese types; the ternary compound $ZnCu_2Al$ crystallizing in the tungsten type; the binary compound $Mg_{17}Al_{12}$ and many ternary compounds such as $Fe_{36}Cr_{12}Mo_{10}$ crystallizing in the α-manganese type.

As examples of ternary compounds crystallizing in binary types, I mentioned $MgNiZn$ crystallizing in the $MgCu_2$ type and $Mg_2Cu_6Al_5$ crystallizing in the Mg_2Zn_{11} type.

In most of such cases the observed isomorphism can be understood on the basis of our present-day knowledge of the influence of the radius ratio of the components, their atomic structure (especially with respect to the number of valence electrons), and their electrochemical differences or similarities. Vice versa, it is frequently possible to make intelligent guesses as to which elements should be alloyed in a q-component system to produce a desired special p-structure type. As an example, I mention the production of the ternary $MgLiSb$ crystallizing (like Mg_2Sn) in the binary CaF_2 type.

THE CONNECTION PRINCIPLE IN COMPOUNDS

Proceeding now to a more detailed discussion of compound structure types, I should first like to stress the power of the connection principle in the realm of the structures of intermetallic phases.

Figure 4 shows the structure of the NaTl type, a rather frequent type. At the top right one sees the whole structure and can recognize that it is a superstructure of the tungsten type. We let black dots represent sodium and white ones thallium, at the bottom we see the position of sodium and thallium drawn separately. We can recognize that in both cases the connection is a three-dimensional one, i.e., a lattice connection. Actually each element considered alone forms the lattice connection of the diamond structure. The geometry of the type is summarized in the top left corner in the form of a "connection map." This map shows that sodium and thallium considered together form a heterogeneous lattice connection with the CN 4, whereas considered alone they form homogeneous lattice connections, also with the CN 4. So far, the structure conforms very well to the connection principle. One further important point has to be discussed. Note the figures written below the connection symbols. They represent the relative values of the shortest distances within the respective connections. In this connection map, as in the others to be shown later, the following procedure is adopted: The shortest distance between different elements is taken as unity, as one would generally expect that

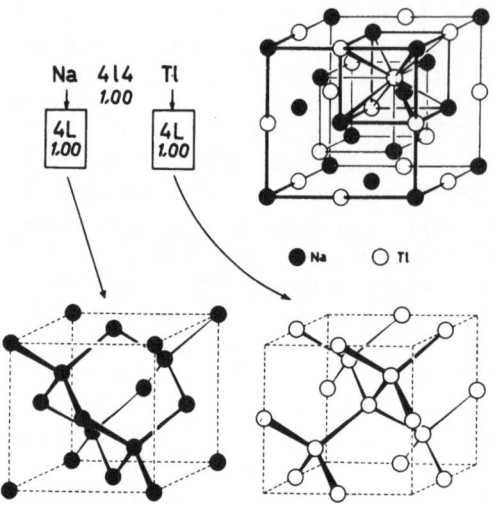

Figure 4. NaTl-type structure and its lattice connections. (Reprinted by permission from *Theory of Alloy Phases*, American Society for Metals, 1956.)

the *different* components of a compound would attract each other most strongly and, therefore, approach each other as closely as possible, i.e., till the atoms which we consider as spherical touch each other.

The shortest distances within the homogeneous connections are then expressed by figures related to the chosen unit value and written below the respective connection symbol. One may now notice that in this NaTl type the distances in the heterogeneous connection are equal to those in the homogeneous connections. This means that either the A atoms or the B atoms in this structure type do not touch each other, because it would be a chance occurrence if both sorts of atoms were of the same size. In addition, one can conclude from the same kind of reasoning that there cannot be an A–B contact. Following Zintl,[4] who discovered this structure type, the explanation for its existence might be the following: The strongly electropositive sodium gives away its valence electron, which is used by the thallium to develop a sort of covalent bond as in the diamond type of structure. Thus, one may visualize this structure as a somewhat negative lattice connection of the B atoms in which positive A atoms are inserted. The connection principle thus receives a somewhat expanded meaning when we are dealing with compound structure types: *Like atoms may tend to approach each other as closely as possible, as their bonds may add to the stability of the compound.* For this reason the connection principle may compete with the space principle, as can be shown by a comparison of the NaTl and the CsCl types. In Figure 5 the CsCl type is drawn for different radius ratios. As can readily be understood, A–B contact is possible within a wide range of different values of the radius ratio. Thus, from the simplified point of view that alloy structures may be visualized as sphere packings, the CsCl type would allow a more dense packing than the NaTl type, except for the special radius ratio equal to 1, in which case the space filling would be the same in both types.

That the connection principle can also compete with the symmetry principle is dramatically shown by the fact that many AB compounds crystallize in the NiAs type, instead of in the NaCl type, which has the higher symmetry. The reason is that the relative position of the octahedral voids in a hexagonal close-packed arrangement of the "negative" partner provides the Ni "ions" with the opportunity to approach each other more closely than in the cubic NaCl-type structure.

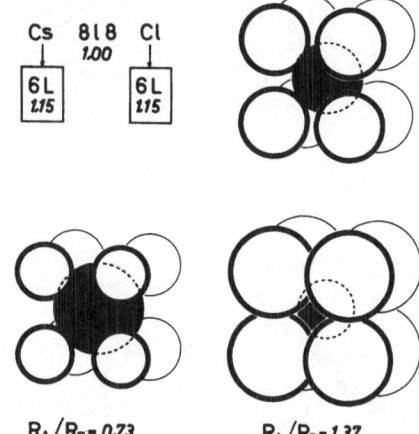

Figure 5. CsCl-type structure drawn for different radius ratios. (Reprinted by permission from *Theory of Alloy Phases*, American Society for Metals, 1956.)

ADDITIONAL STRUCTURE-INFLUENCING FACTORS IN COMPOUNDS

In the discussion of element structures, I mentioned briefly the influence of factors which counteract structure type expectations based on the main geometrical principles. The same factors apply, of course, to the structures chosen by intermetallic compounds. In addition, however, some new factors come into play due to the fact that compounds consist of atoms differing among themselves. These factors are usually called "size factor" and "electrochemical factor." The special influence of these factors depends largely on the differences the atoms composing an alloy may or may not have with respect to size and degree of electronegativity.

SIZE FACTOR

First let us consider the influence of the size factor. In discussing it, difficulties arise from the beginning. It is true that we have some notion of the size of metallic atoms thanks to the work of V. M. Goldschmidt, W. H. Hume-Rothery and G. V. Raynor, but the atomic size of an element A may vary considerably depending on the partners with which an alloy is formed. A collection of many details pertinent to this question was presented in 1956.[1]

Despite these difficulties, which cannot be discussed here in detail, it can be stated that many compounds owe their existence predominantly to the fact that the radius ratio of their components has a value exactly corresponding to that required to build up a structure satisfying the main geometrical principles outlined above. Looking back to the question put forward at the beginning of this lecture, namely, why does a compound KNa_2 exist, I well remember a discussion with a very good friend of mine about the conditions to be satisfied for a compound to have an MgX_2-type structure. I maintained that the main condition is for the radius ratio of the components to be near the value $1.23 = \sqrt{3}/\sqrt{2}$, which can be derived on purely geometrical lines if one believes in the validity of the main geometrical principles. As the radius of potassium is 2.36 and the radius of sodium is 1.92, the radius ratio $2.36/1.92 = 1.255$, i.e., it is the ideal value for a compound with a MgX_2-type structure. Therefore, I said, "I bet that KNa_2 has an MgX_2-type structure." My friend did not believe me and challenged me by asking: "Why don't you do the structure?" I replied that I did not like handling compounds consisting of alkali metals. So we decided that somebody else would do the job for us.

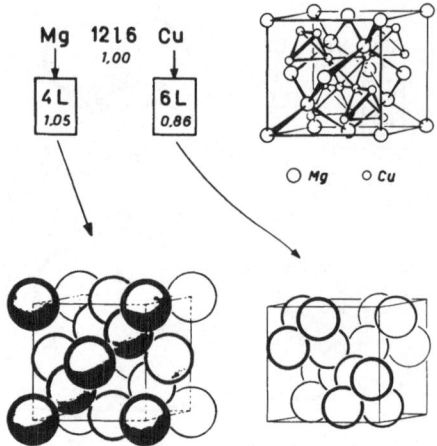

Figure 6. $MgCu_2$-type structure and its lattice connections. (Reprinted by permission from *Theory of Alloy Phases*, American Society for Metals, 1956.)

Fortunately, there are chemists in this world, and so two years later Böhm and Klemm[5] proved the existence of a compound KNa_2 by presenting d values from an X-ray photograph. The issue of the journal arrived just on the evening when I was waiting at the Institute for news of the birth of my first child. To calm myself I took a slide rule and calculated the d values to be expected, assuming that KNa_2 has the $MgZn_2$ structure (as it could be seen immediately that the reported d values were incompatible with a cubic structure of the $MgCu_2$ type). And the assumption turned out to be correct! So, I won the bet and a child in the same night.

I said just now that the MgK_2 types best satisfy the requirements of "the main principles" if compounds AB_2 are considered which have a radius ratio approaching 1.23. I would like to discuss this statement using Figure 6, which represents the cubic version of $MgCu_2$.

Note that the *space principle* is satisfied as indicated by the highest CN's known for AB_2 compounds. Note that the *symmetry principle* is satisfied as the MgX_2 types have either cubic or hexagonal symmetry. Note that the *connection principle* is satisfied as the heterogeneous and both homogeneous connections are three-dimensional ones. In addition, *it is even a Kasper-type structure* as both the magnesium and the X atoms are surrounded by Kasper polyhedra. Thus, there is no ground for astonishment that the MgX_2-type structures are met with most frequently both as binary and as polynary compounds. H. J. Goldschmidt[6] points to the fact that if transition metals are involved, MgX_2 types may occur in which the radius ratio of the components becomes rather irrelevant. This is expressed by the fact that large homogeneity regions exist. In such cases the influence of the ratio between valence electrons and the number of atoms ("electron compounds") is considered to play an important role. Earlier,[7] the influence of the electron concentration was considered only with the question of whether the $MgCu_2$ or the $MgNi_2$ or the $MgZn_2$ type is the favored one. Unpublished work[8] of the author shows also that the radius ratio derived from the elementary radii can be quite unimportant for the formation of an MgX_2-type compound if transition metals are involved. For it was found that a compound Mn_2Cu_3Al crystallizes in the $MgCu_2$ type with aluminum (which has the largest radius) in the copper position. This compound forms a series of solid solutions with a compound $MnInCu_4$ in which the chemically analogous indium (because of its very large radius) occupies the magnesium position together with the manganese. Thus, in addition to the geometrical requirements "electron concentrations" and the formation of special bonds within a homogeneous lattice connection appear to play a decisive role in the formation of MgX_2-type compounds and probably in other types of compounds as well.

Before leaving these MgX_2 types I should like to point out one further interesting feature of their structure. If we again choose as unity the shortest distance between unlike atoms—as can be seen in the connection map—and if we measure the shortest distance between like atoms with the same unit, we observe that the average of the distances between like atoms is smaller than the distance between the unlike atoms. This means that if we visualize the type as made up by spheres that touch each other only, the like atoms touch each other and the unlike do not.

This interesting fact can be expressed[9] by a simple quotient Q. Let d_A be the distance between A atoms, d_B the distance between B atoms, and d_{AB} the A–B distance; if we now calculate the quotient

$$Q = (d_A + d_B)/2d_{AB}$$

this Q value differs from structure type to structure type, and is rather indicative of the bond character which may or may not develop in a certain type. Table III gives some data concerning structure types. As a rule one can say that *the smaller Q is, the less likely is the occurrence of polar bonds.*

The CN's (in a rigid sense) 12 and 6 are apparently the highest that are geometrically possible for AB_2 compounds and two more types with these high CN's are known: the AlB_2 and the $ThSi_2$ type. (Note: Roman B stands for the element boron; an italic B stands for any element in a compound with the stoichiometrical formula AB_2.) However, these types do not satisfy our main principles as well as the MgX_2 types do and they are, therefore, limited to compounds of a less metallic character, in which Hume-Rothery tendencies may have an essential influence. As far as the radius ratio is concerned, however, these types are comparable to the MgX_2 types.

Table III. Some Q Values of Structure Types:
$$Q = (d_A + d_B)/2d_{AB}$$

Structure type	Coordination numbers (CN)	Q
SiO_2	4 and 2	1.82
TiO_2	6 and 3	1.47
CaF_2	8 and 4	1.39
$CuAl_2$	8 and 4	1.06*
$MoSi_2$	10 and 5	1.11
$MgCu_2$	12 and 6	0.96
AlB_2	12 and 6	1.01
$ThSi_2$	12 and 6	1.04
ZnS	4 and 4	1.63
NaTl	4 and 4	1.00
NaCl	6 and 6	1.41
NiAs†	6 and 6	1.19
CsCl	8 and 8	1.16

* Average values of the distances are used.
† As c/a varies with different compositions the Q value given refers here to the NiAs compound.

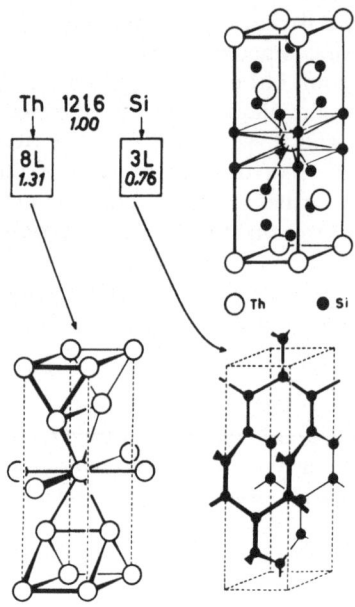

Figure 7. ThSi₂-type structure and its lattice connections. (Reprinted by permission from *Theory
of Alloy Phases*, American Society for Metals,
1956.)

Figure 7 shows the $ThSi_2$ type first described by Brauer and Mitius.[10] Note the dominating $3L$ connection of the silicon atoms which form a sort of three-dimensional graphite structure. Obviously, rather covalent bonds within this $3L$ connection are responsible for the stability of this type, and accordingly the representatives crystallizing in this AB_2 type are those which have chemical compositions with B lying near to the Zintl line. As examples, we can mention: {La, Ce, Pr, Nd, Th, Np, U, Pu}Si_2 and {Pr, Pu}Ge_2.

In a similar way the chemical composition of the representatives of the AlB_2 type (Figure 8) is characterized by components having (1) the proper radius ratio,

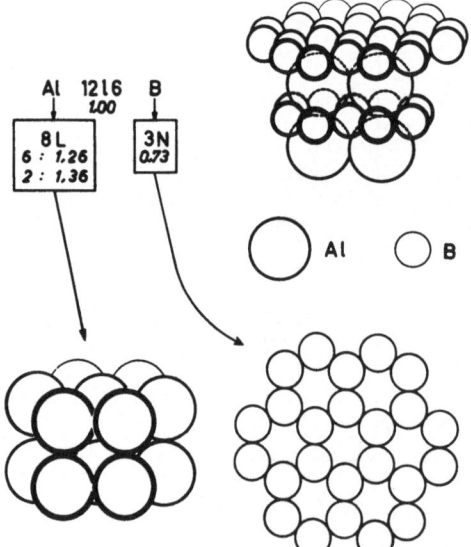

Figure 8. AlB₂-type structure and its connections.
The B atoms form a net connection only.

and (2) a B component lying near the Zintl line. As examples, we can mention: {Mg, Al, Ti, V, Cr}B$_2$; ThAl$_2$; {Ca, La}Ga$_2$; USi$_2$. It is interesting to compare the structures of the two compounds CaAl$_2$ and CaGa$_2$. In both compounds the radius ratio is virtually the same and suitable for the formation of an MgX$_2$ type. Thus, for both cases the MgX$_2$ type was "expected." Whereas CaAl$_2$ was found to have the MgCu$_2$ type, CaGa$_2$ has not. It has the AlB$_2$ type. Obviously, in the latter case a Hume-Rothery tendency plays a decisive role, as gallium shows this tendency already in the elementary state (CN 1 to 7) whereas aluminum itself crystallizes in a cubic close-packed structure.

Apparently no restrictions exist on the chemical character of the components of MgX$_2$ type compounds, as the following examples show: KBi$_2$, KNa$_2$, BiAu$_2$, NaAu$_2$, and (Au, Be)Be$_2$. On the other hand AB_2 compounds occur in the ThSi$_2$ or AlB$_2$-type structures if the B component tends to form bonds of more covalent character (with the small CN 3) than would be possible in a MgX$_2$-type structure (where the CN is 6). (More data on the influence of the size factor on chemical compositions and structure of metallic compounds have been collected in an earlier paper.[1] A rather recent discussion on the size, and other factors involving the stability of MgX$_2$-type structures has been given by Schulze.[11])

ELECTROCHEMICAL FACTORS

The influence of the "electrochemical factor can best be demonstrated by discussing the so-called "Zintl-phases."[12] These are compounds in which one component (an electronegative one) lines to the right of the Zintl line and the other (an electropositive one, which as an ion would have a noble-gas configuration) to the left, and which can be formally explained on the basis of the classically recognized valencies. As an example, I mention Mg$_2$Sn or Mg$_2^{2+}$Sn^{4-}, which crystallizes in the CaF$_2$-type structure. It is characteristic of the Zintl phases that their structure types have Q values larger than those of other structure types chosen by metallic compounds with analogous stoichiometric and radius ratios, but which are composed by elements which do not follow the restrictions outlined above (Table III—the CaF$_2$ and CuAl$_2$ types). This fact indicates that the bonds within the Zintl phases are of a more polar character than in other compounds.

This formal valency concept "explaining" the existence of Zintl phases could be used to predict the existence of many isotopic ternary compounds. As an example I mention the compound Mg^{2+}Li^{1+}Sb^{3-}, which crystallizes as Mg^{2+}Mg^{2+}Sn^{4-}, and I would not be surprised if these two compounds formed a continuous series of quarternary mixed crystals with the CaF$_2$-type structure. In the last decade, the Zintl phases have gained considerable interest as semiconductors.[13]

As already mentioned, several principles and factors may compete with each other and their ratio of importance may change within a series of compounds which have certain features in common. Table IV gives such a series. At the top the compound Li$_2$O is a typical saltlike structure. Going downward we pass the Zintl phases MgLiSb, Mg$_2$Sn, MgLi$_2$Sn. When the strong electropositive alkali and earth-alkali atoms are exchanged by the less electropositive Cu, Ni, and Mn, the metallic character of the compounds becomes increased and the "holes" in the CaF$_2$-type structure are filled up to conform with the space principle, which is a leading one for metallic substances. At the bottom of the table, we end up with the iron structure. Thus, a more or less continuous series of compounds exists (in part forming continuous series of solid solutions) in which the bonding character changes from that of a typical salt (as in Li$_2$O) to that of a typical metal (as in the element iron).

Table IV. CaF$_2$ or "Antifluorite"-Type Compounds and Structurally Related Substances

Chemical formula	Lattice constant a	Element distribution on different positions (only one representative of a face-centered cubic lattice given)				
		$0\,0\,0$	$\frac{1}{4}\frac{1}{4}\frac{1}{4}$	$\frac{3}{4}\frac{3}{4}\frac{3}{4}$	$\frac{1}{2}\frac{1}{2}\frac{1}{2}$	
Li$_2$O	4.61	O	Li	Li		Increasing polar character
Li$_2$S	5.71	S	Li	Li		
Li$_2$Se	6.01	Se	Li	Li		↑
Li$_2$Te	6.50	Te	Li	Li		
MgLiAs	6.21	As	Mg	Li		
MgLiSb	6.61	Sb	Mg	Li		
MgLiBi	6.75	Bi	Mg	Li		
Mg$_2$Sn	6.75	Sn	Mg	Mg		
MgLi$_2$Sn*	6.69*					
	6.75*	Sn	Mg, Li	Mg, Li	Mg, Li†	
MgCuSb	6.15	Sb	Cu		Mg	
MgCuBi	6.26	Sb	Cu		Mg	
MgCuSn	6.22	Sn	Cu		Mg	
MgNiSb	6.04	Sb	Ni		Mg	
MgNi$_2$Sn	6.05	Sn	Ni	Ni	Mg	
(Cu, Ni)$_3$Sb	5.86	Sb	Cu, Ni	Cu, Ni	Cu, Ni†	
Cu$_3$Sb	6.00	Sb	Cu	Cu	Cu	
Cu$_2$MnSn	6.17	Sn	Cu	Cu	Mn	
Cu$_2$MnAl	5.9	Al	Cu	Cu	Mn	
(Cu, Mn)$_3$Al	5.9	Al	Cu, Mn	Cu, Mn	Cu, Mn†	
Cu$_3$Al	5.84	Al	Cu	Cu	Cu	↓
Fe$_3$Al	5.78	Al	Fe	Fe	Fe	Increasing metallic character
Fe	5.72	Fe	Fe	Fe	Fe	

* Two phases of such composition (or near to it) were found to coexist (unpublished work of the author).

† No decision has been reached yet on the actual position of Mg and Li, Cu and Ni, Cu and Mn.

In the Zintl phases, the ratio electrons/atoms is clearly defined and can easily be understood on the basis of classical valency considerations. However, there are several further important groups of compounds in which the number of electrons available for bond formation plays a decisive role for the chemical composition of the compounds to be formed, and in which the size factor loses much of its influence.

I mention the well-known Hume-Rothery phases,[14,15] for example, the CsCl-type (or tungsten-type) compounds in which the electron/atom ratio has a value near to 1.5. (Other well-known Hume-Rothery compounds crystallize in the β-manganese type, in the γ-brass type, and in the ϵ-brass type, which is a disordered hexagonal close-packed structure).

Within the last decade a large number of other structure types have become known, the stability of which also depends mainly on the ratio of valence electrons/atoms or on the ratio of missing d electrons/atoms. It is characteristic for the representatives of these types that their composition is dominated by transition elements. They are usually called "electron compounds" and have at first sight rather complex atomic arrangements. However, applying the points of view developed by Kasper[2] and by Frank and Kasper,[3] we find we can grasp the main features of the structures relatively easily. Among others the σ-phase structure and the α-manganese-type structure should be mentioned here. P. Beck and his co-workers[16] in Urbana were especially successful in exploring the chemistry of these compounds.

There is still another interesting group of types in which the number of electrons available plays an important role. I am referring to those types in which the presence of boron, carbon, nitrogen, or oxygen helps the formation. I mention the E9$_3$ type: Fe$_3$W$_3$C; the D8$_8$ type: Mn$_5$Si$_3$ for which the name Nowotny-phases has been proposed,[17] and the L'1$_2$ type: AlFe$_3$C or $B T_3 M$, where B stands for a B-metal (e.g., magnesium, zinc, and aluminum) and T for a transition element (e.g., iron).

The last type has in recent years been studied especially by Stadelmaier[18] and co-workers. Figure 9 shows its features. In the ideal case it is cubic and its atoms take the positions of the perowskite structure, i.e., CaTiO$_3$. This already indicates that heteropolar forces are involved in its formation. Figure 9 shows three different views of the structure; depending on which sort of atom is placed in the center of the cell. In addition, it shows the connection map. Let us concentrate on the right top drawing: Note that the shortest distance is the one between carbon and the T metal, here taken as nickel. Note the three-dimensional lattice connection made up of carbon and the T metal. It will be a negative one and the B metal helps to build it up by contributing valence electrons. A further discussion would indicate that the type may best be explained as made up of a negative transition-metal framework in which positive B metals and metalloid "ions" are inserted, in much the same way as was discussed at the beginning of this lecture in reference to the NaTl-type structure. I believe Stadelmaier has hitherto found 30 or more representatives of this interesting ternary AlFe$_3$C-type structure.

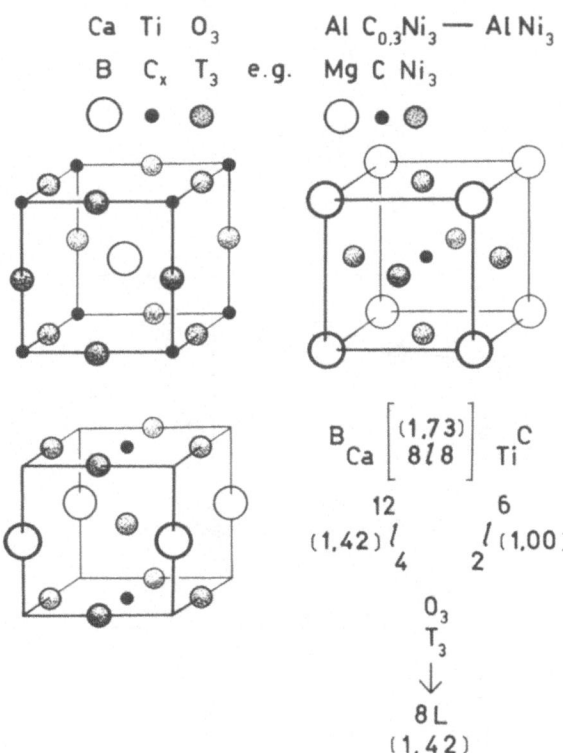

Figure 9. Different views of the perowskite (CaTiO$_3$–AlCFe$_3$) structure corresponding to different choices of the origin. In the connection map, bottom right, the structurally unimportant Ca–Ti (or B metal–carbon) connection is bracketed.

The last group I propose to discuss today, however briefly, is the terrible group of compounds which have the chemical composition T_pB_q in which T means a transition metal and B means a B metal, e.g., aluminum, and in which q is much larger than p. I chose two examples, VAl_{10}, determined by P. J. Brown,[19] and $V_4Al_{23} \approx VAl_6$, determined by J. F. Smith[20] and co-workers.

Figure 10 shows a drawing of the structure as given by Brown.[19] The bottom drawing gives the position of the vanadium (black points). Around these vanadium atoms aluminum atoms are clustered in the form of Kasper polyhedra, as shown in the top drawing.

Another way of looking at the structure would be to search for pertinent connections in the sense discussed before. For this purpose I draw your attention to the vanadium arrangement in the bottom drawing of Figure 10. It is the same as the copper arrangement in the $MgCu_2$ type. Table V contains the atomic distances as given by Brown.[19] There is one remarkably short V–Al distance equal to 2.57.

Let us now look for the heterogeneous connection which results if we consider this short distance of 2.57 only. The result is drawn in Figure 11. It shows a VAl_3 framework made up of vanadium tetrahedra, the edges of which are nearly centered by aluminum. Thus, the structure can be visualized as strongly bonded VAl_3 framework the voids of which are filled by aluminum of normal metallic size.

If we analyze the V_4Al_{23} structure in the same fashion we again find vanadium

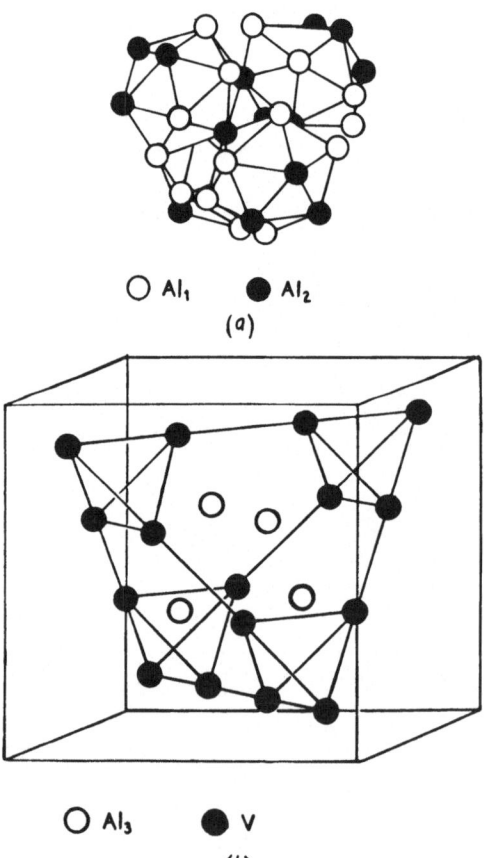

O Al₁   ● Al₂
(a)

O Al₃   ● V
(b)

Figure 10. The structure of $\alpha(V–Al) =$ VAl_{10}, after Brown.[19]

Table V. Interatomic Distances of (V–Al) = VAl$_{10}$, Determined by P. J. Brown[19]

Kind of surrounded atom	Kind and number of surrounding atoms	Distance
V {	6 Al$_1$	2.826
	6 Al$_2$	2.572
Al$_3$	12 Al$_1$	3.083
Al$_1$ {	2 Al$_1$	2.943
	2 Al$_1$	2.740
	1 Al$_1$	2.679
	2 Al$_2$	2.880
	1 Al$_2$	2.679
	2 Al$_3$	3.083
	1 V	2.826
Al$_2$ {	4 Al$_2$	2.882
	2 Al$_1$	2.679
	4 Al$_1$	2.880
	2 V	2.572

The space group is $Fd3m$ with Al$_1$ in 96(g), Al$_2$ in 48(f), Al$_3$ in 16(d) and V in 16(c).

tetrahedra bonded by aluminum. This time, however, chains only are formed, probably due to the relatively smaller amount of aluminum present (see Figure 12).

However, in both cases the difference in the electronic structure of the components has a remarkable and specific effect, and the geometrical properties of space finally determine the queer chemical formulas observed.

SUMMARY

I hope I have been able to convey the meaning of some geometrical principles, and of some physicochemical factors which may counteract them. Both the principles and the factors play important roles in determining the composition of intermetallic compounds. From another point of view, by considering these principles and factors, rules can be recognized which help in predicting the constitution of alloy systems not yet investigated. The principles are: space principle, symmetry principle, and connection principle.

The factors are: size factor, temperature factor, and electrochemical factor (or bond factors). In principle, every structure type can be discussed by considering its known representatives and by examining them to discover which of the principles and which of the factors appear to be the most determining ones. On the basis of such a discussion, certain rules emerge the value of which can be checked by preparing new mixtures of metals which should conform to such rules. In many cases "predictions" have been verified, this adding to our confidence in a certain rule. In other cases compounds are formed that do not conform to the predicted result. If this is the case a new window is usually opened shedding light on a new direction of knowledge concerning the strange relations

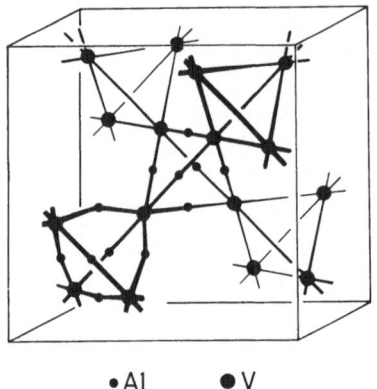

• Al   ● V

Figure 11. VAl₃ lattice connection within the VAl₁₀ structure. The Al positions are near the centers of the tetrahedra edges. In two of the tetrahedra they are marked by small dots. In one of the tetrahedra the slightly bent edges are sketched. As the radii of V and Al are 1.36 and 1.43 Å in the elementary state the actual V–Al distance equal to 2.57 is considerably smaller than the sum of the radii $1.36 + 1.43 = 2.79$. The actual Al–Al distance equal to 2.88, however, is near to the sum $1.43 + 1.43 = 2.86$. For this drawing an origin different from the one used in Figure 10 was chosen in order better to demonstrate the connection feature here discussed.

between chemical composition and the structure of metallic compounds. Slater[21] (1956) once made a comment as follows:

> "I don't understand why you metallurgists are so busy in working out experimentally the constitution of a polynary metal system. We know the structure of the atoms, we have the laws of quantum mechanics, and we have electronic calculating machines, which can solve the pertinent equations rather quickly."

However, no structure of a compound has as yet (1962) been predicted by using such higher methods of approach. On the other hand, many unknown compounds and their structures have been predicted and verified by past experience (expressed as "rules" and by the intuition of persons (like Hume-Rothery[14]) who were intrigued by questions like "Why do the compounds Cu_5Zn_8 and Cu_9Al_4 exist, since they have no similarity in chemical composition, but are isomorphous, and have atomic arrangements of a ridiculous complexity?"

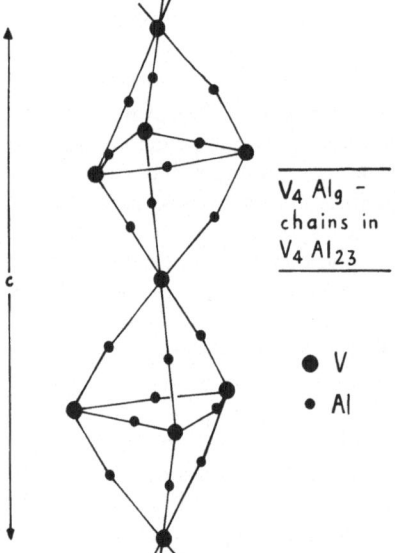

V₄ Al₉ –
chains in
V₄ Al₂₃

● V
• Al

Figure 12. V₄Al₉ chain connection within the V₄Al₂₃ structure. As in the case of Figure 11 the V–Al distances (2.52, 2.58, and 2.60) are rather small, having the average value 2.57. On the other hand the Al–Al distances in the drawn connection (2.83, 2.89, and 2.91) are quite normal, having the average value 2.88. Note that the average values of this connection are exactly the same as the corresponding values of the VAl₃ connection depicted in Figure 11.

REFERENCES

1. F. Laves, "Crystal Structure and Atomic Size," *Theory of Alloy Phases*, Am. Soc. for Metals, Cleveland, 1956, pp. 124–198.
2. J. S. Kasper, "Atomic and Magnetic Ordering in Transition Metal Structures," *Theory of Alloy Phases*, Am. Soc. for Metals, Cleveland, 1956, pp. 264–279.
3. F. C. Frank and J. S. Kasper, "Complex Alloy Structures Regarded as Sphere Packings. I. Definition and Basic Principles," *Acta Cryst.* **11**: 184–191, 1958; II. "Analysis and Classification of Representative Structures," **12**: 483–499, 1959.
4. E. Zintl and G. Woltersdorf, "Gitterstruktur von LiAl," *Z. Elektrochem.* **41**: 877–879, 1935.
5. B. Böhm and W. Klemm, "Zur Kenntnis des Verhaltens der Alkalimetalle zueinander," *Z. anorg. u. allgem. Chem.* **243**: 69–85, 1940.
6. H. J. Goldschmidt, "The Phase Constitutions of Some Niobium-Bearing and Associated Transition Metal Systems," *J. Less Common Metals* **2**: 138–153, 1960.
7. F. Laves and H. Witte, "Der Einfluss von Valenzelektronen auf die Kristallstruktur ternärer Mg-Legierungen," *Metallwirtschaft* **15**: 840–842, 1936.
8. F. Laves, *Z. Kristall.* (to be published).
9. F. Laves and H. J. Wallbaum, "Ueber einige neue Vertreter des NiAs-Typs und ihre kristallchemische Bedeutung," *Z. angew. Mineral.* **4**: 17–46, 1941.
10. G. Brauer and A. Mitius, "Die Kristallstruktur des ThSi$_2$," *Z. anorg. u. allgem. Chem.* **249**: 325–335, 1942.
11. G. E. R. Schulze, "Dichte und Raumerfüllung bei intermetallischen Verbindungen, insbesondere Laves-Phasen," *Z. Kristall.* **115**: 261–268, 1961.
12. F. Laves, "Zintl's Arbeiten über die Chemie und Struktur von Legierungen," *Naturwiss.* **29**: 244–254, 1941.
13. E. Mooser and W. B. Pearson, "The chemical bond in semiconductors," *J. Electronics* **6**, 629–645, 1956, pp. 1–17; *Progress in Semiconductors, Vol. 5*, Heywood Comp., London, 1960, pp. 103–139.
14. W. Hume-Rothery, "Nature, Properties, and Formation of Intermetallic Compounds," *J. Inst. Metals* **35**: 295–361, 1926.
15. H. Witte, "Der Gültigkeitsbereich der Hume-Rotheryschen Regel," *Metallwirtschaft* **16**: 237–245, 1937.
16. P. Beck and co-workers (only few of his papers can be quoted here): 1. K. R. Gupta, N. S. Rajan and P. A. Beck, "Effect of Si and Al on the Stability of Certain Sigma Phases," *Trans. Met. Soc., AIME* **218**: 1960; 2. B. N. Das and P. A. Beck, "Relationship Between the Mu Phase and the Sigma Phase in the Mo–Mm–Co System," *Trans. Met. Soc., AIME* **218**: 1960.
17. E. Parthé, "Contributions to the Nowotny Phases," *Acta Cryst.* **10**: 768–769, 1957; E. Parthé and J. T. Norton, "Crystal Structures of Zr$_5$Ge$_3$, Ta$_5$Ge$_3$, and Cr$_5$Ge$_3$," *Acta Cryst.* **11**: 14–17, 1958.
18. H. H. Stadelmaier, "Ueber ternäre Verbindungen von Uebergangsmetall, B-Metall und Metalloid," *Z. Metallkunde* **52**: 758–762, 1961; "Ternary Carbides of the Transition Metals Ni, Co, Fe, Mn with Zn and Sn," *Acta Met.* **7**: 415–419, 1959.
19. P. J. Brown, "The Structure of α (V–Al)," *Acta Cryst.* **10**: 133–135, 1957.
20. J. F. Smith and A. E. Ray, "The Structure of V$_4$Al$_{23}$," *Acta Cryst.* **10**: 169–172, 1957; see in addition *Acta Cryst.* **13**: 876–884, 1960.
21. J. C. Slater, "Band Theory of Bonding in Metals," *Theory of Alloy Phases*, Am. Soc. for Metals, Cleveland, 1956, pp. 1–12.

THE ZIRCONIUM–IRON SYSTEM

F. N. Rhines and R. W. Gould

University of Florida, Gainesville, Florida

ABSTRACT

By means of a combined X-ray diffraction and metallographic study, two heretofore unrecognized phases, Zr_4Fe and Zr_2Fe, have been identified. Approximate verification of this finding has been made by Dr. David Levinson, using microprobe analysis. A tentative revision of the Zr–Fe phase diagram is proposed.

INTRODUCTION

In the course of a study of the phases present in zircaloy, it was observed that more binary zirconium–iron phases exist than are shown in the phase diagram presented in Hansen's *Constitution of Binary Alloys*.[1] Although the new observations are limited in effect to low temperatures, they can be used to suggest a modification of the zirconium–iron phase diagram as shown in Figure 1.

Two earlier investigators have suggested the existence of zirconium-rich phases other than the $ZrFe_2$ shown in the diagram of Hansen, which is taken largely from the work of Hayes, Robertson, and O'Brien.[2] In 1932, Vogel and Tonn[3] reported one binary intermediate phase Zr_2Fe_3. This finding was challenged by Wallbaum,[4] who asserted that an alloy of the proportion of Zr_2Fe_3 is in fact composed of two intermediate phases, *viz.*, $ZrFe_2$ and a zirconium-rich intermediate of unknown composition. Hayes, Robertson, and O'Brien were unable to confirm the presence of intermediate phases other than $ZrFe_2$.

New evidence has been obtained from X-ray diffraction studies performed upon powders brought to near equilibrium below the α–β transformation of zirconium and metallographic studies performed on both cast binary alloys and the same alloys slowly cooled from above the α–β transformation temperature. Eight binary alloys, ranging from 5 to 55% iron were examined. Two new phases were found, one at about the composition Zr_4Fe the other at about Zr_2Fe. These have been confirmed approximately by Levinson,[5] using electron microprobe analysis.

EXPERIMENTAL PROCEDURES

Grade II crystal bar zirconium and 99.9% pure iron, from the Corey Steel Company, were used in compounding the alloys. Carefully weighed charges of the compositions listed in Table I were melted in a 3-in. cold copper hearth, tungsten electrode arc furnace under a positive pressure of tank helium. Charged at the same time, in a separate depression in the copper hearth, was about 15 g of pure zirconium, which was melted first to purge the furnace atmosphere. Each alloy charge was inverted and remelted several times to ensure uniform alloying. Pilot melts of pure zirconium were made from

---

[1] Superscripts pertain to references at the end of the paper.

Table I. Alloy Preparation

Nominal Composition	Stoichiometric ratio	Weight of charge	Weight after melt	Times melted
Zr + 5%Fe	—	Zr— 24.5908 Fe— · 1.2939 ——— 25.8847	25.7854	4
Zr + 13.27%Fe	Zr_4Fe	Zr— 31.4462 Fe— 4.8153 ——— 36.2615	36.2628	5
Zr + 16.43%Fe	Zr_3Fe	Zr— 31.0652 Fe— 6.1063 ——— 37.1715	37.1510	5
Zr + 19.98%Fe	—	Zr— 23.7190 Fe— 5.9243 ——— 29.6433	29.7190	4
Zr + 23.44%Fe	Zr_2Fe	Zr— 21.7081 Fe— 6.6436 ——— 28.3517	28.3500	5
Zr + 28.9%Fe	Zr_3Fe_2	Zr— 33.0287 Fe— 13.4811 ——— 46.5098	46.4690	3
Zr + 37.97%Fe	$ZrFe$	Zr— 18.0131 Fe— 11.0253 ——— 29.0384	28.9622	5
Zr + 55.3%Fe	$ZrFe_2$	Zr— 12.4834 Fe— 15.3553 ——— 27.8387	27.8075	5

time to time to detect possible oxygen contamination of the alloy, consistently low post-melting hardness being taken to indicate satisfactory purity.

Buttons of the several binary alloys thus prepared were quartered, one quarter being used for X-ray spectrochemical analysis, a second quarter for metallographic examination as-cast, and the third and fourth quarters being heat-treated at 900°C and slowly cooled. Of the heat-treated pieces, one was used for metallographic examination, while the second was partly reduced to powder, which was stress-relieved at 600°C and examined by X-ray diffraction.

Heat treatments were conducted in a fused-silica vacuum furnace capable of maintaining a pressure of less than $0.01\ \mu$ Hg throughout the thermal cycle. To guard further

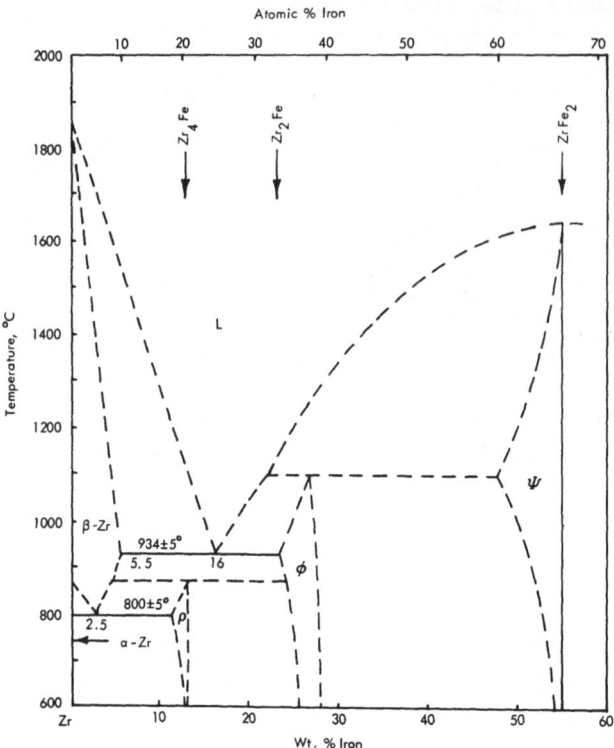

Figure 1. Proposed zirconium-iron phase diagram.

against gaseous contamination, each sample was (1) wrapped in tantalum foil, and (2) enclosed in a 1-in.-diameter, by 4-in.-long silica tube, packed on all sides with fine zirconium turnings. The same procedure was used for both massive samples and powders.

The quarter buttons were first heated slowly to 900°C, at which temperature they were held 65 hr before slow cooling over a period of 24 hr. One quarter of each button was then partly reduced to powder by filing with a new file, in the case of the low iron alloys, or by grinding brittle alloys in a diamonite mortar. The powder, packed as described above, was finally stress-relieved by heating slowly to 600°C, holding for 24 hr at this temperature, and furnace cooling. The absence of gaseous contamination during heat-treatment was demonstrated by lattice parameter measurements made on pilot samples of pure zirconium filings.

METALLOGRAPHY

The several phases of the zirconium–iron system may be identified metallographically by the use of an anodizing etch, originally proposed by Picklesimer.[6] This is best applied after chemical polishing in a solution composed of 45% nitric acid, 45% water, and 10% hydrofluoric acid. The anodizing solution is composed of 60 ml of ethyl alcohol, 35 ml of water, 20 ml of glycerine, 5 ml of phosphoric acid, 10 ml of lactic acid, and 2 g of citric acid. The cathode is aluminum and the sample is made the anode with an applied potential of 23 v. After 20 sec of anodizing the sample is removed from the bath and thoroughly washed before drying. Each phase of the microstructure develops a characteristic color. This method is applicable to all compositions between pure zirconium and about 29%

iron. Photomicrographs of anodized structures, in the eight alloys examined after slow cooling from 900°C, are presented in Figures 2 through 9.

Four phases are distinguishable in these microstructures, namely: α-zirconium, Zr_4Fe (hereafter referred to as ρ), Zr_2Fe (hereafter referred to as ϕ), and $ZrFe_2$ (hereafter referred to as ψ). The α phase, see Figure 1, comes from the eutectoid decomposition of the β phase during slow cooling from 900°C. This results in the formation of a eutectoid, composed of α plus an intermediate phase. Thus, in all of the photographs in which α appears, it is readily distinguished by being the major phase of a two-phase eutectoid structure. It is most abundantly present in Figure 2, where the α + ρ eutectoid constitutes the matrix of the structure. In Figure 3, the α + ρ eutectoid (deep-gray tone) is clearly evident as a two-phased constituent. The quantity of α decreases progressively through Figures 4 and 5, the eutectoid being scarcely resolved in the α particles of Figure 5. No α is identifiable in alloys of high iron content, Figures 6 and 7.

The intermediate phase Zr_4Fe (ρ) appears to form by peritectoid reaction between β and ϕ. Large light-gray particles of ρ are prominent in Figure 2. These appear to be the product of an essentially completed reaction between ϕ and β. Where the quantity of ϕ has been greater, the reaction has not been completed; in Figure 3, ρ can be seen as an envelope around white particles of residual ϕ and everywhere separating the ϕ from the newly formed α plus ρ eutectoid. The same structural characteristic is evident in Figures 4 and 5, except that the relative quantity of ϕ increases progressively while that of α diminishes. Only gray ρ and white ϕ are to be seen in Figure 6, where it may be deduced that the reaction between ϕ and β to form ρ has again gone to near completion. Both ϕ and ρ are to be seen in much the same structural arrangement in Figure 7.

The intermediate phase Zr_2Fe (ϕ) makes its first appearance in Figure 3, where, as

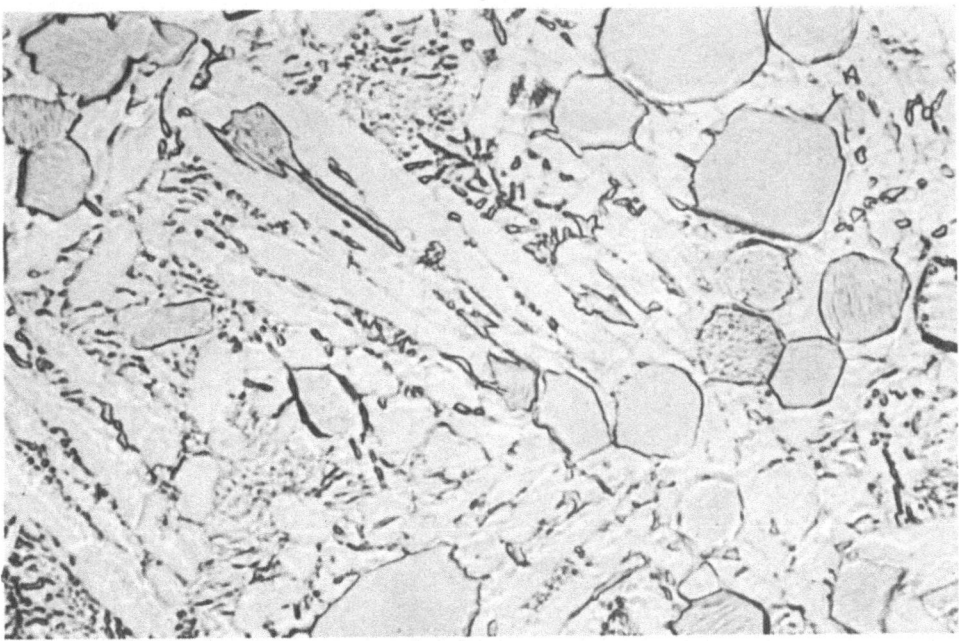

Figure 2. Zr + 5%Fe, 900°C for 65 hr, slow cool to room temperature. Large rounded areas of ρ in a matrix of α + ρ. 500×.

Figure 3. Zr + 13%Fe, 900°C for 65 hr, slow cool to room temperature. White phase is ϕ surrounded by gray ρ. The dark areas are $\alpha + \rho$ eutectoid. The peritectoid transformation, $\beta + \phi$ to form ρ, is clearly evident. 500 ×.

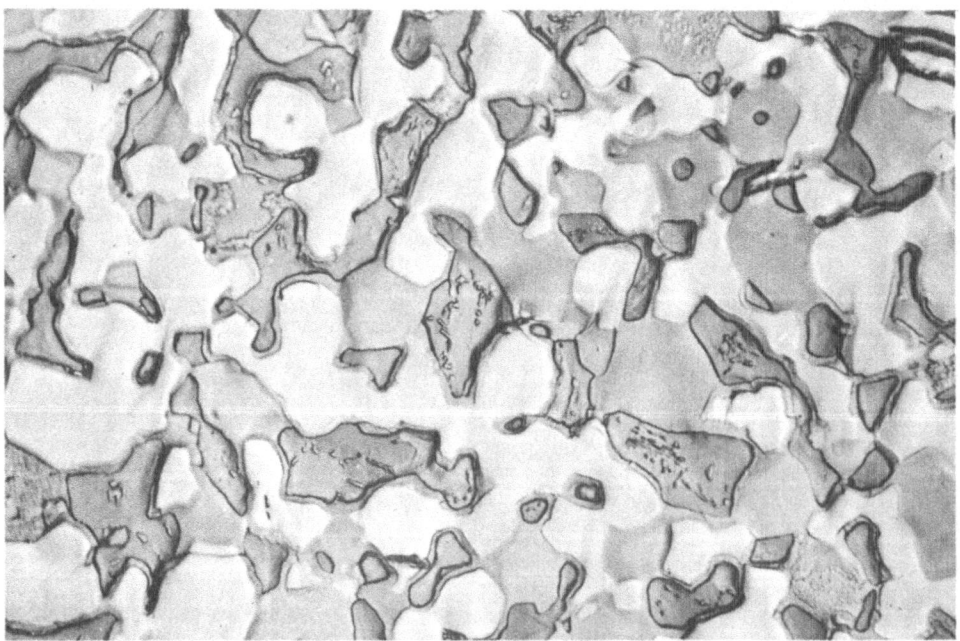

Figure 4. Zr + 17%Fe, 900°C for 65 hr, slow cool to room temperature. White phase is ϕ surrounded by light gray ρ together with darker $\alpha + \rho$ eutectoid. 500 ×.

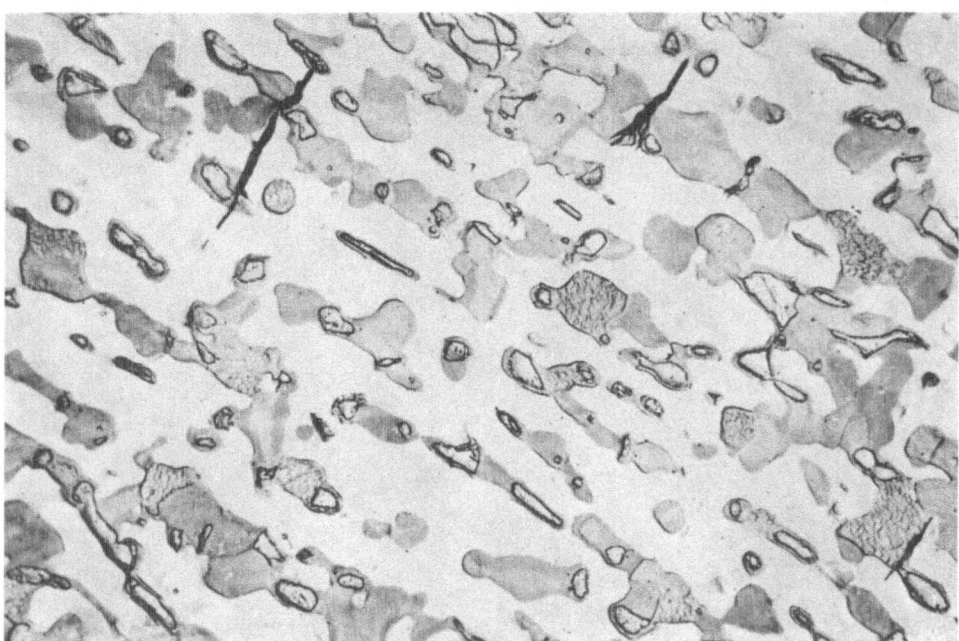

Figure 5. Zr + 20%Fe, 900°C for 65 hr, slow cool to room temperature. Predominant white phase is ϕ, gray phase is ρ, small areas of $\alpha + \rho$ embedded in ρ. 500 ×.

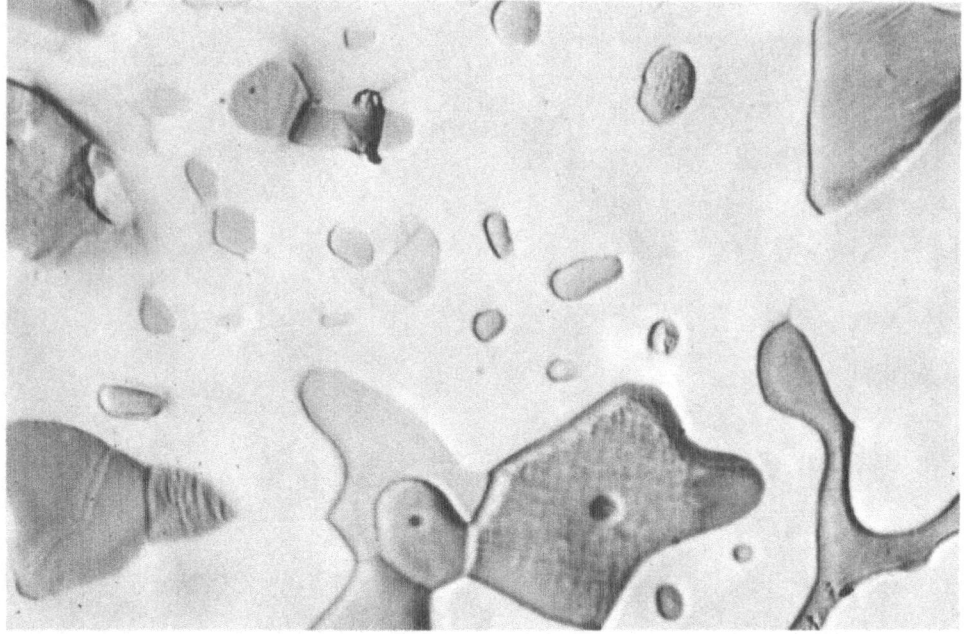

Figure 6. Zr + 23%Fe, 900°C for 65 hr, slow cool to room temperature. Predominantly composed of white ϕ in which darker areas of ρ can be seen. 2000 ×.

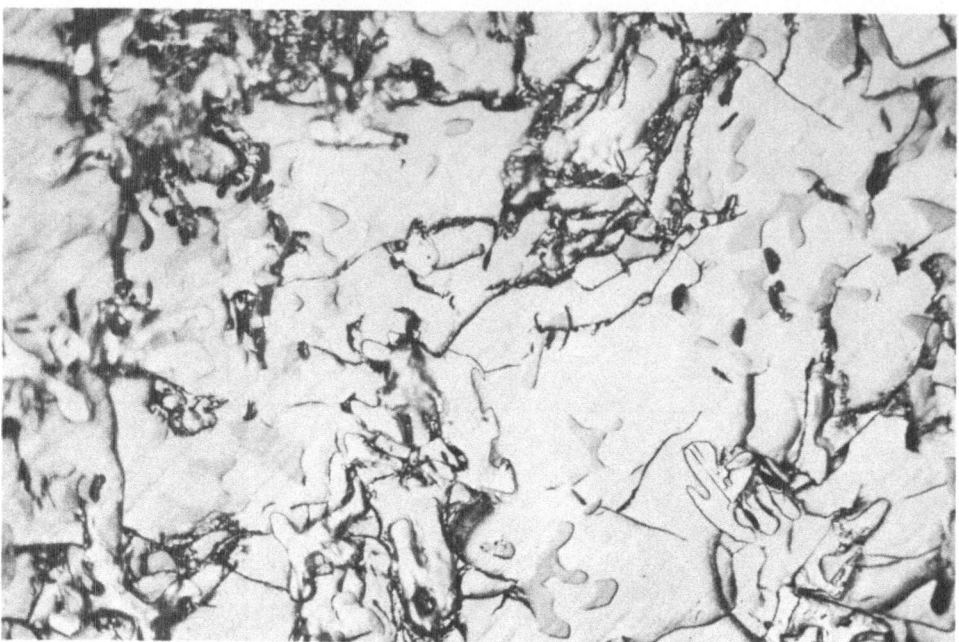

Figure 7. Zr + 29%Fe, 900°C for 65 hr, slow cool to room temperature. Dominant phase is ϕ in which small dark areas of ρ and white ψ are embedded. 500×.

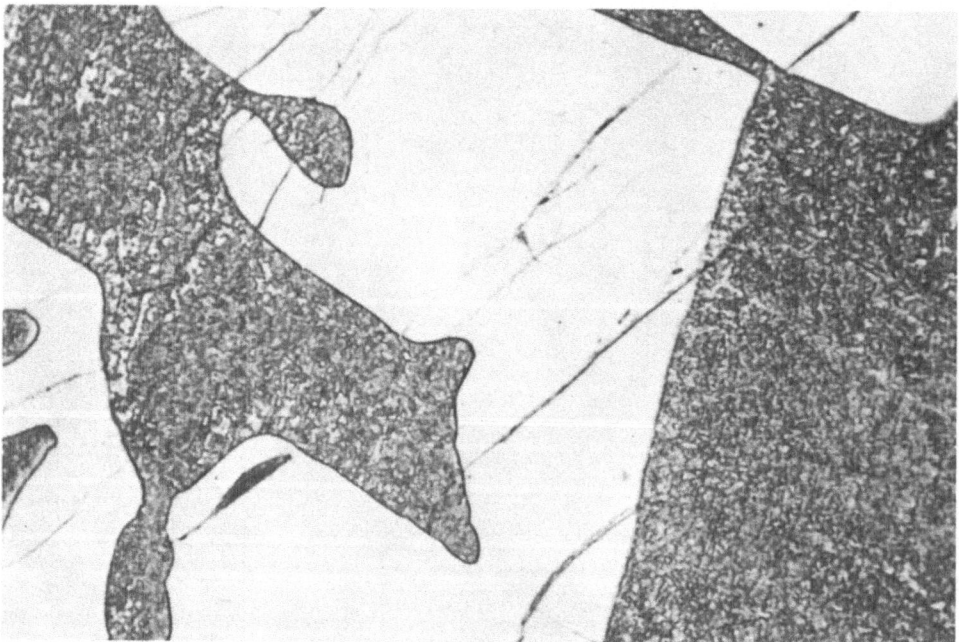

Figure 8. Zr + 38%Fe, 900°C for 65 hr, slow cool to room temperature. White phase is ψ (ZrFe₂) embedded in dark fine-grained polyphased matrix containing ϕ. 2000×.

indicated above, it remains as an unreacted residue of light-colored particles. The quantity of the white ϕ phase increases through Figures 4, 5, 6, and 7, in the latter three of which it forms the matrix.

Finally the intermediate phase ZrFe$_2$ (ψ), which was identified by previous investigators, can be seen in the microstructures of Figures 7, 8, and 9. In these photomicrographs ψ appears as the most nearly white constituent. It constitutes the major part of the structure of the alloy containing 55% iron, which lies very close to the stoichiometric proportions of ZrFe$_2$. As the iron content is diminished, the quantity of ZrFe$_2$ becomes progressively less. It freezes from the melt as an idiomorphic dendrite, which is only partially spherodized by heat treatment at 900°C. Thus, the white branching particles in Figure 7 are ψ, left from the cast structure, while the blocky white phase in Figure 8 is of the same origin. A gray constituent, which is obviously polyphased, is prominent in both Figures 8 and 9. This is presumed to be a mixture of ϕ and ρ which has descended by transformation from the eutectic reaction at 934°C.

X-RAY DIFFRACTION

Annealed filings (less than 200 mesh) prepared in the manner previously described were coated upon petrographic glass slides using a mixture of collodion and amyl acetate. These slides were then examined in a high-angle diffractometer using nickel-filtered copper radiation. In order to present the same surface area to the X-ray beam, each slide was covered with a mask consisting of a $\frac{1}{2}$-in. hole drilled in a thin sheet of tantalum.

The variation of line intensity as a function of alloy composition is plotted in Figures 10 and 11 for each of the reflections observed. Here it will be observed that most of the

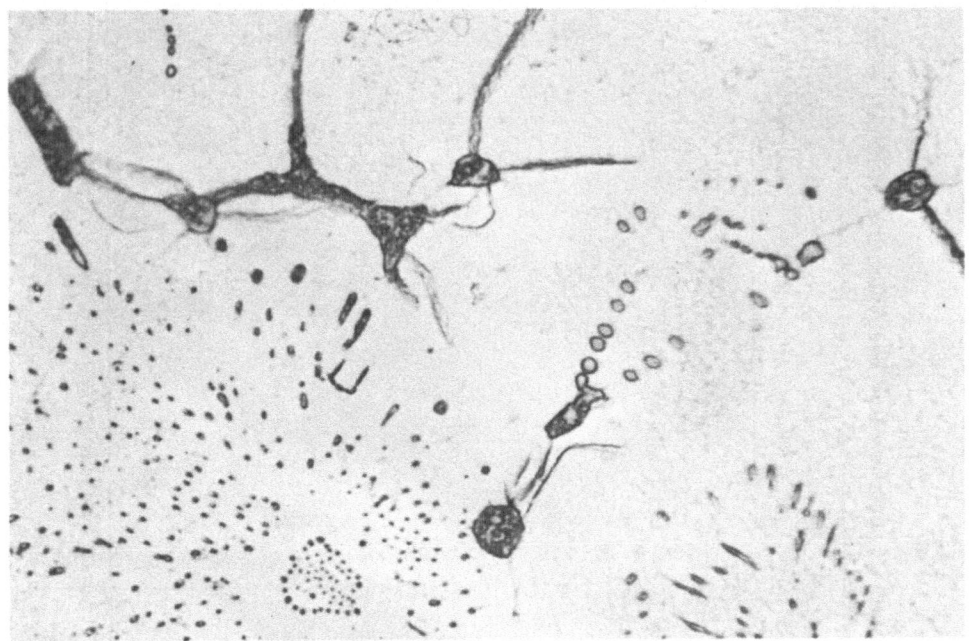

Figure 9. Zr + 55%Fe, 900°C for 65 hr, slow cool to room temperature. Almost entirely composed of ψ with small rounded areas of ϕ. 2000 ×.

Table II. Intensity and Identification of X-Ray Diffraction Reflections for Zr–Fe Alloy Powders Annealed at 600°C

$d(\pm 0.01)$	α-Zr	Zr + 5%Fe	Zr$_4$Fe Zr + 13.3%Fe	Zr$_3$Fe Zr + 16.4%Fe	Zr + 20.0%Fe	Zr$_2$Fe Zr + 23.4%Fe	Zr$_3$Fe$_2$ Zr + 28.9%Fe	ZrFe Zr + 38.0%Fe	ZrFe$_2$ Zr + 55.0%Fe
2.80	33 α	12 α, ϕ	5 α, ϕ	4 α, ϕ	5 α, ϕ	12 ϕ	9 ϕ	9 ϕ	
2.75		3 ρ, ϕ	9 ρ, ϕ	4 ρ, ϕ	4 ρ, ϕ	6 ρ, ϕ	5 ρ, ϕ		
2.63			12 ρ, ϕ	10 ρ, ϕ	10 ρ, ϕ	12 ρ, ϕ	11 ρ, ϕ	5 ϕ	
2.58		24 ρ, ϕ	50 ρ, ϕ	47 ρ, ϕ	38 ρ, ϕ	41 ρ, ϕ	37 ρ, ϕ	12 ϕ	
2.57	32 α								
2.55									
2.50					10 ϕ, ψ	10 ϕ	7 ϕ	13 ϕ	17 ψ
2.47	100 α	59 α	36 α, ϕ	21 α, ϕ	11 α, ϕ	21 ϕ, ψ	18 ϕ, ψ	22 ϕ, ψ	
2.38		7 ρ, ϕ	27 ρ, ϕ	25 ρ, ϕ	30 ρ, ϕ	13 ϕ	12 ϕ		
2.35		10 ϕ, ρ	13 ϕ, ρ	15 ϕ, ρ	23 ϕ, ρ	30 ρ, ϕ	16 ϕ	28 ϕ	
2.29			2 ρ			48 ϕ	39 ϕ		
2.21		2 ρ, ϕ	4 ρ, ϕ	3 ρ, ϕ	4 ρ, ϕ	7 ρ, ϕ	6 ρ, ϕ		
2.16		5 ϕ, ρ	17 ϕ, ρ	20 ϕ, ρ	19 ϕ, ρ	30 ϕ, ρ	28 ϕ, ρ	18 ϕ	5 ϕ
2.14						14 ψ	33 ψ	41 ψ	46 ψ
2.09			5 ϕ		4 ϕ	6 ϕ	3 ϕ	6 ϕ	
2.04					2 ψ	6 ψ	10 ψ	15 ψ	16 ψ
1.90	17 α	10 α	6 α						
1.69			6 ρ, ϕ	6 ρ, ϕ	5 ρ, ϕ	6 ρ, ϕ			
1.66			8 ρ, ϕ	10 ρ, ϕ	5 ρ, ϕ	6 ρ, ϕ			
1.61	17 α	8 α							
1.47	18 α	14 α							
1.44					5 ϕ, ψ	7 ϕ, ψ	5 ϕ, ψ	6 ϕ, ψ	8 ϕ, ψ
1.42		3 ρ	5 ρ	5 ρ	2 ρ				
1.40	3 α	5 ρ, ϕ	10 ρ, ϕ	12 ρ, ϕ	8 ρ, ϕ	11 ρ, ϕ			
1.37	18 α	13 α	7 α						

Note: The numbers refer to peak minus background intensity.

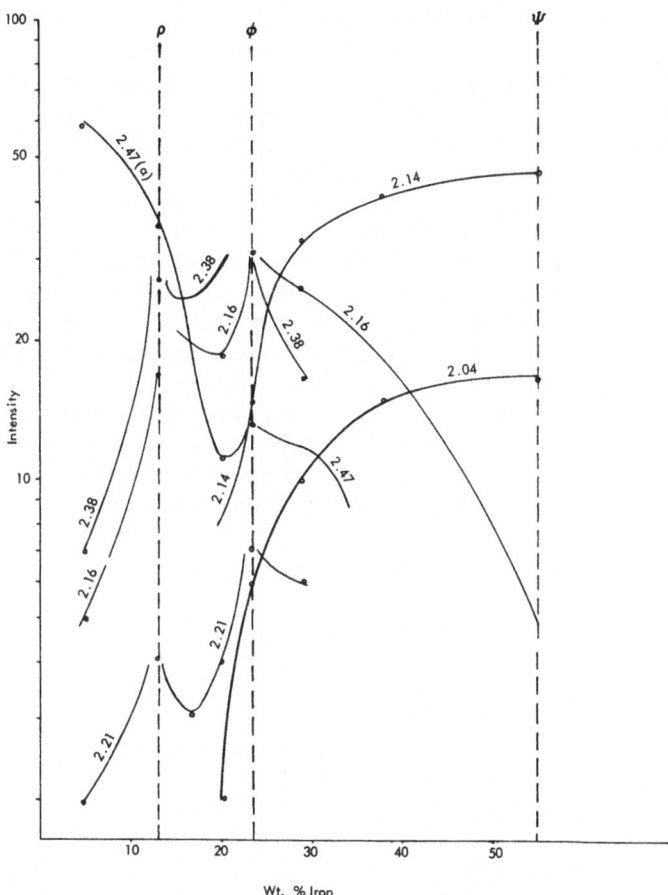

Figure 10. X-ray diffraction intensity of several important reflections plotted against iron content of alloys.

reflections exhibit more than one intensity maximum as the composition of the alloy is varied. This indicates that such reflections represent the superposition of lines from two or more phases. This method of plotting has been used to identify the composition of the phases which contribute to each reflection. It has thus been possible, in Table II, to designate by Greek letter symbols those phases which could have caused the recorded reflections and in Table III to assign certain reflections to the two new phases. In order to obtain a complete interpretation of the results it is necessary to compare the X-ray diffraction patterns with the metallography. This procedure leads to a unique identification of the phases which are present over each composition range as is indicated in the phase diagram, Figure 1.

REMARKS

The proposed phase diagram is incomplete. From the present research it has been possible to determine the approximate compositions of two new phases and to identify, crudely, the temperature range in which each of these phases occurs. The metallographic evidence shows that ρ is generated by reaction between β and ϕ, which must occur below 900°C and above 800°C. It indicates also that ϕ is stable above 900°C. Photomicrographs of cast alloys, not reproduced here, suggest that ϕ forms by peritectic reaction of ψ with the liquid phase; this means that ϕ must be stable above 934°C, but not up to the melting point of ψ, near 1600°C. For lack of evidence to the contrary, it is assumed that both ρ and ϕ are stable to room temperature. There is some suggestion of precipitates forming within particles of both phases. This observation has been used as a basis for indicating a range of solid solubility in each phase.

It should be noted further that the alloys which have been examined were mostly in a nonequilibrium state. This is clearly evident, since most of the alloys exhibited as many as three readily identifiable solid phases. Such lack of equilibrium does not vitiate the finding of the additional phases, as is proved by the fact that the quantities of the intermediate phases increase with slower cooling from high temperature. There is little doubt, therefore, that ρ and ϕ are, in fact, the equilibrium phases.

In the course of the X-ray diffraction study, some additional reflections were observed in the as-cast 20% alloys and in this alloy again after certain heat treatments. There is a possibility that this may indicate the presence of yet another intermediate phase in the general neighborhood of 20% iron. The nature of the evidence is such however that the proposal of the existence of such a phase at this time is premature.

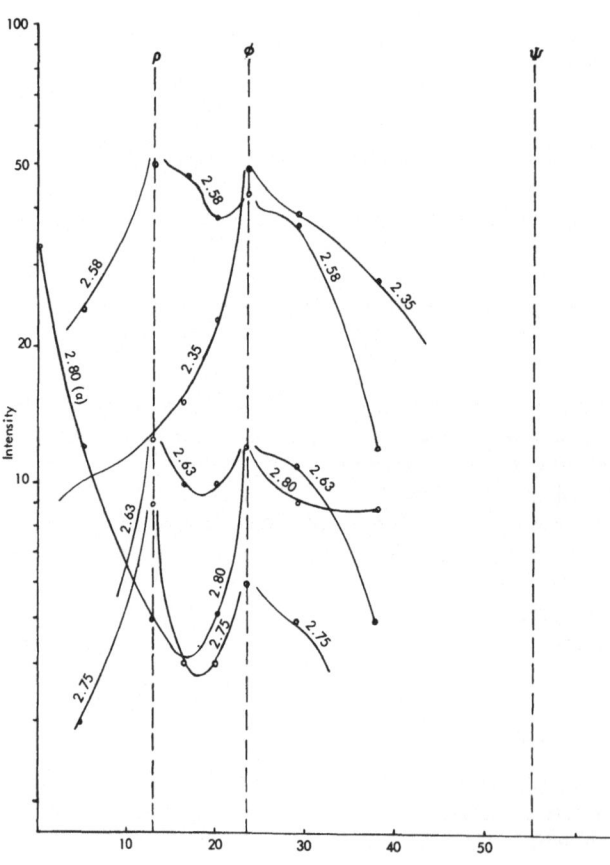

Figure 11. X-ray diffraction intensity of several important reflections plotted against iron content of alloys.

Table III. *d*-Spacings and Intensities for the Intermediate Phases in the Zr–Fe System

Zr + 13%Fe ρ Zr₄Fe		Zr + 23%Fe φ Zr₂Fe		Zr + 55%Fe ψ ZrFe₂	
d	*I*	*d*	*I*	*d*	*I*
2.75	mw	2.80	m	2.49	43
2.63	mw	2.75	w	2.13	100
2.58	s	2.63	mw	2.03	30
2.38	ms	2.58	ms	1.62	w
2.29	vw	2.55	w	1.44	22
2.21	vw	2.50	ms	1.36	30
2.16	m	2.47	w	1.25	30
1.69	w	2.38	ms	1.25	30
1.66	w	2.35	s	1.19	w
1.42	w	2.21	w	1.12	27
1.40	w	2.16	ms		
		2.09	vw		
		1.69	w		
		1.66	w		
		1.44	w		
		1.40	w		

REFERENCES

1. D. Hansen, *Constitution of Binary Alloys*, second edition, McGraw-Hill Book Co., Inc., New York, 1958.
2. E. T. Hayes, A. H. Robertson, and W. L. O'Brien, *ASM Trans.* **43**: 888–904, 1951.
3. R. Vogel and W. Tonn, *Arch. Eisenhüttenw.* **5**: 387–389, 1931, 1932.
4. H. J. Wallbaum, *Arch. Eisenhüttenw.* **14**: 521–526, 1940–1941.
5. D. Levinson, Armour Research Foundation, Chicago, Illinois, private communication, 1961.
6. M. L. Picklesimer, Oak Ridge National Laboratory, Oak Ridge, Tennessee, private communication.

DISCUSSION

F. Laves (Polytechnic Institute, Zurich, Switzerland): Is it possible to increase the length of annealing time to see if other phases will come out?

R. W. Gould: We had one specimen heat-treated that had been in the furnace for two weeks, and it showed a two-phase structure in the 13% alloy. Unfortunately, it did not come out single phase in two weeks. It was two phases; however, one phase predominated. I assumed it to be predominantly 30% ρ. We haven't had a chance to correlate that report, but expect to do this and run some month-long heat treatments. It is going to take that long to reach complete equilibrium.

R. L. Beck (University of Denver): Could oxygen possibly have stabilized the ε-phase?

R. W. Gould: This may be. We don't deny this. We took extreme precautions to eliminate oxygen contamination. It may be pointed out that the buttons were made in a cold hearth vacuum arc melting furnace and we used both hardness and lattice parameter of α-zirconium. We ran pilot melts of α-zirconium and measured the lattice parameter to see if there was any shifting to detect any oxygen pickup while we were melting. And we ran pure zirconium powders through our powder annealing technique again to detect any oxygen pickup. We detected none, but I don't throw out the possibility that this may be an oxygen-stabilized phase.

X-RAY DIFFRACTION STUDIES ON THE TITANIUM–NICKEL SYSTEM

John V. Gilfrich

U.S. Naval Ordnance Laboratory, White Oak, Silver Spring, Md.

ABSTRACT

X-ray diffraction studies were made on the Ti–Ni system around the stoichiometric composition of the intermetallic compound TiNi to clarify some confusion which has existed about the phase diagram in this region, and to explain some anomalies in the physical properties of this material. Wrought and cast samples were examined at room temperature both before and after heat treatment and at temperatures both above and below ambient. The compound TiNi does exist at room temperature. The phase purity of the particular sample was found to be greatly affected by such factors as minor variations in composition, heat treatment, and method of sample preparation. Some confirming metallographic and physical property data will also be presented.

INTRODUCTION

Over the past ten years, investigations of intermetallic compounds have received considerable emphasis, mostly because many of them possess semiconducting properties. Intermetallic compounds also have unusual physical and mechanical properties for structural applications. One inherent property of these compounds that has limited their structural use is their extreme brittleness.

The definition of an "intermetallic compound" is a difficult task and one which evokes considerable disagreement among workers in the field. Some agree with the *Metals Handbook*[1] which defines it as "an intermediate phase in an alloy system, having a narrow range of homogeneity and relatively simple stoichiometric proportions, in which the nature of the atomic binding can vary from metallic to ionic." Others[2,3] feel that in order to classify an intermediate phase as an intermetallic compound, identical kinds of atoms must occupy identical points on the lattice. And still others, like Westbrook,[4] consider "all intermediate phases in binary and higher order metal systems whether ordered or disordered."

An extensive investigation[5] was initiated to develop an alloy system based on an "intermetallic compound," as defined by Westbrook. As a result of this study a new series of engineering alloys was developed, based on the compound TiNi, which showed remarkable and unusual physical and mechanical properties. These alloys, called Nitinols, have the characteristic high melting point of the intermetallic compounds, but have sufficient room-temperature ductility to make their application to structural components feasible. They are nonmagnetic, corrosion and abrasion resistant, of moderate density and high strength, and can be hardened to values approaching those of tool steels ($R_c = 62$) by suitable composition and heat treatment. Details of these engineering properties are available elsewhere.[5,6]

---

[1] Superscripts pertain to references at the end of the paper.

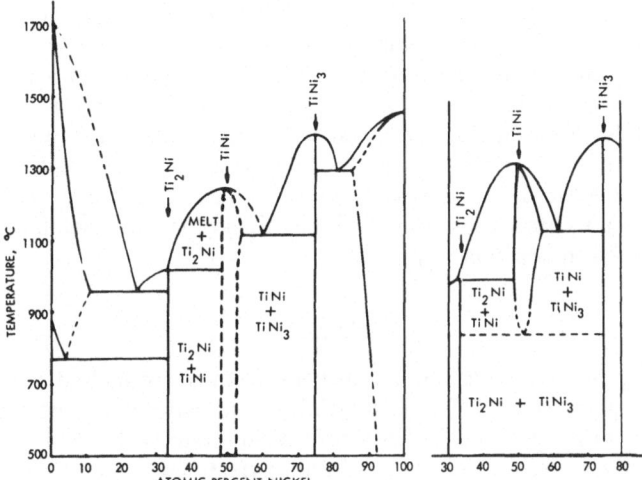

Figure 1. Two versions of the nickel–titanium constitution diagram.

Hansen[7] gives two constitutional diagrams for the titanium–nickel system, which disagree mainly in the region of interest around 50 at. % as shown in Figure 1.* The stoichiometric intermetallic compounds Ti_2Ni and $TiNi_3$ existing on either side of this ductile alloy were investigated[5] and it was discovered that they were, in fact, typical intermetallic compounds; that is, they were hard, abrasion resistant and brittle at room temperature, and they existed at a discrete composition.

Duwez and Taylor[8] and Poole and Hume-Rothery[9] using powdered samples have shown the existence of a CsCl type of cubic structure in the equiatomic alloys, but have stated that this compound is not stable at room temperature. On the other hand, Margolin *et al.*[10] were unable to detect any dissociation of TiNi into Ti_2Ni and $TiNi_3$ in bulk material. In addition, Philip and Beck[11] and Pietrokowsky and Youngkin[12] in studying the ordering phenomena in the body-centered cubic structure of transition metal alloys all worked with TiNi at room temperature. This contradictory evidence explains the inclusion of two alternative constitution diagrams in Hansen's compendium and casts some uncertainty on the stability of the body-centered cubic structure.

The present X-ray investigation was initiated to resolve this uncertainty and to examine the system in enough detail to explain some of the unusual mechanical behavior, severe dimensional changes, and acoustical damping which occur in the alloy.

EXPERIMENTAL

The X-ray studies were all conducted using a standard Norelco diffractometer equipped with a krypton-filled geiger tube for detecting the zirconium-filtered molybdenum K_α radiation used in the investigation. Calculations based on the wavelength absorption coefficient of the alloy, and the range of 2θ angles employed showed that the effective depth of penetration of the X-rays varied from 0.002 to 0.004 in. Certain special sample holders were available for X-ray patterns run at temperatures above and below ambient. The sample holder for low temperatures (slightly below 0°C) and for temperatures up to 85°C was a copper block machined to take the sample and to provide a

* Figures 1, 2, 3, 9, 10, and 11 are taken from Buehler and Wiley, NOLTR 61–75, Reference 5.

chamber in good thermal contact with the sample through which the cooling or heating media could be passed. For temperatures above ambient, heated water maintained at constant temperature was passed through the holder. For low temperatures, ice water, alcohol cooled by dry ice, or nitrogen gas cooled by liquid nitrogen was circulated through the system, and dry, precooled nitrogen gas was directed over the sample surface to prevent condensation or frost from forming. For temperatures above 85°C, a high-temperature attachment for the diffractometer was used similar to the one designed by Mauer and Bolz.[13] This attachment was modified slightly so that bulk samples in the form of sheet could be used in addition to powder.

RESULTS

The first samples to be run were filings made from an arc-melted button of stoichiometric TiNi (55.06 w/o Ni). The X-ray diffraction pattern of this sample, as filed, showed one very broad peak in the low-angle region, due to the cold work necessary to file this very ductile material. Numerous annealing treatments were attempted at temperatures up to 1000°C and for times as short as 5 min, all leading to diffraction patterns which corroborated the findings of Duwez and Taylor, and Poole and Hume-Rothery; that is, the sample consisted of a mixture of Ti_2Ni and $TiNi_3$ with no TiNi. This sample was then run at temperatures up to 1000°C in the high-temperature diffractometer attachment, but the results did not change. The patterns became successively less sharp as the temperature increased but they indicated that the sample was remaining a mixture of the two phases rather than TiNi.

A preliminary run was next made on a piece of an arc-melted button of 55.06 w/o Ni composition. This piece was irregular in shape but it gave a very satisfactory pattern indicating phase-pure TiNi, a CsCl type of body-centered cubic with a lattice parameter of about 3 Å and a rather strong (100) superlattice peak.

A number of samples of the stoichiometric composition were available which had been arc-cast and then hot-rolled at various temperatures from 600 to 1100°C. A sample of the material rolled at each of the temperatures (600, 700, 950, 1000, and 1100°C) was ground flat, metallographically polished and then deep-etched to remove the worked layer. The first one of these samples gave a diffraction pattern indicating a mixture of Ti_2Ni and $TiNi_3$ with a minor amount of TiNi. A metallographic examination of the sample is shown in Figure 2(a) which shows a martensitic-like structure in the surface layer. Figure 2(b) shows the true base structure of this sample showing a portion of the martensitic layer removed by careful surface preparation. The lower part of this photo shows the true structure, the upper part is masked by the remaining overlay of the martensitic-like material.

All five of these samples were then repolished carefully using diamond paste and very lightly etched. Metallographic examination of them showed the true base structure as indicated by the lower part of Figure 2(b). X-ray diffraction scans of all samples showed that they were without exception the body-centered cubic CsCl type of structure ascribed to TiNi.

To determine the possibility of a dissociation of TiNi into Ti_2Ni and $TiNi_3$ in bulk material, seven samples were prepared, varying in composition from 50 to 60 w/o Ni, by arc-casting them into a mold which had a carefully prepared flat bottom. This produced a button which had one surface flat enough to be used in the diffractometer without the necessity of any surface preparation. X-ray diffraction patterns were run at room temperature on each of the seven samples five times, after the following heat treatments in an argon atmosphere: (1) As-cast, (2) After 750°C for 24 hr, (3) After 750°C for 72 hr,

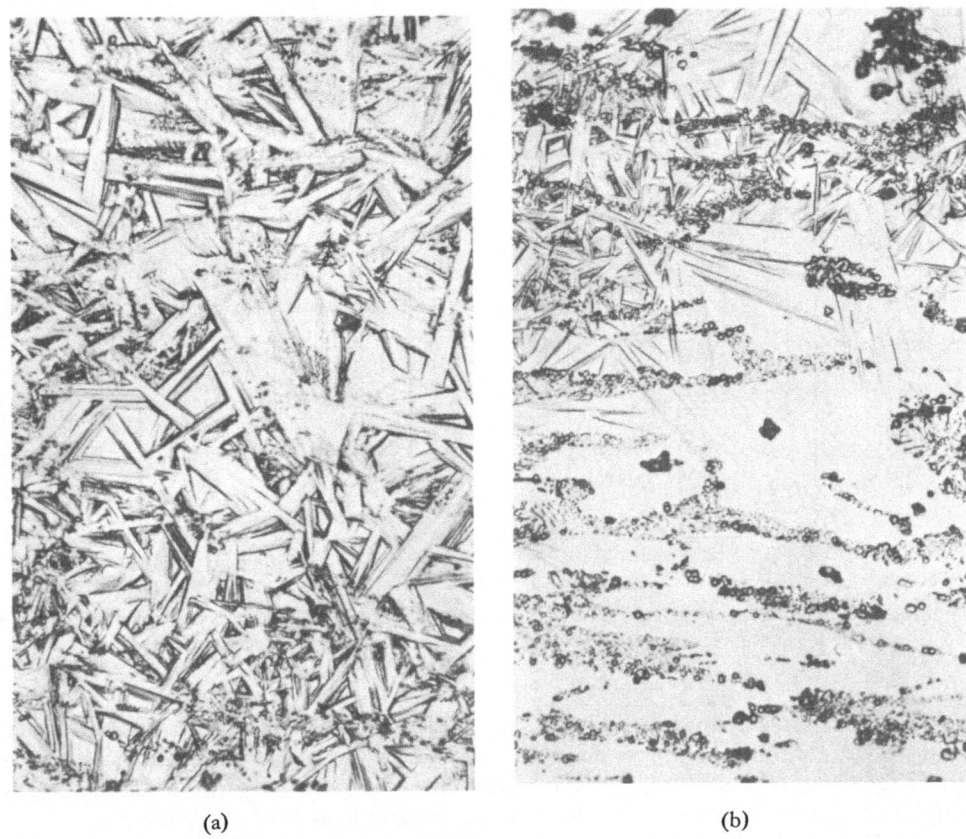

(a)   (b)

Figure 2(a). Photomicrograph showing the superficial layer of martensitic structure formed upon the surface during abrasive cutting of the sample. 500×. (b). Photomicrograph showing a portion of the martensitic layer removed by careful surface preparation. Lower part of photo shows true base structure; upper part is masked by remaining overlay of martensite. 500×.

(4) After 750°C for 144 hr, and (5) After 750°C for 144 hr plus 800°C for 120 hr. The results are shown in Figure 3, which gives a qualitative estimate of the relative amounts of the three phases present in each sample. It must be borne in mind that there is a great deal of preferred orientation in these samples, and therefore, small changes in percentage of a phase present at a low level are not really significant. From Figure 3 it can be seen that TiNi does, in fact, exist at room temperature, in a stable or metastable condition in samples containing more than 54 w/o Ni, even after a total of 264 hr at temperatures of 750 and 800°C. It is also evident from these data that in samples containing 54 w/o Ni or higher, there are three phases coexisting, seeming to contradict the "phase rule" for equilibrium alloy systems. Prolonged heating at 750 and 800°C does not seem to have any significant effect in promoting any greater equilibrium condition. The samples containing less than 54 w/o Ni contained only the two compounds Ti_2Ni and $TiNi_3$. The 54 w/o Ni sample, as cast, contained only these two compounds but were converted to predominantly TiNi material by the anneal.

Concurrently with the X-ray diffraction studies just mentioned, it was found[5] through hardness, vibration damping, and dimensional changes that compositions close

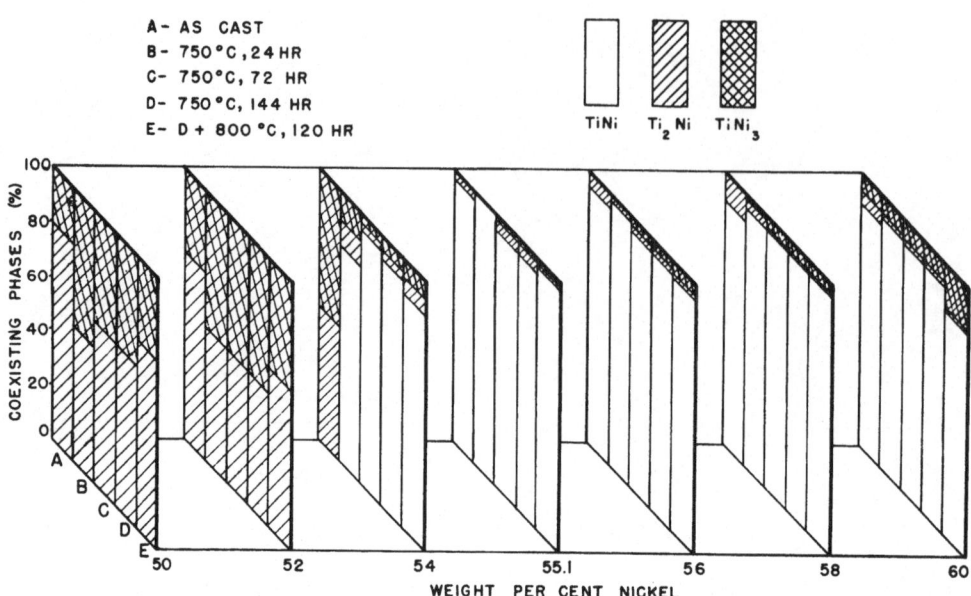

Figure 3. Shows phases existing in several Ni–Ti alloys from 50 to 60 w/o Ni. Also shown are the changes in the quantities of the phases with thermal treatment.

to stoichiometric TiNi possessed some very unusual properties, which appeared related to the phase equilibria of the system. Very briefly, these properties indicated that between room temperature and 100°C, drastic transitions were occurring in the sample. X-ray diffraction patterns were run on a stoichiometric sample at temperatures between −10 and 85°C using the controlled temperature sample holder previously described. Figure 4 shows a portion of the patterns indicating the phase relationship between the TiNi which is the predominant phase and the Ti_2Ni which is present in a minor amount. The room-temperature pattern (Figure 4A) shows two TiNi peaks and three Ti_2Ni peaks, all fairly well resolved. As the temperature is increased the amount of Ti_2Ni decreases until, at 82°C (Figure 4C) only one of its peaks is barely detectable. As the temperature is decreased down to −10°C (Figures 4D, E and F), the relative amount of the second phase increases as shown by the intensifying of the Ti_2Ni peaks. When the sample returns to room temperature, the pattern remains the same as at −10°C but by heating the sample by immersion in warm water, the room-temperature condition as shown in Figure 4A can be recovered. If the sample is cooled to −196°C by being placed in liquid nitrogen, the Ti_2Ni peaks become even stronger, the peak at $27°2\theta$, for example, becoming considerably more intense than the neighboring TiNi peak. Again, by heating in warm water, the room temperature condition can be recovered.

In all samples containing TiNi, with a few exceptions, the (100) superlattice line was observed. The exceptions to this occurred where the preferred orientation was such that even the (200) reflection was missing or very weak. In an attempt to study the order-disorder phenomenon, and to obtain additional data at higher temperatures, four additional samples were prepared in two forms. These had compositions of 54, 55.06, 55.4, 56 w/o Ni, balance Ti, and were prepared as arc-cast buttons with one flat surface suitable for use in the diffractometer, and as hot-rolled sheet approximately 0.020 in. in thickness, for use in the high-temperature diffractometer attachment. The buttons were run as

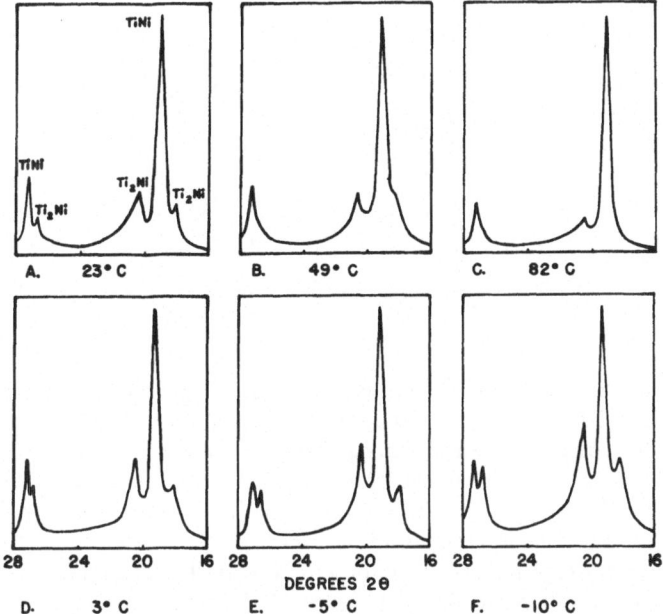

Figure 4. X-ray diffractometer scans of stoichiometric TiNi at various temperatures.

cast and the sheets were cut to fit the sample holder and diamond polished to remove the oxidized surface. Portions of the diffraction pattern of these samples are shown in Figures 5, 6, 7, and 8.

The 54 w/o Ni sample, in the as-cast condition (Figure 5A) showed a mixture of Ti_2Ni and $TiNi_3$. The hot-rolled sheet of this composition, at room temperature, was the same (Figure 5B). By increasing the temperature of this sample up to about 90°C, the diffraction pattern shows TiNi beginning to form in the surface layer being studied (Figure 5C). On attaining 125°C, the sample is completely TiNi (Figure 5D). When the sample returns to room temperature, the TiNi virtually disappears and the sample is almost completely Ti_2Ni and $TiNi_3$ (Figure 5C). This transformation is reversible up to this temperature level and can be repeated any number of times with, however, considerable temperature hysteresis. In this composition TiNi is the stable phase up to 1000°C and after having been heated above about 800°C, is the phase present at room temperature; the transformation TiNi to Ti_2Ni + $TiNi_3$ is no longer reversible.

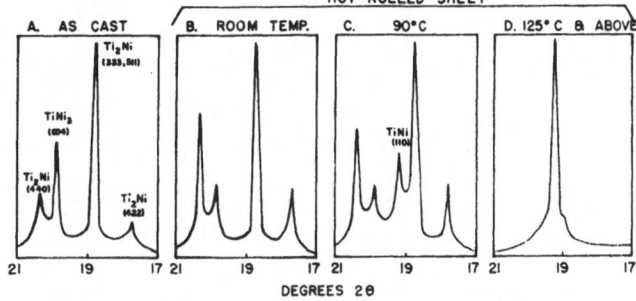

Figure 5. Partial X-ray diffractometer scans of 54% Ni–Ti, as arc-cast, and as hot-rolled sheet at various temperatures.

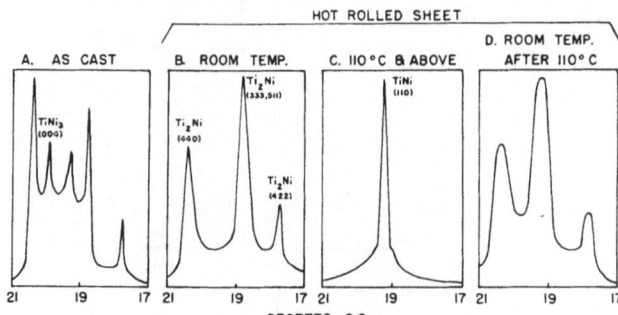

Figure 6. Partial X-ray diffractometer scans of 55.06% Ni–Ti, as arc-cast, and as hot-rolled sheet at various temperatures.

The stoichiometric alloy (55.06 w/o Ni), as cast, (Figure 6A), is a mixture of all three phases, the TiNi being present in a fairly small amount. The sheet sample at room temperature (Figure 6B) on the other hand, gives a diffraction pattern which looks like single-phase Ti_2Ni. Gross mechanical properties belie this observation because Ti_2Ni is very brittle and this sample can be worked. The Ti_2Ni must be therefore a surface layer. Heating to 110°C converts the sample entirely to TiNi (Figure 6C). After returning to room temperature, the sample seems to be a mixture of TiNi and Ti_2Ni and the diffraction peaks are considerably broadened (Figure 6D). Again, the TiNi is stable up to 1000°C and TiNi is the stable room-temperature phase after having been heated to about 800°C or above.

The sample containing 55.4 w/o Ni, as cast, is also a three-phase alloy, the predominant phase, this time being TiNi (Figure 7A). The sheet material, at room temperature is also three-phase; however, the TiNi is a minor component (Figure 7B). The TiNi reflection begins to become more intense at about 65°C and becomes the only peak observable at 100°C (Figure 7C). On cooling to room temperature, the sample remains predominantly TiNi but minor amounts of the other phases are present. Heating to 800°C causes the sample which is single-phase TiNi at lower temperatures to develop a minor amount of $TiNi_3$ (Figure 7D), which remains present in the sample at room temperature.

In the as-cast 56 w/o Ni sample (Figure 8A) single-phase TiNi is demonstrated. The sheet sample at room temperature is identical. On heating to 650°C a minor amount of $TiNi_3$ is developed in the sample (Figure 8B), which remains regardless of further treatment.

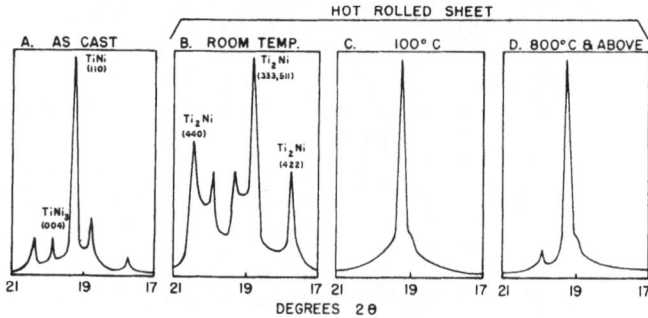

Figure 7. Partial X-ray diffractometer scans of 55.4% Ni–Ti, as arc-cast, and hot-rolled sheet at various temperatures.

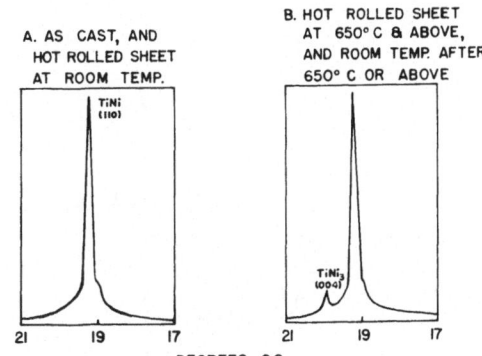

Figure 8. Partial X-ray diffractometer scans of 56% Ni–Ti, as arc-cast and as hot-rolled sheet at various temperatures.

Figure 9 shows the data from Figure 3 plotted as quantity of coexisting phases *vs.* weight per cent nickel and related to the two alternative constitution diagrams. It can be seen that below 52 w/o Ni the data seem to agree with the diagram which shows the TiNi dissociating into Ti$_2$Ni and TiNi$_3$ at low temperatures and, above 54 w/o Ni the data agree with the other diagram with increasing amounts of TiNi$_3$ as the nickel content is

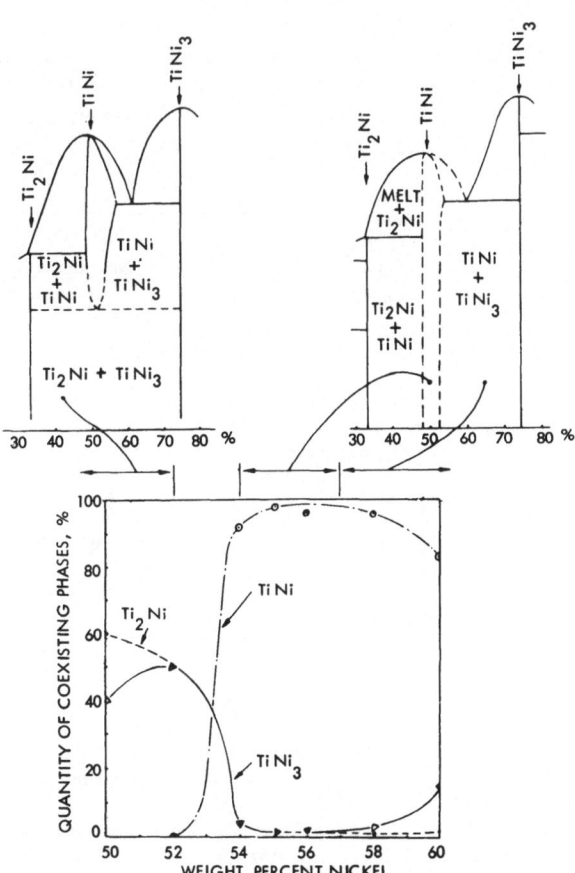

Figure 9. Room-temperature phase relationship of nickel–titanium alloys from 50 to 60 w/o Ni.

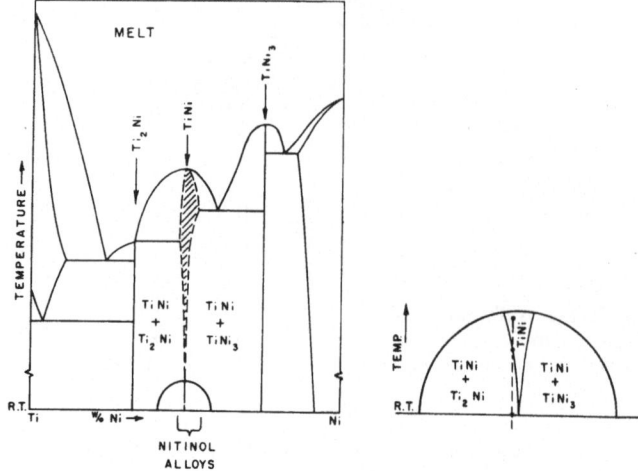

Figure 10. Equilibrium diagram for titanium–nickel system.

increased. The region between 52 and 54 w/o Ni is ambiguous as there seems to be a significant quantity of all three phases present.

Combining these data with those seen in Figures 5 through 8, we obtain the true phase diagram as shown in Figure 10. Purdy and Parr[14] in studying the TiNi system show a phase diagram essentially the same in the region of interest between the compounds Ti_2Ni and $TiNi_3$. They state, however, that TiNi in alloys containing from 35 to 48% Ni (atomic) completely transforms to a previously unreported π-phase, which they

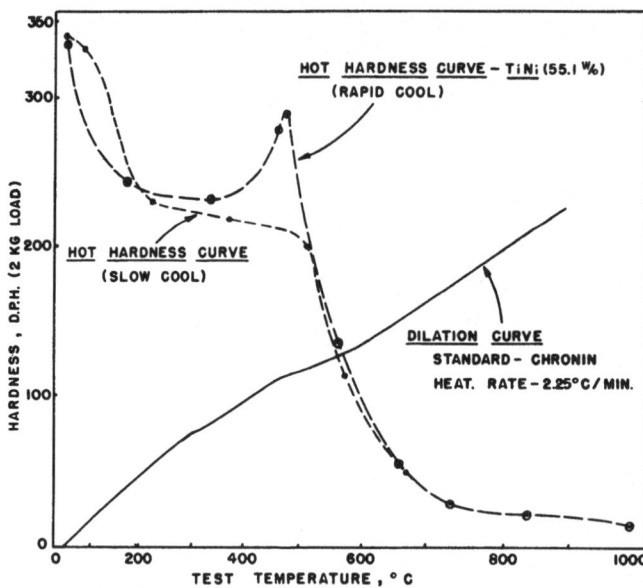

Figure 11. Curves of hot hardness and differential dilation plotted on the same temperature axis. Note changes in slope of the dilation curve at a temperature corresponding to the secondary hardening peak.

tentatively index as hexagonal, $a = 4.572$ Å, $c = 4.660$ Å, $c/a = 1.02$, at room temperature. We have been unable to confirm this result, finding only X-ray patterns due to TiNi, Ti_2Ni, $TiNi_3$, or mixtures of these in alloys covering this range of composition.

In all of our samples, the (100) superlattice line is quite evident except, as was mentioned previously, in cases where because of preferred orientation the (200) peak was missing or weak. Attempts to determine the disordering temperature were unsuccessful because the superlattice line was present up to 1000°C, the limit of the investigation. The only change observed in this peak was in the neighborhood of the recrystallization temperature, approximately 650°C, where the intensity changed in direct relation to the intensity change in the (200) fundamental line brought about by recrystallization. This is not what was expected since the hot hardness curves run on the stoichiometric composition (Figure 11) show a secondary hardening phenomenon at a temperature slightly below 500°C and this was thought to be related to the order–disorder transformation. Significant slope changes can be seen in the differential dilation curve on this same figure, at a temperature corresponding to the peak in the hardness curves. At the present time, the explanation for this effect is unknown, but in the light of the X-ray data, it does not seem to be a simple order–disorder phenomenon, as was previously thought.

CONCLUSIONS

We have determined that TiNi, a CsCl type of body-centered cubic intermetallic compound does exist at room temperature, in a stable or metastable form, in bulk material. Since even small amounts of energy can effect the transformation of TiNi at room temperature, it is doubtful that this compound can exist in powder or filings. The presence of minor amounts of Ti_2Ni, $TiNi_3$, or both, in alloys which are predominantly TiNi make it possible to control the physical and mechanical properties over a wide range. The true phase diagram around 50 at. % nickel indicates a fairly wide range of homogeneity for single-phase TiNi at elevated temperatures; however, this homogeneous range becomes quite narrow at lower temperatures and approaches a discrete composition at room temperature.

REFERENCES

1. *Metals Handbook*, eighth edition, American Society for Metals, Cleveland, 1961.
2. C. S. Barrett, *Structure of Metals*, McGraw-Hill Book Co., Inc., New York, 1952.
3. W. Hume-Rothery, *Atomic Theory for Students of Metallurgy*, The Institute of Metals, London, 1948.
4. *Mechanical Properties of Intermetallic Compounds*, J. H. Westbrook, ed., John Wiley & Sons, Inc., New York, 1959.
5. W. J. Buehler and R. C. Wiley, The Properties of TiNi and Associated Phases, NOLTR 61-75, August, 1961.
6. W. J. Buehler and R. C. Wiley, *Trans. Quart. ASM* 55: 269–276, 1962.
7. M. Hansen, *Constitution of Binary Alloys*, second edition, McGraw-Hill Book Co., Inc., New York, 1958.
8. P. Duwez and J. L. Taylor, *AIME Trans.* 188: 1173–1176, 1950.
9. D. M. Poole and W. Hume-Rothery, *J. Inst. Metals* 83: 473–480, 1955.
10. H. Margolin, E. Ence, and J. P. Nielsen, *AIME Trans.* 197: 243–247, 1953.
11. T. V. Philip and P. A. Beck, *AIME Trans.* 209: 1269–1271, 1957.
12. P. Pietrokowsky and F. G. Youngkin, *J. Appl. Phys.* 31: 1763–1766, 1960.
13. F. A. Mauer and L. H. Bolz, Measurement of Thermal Expansion of Cermet Components by High Temperature X-ray Diffraction, WADC Technical Report 55–473, December, 1955.
14. G. R. Purdy and J. G. Parr, *Trans. Met. Soc. AIME* 221: 636–639, 1961.

DISCUSSION

F. Laves (Polytechnic Technical Institute, Zurich, Switzerland): How long a time elapsed between room temperature where $TiNi_3$ and Ti_2Ni existed and at 300° where TiNi existed?

J. F. Gilfrich: This was on the order of a few minutes. The heating rate on the high-temperature diffractometer attachment is quite rapid in this temperature range and the sample could be heated from room temperature to 100°C in certainly no more than 2 min. The diffraction plot was run immediately and the sample was single-phase TiNi. This may be a diffusion transformation such as exists in the copper–manganese and the terbium–gold systems. But there is a great deal of work still remaining to be done on the system, and we have no specific data.

W. J. Wittig (Haynes-Stellite Co.): What was the arc-cast mold that these samples were cast into to make them suitable for use with the diffractometer?

J. F. Gilfrich: It was simply a conventional, water-cooled copper hearth as is used in an arc-melting furnace. In our laboratories these are usually made with a concave bottom. This one was made specifically with a machined flat bottom and it was not polished. The balance of the buttons which we used in the diffractometer were of cubical cast surface having impressions in them, etc., but nevertheless providing a more or less flat surface which could be held in the plane of the diffractometer and used with at least some idea that the lattice parameters which had been determined were reasonably correct. In an attempt to determine the limits of the homogeneous field by lattice parameter measurements, we ran into difficulty, and I think this was because we at no time really attained equilibrium in these samples, because our lattice parameter data scattered all over the place. We could not use any of the standard methods or any technique to try to determine the phase boundaries.

M. Semchyshen (Climax Molybdenum Co.): Did you attempt to influence the approach to equilibrium by deforming the samples?

J. F. Gilfrich: Not specifically, no. The only thing that comes anywhere close to that is the hot-rolled material which I examined in the initial stages. The matrix was very definitely TiNi; however, particles were still present which could be determined to be Ti_2Ni and $TiNi_3$ so that even under those circumstances the three-phase condition existed and we were not at equilibrium.

HELIUM PATH DIFFRACTOMETRY AND ITS APPLICATION TO DETERMINATION OF RETAINED AUSTENITE AND MACROSTRESS IN STEEL

Robert A. McCune

University of Denver, Denver, Colorado

ABSTRACT

The helium X-ray path has long been used in spectrography to increase the intensity of the long-wavelength X-rays. The same principle has been applied to diffractometry. Up to threefold intensity increase is observed with chromium K_α radiation with very little increase in background. The peak to background ratio, therefore, is improved by almost the same factor as the increase in intensity. Application of the technique is illustrated by analysis of gauge block steel for retained austenite and macrostress.

X-ray diffractometry has been practiced almost exclusively using copper radiation. The marked low intensity of other radiations with the G–M diffractometer made copper radiation the standard. However, the X-ray emission spectrographer had to detect, with the utmost sensitivity, radiations of all wavelengths. The scintillation X-ray detector and the helium path spectrometer were developed and made available.

The advantage of the scintillation X-ray detector over other detectors is shown in Figure 1. The NaI scintillation detector is the most generally useful detector having a high spectral sensitivity throughout the important wavelength region.[1] Many diffractionists adopted this detector to gain intensity with copper radiation. They found that now the harder radiations, such as from molybdenum, could be detected with good sensitivity. The detection of the softer radiations such as Co K_α and Cr K_α was improved slightly. The intensities were still too weak for practical diffractometry. However, the helium path was used in spectrometry to enhance the intensities of the softer radiations. This technique should work with diffractometry. Table I compares transmission of X-rays of various wavelength in dry air and helium through a 30-cm path of a Norelco diffractometer. Note the threefold improvement in transmission of Cr K_α in helium *vs.* air.

In our first application of Cr K_α radiation we were interested in raising the intensity level but not necessarily in reaching the ultimate. Since the scatter shield has a diameter of 20 cm, excluding air from this area would leave an air path of only 10 cm. Transmission of Cr K_α through 10 cm of air and 20 cm of helium is calculated to be 69.4%. By cementing 1-mil mylar film about the opening of the scatter shield and flowing helium into the center hub through a small hose nozzle, an increase in intensity of more than 100% was obtained.

The maximum available intensity must be realized when one wishes to determine retained austenite in steels. One can reduce the background and thereby increase the peak to background ratio by using crystal-monochromated radiation. However, the intensities

---

[1] Superscripts pertain to references at the end of the paper.

Table I. Transmission of X-rays in Air and in Helium in a 30-cm Path Length Diffractometer

Target	Wavelength	μ/ρ		$-(\mu/\rho)\rho X(10)^3$		$\% \, I_x/I_0$	
		Air	He	Air	He	Air	He
Mo	0.711	1.2	0.18	43.4	0.89	95.8	99.9
Cu	1.542	9.4	0.37	339.7	1.85	71.2	99.8
Co	1.790	15.0	0.52	542.1	2.60	58.2	99.7
Fe	1.973	19.0	0.64	686.7	3.19	50.3	99.7
Cr	2.291	30.3	0.86	1095.1	4.29	33.5	99.6

$X = 30$ cm $\begin{cases} \text{Air . . . } \rho = 1.2047 \times 10^{-3} \quad \text{and} \quad \rho X = 36.1410 \times 10^{-3} \\ \text{Helium . } \rho = 0.1664 \times 10^{-3} \quad \text{and} \quad \rho X = 4.9920 \times 10^{-3} \end{cases}$

are so reduced that long counting times or exposures must be used to obtain a measurable austenite line. However, by replacing all of the air in the X-ray path with helium, a 300% increase in intensity is obtained with only a 10 to 20% increase in background. Thus, the peak to background ratio is increased by a factor of nearly three with a similar increase in signal level. To achieve a near 100% helium X-ray path a special 1-mil polyethylene bag was obtained. The bag was made from 54-in.-circumference material sealed off to form a 24-in.-deep bag. The thermally sealed edge was further sealed with

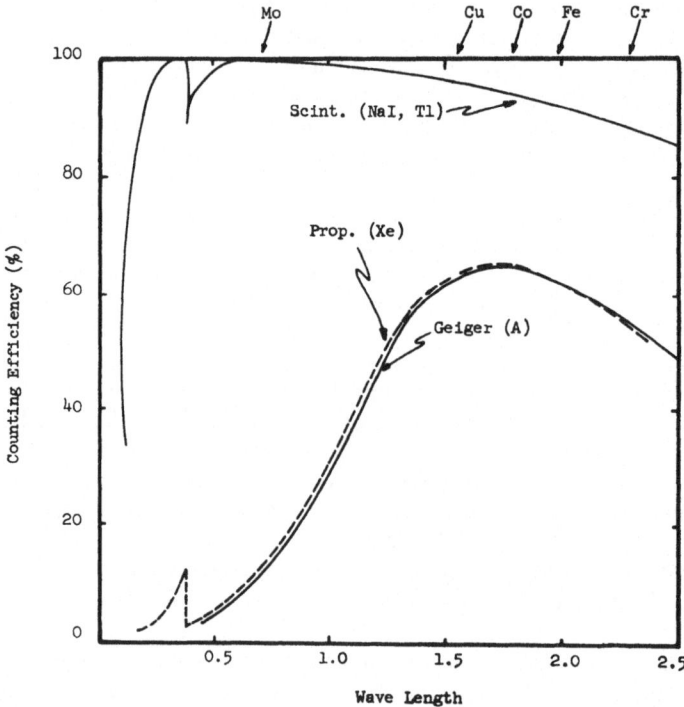

Figure 1. X-ray detector counting efficiency as a function of wavelength.

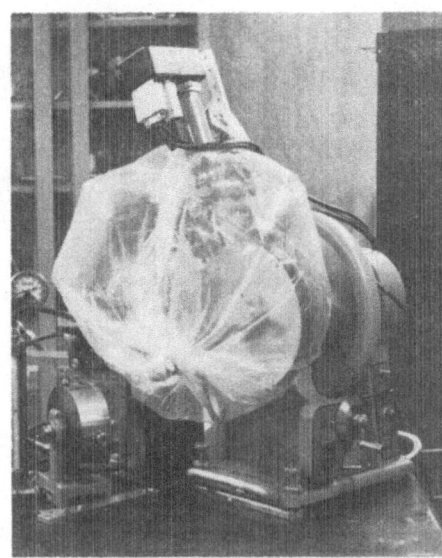

Figure 2. Polyethylene bag in place on diffractometer for a helium X-ray path.

black pressure-sensitive electrical tape. Holes were punched near the sealed edge for the
scatter slit arm and diagonally opposite for the detector. The receiving slit and detector
arm were placed in the bag and a small hole punched to take the support arm of the
detector. With opposite ends of the bag in place and holding various spots of the bag
to the center hole of the diffractometer, one can readily determine where to punch the
hole for the sample-holder shaft. This hole should be made slightly smaller than the
shaft and the plastic stretched over the shaft. All places where a part (except the sample-
holder shaft) goes through the plastic, the hole is sealed to the part with the black tape.

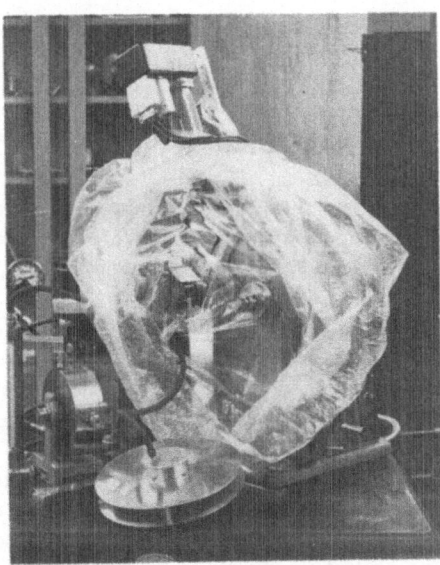

Figure 3. Helium bag open to show positioning of polyethylene around diffractometer parts.

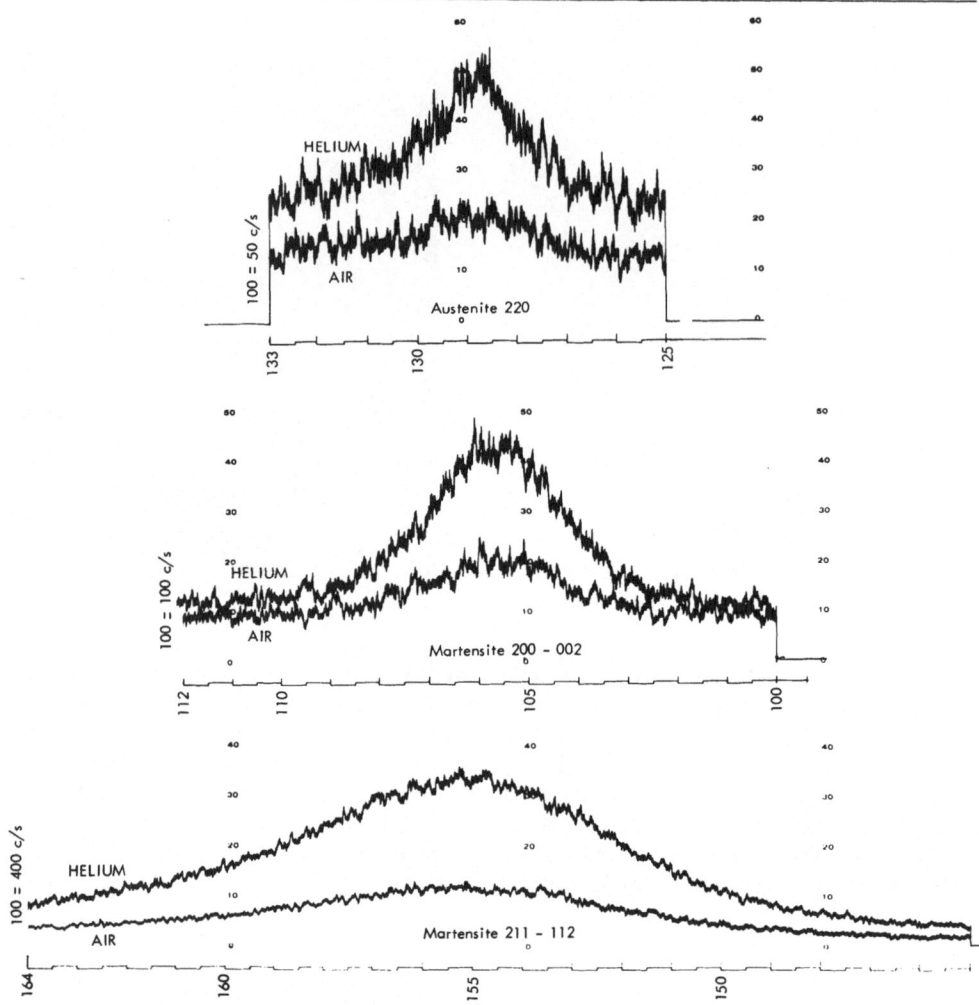

Figure 4. Diffractometer traces for various steel reflections with air and helium X-ray paths. Chromium radiation—40 kvp, 10 ma; slits—4°–0.006″–4°; scintillation detector; filter Va—PHD; scan speed—$\frac{1}{2}$ deg/min; paper speed—$\frac{1}{2}$ in./min; time constant—4 sec.

After the sample shaft is pushed through its hole in the plastic, a snug fitting O-ring is placed in the shaft, followed by the steel positioning washer and the 2 to 1 drive "dog." When these parts are pushed together and the "dog" fastened in position, the O-ring will form an effective seal.

By carefully placing the divergent slit scatter shield over the shutter housing, the plastic can be positioned in front of the shutter opening and the excess plastic kept from bunching between the shutter opening and the slit. A rubber tubing from the flowmeter or helium supply can be brought through the plastic below the divergent slit. With this arrangement, the regular scatter shield can be used. The helium hose can be attached to a small hose nozzle fastened to the hub of the shield.

Many of the details of positioning the bag can be seen in Figure 2. After the sample has been positioned and the scatter shield replaced, the bag can be closed by twisting the neck through several turns and fastening it with a rubber band. If the possible leaks have been sealed, 3 to 4 liters/min of helium should keep the bag inflated as shown in Figure 3. There is no waiting time for the air to be displaced when the following procedure is used. After the sample has been positioned and the scatter shield replaced, the bag is collapsed. The flow rate of the helium is then increased to about 10 to 12 liters/min and the neck of the bag slowly twisted. As soon as the bag is fully inflated the flow rate is decreased to 3 to 4 liters/min, and the neck of the bag is secured with a rubber band.

No attempt is made here to describe a procedure for determining retained austenite or macrostress in steels as these are well documented. Any of the published methods can be adapted to take advantage of the longer-wavelength chromium X-rays and the diffractometer when the helium path is used. Figure 4 compares the diffractometer traces with an air path and with a helium path obtained for austenite and martensite in a very fine-grain, high-carbon steel sample. This sample contained about 10% austenite. As one can readily see, the integrated area under the helium path curves are real, whereas the area under the air path curves would be highly questionable. Cullity[2] states that the minimum detectable amount of austenite is 5 to 10 vol. % with ordinary filtered radiation. We were able to measure down to 2 vol. % with a standard error of 0.4%. This is with $\frac{1}{2}$-deg/min scanning speed or a total of 78 min to scan the three peaks. This is a compromise between accuracy and instrument time. Should one have the time to scan at slower speeds or use a step scanning and the counting rate computer, both the minimum detectable amount and the error can be substantially reduced.

In determining the macrostress in the martensite, we used the 211–112 unresolved doublet. The curve (or peak) is assumed to be parabolic. Five fixed count readings at half-degree intervals across the top of the curve were taken. The 2θ position of the peak (or parabolic axis) was determined by least-squares analysis. It is obvious that the air path curve is too flat for this type of treatment. Our estimated error in the macrostress was ± 1800 psi compared to ± 2880 psi by other techniques. The time required to obtain the data averaged 24 min/peak.

Sometimes it is necessary or desirable for various reasons to use chromium radiation with the Philips-type Debye–Scherrer camera. The exposure time can be reduced by one-half, and the fogging due to air scatter eliminated by filling the camera with helium. This is accomplished simply by removing the sample-aligning screw after the alignment is complete. A hose nozzle with matching threads is inserted in the screw hole. A camera so fitted is shown in both Figures 2 and 3. After the film is loaded and the camera positioned on the track, the helium hose is connected and the flow rate adjusted to about 1 liter/min. (Caution: too fast a flow rate will blow the camera cover off.) This procedure has been used to obtain excellent patterns of many of the rare earth elements.

REFERENCES

1. W. Parrish, *Norelco Rept.*, **3**: 24–36, 1956.
2. B. D. Cullity, *Elements of X-ray Diffraction*, Addison-Wesley Publishing Co., Inc., Reading, Mass., 1956.

DISCUSSION

R. C. Larsen (Sundstrand Aviation): Were the helium losses due to an outlet flow or leakage?

R. A. McCune: I used 4 liters/min to keep the bag inflated because there was that much leakage. When we took greater pains in sealing the bag, we were able to keep the bag inflated with a flow rate of about 1 liter/min.

R. C. Larsen: Did you use a 6-mil receiving slit?

R. A. McCune: Yes.

R. C. Larsen: Did you think of using a wider receiving slit to increase your intensity?

R. A. McCune: Yes, I did. I used a 4° divergent, a 6-mil receiving, and a 4° scatter slit, but I didn't need the wider slits for this problem.

THE CRYSTAL STRUCTURE OF ThPd$_4$

J. R. Thomson*

Imperial College of Science and Technology, London, England

ABSTRACT

The structure of ThPd$_4$ has been determined by X-ray powder methods. It has the simple cubic Cu$_3$Au (LI$_2$) structure $a = 4.110 \pm 0.002$ Å at 80 at. % palladium. Interatomic distances and the range of stability of this compound and of ThPd$_3$ are discussed briefly.

The crystal structure of Th$_2$Pd has been reported by Ferro[1] and that of ThPd$_3$ was given by Dwight.[2] During a current investigation of the constitution of alloys of thorium and palladium, several other compounds were identified and the structure of an intermetallic compound observed at 80 at. % Pd is the subject of the present paper.

The alloys were prepared by arc-melting the component metals into 1-g buttons in a zirconium-gettered argon atmosphere. Weight losses on melting were negligible and nominal compositions have been accepted. All alloys were homogenized in vacuo for 4 days at 1000°C. The alloys were brittle and powders for X-ray studies were prepared by crushing in air; these gave very sharp X-ray patterns without a strain-relieving anneal. Powder patterns were obtained with a Guinier-type focusing camera using copper radiation and a quartz monochromator ($\lambda K_{\alpha 1} = 1.54050$ Å). The line intensities were estimated visually and calculated intensities were determined from the expression

$$I_c \propto \frac{1 + \cos^2 2\theta \cos^2 2\alpha}{(1 + \cos^2 2\alpha)\sin^2\theta \cos\theta} \cdot pF^2$$

where I_c is the calculated intensity, θ the Bragg angle, α the angle which the X-ray beam from the anticathode makes with the monochromator (for Cu K_α radiation $\alpha = 13°\ 20'$), p the multiplicity, and F the structure factor.

The pattern from the alloy of 80 at. % Pd was indexed as simple cubic $a = 4.110 \pm 0.002$ Å; observed and calculated values of sin$^2\theta$ are given in Table I. Intensities were calculated on the basis of a Cu$_3$Au-type lattice L1$_2$ (space group $O_h^1 - Pm3m$) with one in five of the thorium atom sites occupied by palladium atoms. Good agreement with the observed intensities was obtained as shown in Table I. For convenience, the compound has been designated ThPd$_4$ since it occurs at approximately 80 at. % Pd. It exists over only a narrow range of composition since the X-ray film of 78 at. % Pd showed the presence of ThPd$_3$ + ThPd$_4$, the patterns being of approximately equal intensity, while the film of 81 at. % Pd showed that only the phase richer in palladium was present. The variation of lattice parameter and cell volume with composition is given

---

* Formerly J. R. Murray.
[1] Superscripts pertain to references at the end of the paper.

Table I. Comparison of Observed and Calculated $\sin^2\theta$ Values and Line Intensities for ThPd$_4$

hkl	$(\sin^2\theta)_{obs.}$	$(\sin^2\theta)_{calc.}$	$I_{obs.}$	$I_{calc.}$
100	0.03502	0.03512	w	104
110	0.07009	0.07024	w	83
111	0.10548	0.10536	vvs	1000
200	0.14050	0.14048	vs	516
210	0.17560	0.17560	w	50
211	0.21077	0.21072	w	37
220	0.28094	0.28096	s	338
300	0.31630	0.31608	vw	21
221	0.31630	0.31608	vw	5
310	0.35135	0.35120	vw	18
311	0.38639	0.38632	s	397
222	0.42134	0.42144	m	119

vvs = very, very strong,   vs = very strong,   s = strong,   m = medium,   w = weak, vw = very weak.

in Table II. The temperature range over which ThPd$_4$ is stable was not studied in detail but material which had been annealed for 1 day at 1200°C still gave the L1$_2$ pattern.

Visual estimation of the line intensities is insufficiently sensitive to distinguish between the possible defect structures for this lattice, e.g., thorium sites vacant or occupied by palladium atoms since both give only a slight alteration in the relative intensities between the face-centered cubic lines and the superlattice lines. Random substitution for thorium atoms is suggested as this seems the most reasonable and allows for the increase of volume of the unit cell with increasing thorium composition as observed.

In the present investigation, it has been established that the compound ThPd$_3$ which has the hexagonal TiNi$_3$ structure (DO$_{24}$) can also exist over a narrow range of composition shown in Table III.

The stronger lines of the adjacent compound were also present on the film of the 70 at. % Pd alloy. While the lattice parameters of the 70 and 75 at. % Pd alloys agree within the experimental error, those for 77 and 78 at. % Pd are significantly different and lead to marked changes in the relative separation of several pairs of lines. The lattice parameters obtained on annealed powders by Dwight[2] were $a = 5.856$ Å and $c = 9.826$ Å (though he does not state his annealing temperature) and show reasonable agreement with those obtained at the thorium limit of composition in the present investigation.

Table II. Variation of Lattice Parameter and Cell Volume with Composition

At. % palladium	a, Å	Volume of unit cell, Å^3	Closest Th–Pd and Pd–Pd neighbors
78	4.126 ± 0.002	70.24	2.918
79	4.119 ± 0.002	69.88	2.913
80	4.110 ± 0.002	69.43	2.905

Table III. Composition Range of ThPd$_3$

At. % palladium	$a,$ Å	$c,$ Å	c/a	Closest Th–Pd and Pd–Pd neighbors
70	5.860 ± 0.003	9.806 ± 0.004	1.67	2.930
75	5.858 ± 0.003	9.814 ± 0.004	1.67	2.929
77	5.851 ± 0.003	9.657 ± 0.004	1.65	2.925
78	5.846 ± 0.003	9.643 ± 0.004	1.65	2.923

Since the range of composition over which ThPd$_3$ exists is on the palladium side of the ideal composition, it is suggested that this too is achieved by random replacement of some thorium atoms by palladium atoms. This would also account for the observed volume contraction with increasing palladium content. In an investigation of the U–Pd system, Catterall[3] observed that UPd$_3$, which is isostructural with ThPd$_3$, could exist over a narrow range of composition with little variation in the a parameter but with

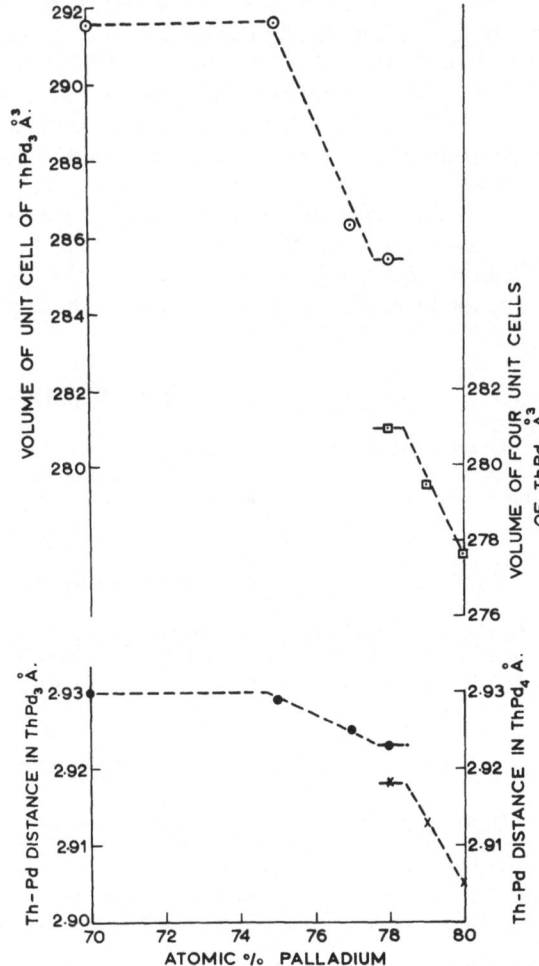

Figure 1. Comparison of unit cell volumes and Th–Pd nearest neighbor distances in ThPd$_3$ and ThPd$_4$.

the c parameter varying sufficiently for c/a to be 1.67 with excess uranium present and 1.65 with excess palladium.

In the $ThPd_3$ lattice, the closest Th–Pd and Pd–Pd distances are given by $\frac{1}{2}a$ and it will be seen in Figure 1 that these vary between 2.92 and 2.93 Å. The Th–Th interatomic distances depend on a and c and are appreciably greater than those observed in the pure metal. In $ThPd_4$, the nearest Th–Th neighbors are at a, which is again larger than the thorium atomic diameter, while the closest Th–Pd and Pd–Pd distances are equal to each other and increase significantly with increasing thorium composition. Assuming that the atomic radii (for CN 12) in the pure metals are 1.37 and 1.80 Å for palladium and thorium, respectively, a considerable contraction in the Th–Pd distance occurs in both $ThPd_4$ and $ThPd_3$.

The variation in Th–Pd distance and in the volumes of the unit cells with composition in the range 70–80 at. % Pd is shown in Figure 1. Since the unit cell of $ThPd_3$ contains 4 moles while that of $ThPd_4$ has 1 mole, the volume of four unit cells of $ThPd_4$ has been plotted for comparison purposes. It can be seen that there is a significant decrease in volume on passing from $ThPd_3$ to $ThPd_4$ and also a shorter Th–Pd distance in $ThPd_4$.

The detailed coordination around the thorium and palladium atoms differs in the two structures. Although there are two sets of thorium and two sets of palladium positions in $ThPd_3$, the coordination around each thorium atom is $(6 + 6)Pd + 6Th = 18$, and that around each palladium atom is $(4 + 4)Pd + (2 + 2)Th = 12$. In the ideal $L1_2$ structure the coordination around each thorium atom is $12Pd = 12$ while that of palladium is $4Th + 8Pd = 12$. When the structure occurs in alloy systems where the size difference between the atoms is appreciable, the coordination around the larger atom is increased to 18 since it has six neighbors of its own kind at a. In the present system, a is less than 1.2 times the atomic diameter of thorium. In the nonideal structure proposed for $ThPd_4$ the average coordination for thorium would be $4.8\ Th + (12 + 1.2)Pd = 18$ and for palladium $3.2\ Th + (8 + 0.8)Pd = 12$, at 80 at. % Pd. There is therefore little difference in the total coordination around the thorium or palladium atoms in $ThPd_3$ and $ThPd_4$.

ACKNOWLEDGMENT

Grateful acknowledgment is made to Professor J. G. Ball under whose supervision this work was carried out, and also to the Atomic Energy Research Establishment, Harwell, for financial support and for allowing the author to make use of some of their experimental facilities.

REFERENCES

1. R. Ferro and R. Capelli, *Acta Cryst.* **14**: 1095, 1961.
2. A. E. Dwight, J. W. Downey, and R. A. Conner, Jr., *Acta Cryst.* **14**: 75, 1961.
3. J. A. Catterall, J. D. Grogan, and R. J. Pleasance, *J. Inst. Metals* **85**: 63, 1956–7.

DISCUSSION

H. M. Otte (RIAS): Is $ThPd_4$ an ordered structure?

J. R. Thomson: I think the question is, are there really any effects of disorder? This is not the case. The as-cast 80 at. % alloy is two-phase. We chose the $ThPd_3$ and the $ThPd_4$ because it's obviously not in equilibrium. When we annealed, we saw no evidence of the ordered phase.

H. M. Otte: Is the $ThPd_3$ structure ordered?

J. R. Thomson: We have never observed disorder as the temperature decreased. As far as I know, there is no order–disorder change.

Henry Chessin (U.S. Steel Corp.): If there was a question of order–disorder, wouldn't you see this by comparing your observed and calculated intensities?

J. R. Thomson: Yes, I have. I calculated these for $ThPd_3$ and $ThPd_4$, and I have not observed any variations between my observed and calculated. I have no evidence whatsoever of any disorder.

THE EFFECT OF COLD-WORK ON THE
X-RAY DIFFRACTION PATTERN OF A
COPPER–SILICON–MANGANESE ALLOY

D. O. Welch and H. M. Otte

RIAS
Baltimore, Maryland

ABSTRACT

Plastic deformation of metals produces a state characterized by the presence of residual elastic strains, small domains which diffract X-rays coherently, and often stacking faults; these effects may be studied with X-ray diffraction techniques. Changes in the lattice parameter, shifts in the relative positions of diffraction lines, and the broadening of diffraction lines were used to study the state of cold-work resulting in Cu–6.6 at.%Si–1.2 at.%Mn after deformation by filing, wire-drawing, and uniaxial tension at room temperature.

Both filing and wire-drawing produce large root-mean-square strains and stacking faults, whereas deformation by tension up to 22% extension fails to produce any clear evidence of faulting or root-mean-square strains. Tensile deformation causes fragmentation of coherent domains to an average dimension of 250 Å after 22% extension, and results in a radial, tensile, residual macrostrain arising from a smaller rate of work hardening in the surface layers than in the interior. Wire drawing also results in a residual macrostrain system. Deformation appears to enhance diffusion and promote solute clustering at room temperature.

INTRODUCTION

The interpretation of the changes in position, shape, and width of diffraction lines in the Debye–Scherrer (D–S) X-ray patterns from metals and alloys after deformation and/or heat treatment forms an important part of investigations into the properties and behavior of these materials. As is well known, deformation introduces not only additional dislocations and point defects into the material, but also causes related fragmentation of coherently diffracting domains together with residual elastic stresses and strains. In addition, stacking faults may frequently be formed. Information concerning the magnitudes of the domain size, residual strains, and stacking fault density may be obtained by an appropriate analysis of the diffraction lines. In performing such an analysis, the physical condition of the specimen must be taken into consideration. Thus, whereas for filings the line profile is generally obtained from a sufficiently large number of randomly oriented crystallites, for "bulk" samples such as drawn wires or tensile specimens, this is frequently not the case. Since many physical and mechanical properties are, in fact, measured on bulk samples, it becomes inportant to make X-ray measurements on these also rather than on filings.

The object of the present investigation is to compare the effect of deformation (at room temperature) by filing, wire-drawing, and uniaxial tension on the residual strains, domain size, and stacking fault density of a Cu–Si–Mn alloy. At the outset it is evident that not only is the mode of deformation different in the three processes (filing, wire-drawing, straining in tension), but the samples differ also in their physical state when examined. An attempt to evaluate the influence of this on the determination of residual strains, stacking fault density, and domain size must first be made. The approach to the problem is different to any that has been made by previous investigators and involves certain new features. In order to bring these out clearly, a short summary of earlier work is presented in the next section.

BRIEF EVALUATION OF THE LITERATURE

Residual Elastic Strains

For *cold-worked filings* these will simply produce a broadening of the diffraction peak since on the average there will be as many particles (in the filings) oriented with their reflecting planes in compression as there will be with the corresponding planes in tension; the maximum will be around zero strain. The broadening is essentially symmetrical and produces no peak displacement. The root-mean-square strain $\bar{\epsilon}_{hkl}$ in the *hkl* direction (cubic crystals) can be obtained from the integral breadth $(B_s)_{hkl}$ of the *hkl* diffraction peak, by means of the relation[1]

$$(B_s)_{hkl} = 2\bar{\epsilon}_{hkl} \tan \theta \text{ in radians} \tag{1}$$

where θ is the Bragg angle for the *hkl* planes.

For any given particle (in the filings) it is hypothetically possible to determine the three principal axes of strain. The residual elastic strain (or stress) may be associated with what are sometimes called the macrostresses, and are long range in nature compared with the microstresses which are to be associated with local distortions around imperfections such as dislocations and point defects. By this definition, microstresses are not expected to contribute to any great extent directly to either line broadening or line shape.

In a *bulk crystalline sample* residual elastic strains are very important and have formed the subject of innumerable investigations since about 1887.[2] The *Proceedings of the Society for Experimental Stress Analysis* are devoted to the furtherance of the many methods and techniques available for the determination of stresses and strains in materials; books on the subject are also available.[3-5]

Although the X-ray method of measuring residual strains rates high among the nondestructive methods, its widespread application has been prevented by the failure of various investigators to obtain not only consistent results, but also agreement with other methods. The status of the field for the X-ray diffraction techniques, of primary interest in the present work, was reviewed in 1952[6] and it would seem that little clarification of the confusion has occurred since.

Elastic stresses may arise either from deformation, thermal gradients, or metallurgical (structural) changes in the material. The stresses produce a uniform contraction or expansion of the diffracting atomic planes and this ideally causes only a shift in the position of the diffraction peak. By measuring this shift, the lattice strain is readily calculated, and, depending on the type and distribution of the stresses assumed or expected, the stress may be computed directly from classical elasticity theory with or without an additional measurement in some other direction. Difficulties arise, however, from two main sources: the anisotropic strain distribution in the individual grains making up the aggregate of the polycrystalline sample, and the selective nature of the X-ray diffraction process. The situation is further complicated when the material is elastically anisotropic.

Two oversimplified cases can be solved: (1) when the stress acting on each grain is

[1] Superscripts pertain to references at the end of the paper.

equal and the same as the stress acting on the aggregate (isotropic strain)[7]; and (2) when the strain is the same for all grains and the aggregate as a whole obeys elastic laws for isotropic bodies (isotropic strain).[8] The experimental results do not agree with the results predicted from either of these two theories, but in certain cases agree quite well with the average of the two theoretical values.[6,9] Attempts to overcome the selective nature of the X-ray diffraction process have been made by oscillating the sample through a small angle. It is questionable, however, how effective or successful this is in all cases.

The results of Greenough[10–13] on wires of nickel, magnesium, copper, aluminum, and mild steel indicate that the residual strains calculated from different diffraction lines may be either compressive or tensile. Similar results for mild steel were found by Finch[14] and by Garrod.[15] Greenough[10] found fairly good agreement between the experimental results and those predicted from an intergranular stress system model. In Greenough's model the residual stress system is caused by the difference in yield stress of grains oriented differently with respect to the applied stress. Since only those grains with appropriate orientation will contribute to a particular diffraction peak, they will be, on the average, either in tension or in compression, and the sign of the strains should not vary with depth from the surface.

A second and distinctly different theory is that the grains at the surface of a specimen are deformed more than the grains in the interior causing a residual compressive, longitudinal, surface stress balanced by an interior tensile stress. Corroborating data have been obtained on mild steel by Bollenrath, Hauk, and Ostwald,[16] who attribute the nonuniform deformation to the surface grains having a lower flow stress than the grains in the interior. Similar data have been obtained on nickel by Kolb and Macherauch[17] who show some evidence that it is not an initially different flow stress, but a different rate of work hardening which results in nonuniform deformation.

The variation of the residual strains with distance from the surface should provide evidence as to the validity of either of the two theories, since if an intergranular stress system is correct, the residual strains should not change on etching. If, however, a macroscopic stress system is correct, there should be a marked variation of residual strain with depth from the surface. Bollenrath, Hauk, and Ostwald[16] find a change on etching consistent with the macrostress theory. Wood[18] and Greenough[13] find no change in residual strain upon etching. Furthermore, Donachie and Norton[19] have reported X-ray results which are in agreement with a macroscopic stress and which do not agree with an intergranular stress model; they observed no change when the specimen was progressively etched away. The current controversy is thus clearly illustrated.

There is another aspect to consider, which, though not of direct consequence in the present investigation, is of importance when comparing X-ray measurements with residual stresses obtained by certain destructive methods. In these, dimensional changes are recorded when layers of determined thickness are removed mechanically, chemically, or electrolytically, and elasticity theory used to compute the stress that must have existed in the layer removed.[5] The assumption is made that all of the dimensional changes may be attributed to stress relaxation, and that no significant dislocation rearrangements occur. It is known however, that this is not true and that a certain amount of reverse plastic flow occurs whenever a stressed specimen is unloaded.[20,21] Thus these methods are bound to overestimate the residual stresses.

Layer Faults

These produce peak shifts as well as changes in the line profile and width of the powder patterns from cubic crystals.[22–26] Types of stacking faults in the fcc lattice are illustrated in Figure 1; the relative positions of the planes are denoted by *A*, *B*, and *C*.

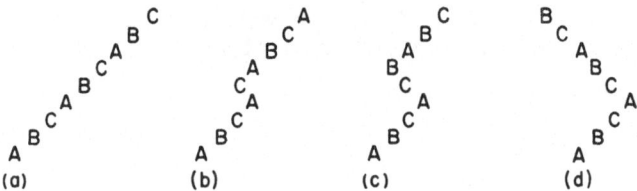

Figure 1. Stacking faults in the face-centered cubic lattice. (*a*) Stacking sequence of (111) planes in normal face-centered cubic lattice. (*b*) Deformation (intrinsic) fault. (*c*) Double deformation (extrinsic) fault. (*d*) Twin fault.

Deformation faults give rise to a symmetrical broadening and a peak shift, while twin faults produce an asymmetrical broadening and a negligibly small peak shift. Extrinsic faults produce initially both an asymmetrical broadening and a peak shift.[56,57] The peak shift is in the opposite direction to that produced by intrinsic faults, and may cancel any peak shifts due to the latter in special cases.[58] For layer faults, there is in addition a change in interplanar spacing at the stacking fault and in certain instances the effect from this is measurable.[26] The fractional change in interplanar spacing for a set of *hkl* planes as observed in the powder pattern for a face-centered cubic structure containing both stacking faults and spacing faults is

$$(\Delta d/d)_{hkl} = \alpha\,(G_{hkl} + J_{hkl}\varepsilon) \tag{2}$$

where α is the probability that a (111) plane is a deformation faulted plane, ε is the fractional change in interplanar spacing at the stacking fault, and G_{hkl} and J_{hkl} are constants (values are listed in Table I).

The line broadening caused by layer faults is independent of the order of the reflection in the same way as line broadening caused by small coherent domain size. If it is assumed that the stacking faults extend completely across the coherent domain, as the results of Warren suggest,[27] the apparent domain size determined from line broadening is related to the true domain size, the faulting probability, and the spacing fault magnitude by

$$(1/D_{\mathrm{eff}})_{hkl} = \frac{1}{D_{hkl}} + \frac{1.5\alpha + \beta}{a}\,V_{hkl} + \frac{\alpha\varepsilon}{a}\,W_{hkl} \tag{3}$$

where $(D_{\mathrm{eff}})_{hkl}$ is the apparent domain size* in the *hkl* direction, β is the probability that a (111) plane is a twin-faulted plane, a is the unit cell edge length, and V_{hkl} and W_{hkl} are constants given in Table I.

Stacking faults are readily produced during deformation, particularly in many metals and alloys with a fcc structure. The resulting peak shifts in the diffraction pattern can be of the same order of magnitude as that possible from the elastic straining of the lattice. However, an important difference exists. Whereas elastic strains should produce line displacements that are in the same direction though not necessarily of the same magnitude for all reflections, displacements due to stacking faults are not only different for different reflections, but also in opposite directions.

Since, in the case of filings, the residual elastic stresses produce no net peak shift, it is possible to interpret all such shifts as arising from layer faults. Agreement with the expected direction and magnitude of the shift is often good quantitatively and always good qualitatively.

* D_{eff} from integral breadth is slightly larger than that from Fourier analysis (see the Appendix), and hence equation (3) is only approximately valid for integral breadths.

Table I. Values of Constants for Spacing Faults and Stacking Faults in Face-Centered Cubic Metals and Alloys (After Wagner, *et al.*[26])

hkl	G_{hkl}	J_{hkl}	V_{hkl}	W_{hkl}
111	−0.0345	+0.209	$\sqrt{3}/4$	−0.79
200	+0.0689	−0.167	1	+3.63
220	−0.0345	+0.167	$1/\sqrt{2}$	−5.13
311	+0.0125	+0.038	$3/2\sqrt{11}$	+3.28
222	+0.0172	+0.209	$\sqrt{3}/4$	+1.57
400	−0.0345	−0.167	1	−7.26
331	−0.0073	−0.167	$7/2\sqrt{19}$	−2.70
420	+0.0069	+0.284	$1/2\sqrt{5}$	+0.81
422	0	−0.167	$2/\sqrt{3}$	+2.96
333+511	+0.0029	−0.0104	$17/16\sqrt{3}$	−0.065

Except for a few cases, including the present investigation, in which all the diffraction peaks available in the spectrum with the selected radiation have been measured, the usual procedure is to measure only the first two or three low-angle diffraction peaks for a determination of the stacking fault density. Residual strains, on the other hand, are invariably determined from the highest-angle peaks available. A simple consideration of the line shifts involved shows that for solid polycrystalline materials it is essential to measure as many diffraction peaks as possible when the presence of both residual strains and layer faults is suspected. The importance of this will be brought out in the results to be presented. A number of investigators,[28-30] have recently measured stacking faults on deformed bulk samples without taking into consideration possible peak shifts from residual elastic stresses. This has already been briefly discussed elsewhere.[59] Conversely, investigators studying residual elastic strains invariably ignore possible peak shifts that may arise if stacking faults are present. The kind of apparently anomalous results that can arise has been pointed out.[31]

Coherent Domain Size

In addition to the line broadening that may arise from strain distribution and layer faults, broadening will also be produced when the size of the crystalline domains which are coherently diffracting X-rays becomes less than about 10,000 Å. The integral breadth $(B_L)_{hkl}$ of an *hkl* diffraction peak broadened by a particle size D_{hkl} in the *hkl* direction is given by[32]

$$(B_L)_{hkl} = \lambda/D_{hkl} \cos\theta \text{ in radians} \tag{4}$$

where λ is the wavelength of the X-rays. For cold-worked filings the broadening caused by small particle size is always present. In deformed bulk samples, Wood and Rachinger[54] have shown that domain (or crystallite) size is the principal source of broadening. They deformed several bcc metals in tension, compression, by cold-rolling, swaging, and drawing, until maximum line broadening was obtained; this broadening they could then explain solely on the basis of limiting domain size in the region 250–400 Å.

PRESENT INVESTIGATION

The foregoing considerations show that if the peak shifts and broadening in deformed polycrystalline aggregates are to be properly evaluated, then, in general the peak shift, or corresponding fractional change in the spacing of (hkl) planes should be expressed by:

$$(\Delta d/d)_{hkl} = \epsilon_{hkl} + \alpha(G_{hkl} + J_{hkl}\epsilon) \tag{5a}$$

$$\epsilon_{hkl} = \sigma_L\left[-\frac{c_{11} + 4c_{12} - 2c_{44}}{4(c_{11} + 2c_{12})(c_{11} - c_{12} + 3c_{44})} + \tfrac{1}{2}s_{12} + \tfrac{1}{2}(s_{11} - s_{12} - \tfrac{1}{2}s_{44})\Gamma\right] \tag{5b}$$

$$\Gamma = \frac{h^2k^2 + k^2l^2 + h^2l^2}{(h^2 + k^2 + l^2)^2} \tag{5c}$$

where σ_L is the stress parallel to specimen axis, c_{ij} and s_{ij} are elastic constants, and ϵ_{hkl} is the component of elastic strain in the hkl direction. By measuring the change in spacing of a number of (hkl) planes, it becomes possible to separate the components of equation (5a). For filings, ϵ_{hkl} is zero.

The line broadening due to cold work will be a function of the root-mean-square strain $\bar{\epsilon}$ and an effective domain size D_{eff} which incorporates the contribution from the stacking faults. If the diffraction line profile may be described by a Cauchy distribution curve, the integral breadth corrected for instrumental broadening is:[33]

$$B = (\lambda/D_{eff})\cos\theta + 2\bar{\epsilon}\tan\theta \tag{6}$$

If, however, the diffraction profile may be fitted by a Gaussian distribution curve, then[33]

$$B^2 = (\lambda/D_{eff})^2\cos^2\theta + (2\bar{\epsilon}\tan\theta)^2 \tag{7}$$

The two terms on the right-hand side of equations (6) and (7) can be separated by utilizing the difference in their dependence upon the Bragg angle θ.

In order to obtain successful averaging of the diffraction from individual crystallites, these should be small and randomly oriented. The condition of random orientation may be more nearly met by rotating the specimen. This, of course, is only possible for cylindrical specimens in which the strain distribution is radial and perpendicular to the axis.

Material and Specimen Preparation

An alloy of copper, with 6.6 at.%Si and 1.2 at.%Mn (available commercially as Everdur) was selected as one which was known to fault profusely upon deformation,[34] and thus permit a determination of the influence of stacking faults on the various measurements.

Filings, both as cold-worked at room temperature, and after annealing at 700°C in vacuo for 2 hr and water quenching, were compacted, using Duco cement as a binder, into cylindrical specimens 2.2 mm in diameter.

Wire specimens, one 3.13-mm diameter annealed at 700°C for 2 hr and water quenched, and one after annealing, given 40.6% reduction in area at room temperature to 2.40-mm diameter were also prepared.

The tensile specimens conformed to standard ASTM specifications for $\tfrac{1}{8}$-in. (3-mm) diameter test piece[35]: gauge length $\tfrac{1}{2}$ in. (12 mm). After machining they were given the normal annealing treatment (2 hr at 700°C in vacuo) and air cooled.

Diffraction Equipment and Technique

The X-ray diffraction patterns were obtained by means of a Debye–Scherrer method utilizing a diffractometer.[36] A General Electric XRD–5 spectrogoniometer (SPG) was used with the bisecting mechanism removed so that it was possible to measure both halves of a diffraction ring. Copper K_{α_1} radiation was used throughout the investigation and was obtained with a Siemens curved quartz crystal monochromator, Johansson focusing, bent elastically to a radius of 25 cm; an adjustable slit at the focal point of the monochromator cut out the K_{α_2} component of the doublet. Diffraction peaks were scanned with a krypton-filled Geiger counter at a rate of 10 min/deg 2θ, and the amplified counter output was recorded simultaneously on a chart recorder and a digital printer. The position of each diffraction peak was taken to be the average of the positions of the "positive" and "negative" halves of the Debye–Scherrer ring. This procedure eliminated any displacement from the true-line position caused by finite scanning rate and time constant of the recording circuit as well as small errors in the alignment of the beam with the diffractometer scale zero position.[36] A specimen spinner (150 rpm) was used; this is a standard procedure in the D–S method to improve sampling, especially in the bulk specimens. The temperature of the room was maintained at $24 \pm 1°C$; the temperature fluctuations in the specimen were determined and found to be a little greater than $\pm \frac{1}{2}°C$. The position of the center of gravity of diffraction peaks could be determined reproducibly to within $\pm 0.005°2\theta$.

Tensile Specimen Holder

This holder permitted X-ray diffraction patterns to be obtained from a cylindrical specimen while the specimen was loaded under uniaxial tension. The holder fitted on the SPG platform and had a motor attached that made it possible to spin (at 170 rpm) the loaded specimen around the coincident specimen-diffractometer axis in order to improve sampling and thus eliminate or reduce the selective diffraction generally associated with bulk specimens. The D–S geometry was thus maintained. The holder could be easily withdrawn from the platform and placed under a traveling microscope to measure the extension of the test piece when the load was applied or removed. The vernier on the traveling microscope could be read to 0.01 mm.

Analysis of the Data

Determination of the Line Position. When the line profile is symmetrical, the peak position (PP) and center of gravity (CG) coincide, and a distinction is not necessary. However, this is only rarely the case. When the line is asymmetrical due to geometrical factors, a choice arises: here it was decided to measure the CG since lattice parameter values from this gave a linear relationship with the Nelson–Riley Function (NRF)[37] for all the lines measured.[36]

Strain and particle-size distributions are not expected to introduce any asymmetry, but twin faults will.[22,23,25] Thus, for a symmetrical line (from an annealed material) broadened by deformation and twin faults, the change in PP will be due to the deformation faults and the difference between the CG and the PP will be related to the twin fault density,[38] which can alternately also be obtained from an appropriate Fourier analysis of the asymmetry of the line profile[23,25] and otherwise.[22] The Fourier analysis will also yield a value for the deformation fault probability.

In practice the fault probability is not the same for all the coherently diffracting domains. There appears to be a tendency for the smaller particles in the filings to have a higher fault probability.[39] In any case, it is not unreasonable to expect a distribution of deformation fault densities (respectively, probabilities) among the domains, and this would, of course, create an asymmetry in the diffraction line profile in addition to that produced by twin faults. Under these conditions the shift of the CG would be a more correct measure of the deformation fault density, whereas the shift of the PP would tend to give an underestimate. On the other hand, if twin faults are present, the shift of the CG would tend to give an overestimate of the actual deformation fault density.

In the present work twin fault densities have been ignored, and all the CG shift attributed to deformation faults when present.

In the heavily cold-worked materials, for which the lines are very broad with extremely long tails that overlap frequently with other lines, the CG has been determined from that part of the profile above the general raised background. The error introduced by this procedure is difficult to evaluate because of the extreme difficulty in determining sufficiently accurately the position of such long overlapping tails. It is unlikely that the errors will be so large as to affect any of the conclusions reached in this investigation, though some of the numerical values quoted would perhaps have to be corrected.

Lattice Parameter and Peak Displacements. The apparent lattice parameter values, calculated for each of the eight diffraction peaks (111) through (420) obtained from a specimen were plotted against the extrapolation function of Nelson and Riley.[37] For annealed metals and with the diffraction geometry employed in this investigation, a Nelson–Riley plot yields a straight line through all the experimental points, and the extrapolated lattice parameter is (except for one or two minor corrections) the true value, independent of specimen diameter,[36] absorption, etc. For cold-worked specimens any displacements of the experimental points from a straight line can be regarded as caused by the deformation and not from instrumental error.

Figure 2 shows the direction and relative magnitude of the displacements to be expected in powder patterns on theoretical grounds if only deformation faults are present and Paterson's analysis[22,23] is assumed applicable. On this basis a straight line can be drawn through observed data by selecting its position so that the relative displacements are in accordance with the theoretical predictions. In practice, however, discrepancies generally arise for various reasons when all displacements are taken into consideration, and therefore in the present work the straight line was positioned so that it gave the following relative displacements, Δa_{hkl}, calculated from the theory:

$$2\Delta a_{111} = - \Delta a_{200} \tag{8a}$$

and

$$\Delta a_{331} \cong - \Delta a_{420} \tag{8b}$$

The anticipated relative peak displacements* from an anisotropic material, such as the Cu–Si–Mn alloy, when loaded uniaxially and X-rayed radially (i.e., perpendicular to the axis) are illustrated in Figure 3. If the unloaded values of the lattice parameters are not known, then a straight line drawn through the points will show displacements of the points as in the case of stacking faults, but the directions and magnitudes of the displacements will be different. In particular, the displacements of the first and second orders of a reflection will be in the *same* direction whereas for stacking faults they will be in *opposite* directions. This distinction becomes especially important when the initial lattice parameter is either not known at all or known only approximately and/or when the presence of stacking faults is suspected. Further mention of this will be made in the discussion.

The extrapolated lattice parameter in Figure 3 for the stressed specimen may be taken as an average value independent of anisotropy and comparable to the measurement of Young's modulus on a polycrystalline sample. In the present work, such an extrapolated value was of interest. When Figures 2 and 3 are combined, i.e., when we have an alloy in which both stacking faults and residual stresses have arisen, it becomes difficult to position the straight line correctly *a priori*, though, of course, in principle its position

---

* Calculated assuming an average of isotropic stress and isotropic strain, henceforth referred to as the elastic anisotropy effect.

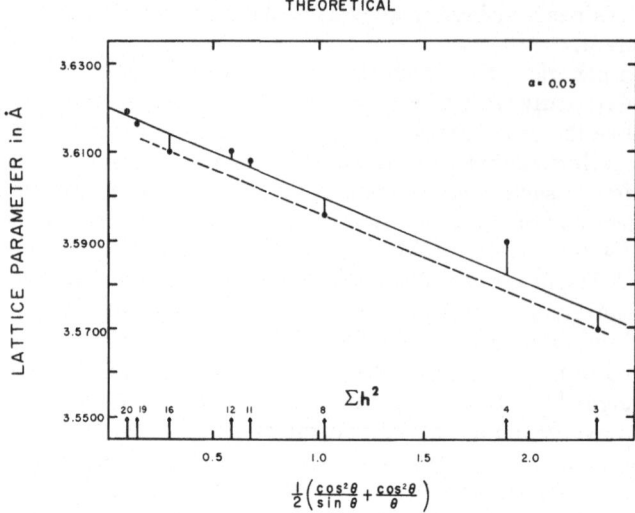

Figure 2. Theoretical Nelson–Riley plot according to the theory of Paterson for a face-centered cubic metal containing deformation stacking faults.

can be ascertained by first determining the values of ϵ_{hkl}, (and ε) by solving equations (5), using the displacements of the points calculated from two (or three) reflections. In practice this is not feasible because of experimental errors and the sensitivity of the values obtained to these errors. Therefore, the rather arbitrary procedure was adopted to position the line as shown in Figure 3. The consequence of doing this has a relatively small effect on the extrapolated lattice parameter, but makes the residual strain displacements from the first and second orders appear to have slightly different magnitudes, instead of the same.

Because of the comparatively large diameter of the specimens used in the present work, absorption effects cause the straight line in the NRF plot to have a steep slope. Since absorption is not of interest here, it can be essentially eliminated simply by re-plotting only the displacements of the points from the straight line, which should be horizontal in the absence of absorption (specimen eccentricity, etc.). This has been done, for example, in Figures 7 and 8.

Line Broadening. The integral breadth was used as the measure of line broadening and was obtained by dividing the total diffracted intensity of a diffraction line by the peak

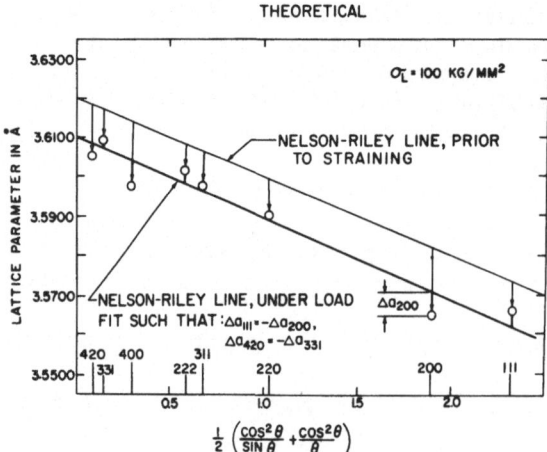

Figure 3. Theoretical Nelson–Riley plot for Cu–6.6 at. %Si–1.2 at. %Mn stressed in uniaxial tension.

intensity. The integral breadths were corrected for instrumental broadening by the expression[40]

$$B^2 = B_E^2 - b^2 \tag{9}$$

where B is the integral breadth caused by properties of the diffracting crystal, B_E is the experimental integral breadth, and b is the integral breadth caused by instrumental broadening, obtained from annealed specimens.

Equation (9) is strictly correct only when all the functions in the fundamental expression for line broadening are Gaussian. The diffraction profiles from both annealed and cold-worked materials were found to be predominantly Gaussian. Appreciable particle-size broadening should result in a Cauchy distribution curve,[22] but in the present work this was never pronounced, even when the particle size was small.

RESULTS

The observations on the filings, the tensile specimens and the drawn wires will be presented separately. In the case of the bulk samples, especially the drawn wires, strong textures (or preferred orientations) may develop. Such textures combined with some vertical divergence in the incident X-ray beam can lead to apparent line shifts in the D–S pattern from solid, polycrystalline specimens[41] and is a possible source of error in residual strain determinations. With the present technique this can be detected by checking for anomalously high or low intensities of the diffraction peaks. All the results quoted here were from specimens which showed no strong texture, and therefore possible peak shifts from this source need not be considered.

Filings

Peak Displacements and Lattice Parameter Changes. The results shown in Figure 4 are reproduced from an earlier paper,[36] in which, however, no calculations of stacking fault probabilities or of line broadening were reported. The direction of the displacements is in agreement with theory, Figure 2, but the magnitudes do not all yield the same stacking fault probability, as shown in Table II. These differences could be partly reconciled by assuming the existence of spacing faults. Similar discrepancies have also been noted in alpha brass[25,42,43] and in alpha copper–tin alloys.[42]

The lattice parameter of the cold-worked filings appears to be slightly smaller

Table II. Stacking Fault Probabilities Determined from the Peak Displacements of Various *hkl* Diffraction Lines for Cu–6.6 at. %Si–1.2 at. %Mn after Room-Temperature Cold-Work by Filing and by Wire-Drawing (40.6% Reduction in Area)

hkl	α(filing)	α(wire-drawing)*
111	0.025	0.049
200	0.025	0.049
220	0.025	0.038
311	0.013	0.048
222	0.006	0.011
400	0.002	0.037
331	0.046	0.065
420	0.046	0.065

* Neglecting effect of residual stresses.

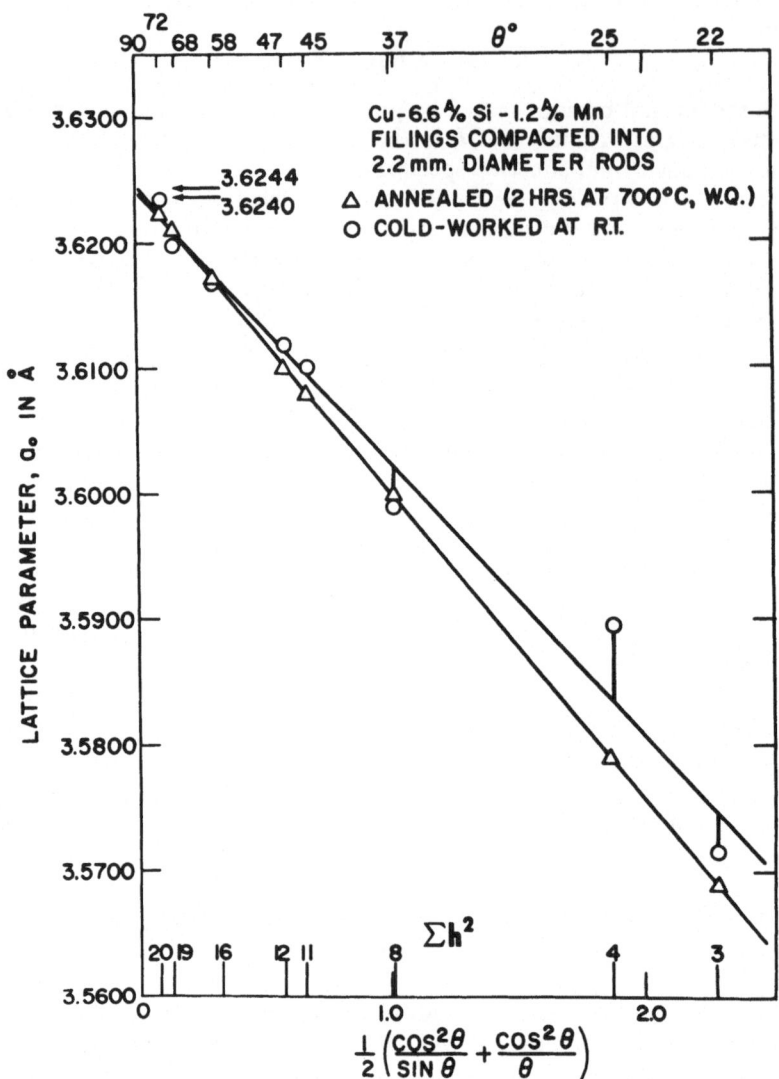

Figure 4. Nelson-Riley plot of lattice parameters for annealed and cold-worked filings of Cu–6.6 at. %Si–1.2 at. %Mn. (After Otte[36].)

(by 0.0004 Å) than that for the annealed filings, and both are smaller than the value of an annealed solid specimen. These differences have already been discussed elsewhere.[36]

Line Broadening. The integral breadth of diffraction lines from the filings is shown in Figure 5. The integral breadths from the annealed filings were employed to correct for instrumental broadening using equation (9). Since broadening caused by nonuniform strain is proportional to $\sin^2\theta$ (equation 7), an extrapolation to $\sin^2\theta$ equal to zero for lines through points of multiple orders yields the broadening caused by particle size and stacking faults. From this extrapolation and by use of equation (4) the effective particle size in the $\langle 111 \rangle$ direction is calculated to be 142 Å and in the $\langle 100 \rangle$ direction 92 Å. These values are indicative of an appreciable contribution from the stacking faults

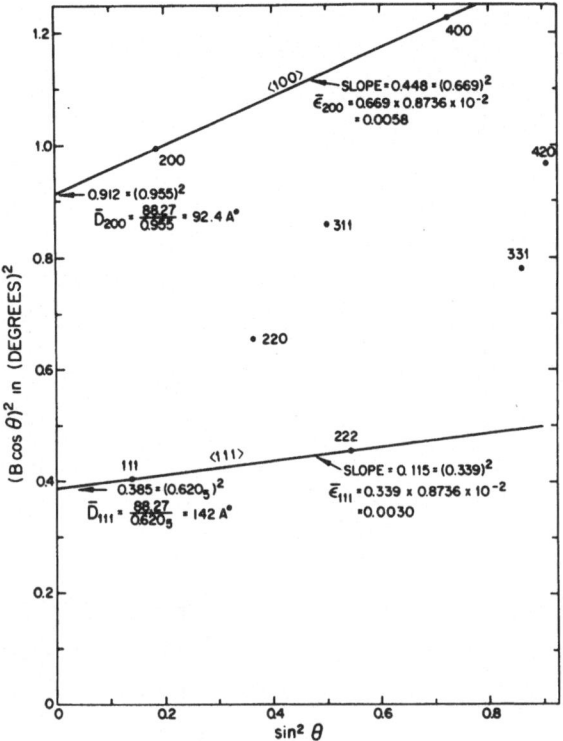

Figure 5. Diffraction line broadening of Cu–6.6 at. %Si–1.2 at. %Mn resulting from filing at room temperature.

to the particle size broadening. The extent of this contribution has recently been discussed in detail.[60]

The strain estimated from the broadening is also indicated in Figure 5. The order of increasing strain among the various *hkl* directions agrees qualitatively with the order predicted from the variation of Young's modulus with *hkl*.[44] No good quantitative fit could be obtained, however. The strains observed are the same order of magnitude as the fracture strength divided by Young's modulus.

Tensile Specimens

Two similar specimens, *A* and *B*, were examined. The D–S pattern of both in the annealed condition was obtained. Specimen *A* was then given approximately 1% extension, the exact amount being measured with the traveling microscope, and the D–S pattern obtained with the specimen under load. The load was removed and another D–S pattern obtained. This procedure was repeated until the specimen had been given 7.35% extension, at which point it was unloaded and allowed to age for 5 months at room temperature. After aging, the specimen was given further load–unload cycles with D–S patterns for each until an extension of 22.3% was reached. The specimen, unloaded, was aged for 45 hr at 300°C, and then given one load–unload cycle. Specimen *B* was loaded in continuous steps. The amount of strain in each step was determined with the traveling microscope. D–S patterns were obtained after each additional loading. A standard stress–strain curve was obtained with a Baldwin testing machine and is reproduced in Figure 9a.

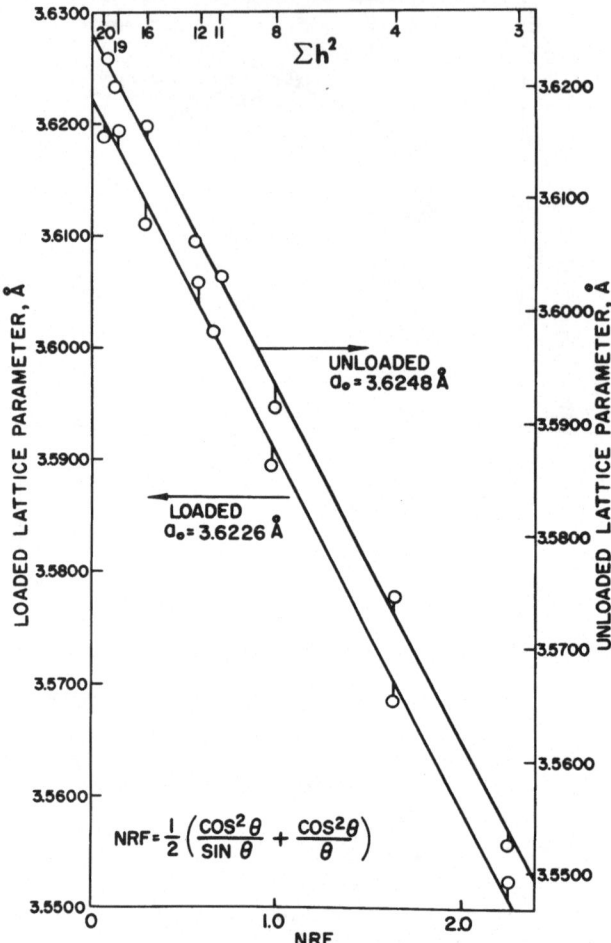

Figure 6. Nelson–Riley plot of lattice parameters for Cu–6.6 at. %Si–1.2 at. %Mn after 17% extension by uniaxial tension at room temperature.

Peak Displacements and Lattice Parameter Changes. A typical NRF plot for tensile specimen *A*, given the load–unload cycles is shown in Figure 6. The displacements are reproducible and have been replotted, to show up more clearly, in Figures 7 and 8. The agreement between the observed peak displacements and those resulting from elastic anisotropy (Figure 3) is better than the agreement with peak displacements to be expected from stacking faults (Figure 2).

The lattice parameter changes accompanying the tensile deformation are shown in Figure 9b for both specimens *A* and *B*. The lattice parameter of specimen *B* which was continuously loaded, decreased to a limiting value which was reached after about 7% extension and then remained virtually constant up to the largest extension (20%) for which results are reported here. The curves for both specimens coincide up to the point where specimen *A* was aged at room temperature. This indicates that the unloading cycle had no measurable effect on the results obtained under load, at least for the first few cycles.

The decrease in the extrapolated lattice parameter for the specimens (both *A* and *B*) under load corresponds to a radial compressive strain. The extrapolated lattice parameter

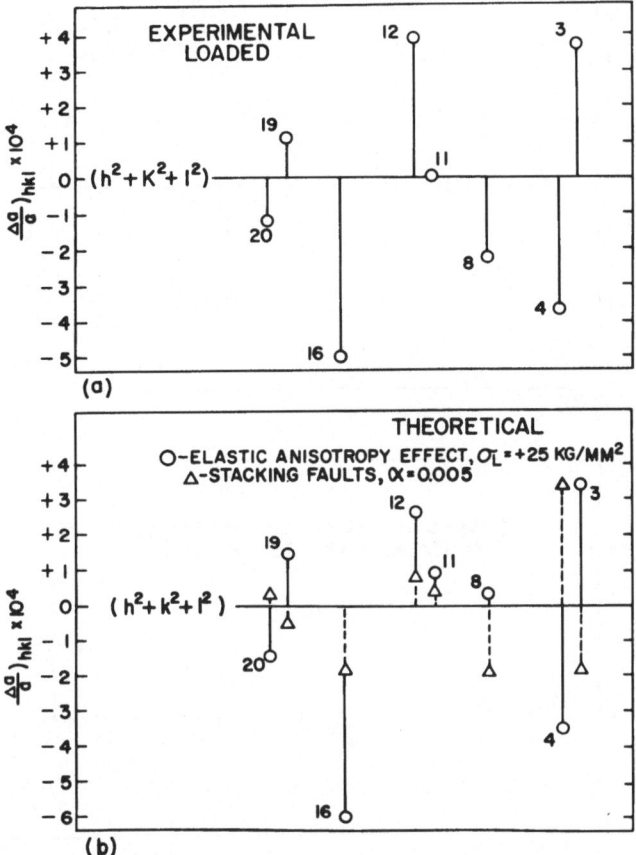

Figure 7. Comparison of experimental peak displacements with theoretical peak displacements. (a) Experimental differential displacements of lattice parameters from a straight-line fit in Nelson–Riley plot for Cu–6.6 at. %Si–1.2 at. %Mn after 17.5% extension, under load. (b) Theoretical differential displacements caused by: (i) Elastic anisotropy and radial Poisson contraction near specimen surface produced by tensile stress, σ_L, parallel to specimen axis. (Average of isotropic stress and isotropic strain.) (ii) Deformation stacking faults.

for the unloaded specimen (A) was also smaller, though only very slightly than that obtained for the initially annealed condition. The displacements of all peaks were, however, reversed compared with those of the specimen under load and indicated a residual radial tensile strain.

The room-temperature aging treatment of specimen A left the lattice parameter in the unloaded condition essentially unchanged, but lowered that obtained under load. The 300°C aging (specimen A) produced a decrease in the lattice parameter and caused the peak shifts due to the residual strains to be greatly diminished but not to disappear entirely. Upon loading, the lattice parameter was now correspondingly lower, but with subsequent unloading it increased slightly above that obtained immediately after aging (Figure 9b).

Line Broadening. The broadening resulting during the tension test appears to be predominantly a particle-size type of broadening with little or no contribution from

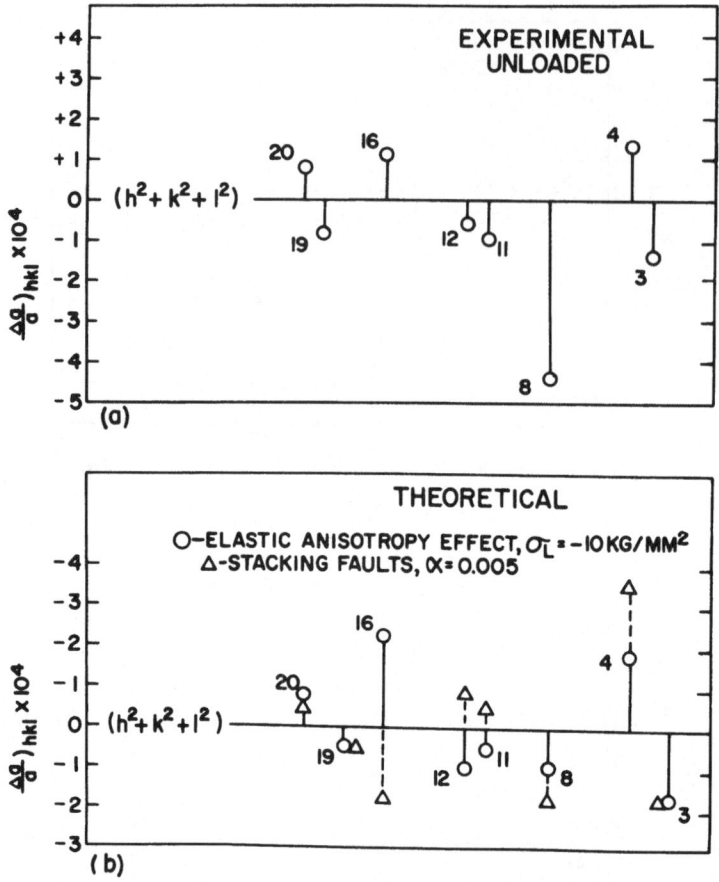

Figure 8. Comparison of experimental peak displacements with theoretical peak displacements. (a) Experimental differential displacements of lattice parameters from a straight-line fit in Nelson–Riley plot for Cu–6.6 at.%Si–1.2 at.%Mn after 17.5% extension, unloaded. (b) Theoretical differential displacements caused by: (i) Elastic anisotropy and radial Poisson expansion near specimen surface produced by compressive stress, σ_L, parallel to specimen axis. (Average of isotropic stress and isotropic strain.) (ii) Deformation stacking faults.

strain broadening. A typical plot of the broadening is shown in Figure 10. From the fact that the two orders of (111) and of (200) are broadened equally we can infer the absence of strain broadening. Table III shows the effective particle sizes corresponding to the broadening in Figure 10 and the stacking fault probabilities which would correspond to these particle sizes if the broadening had been attributed to stacking faults. These probabilities were obtained by using equation (3) with $1/D_{hkl} = 0$, $\beta = 0$, and $\varepsilon = 0$. The observed peak displacements do not suggest the presence of such a high stacking fault probability and the broadening must be in fact a true small domain size effect. The broadening is much smaller than that obtained by filing. The maximum integral breadth resulting from filing is 2.32° 2θ compared with 0.43° 2θ resulting from tensile extension.

The variation of the average coherently diffracting domain size with extension is shown in Figure 11. Aging does not appear to affect the broadening, indicating that the recovery accompanying the aging is small. Any stacking faults in this alloy appear to be

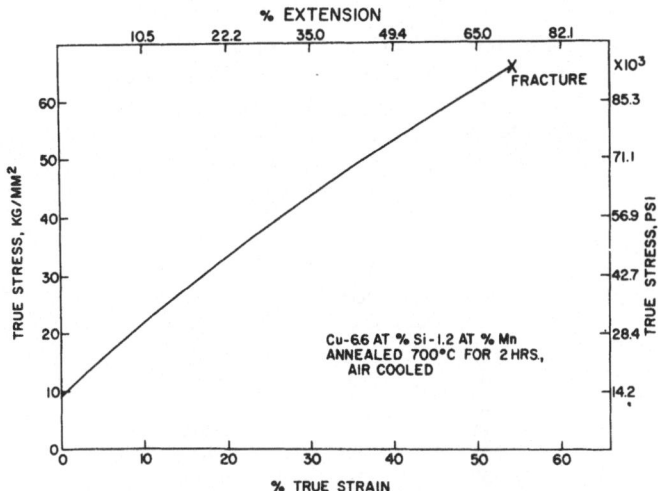

Figure 9a. Mechanical stress–strain curve for Cu–6.6 at. %Si–1.2 at. %Mn loaded in tension. Extension = $(l - l_0)/l_0$; true strain = $\ln(l/l_0)$; l_0 = original length; l = final length.

completely removed by a 300°C aging,[45] so if any were present due to the tensile deformation their removal by a 300°C aging could not be detected.

Drawn Wire

Peak Displacements and Lattice Parameter Changes. The Nelson–Riley plot for a specimen given 40.6% reduction in area by drawing through a die is shown in Figure 12. From this plot it can be seen that the lattice parameter has slightly increased, indicating the presence of a residual radial tensile strain at the surface. To investigate this possibility further, the surface of the specimen was electropolished in several steps, and the lattice parameter as a function of surface removal obtained. The result is shown in Figure 13. These data indicate that the surface of the wire has a radial tensile strain and the interior

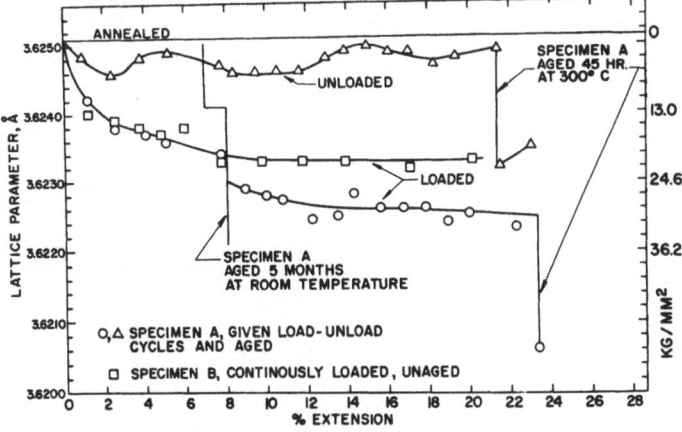

Figure 9b. Lattice parameter changes in Cu–6.6 at. %Si–1.2 at. %Mn caused by tensile extension and aging. The longitudinal stress, right-hand scale, is calculated from the change in a_0 and the bulk elastic constants ($\nu/E = 2.35 \times 10^{-5}$ mm²/kg).

Table III. Effective Coherent Domain Size and Stacking Fault Probability Calculated Assuming Broadening Caused Entirely by Stacking Faults for Cu–6.6 at. %Si–1.2 at. %Mn after 14.3% Extension by Uniaxial Tension

hkl	D_{eff}, Å	α
111	420	0.013
200	315	0.007
220	380	0.009
311	315	0.017
222	400	0.013
400	295	0.008
331	269	0.011
420	242	0.044

a radial compressive strain. This result is not general for wire drawing since it depends to a large extent on the amount of deformation. The peak displacements in Figure 12 indicate that a large number of stacking faults are produced by 40% reduction in area. The values of α corresponding to the various peak displacements are given in Table II.

As with the filings, discrepancies between the stacking fault probabilities from different *hkl* reflections could be accounted for only partly by means of a spacing fault.

Line Broadening. The line broadening resulting from the wire drawing is shown in Figure 14. The data were analyzed in the same manner as those for the filings. The stacking fault probability α obtained from the particle-size type of broadening in the $\langle 111 \rangle$ and $\langle 200 \rangle$ directions is 0.033, in relatively poor agreement with the value of α obtained from the (111) and (200) peak displacements. The stacking fault broadening was subtracted from the total broadening and the resulting strain component of broadening is given in Figure 15. The root-mean-square strains for the $\langle 111 \rangle$ and $\langle 200 \rangle$ directions are 0.73×10^{-2} and 1.29×10^{-2}, respectively. These strains are larger than those resulting from filing.

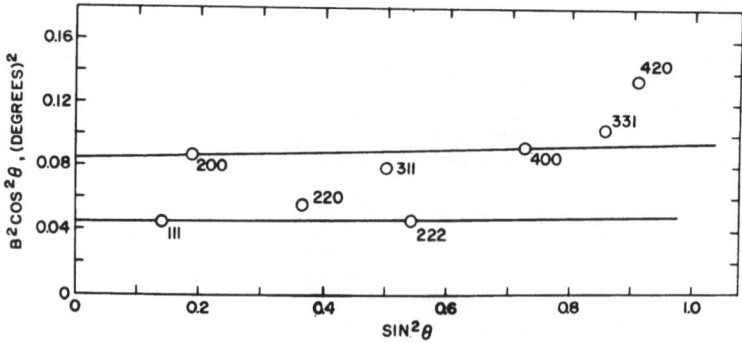

Figure 10. Diffraction line broadening of Cu–6.6 at. %Si–1.2 at. %Mn after 14.3% tensile extension, loaded.

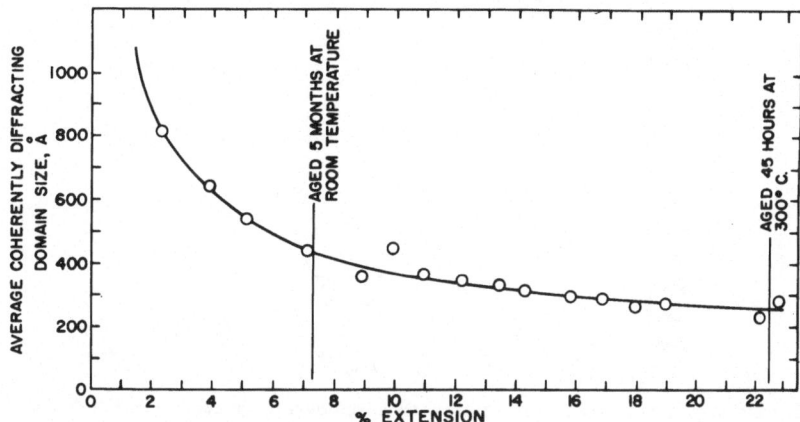

Figure 11. Coherent domain size resulting after uniaxial extension of Cu–6.6 at. %Si–1.2 at. %Mn. Specimen under load-curve virtually the same for unloaded specimen.

DISCUSSION

The most interesting result in the present investigation is probably the observation that little evidence of stacking faults could be found in the tensile specimen, whereas in the filings and the drawn wires their presence was readily detected.

The explanation for this probably is that up to 20% elongation, the deformation is insufficient to produce a significant density of stacking faults at the surface, even though the stacking fault energy is probably in the neighborhood of 1–8 ergs/cm² and the

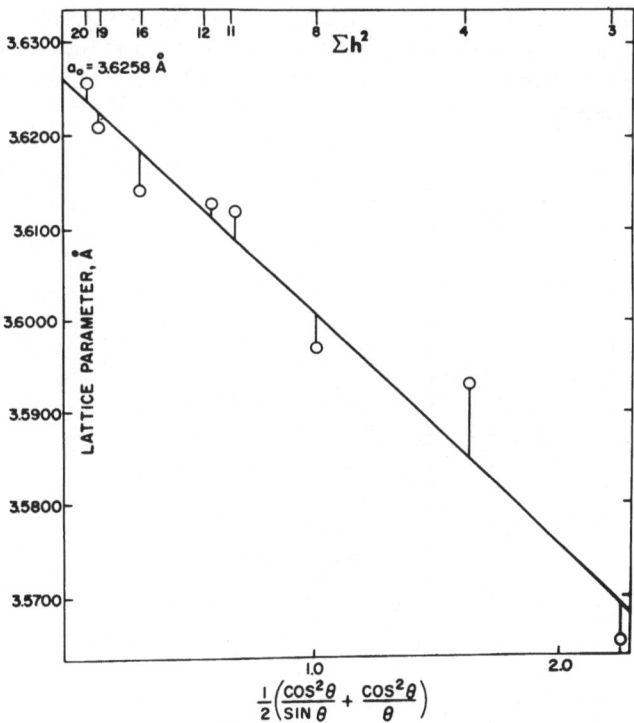

Figure 12. Nelson–Riley plot for Cu–6.6 at. % Si–1.2 at. %Mn after 40.6% reduction in area by wire-drawing.

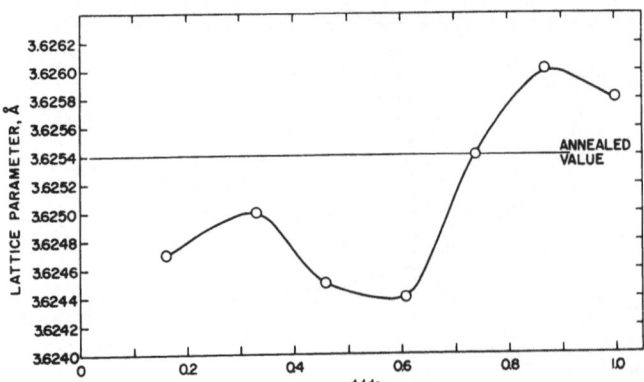

Figure 13. Lattice parameter as a function of surface removal for specimen of Cu–6.6 at. %Si–1.2 at. %Mn given 40.6% reduction in area by wire-drawing.

separation distance between the partial dislocations between 17 to 90 atom spacings (as estimated from investigations[46,51] on alloys similar to the present one).

The surface layers of the tensile specimen show mainly "fragmentation" or reduction in the size of coherently diffracting domains. This implies a certain increase in the number of dislocations required at the domain boundaries in addition to any that may form within the domains. None of these dislocations appear to cause any appreciable amount of strain broadening. This may be due either to the increase in dislocation density being insufficient to be detectable, or due to the localized strain around the dislocations, even cumulatively, not being sufficiently large to be measurable. The former situation is more probable since no stacking faults could be detected either. Under favorable conditions the stacking fault ribbons tend to widen at the surface;[47] furthermore, the results on the drawn wires show that appreciable dissociation can in fact occur in the surface layers.

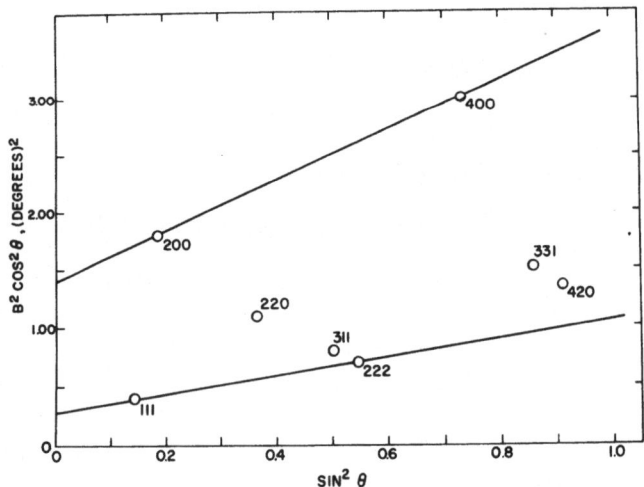

Figure 14. Line broadening for a specimen of Cu–6.6 at. %Si–1.2 at. %Mn given 40.6% reduction in area by wire-drawing.

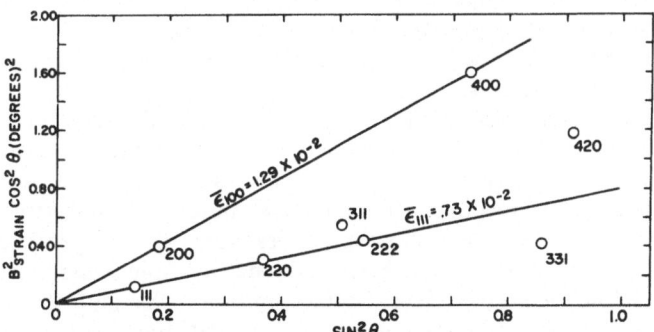

Figure 15. Strain component of line broadening for a specimen of Cu–6.6 at. %Si–1.2 at. %Mn given 40.6% reduction in area by wire-drawing.

Comparison of Figures 9a and 9b reveals clearly that although the tensile specimen as a whole work-hardens appreciably, the surface layers show only very slight work hardening, and fragmentation of the surface layers appears to reach a limiting domain size (Figure 11). Thus, in uniaxial tensile deformation of a polycrystalline specimen the major contribution to the overall work hardening of the material at least up to 20% extension is made by the interior, and the contribution of the surface layers after the initial stages is relatively small. The apparent absence of work hardening in the surface layers has been also found by other investigators.[6,18,19]

Residual Strains

A consequence of the above behavior is that upon unloading after a given amount of extension, one should expect the surface layers to be in radial tension (and longitudinal compression). This is generally observed.[6,18] It will be seen from the values of the lattice parameter in Figure 9b, however, that for the Cu–Si–Mn alloy this does not appear to be true except, perhaps, after aging the specimen at 300°C, although, as Figure 6 shows, the relative line shifts are in the expected direction for radial tensile strain. The result can be explained if it is assumed that after the first straining (and for all subsequent cycles), the "reference" lattice parameter should not be with respect to that measured from the annealed condition of the specimen, but should be with respect to some lower value. This situation may arise if, as a consequence of the deformation produced by the first loading, and to a lesser extent by subsequent loadings, enhanced diffusion permitted a certain amount of clustering (or additional clustering, if some is present already) to occur. When the addition of solute to the matrix causes an expansion of the (matrix) lattice, then any process which results in the clustering of these solute elements will be accompanied by a contraction of the lattice parameter, and this is detectable.[36,43,45] The observations in Figure 9b are consistent with such a possibility.

As a further test of this interpretation the 300°C aging treatment was performed. At this temperature substantial clustering* occurs, whereas the amount after 5 months at room temperature is negligible. The resulting decrease in lattice parameter to an aged a_0 of 3.6232 Å is clearly apparent in Figure 9b and after one load–unload cycle, the lattice

---

* No evidence of any precipitation could be found.

parameter now is slightly higher than the as-aged a_0. This is in agreement with an expected residual radial tensile strain. Thus we may conclude that in the case of the annealed specimens the decrease in lattice parameter due to clustering is larger than the increase due to residual strains.

The preceding interpretation may also be applied to the results of Donachie and Norton,[19] who introduced what they termed a "metastable" lattice parameter, for which, however, they had no explanation. Their metastable lattice parameter is most probably the result of carbon (and nitrogen) clustering to dislocations to form Cottrell atmospheres.

The present results favor the macroscopic stress theory of residual stresses rather than the intergranular stress theory. On the basis of the macroscopic stress theory, it is anticipated that removal of the surface layers should produce a change in the internal strain. Such a change has been reported by some investigators.[16,17] Explanations why others were unable to find it are possible. Thus Wood[18] used cylindrical specimens as in the present investigation, but it seems certain that he did not correct for absorption when he compared his results after etching. Less absorption in the smaller diameter specimens would tend to increase the lattice parameter, whereas the relief of residual stress would tend to decrease it in the radial direction in which he presumably made his measurements. The negative results on aluminum obtained by Donachie and Norton[19] could be attributed to the large scatter in their results for this material. In the ingot iron they observed a change in lattice parameter upon deformation, as already mentioned, and this could vary with depth, and in such a way that it compensates for changes in lattice parameter on relieving the stress by surface removal.

In the case of mild steel, Smith and Wood[48] measured the change in the lattice parameter (the lattice strain) for a series of load–unload cycles, increasing the applied load (or stress) for each cycle as in the present work. Contrary to the present observations, however, they reported that the lattice strain under load decreased with increasing *plastic* strain. One of the possibilities mentioned above could account for the observation. The residual lattice strain, on the other hand, progressively increased, reflecting continued work hardening of the material; no "undulating" of the residual lattice strain, as in Figure 9b, was observed, though Wood[49] had earlier reported such an effect, for a number of materials, including iron and copper. The magnitude of the effect was surprisingly large, larger than observed for the Cu–Si–Mn alloy, Figure 9b, and yet no similar observations appear to have been recorded by anyone since. He interpreted the effect as a periodic recovery during continued working, since it was accompanied by a similar periodic behavior of the line broadening. This does not appear to be the case for the present alloy; the effect, if significant here, is not clearly understood.

The drawn wire for which the results are shown in Figure 12 was selected for comparison with the tensile test results; the relative line shifts here are predominantly due to stacking faults. On removing the surface layers by electropolishing the lattice parameter changes, Figure 13, reflecting the change in the radial residual strain pattern. The stacking fault density also decreases. A complete interpretation of all the data has not yet been possible. It is clear, however, that in neither tensile deformation nor wire drawing is the residual strain pattern uniform through the cross section. For the wire, this is in agreement with observations by destructive methods.[50]

Stacking Faults

From Figures 6 and 8 it is evident that for the tensile specimen the displacements of the first three diffraction peaks (lines 3, 4, and 8) are in the direction to be expected from stacking faults, and that, unless the displacements of the next three peaks (lines 11, 12, and 16), in particular that of line 16, are also checked, the erroneous conclusion would be drawn that upon unloading stacking faults had formed. The effect becomes more pronounced the greater the anisotropy of the material. It so happens also that in general the stacking fault energy is lower the greater the anisotropy. Thus if stacking faults are known (from electron microscopy, observations on filings, or otherwise) to form upon deformation in a material, added caution is required when interpreting the peak displacements from cold-worked bulk samples. This precaution is sometimes overlooked and the relative peak displacement of lines 3, 4, and 8 is used not only as evidence for the presence of stacking faults, but in a number of cases to actually obtain a stacking fault probability. The present work shows that if the stacking fault density is sufficiently large ($\alpha = 0.01$ approximately), then the peak displacements from the faults are likely to predominate over those from the residual strains. Otherwise it is probable that peak shifts due to residual strains are being in fact observed.

Aging

The phase diagram for the copper–silicon–manganese system by Smith[52] and evidence reported by Dreyer[53] indicates that a solid solution with the composition Cu–6.6 at. %Si–1.2 at. %Mn may be supersaturated with respect to Mn_2Si_3 or Mn_5Si_3. However, so far no evidence of any precipitate has been obtained even after treatment considered to be highly favorable for precipitation. A certain tendency to solute clustering is, however, very likely. Thus the lattice parameter of the annealed filings is lower than that of annealed solid specimens by 0.0010 Å; this has been attributed to solute clustering promoted by vacancies quenched in from the annealing temperature.[36]

The decrease in lattice parameter after aging tensile specimen A at 300°C for 45 hr is attributed to solute clustering. The difference between the unloaded and loaded lattice parameter suggests that, if no change in modulus has occurred, the yield stress following aging at 300°C is the same as that prior to aging. Thus either no strengthening accompanies this aging treatment or sufficient recovery has occurred to cancel any strength increase. On the other hand, after the room temperature aging, there appears to be a definite increase in yield stress (assuming again no modulus change) since, while the lattice parameter of the unloaded specimen remains essentially the same after aging, the lattice parameter decreases more under load than it did prior to aging. Direct evidence of age hardening in this alloy has been reported.[45]

CONCLUSIONS AND SUMMARY

Deformation by filing, tensile straining, and wire drawing has been examined in a Cu–6.6 at. %Si–1.2 at. %Mn alloy which has a low stacking fault energy. Evaluation of the strain distribution, domain size, and stacking fault probability shows important differences in the different modes of deformation.

1. *Filing* produces a state characterized by a root-mean-square strain, on the order of the fracture strength divided by Young's modulus, and by the presence of stacking faults, with a stacking fault probability of 0.025. Fragmentation of coherent domains makes no significant contribution to line broadening, as determined by integral breadth, all particle-size type of broadening being accounted for by faulting.

2. Deformation by *tensile straining* gives rise in the surface of the specimen to a radial residual strain that is tensile, but which, on the basis of the extrapolated lattice parameter, appears to be compressive with respect to the lattice parameter in the annealed condition. The effect is explained as the consequence of a small decrease in lattice parameter due to solute clustering by enhanced diffusion through deformation. After aging at 300°C for 45 hr, the decrease is more pronounced, and with respect to the lattice parameter so

obtained, the extrapolated lattice parameter of the aged specimen after a load–unload cycle now shows the expected increase indicative of a residual radial tensile strain. By inference, the longitudinal residual strain is compressive. There appears to be little work hardening in the surface layers after a certain critical strain has been obtained whereas the specimen as a whole work-hardens rapidly. No evidence exists for extensive formation of stacking faults by tensile straining up to about 20% (which is well below the fracture strain). There is essentially no root-mean-square strain contribution to the diffraction line broadening and the minimum average particle size calculated from the broadening is 250 Å.

3. Reduction in area by 40.6% by *wire drawing* gives rise to a radial tensile residual strain at the surface, and a radial compressive strain in the interior. Thus, whereas in a tensile specimen the interior appears to work-harden more than the surface layers, in wire drawing the reverse situation may prevail. Root-mean-square strains after 40.6% reduction in area are somewhat larger in magnitude than those introduced by filing. The stacking fault probability, 0.049, is the highest of the three types of deformation. The particle size type broadening is, as in the case of filings, predominantly due to faulting.

APPENDIX: DERIVATION OF THE RELATIONSHIP BETWEEN PARTICLE SIZE AND STACKING FAULT BROADENING OBTAINED BY FOURIER ANALYSIS AND THAT OBTAINED BY INTEGRAL BREADTH

The following derivation will make use of the notation and results of Warren.[22] It is assumed that the peaks are broadened only by particle size and intrinsic stacking faults, hence the peaks are symmetrical. Using Warren's equation (44) with $B_n = 0$ (no asymmetry), we obtain as the diffracted power per unit length of a diffraction circle, $P'_{2\theta}$, in terms of a position in reciprocal space

$$P'_{2\theta}(h_0) = \frac{G(u + b)}{|b'_3|} \sum_{n=-\infty}^{+\infty} A_n \cos 2\pi n(h'_3 - l' - \delta) \tag{A-1}$$

(Each diffraction peak is considered as a $00l'$ reflection in terms of orthorhombic axes $a'_1 \, a'_2 \, a'_3$ and corresponding reciprocal axes $b'_1 \, b'_2 \, b'_3$. G is a constant.) The summation of the Fourier series is normalized by the relation

$$\int_{l'-1/2}^{l+1/2} \sum_{n=-\infty}^{\infty} A_n \cos 2\pi n(h'_3 - l' - \delta) \, dh_3 = 1 \tag{A-2}$$

For small values of n, we have from Warren's equation (45) with $\beta = 0$:

$$A_n = 1 - \frac{|n|}{|b'_3|}\left[\frac{1}{D} + \frac{1.5\alpha}{a(u+b)h_0} \sum_b |L_0|\right] \tag{A-3}$$

($L_0 = h + k + l$) from which by Warren's equation (47),

$$\frac{1}{(D_{\text{ett}})_{F.A.}} = \left[\frac{1}{D} + \frac{1 \cdot 5\alpha}{a(u+b)h_0} \sum_j |L_0|\right] = \frac{1}{D} + \frac{3\alpha}{2a}V_{hkl} \tag{A-4}$$

in the notation of reference 26.

Now the broadening $\beta_{SF}(h_3)$ of those components affected by faulting is approximately $9\alpha/4$ according to Paterson.[23] In terms of spectrometer coordinates,[22]

$$\beta_{SF}(2\theta) = \frac{\lambda}{\cos\theta} \cdot \frac{|L_0|}{3ah_0} \cdot \frac{9\alpha}{4}$$

where $h_0^2 = h^2 + k^2 + l^2$. Averaging over all components gives

$$\langle \beta_{SF}(2\theta) \rangle = \frac{\lambda}{\cos\theta} \frac{3\alpha}{4a} \left[\frac{1}{k_0(u+b)} \sum_b |L_0| \right] = \frac{\lambda}{\cos\theta} \left(\frac{3\alpha}{4a} V_{hkl} \right)$$

using the notation of reference 26. Thus if the line profiles due to particle size D_P and due to stacking faults are predominantly Cauchy, we can add the broadening due to these two effects and obtain

$$\langle \beta(2\theta) \rangle = \frac{\lambda}{D_P \cos\theta} + \frac{\lambda}{\cos\theta} \left(\frac{3\alpha}{4a} V_{hkl} \right) = \frac{\lambda}{(D_{ett})_\beta \cos\theta}$$

so that

$$\frac{1}{(D_{ett})_\beta} = \frac{1}{D_P} + \frac{3\alpha}{4a} V_{hkl} \tag{A-5}$$

When $D_P = \overline{D} = 0$, we obtain

$$(D_{ett})_\beta = 2(D_{ett})_{F.A.}$$

When the line profiles are predominantly Gaussian, a similar expression (but with a slightly larger factor) can be obtained. For $\alpha = 0$, we have,[55]

$$\frac{(D_{ett})_\beta}{(D_{ett})_{F.A.}} = \frac{\overline{D^2}}{(\overline{D})^2} \geqslant 1$$

Thus, within a factor of about 2, the particle size determined from Fourier analysis and from integral breadth are essentially the same order of magnitude.

ACKNOWLEDGMENTS

The authors are indebted to W. W. Bender for his interest and encouragement and to W. G. Montague for assistance with the experimental work. We are also deeply indebted to Dr. C. N. J. Wagner for his very helpful suggestions and discussion, and for his having read an earlier draft of this paper. This project was sponsored by the Office of Naval Research.

REFERENCES

1. A. R. Stokes and A. J. C. Wilson, Proc. Phys. Soc. 56: 174, 1944.
2. N. Kalakoutzky, Revue d'artillerie 31: 289, 389, 485, 1888; 32: 5, 165, 1888.
3. "Internal Stresses in Metals and Alloys," Institute of Metals Symposium, London, 1948.
4. M. Hetenyi, ed., John Wiley & Sons, Inc., Handbook of Experimental Stress Analysis, New York, 1950.
5. W. R. Osgood, ed., Residual Stresses, Reinhold Publishing Corp., New York, 1954.
6. G. B. Greenough, Progr. in Metal Phys. 3: 176, 1952.
7. A. Reuss, Z. angew. Math. Mech. 9: 49, 1929 (quoted in ref. 6).
8. W. Voigt, Lehrbuch der Kristallphysik, B. G. Teubner, ed., 1910 and 1928 (quoted in ref. 6).
9. H. Neerfield, Mitt. Kaiser-Wilhelm-Inst. Eisenforsch. Düsseldorf 24: 61, 1942 (quoted in ref. 6).
10. G. B. Greenough, Proc. Roy. Soc. (London) A197: 556, 1949.

11. G. B. Greenough, *Nature* **160**: 258, 1947.
12. G. B. Greenough, *Nature* **166**: 509, 1950.
13. G. B. Greenough, *Metal Treatment* **16**: 58, 1949.
14. L. G. Finch, *Nature* **163**: 402, 1949.
15. R. I. Garrod, *Nature* **165**: 241, 1950.
16. F. Bollenrath, V. Hauk, and E. Ostwald, *Z. Ver. dtsch. Ing.* **83**: 129, 1939.
17. K. Kolb and E. Macherauch, *Phil. Mag.* **7**: 415, 1962.
18. W. A. Wood, *Proc. Roy. Soc.* (London) *A***192**: 218, 1948.
19. M. J. Donachie and J. T. Norton, *Trans. AIME* **221**: 962, 1962.
20. J. D. Meakin and H. G. F. Wilsdorf, *Trans. AIME* **218**: 737, 1960.
21. R. A. Ekvall and N. Brown, *ONR Report*, Feb. 16, 1962.
22. B. E. Warren, "X-ray Studies of Deformed Metals," *Progr. in Metal Phys.* **8**: 147, 1959.
23. M. S. Paterson, *J. Appl. Phys.* **28**: 805, 1952.
24. B. E. Warren and E. P. Warekois, *Acta Met.* **3**: 473, 1955.
25. C. N. J. Wagner, *Acta Met.* **5**: 427, 1957.
26. C. N. J. Wagner, A. S. Tetelman, and H. M. Otte, *J. Appl. Phys.* **33**: 3080–3086, 1962.
27. B. E. Warren, *J. Appl. Phys.* **32**: 2428, 1961.
28. G. F. Bolling, T. B. Massalski, and C. J. McHargue, *Phil. Mag.* **6**: 491, 1961.
29. C. J. McHargue, *Acta Met.* **9**: 851, 1961.
30. M. J. Klein, J. L. Brimhall, and R. A. Huggins, *Acta Met.* **10**: 13, 1962.
31. H. M. Otte, *Acta Cryst.* **13**: 1064, 1960 (Intl. Union Cryst. Fifth Intl. Cong. and Symp., paper 10–16, Abstracts, p. 89).
32. A. Taylor, *X-ray Metallography*, first edition, John Wiley & Sons, Inc., New York, 1961, p. 679.
33. A. Taylor, *X-ray Metallography*, first edition, John Wiley & Sons, Inc., New York, 1961, p. 788.
34. C. S. Barrett, *Trans. AIME* **188**: 123, 1950.
35. *Metals Handbook*, ASM, 1948, p. 88.
36. H. M. Otte, *J. Appl. Phys.* **32**: 1536, 1961.
37. J. B. Nelson and D. P. Riley, *Proc. Phys. Soc.* (*London*) **57**: 160, 1945.
38. J. B. Cohen and C. N. J. Wagner, *J. Appl. Phys.* **33**: 2073–2077, 1962.
39. T. R. Anantharaman, *Acta Met.* **9**: 903, 1961.
40. A. Taylor, *X-ray Metallography*, first edition, John Wiley & Sons, Inc., New York, 1961, p. 686.
41. H. M. Otte, *J. Appl. Phys.* **33**: 2892, 1962.
42. J. C. Helion, MS Thesis, Yale University, 1962.
43. H. M. Otte, *J. Appl. Phys.* **33**: 1436, 1962.
44. J. R. Neighbors and C. S. Smith, *Acta Met.* **2**: 591, 1954.
45. M. S. Wechsler, J. M. Williams, and H. M. Otte, *J. Metals* **14**: 81, 1962; *J. Phys. Soc. Japan* **18**: (Supp.) (in press) (*Proc. Int. Conf. Cryst. Lattice Defects, Japan*, 1962).
46. P. R. Swann and J. Nutting, *J. Inst. Metals* **90**: 133, 1961.
47. R. Gevers, S. Amelinckx, and P. Delavignette, *Phil. Mag.* **6**: 1515, 1961.
48. S. L. Smith and W. A. Wood, *Proc. Roy. Soc.* (*London*) **179**: 450, 1942.
49. W. A. Wood, *Proc. Roy. Soc.* (*London*) **172**: 231, 1939.
50. W. M. Baldwin, Jr., *Am. Soc. Testing Materials* **49**: 1, 1949.
51. P. Haasen and A. King, *Z. Metallkunde* **51**: 722, 1960.
52. C. S. Smith, *Trans. AIME* **89**: 164, 1930.
53. K. L. Dreyer, *Metall* **7**: 186, 1953.
54. W. A. Wood and W. A. Rachinger, *J. Inst. Metals* **75**: 571, 1949; discussion, p. 1120.
55. C. N. J. Wagner, Private communication.
56. C. A. Johnson, *Acta Cryst.* **16** (in press).
57. B. E. Warren, *J. Appl. Phys.* **34** (in press).
58. H. M. Otte, *J. Phys. Chem. Solids*: **24**, 169, 1963.
59. H. M. Otte, D. O. Welch, and G. F. Bolling, *Phil. Mag.* **8**: 345, 1963.
60. R. P. I. Adler and C. N. J. Wagner, *J. Appl. Phys.* **33**: 3451, 1962.

LATTICE SPACINGS IN SOME TRANSITION METAL TERMINAL SOLID SOLUTIONS

Henry Chessin, Sigurds Arajs, and D. S. Miller

United States Steel Corporation
Monroeville, Pennsylvania

ABSTRACT

The lattice parameter–composition curves for several nickel solid solutions and for some chromium and iron solid solutions are discussed. It is shown that the size effect may be the predominating influence on the change of lattice parameters in these systems. This is demonstrated by comparing observed and calculated data employing various methods. A new scheme for evaluating the atomic size in solid solutions is proposed, based on regarding the atom as an incompressible core surrounded by a smeared-out compressible volume. The suggestion that classical elasticity theory may be used as a basis for understanding the size effect in solid solutions is justified by examination of the Ag–Pd system for additions of Ag from 0 to 100 at. %.

INTRODUCTION

The investigation of accurate lattice parameters has been directed into two fields of endeavor:

1. Efforts to determine the systematic crystallographic similarities between related phases.
2. The possibility of using lattice parameter changes with changes of composition in solution as a probe to help clarify the understanding of solid solution formation.

This paper will be concerned with the latter topic and will consider in particular the changes in the lattice parameters on formation of terminal solid solutions of binary transition metals. The work reported here is part of a program we have undertaken to investigate various physical and structural properties of the transition metals in solution by means of X-ray diffraction and magnetic measurements. The results of some of these studies have been already reported elsewhere.[1-3]

ATOMIC SIZE EFFECT IN SOLID SOLUTIONS

The formation of substitutional solid solutions considered here gives rise to an increase or decrease from the lattice parameter of the solvent. This results in a shift of the X-ray interference maxima given by Bragg's law and hence may be observed by careful measurements of the positions of the diffraction lines on a suitable X-ray diagram. It must be made clear that such measurements yield the average constants of the unit cell[4] and throughout the course of this paper we refer to these average measurements. In

---

[1] Superscripts pertain to references at the end of the paper.

principle, it is possible to evaluate the individual atomic sizes in solid solution as opposed to average atomic sizes by means of diffuse X-ray scattering as shown by recent workers.[5-8]

The systematic examination of terminal solid solutions has shown that three main factors must be considered to explain the lattice parameter changes when a solute atom is added to or substituted for the atoms of the solvent. These factors are the atomic size factor, the electronegative valence effect, and the relative valency effect. It is assumed by most workers in this field of research that these three factors operate independently and, consequently, that an analysis of any alloy system may be complicated by the fact that these three factors may be operating simultaneously. Nevertheless, it has been possible by careful selection of alloy systems to choose examples so that one factor is predominant, the others playing a relatively small role. It is the purpose of this paper to discuss the role of the size factor in explaining lattice spacings in transition metal solid

Table I. Structure, Volume, and Radius Data for Transition Elements
(Source: Pearson[22])

Atom		Structure Type	a, (Å)	c, (Å)	Atoms per unit cell	Volume per unit cell, (Å³)	Volume per atom, (Å³)	r^3, (Å³)	r, (Å)
Sc	(21)	A3 (hcp)	3.3080	5.2653	2	49.8984	24.9492	5.9562	1.8127
Ti	(22)	A3 (hcp)	2.9504	4.6833	2	35.3057	17.6529	4.2143	1.6153
V	(23)	A2 (bcc)	3.0282		2	27.7686	13.8843	3.3146	1.4910
Cr	(24)	A2 (bcc)	2.8846		2	24.0025	12.0013	2.8651	1.4203
α–Mn	(25)	A12 (bcc)	8.9139		58	708.2772	12.2117	2.9153	1.4285
Fe	(26)	A2 (bcc)	2.8663		2	23.5486	11.7743	2.8109	1.4113
Co	(27)	A3 (hcp)	2.5073	4.0698	2	23.2848	11.6424	2.7794	1.4060
Ni	(28)	A1 (fcc)	3.5238		4	43.7556	10.9389	2.6115	1.3771
Cu	(29)	A1 (fcc)	3.6147		4	47.2299	11.8075	2.8189	1.4126
Y	(39)	A3 (hcp)	3.6451	5.7305	2	65.9393	32.9696	7.8709	1.9892
Zr	(40)	A3 (hcp)	3.2312	5.1477	2	46.5450	23.2725	5.5559	1.7711
Nb	(41)	A2 (bcc)	3.3007		2	35.9593	17.9797	4.2923	1.6252
Mo	(42)	A2 (bcc)	3.1468		2	31.1611	15.5806	3.7024	1.5470
Tc	(43)	A3 (hcp)	2.735	4.388	2	28.425	14.212	3.393	1.5027
Ru	(44)	A3 (hcp)	2.7057	4.2816	2	27.1455	13.5725	3.2402	1.4798
Rh	(45)	A1 (fcc)	3.8043		4	55.0585	13.7646	3.2861	1.4867
Pd	(46)	A1 (fcc)	3.8907		4	58.8956	14.7239	3.5151	1.5205
Ag	(47)	A1 (fcc)	4.0855		4	68.1923	17.0481	4.0699	1.5966
La	(57)	A3 (hcp)	3.770	12.159	8	149.663	18.7079	4.4662	1.6468
Hf	(72)	A3 (hcp)	3.1946	5.0511	2	44.6428	22.3214	5.3288	1.7467
Ta	(73)	A2 (bcc)	3.298		2	35.8717	17.9358	4.2819	1.6239
W	(74)	A2 (bcc)	3.1650		2	31.7045	15.8523	3.7845	1.5584
Re	(75)	A3 (hcp)	2.760	4.458	2	29.4097	14.7049	3.5105	1.5198
Os	(76)	A3 (hcp)	2.7353	4.3191	2	27.9857	13.9929	3.3406	1.4949
Ir	(77)	A1 (fcc)	3.8389		4	56.5749	14.1437	3.3766	1.5002
Pt	(78)	A1 (fcc)	3.9239		4	60.4164	15.1041	3.6058	1.5334
Au	(79)	A1 (fcc)	4.0786		4	67.8477	16.9619	0.0494	1.5939
Zn	(30)	A3 (hcp)	2.6649	4.9468	2	30.4241	15.2121	3.6316	1.5371
Ga	(31)	ortho-rhombic	$a = 4.5198$ $b = 7.6602$	4.5258	8	156.6948	19.5869	4.6760	1.6722
Ge	(32)	A4 (dia)	5.6575		8	181.0813	22.6352	5.4038	1.7548
Si	(14)	A4 (dia)	5.4305		8	160.1472	20.0184	4.7790	1.6844

solutions. By the size effect, we mean the local strains introduced into the solution lattice by virtue of the difference in atomic diameters between solvent and solute atoms. Various workers have chosen different definitions of the atomic size and then have applied an analog of Vegard's rule to this definition. We proceed to discuss these differences. We shall show our scheme employing an analog of Vegard's law invoking still another concept of atomic size, show how this leads to a further understanding of lattice spacings in solid solutions based on elasticity theory, and then employ this theory for further elucidation.

Efforts to determine a quantitative measure of the change in lattice constants due to the distortions around the solute atom have relied on the so-called Vegard's law. In this law, it is assumed that the atomic sizes of the pure constituents of the solid solution have fixed values. These ideas are not easy to express quantitatively because of the difficulty in deciding how to express the size of an atom. Some workers, notably Hume-Rothery[9] and Raynor[10], have taken the atomic diameter as defined by the closest distance of approach in the crystal of an element as a measure of the size of the atom concerned. The atomic sizes thus dealt with are essentially those defined by assuming hard spheres in contact in the crystalline lattice. Other workers, notably Goldschmidt[11] and Pauling,[12] maintain that these interatomic distances also depend on the crystal structure of the element, that is to say, on the coordination of the element in the particular structure. In the generally accepted scheme of Goldschmidt, all radii of the elements are calculated for coordination number 12 so that these Goldschmidt radii may be compared favorably in a given alloy system. Pauling considers the dependence of the radius on the type of the bond defined as the number of shared electron pairs (corresponding to the single, double, and triple bonds in ordinary covalent molecules and crystals) and then considers separately the effect of the resonance in stabilizing the crystal and decreasing the interatomic distance. Presumably, once these atomic sizes are so defined, it should then be possible to predict the interatomic distance in a solid solution invoking Vegard's law. We have chosen to define the atomic radius in terms of the elemental volume per atom which is in turn defined as the elemental unit cell divided by the number of atoms required to specify that unit cell. The reasons for this choice will be made evident later.

Table I lists the data pertinent to this paper for the transition elements of the three long periods. The radii shown in the last column were computed using the data from the seventh column and assuming the relationship

$$\tfrac{4}{3}\pi r^3 = \frac{\text{Volume of unit cell}}{\text{Number of atoms in unit cell}}$$

This seemingly drastic assumption will be justified later. Figure 1 is a plot of the volume per atom *vs.* atomic number of the transition elements and this should be compared with Figure 2,[4] the corresponding plot of atomic diameters computed as the distances of closest approach. Even though some elements have crystal structures in which the atoms in the unit cell are not all equal in size, there is no prior knowledge that they retain this property when alloyed and hence we have used the average volume per atom in Figure 1 and Table I. Within each long period there is a progressive decrease and then increase of atomic diameters or volumes per atom. However, in the first long period, there are some detailed differences between Figures 1 and 2. From Ti to Cr the atomic volumes and diameters exhibit comparable decreases, but there is a slight increase in volume at α–Mn while the atomic diameter shows a distinct minimum. On passing from Mn to Cu the volumes per atom show a distinct minimum at Ni while the minimum in the closest-approach distance appears at Fe as shown in Figure 2. The significance of this difference will be made clear

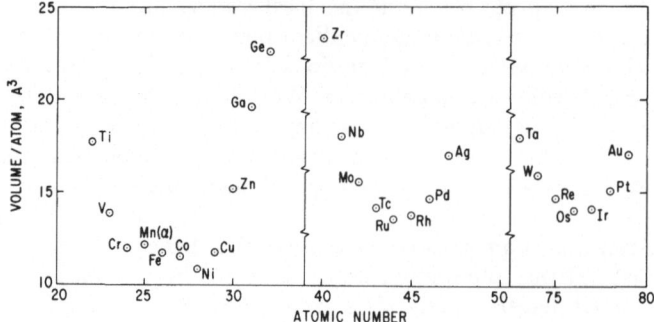

Figure 1. Volume per atom of the transition elements.

later. In the second long series, the atomic diameters and volumes diminish in the same fashion both reaching a minimum at Ru and then increasing. In the third long period, there is again a minimum in atomic volume and diameter at Group VIII A (Os). Thus it can be seen that the atomic diameters and volumes of the elements at the end of the second and third transition series follow different schemes from those in the first.

SIDE EFFECT CALCULATIONS

During the course of our studies of the lattice parameter relationships in iron-rich iron–vanadium and iron–germanium alloys, we proposed a simple scheme which helped explain the change of lattice constant with small amounts of V and Ge in Fe. Since this scheme appears to lay the basis for a physical picture of the size effect, we proceed to outline this concept. The lattice constant–composition curves for the terminal solid solutions of Fe–V and Fe–Ge observed in our studies are shown in Figures 3 and 4. We regard the crystalline lattice as a packing of atoms in their usual characteristic ways in a metal (i.e., body-centered cubic, close-packed hexagonal or cubic, etc.) and assume these spheres are hard incompressible cores each of volume V_h in contact with each other as defined by the unit cell. In such a structure a considerable amount of space in the unit cell is not occupied by the spheres, and we assume that this space is not empty but rather

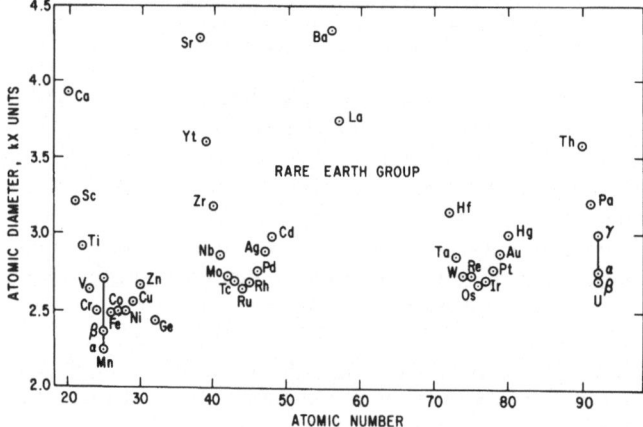

Figure 2. Atomic number.[4]

occupied by the compressible parts of the atoms, that is, a portion of volume V_s of each atom is smeared out through the space not filled by the hard spherical cores. It is necessary, then, to calculate the maximum proportion of the available volume in the unit cell which may be filled by hard spheres arranged on various structures. It can be shown that these volume fractions are:

Face-Centered Structure

$$w_{\text{fcc}} = 0.74$$

Body-Centered Structure

$$w_{\text{bcc}} = 0.68017$$

Diamond Structure

$$w_{\text{dia}} = 0.34008$$

An illustration of this scheme will be given for the iron–germanium system. The lattice parameter of germanium in the diamond lattice is $a_{\text{Ge}} = 5.6575$ Å at 298°K.[13,14] The unit cell of this structure contains 8 atoms (spheres). The total volume $(V_{\text{Ge,dia}})$ of a germanium atom in the diamond structure consists of a hard incompressible volume $(V_{h,\text{Ge}})$ and a smeared out volume $(V_{s,\text{Ge}})$. These volumes are

$$V_{h,\text{Ge,dia}} = \tfrac{1}{8}a_{\text{Ge}}{}^3 w_{\text{dia}} = 7.6978 \text{ Å}^3 \tag{1}$$

and

$$V_{s,\text{Ge,dia}} = \tfrac{1}{8}a_{\text{Ge}}{}^3(1 - w_{\text{dia}}) = 14.9374 \text{ Å}^3 \tag{2}$$

In the case of a dilute solution of germanium in iron, this latter volume must be compressed into a hypothetical body-centered cubic lattice with an available volume 0.32 of the unit cell volume as compared to an available volume 0.66 of the unit cell volume in the diamond structure. Thus, the volume $V_{s,\text{Ge,bcc}}$ of a germanium atom smeared out in the body-centered unit cell is

$$V_{s,\text{Ge,bcc}} = \frac{1 - w_{\text{bcc}}}{1 - w_{\text{dia}}} V_{s,\text{Ge,dia}} = 7.2393 \text{ Å}^3 \tag{3}$$

The above argument is supported somewhat by the relatively large compressibility of germanium (1.44×10^{-12} cm^2/dyne at room temperature).[15] Hence, the volume of a germanium atom in the body-centered cubic structure is

$$V_{\text{Ge,bcc}} = V_{h,\text{Ge,dia}} + V_{s,\text{Ge,bcc}} = 14.9371 \text{ Å}^3 \tag{4}$$

The volume of an iron atom in the body-centered structure with $a_{\text{Fe}} = 2.8663$ Å at 298°K is

$$V_{\text{Fe,bcc}} = \tfrac{1}{2}a_{\text{Fe}}{}^3 = 11.7743 \text{ Å}^3 \tag{5}$$

Now we can apply the analog of Vegard's law to calculate the atomic volume of iron–germanium solid solutions of the body-centered cubic structure,

$$V_{\text{Fe–Ge,bcc}} = (1 - x)V_{\text{Fe,bcc}} + xV_{\text{Ge,bcc}} \tag{6}$$

where x is the atomic fraction of germanium. For 1 a/o germanium this method gives the volume of 11.8060 Å^3 and the lattice constant, taken as a cube root of the volume, is 2.8689 Å. This value should be compared with the observed value 2.8684 Å obtained

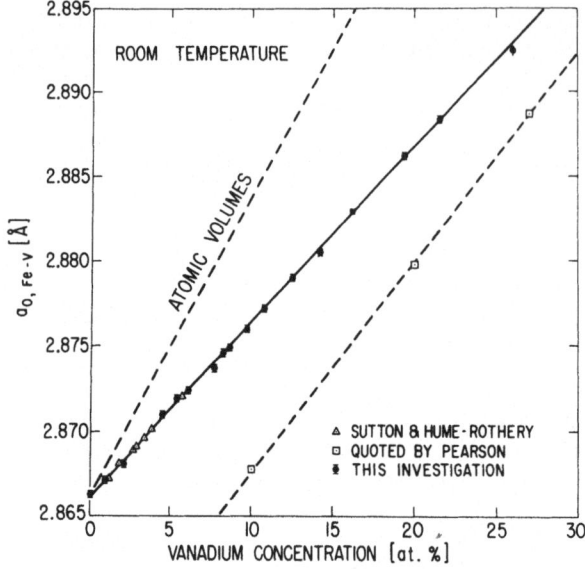

Figure 3. Lattice spacings of iron–vanadium alloys.

from Figure 4. We believe, in view of the agreement between observed and calculated values, that the above scheme satisfactorily describes the lattice distortion of the dilute solid solution of germanium in iron. When the constituents of a solid solution have the same lattice type, it is evident from the foregoing analysis that the application of the analog of Vegard's law should yield the same value of the lattice parameter regardless of the definition of atomic size chosen. This is shown in Figure 3 for the Fe–V system.

This scheme that we propose bears a resemblance to the model of distortion of a crystal lattice by a substitutional or interstitial atom as the dilation center in an infinite elastic isotropic continuum. Many authors have addressed themselves to this problem.[16–19]

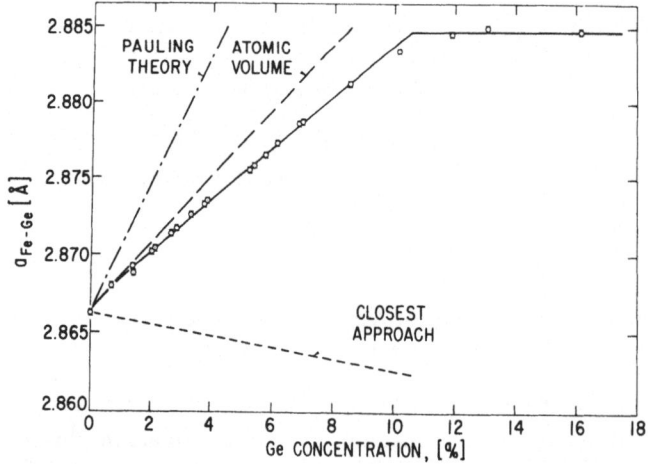

Figure 4. Lattice parameters of iron–germanium solid solutions at 298°K.

Of these we shall employ the treatment of Friedel which is valid for dilute solutions. Friedel regards the substitution of a solute atom as equivalent to removal of spherical hole from the matrix, substitution of a sphere of a different radius, and the resulting disturbed spherical medium reaching an equilibrium radius. This concept therefore defines the atomic radius or size as $\frac{4}{3}\pi r_i^3$ equal to atomic volume. Thus, the atomic size defined in the simple scheme above is precisely the same as that defined by Friedel. Applying classical elasticity theory, Friedel[19] arrives at the relationship

$$r_0 = r_1 + (r_2 - r_1)\left(\frac{\alpha + \chi_1/\chi_2}{\alpha + 1}\right)c \tag{7}$$

$$\alpha = \frac{(1 + \nu)\chi_1}{2(1 - 2\nu)\chi_2} \tag{8}$$

where

$$r_i = \left(\frac{\text{volume per atom}}{\frac{4}{3}\pi}\right)^{1/3}$$

$i = 1$ for the solvent; $i = 2$ for the solute; $i = 0$ for the solution; χ is the compressibility; c is the atom fraction solute; and ν is the Poisson ratio of the solvent.

It was felt that it would be a worthwhile venture to evaluate the size effect using these various approximations listed above. Since Pearson,[20] and Pearson and Thompson[21] show that the size effect appears to be the predominant factor in determining the lattice distortion at very low solute concentrations in terminal solid solutions of the first long period transition metals and similarly in Ni solid solutions, these systems were employed for the test. The lattice parameter changes in solid solutions of Ni are shown in Figure 5.[21] The top curve shows the lattice parameter changes expected initially applying Vegard's law to the distances of closest approach in the elemental structures while the middle curve

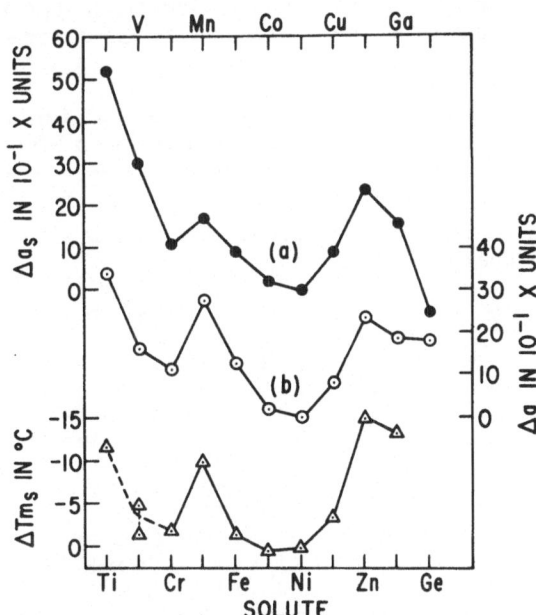

Figure 5. Initial change in lattice parameters. Nickel solid solutions (a) calculated (b) observed. (Pearson and Thompson.)

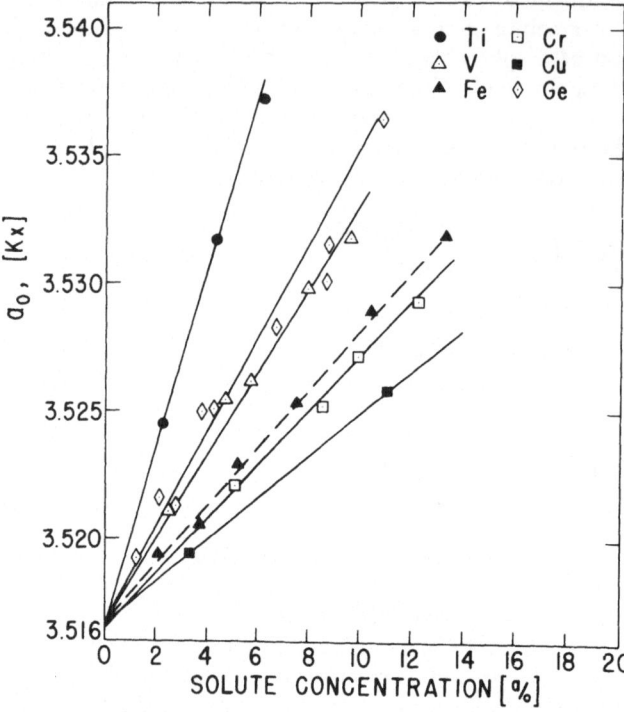

Figure 6. Nickel solid solutions.

is plotted from the observed data. It can be seen that the shapes of both curves are similar and, by and large, the magnitudes of the changes in lattice constants are similar. A closer examination of these solid solutions for the size effect relationships is therefore in order and will be conducted in this paper. It is significant, from an inspection of Figures 2 and 5 that the calculated data from the closest approach distances predict a decrease in lattice parameters in the Fe–Ge and Ni–Ge systems while the observed lattice parameters

Table II. Observed and Calculated Initial Change in Lattice Constants for Several Transition Metal Solid Solutions

Solvent	Solute	Δa, Å				
		Observed	Simple scheme	Friedel's relation	Goldschmidt radii	Pauling radii
Cr	V	1.5×10^{-2}	2×10^{-2}	2×10^{-2}		
Cr	Mo	2.6×10^{-3}	2.8×10^{-3}	2.9×10^{-3}		
Ni	V	1.65×10^{-3}	2.25×10^{-3}	2.9×10^{-3}	2.9×10^{-3}	4×10^{-3}
Ni	Ti	3.5×10^{-3}	3.5×10^{-3}	5.2×10^{-3}	5.6×10^{-3}	7×10^{-3}
Ni	Cr	7×10^{-4}	4×10^{-4}	11×10^{-4}	11×10^{-4}	Pauling's theory
Ni	Fe	9×10^{-4}	2×10^{-4}	8×10^{-4}	9×10^{-4}	not capable of
Ni	Cu	8×10^{-4}	9×10^{-4}	8×10^{-4}	10×10^{-4}	predicting these
Ni	Ge	2.0×10^{-3}	2.5×10^{-3}	6.2×10^{-3}	3.5×10^{-3}	small changes
Fe	Ge	2.1×10^{-3}	2.6×10^{-3}		3.0×10^{-3}	
Fe	V	9×10^{-4}	17×10^{-4}		17×10^{-4}	

Table III. Comparison of Calculated Initial Change in Lattice Constants for Several Transition Metal Solid Solutions

0 = Observed; 1 = Atomic volume (as proposed here); FR = Friedel's relation; GR = Goldschmidt radii; PR = Pauling radii

Solvent	Solute	$\Delta a_1 - \Delta a_0$, Å	$\Delta a_{FR} - \Delta a_0$, Å	$\Delta a_{GR} - \Delta a_0$, Å	$\Delta a_{PR} - \Delta a_0$, Å
Cr	V	5×10^{-3}	5×10^{-3}		
Cr	Mo	2×10^{-4}	3×10^{-4}		
Ni	V	6×10^{-4}	12×10^{-4}	12×10^{-4}	23×10^{-4}
Ni	Ti	0	17×10^{-4}	21×10^{-4}	35×10^{-4}
Ni	Cr	-3×10^{-4}	4×10^{-4}	4×10^{-4}	
Ni	Fe	-7×10^{-4}	-1×10^{-4}	0	
Ni	Cu	10^{-4}	0	2×10^{-4}	
Ni	Ge	5×10^{-4}	42×10^{-4}	15×10^{-4}	
Fe	Ge	5×10^{-4}		9×10^{-4}	
Fe	V	8×10^{-4}			

increase in both systems (Figures 4 and 6). We have already resolved this apparent anomaly for Fe–Ge when the atomic volume scheme is used and we shall show shortly that this is so for Ni–Ge.

Of the several methods which have been proposed to explain the lattice parameter size effect relations in solid solutions we have chosen four:

a. Our scheme based on the hard-core volume and compressible shell surrounding the hard core. This concept was explained using Fe–Ge as an example.

b. Goldschmidt atomic radii, which, on empirical grounds, applies a percentage correction to the distance of closest approach to evaluate the radii in a hypothetical structure having a coordination number 12.

c. Pauling radii.

d. Classical elastic continuum theory.

Following the suggestion outlined above for the Ni solid solutions we have first plotted the lattice parameter–composition curves of these solutions using the experimental data reported by Pearson.[22] These curves are shown in Figure 6. From these we evaluated the change in lattice parameter per unit concentration of solute (atomic percent) and then applied the methods (a) to (d) invoking the analog of Vegard's law for methods (a), (b), and (c). The results of this analysis are shown in Figure 7 and Table II; in Figure 8 and Table III the discrepancies between the observed and various calculated values are shown. The constituent radii r_1 and r_2 were taken from Table I and χ_1, χ_2 and ν were taken from Köster and Franz.[15]

Recently, Dr. Richmond of this Laboratory has examined the simplifying assumptions made by Friedel and has rederived the relationship for lattice parameters in solid solutions under more rigorous assumptions but employing the same physical picture and classical elasticity theory. The results of this work will be the subject of a separate publication; however, it is of interest here to report a calculation carried out. Since Dr. Richmond feels that his derivation should hold completely across the composition diagram and Ag is completely soluble in Pd and, furthermore, this alloy system satisfies certain conditions placed on the slope of lattice parameter *vs.* composition, it was decided to see how closely his relationship would match the experimental data. The results of this calculation are compared with the observed data in Table IV and are plotted in Figures

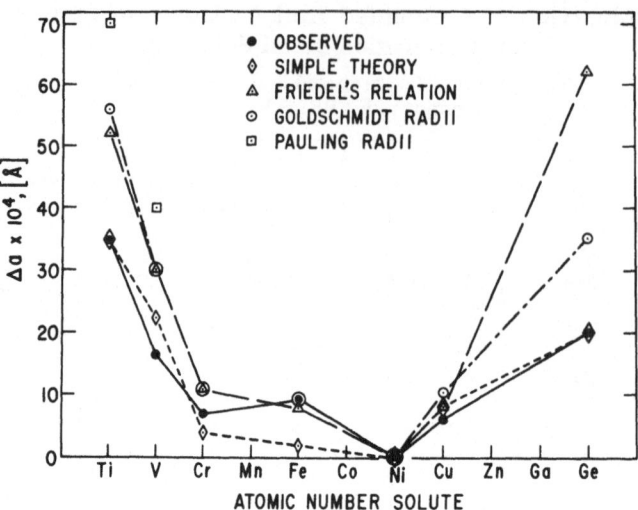

Figure 7. Nickel solid solutions.

9 and 10. It can be seen that the maximum discrepancy is about 1 part in 2000 while the average discrepancy is 1 part in 20,000. Coles[23] points out that other physical properties show that the $4d$ band holes are just filled at a silver content slightly greater than 60 a/o and that the change in slope reflects this change in electronic structure. Our analysis shows that there is a sharp change, and extremum in slope, of the lattice spacing–composition

Table IV. Lattice Parameters of the Ag–Pd System*

a/o Ag	a, kX	a, Å	a_{calc}, Å	Δa_{obs}, Å	Δa_{calc}, Å
0	3.8829	3.8907	3.8907	0	0
9.86	3.8991	3.9070	3.9066	0.0163	0.0159
19.02	3.9154	3.9233	3.9219	0.0326	0.0312
28.88	3.9324	3.9403	3.9390	0.0496	0.0483
33.60	3.9409	3.9489	3.9474	0.0582	0.0567
39.28	3.9512	3.9592	3.9578	0.0685	0.0671
44.00	3.9596	3.9676	3.9666	0.0769	0.0759
49.13	3.9686	3.9766	3.9763	0.0859	0.0856
53.80	3.9770	3.9850	3.9854	0.0943	0.0947
59.37	3.9881	3.9962	3.9964	0.1055	0.1057
69.43	4.0077	4.0158	4.0170	0.1251	0.1263
79.64	4.0296	4.0377	4.0388	0.1470	0.1481
90.92	4.0561	4.0643	4.0643	0.1736	0.1736
100.00	4.0773	4.0855	4.0855	0.1948	0.1948

* Observed lattice parameters taken from Coles.[23]

$$\Delta a = a_{Ag–Pd} - a_{Pd}$$

$$\sum \frac{|a_{obs} - a_{calc}|}{\Sigma |a_{obs}|} = 0.00021$$

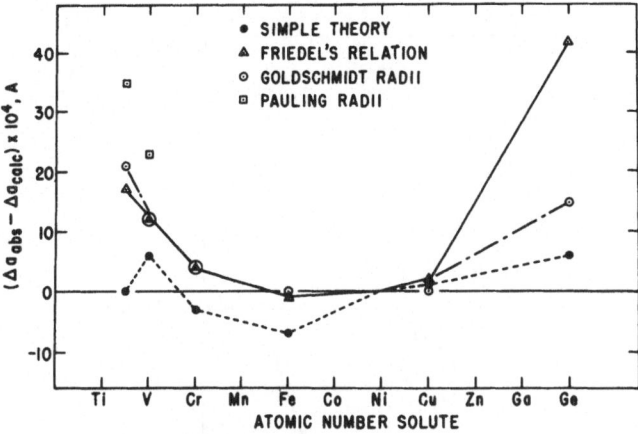

Figure 8. Nickel solid solutions.

curve and serves to illustrate the importance of evaluating the size effect in alloys in order to infer other physical properties from lattice parameter measurements.

CONCLUSIONS

An inspection of Figures 7 and 8 reveals that for the nickel solid solutions all methods for calculating the size effect predict the correct sign of the lattice constant change. Within the first long period and extending into the B subgroup elements, the poorest agreement

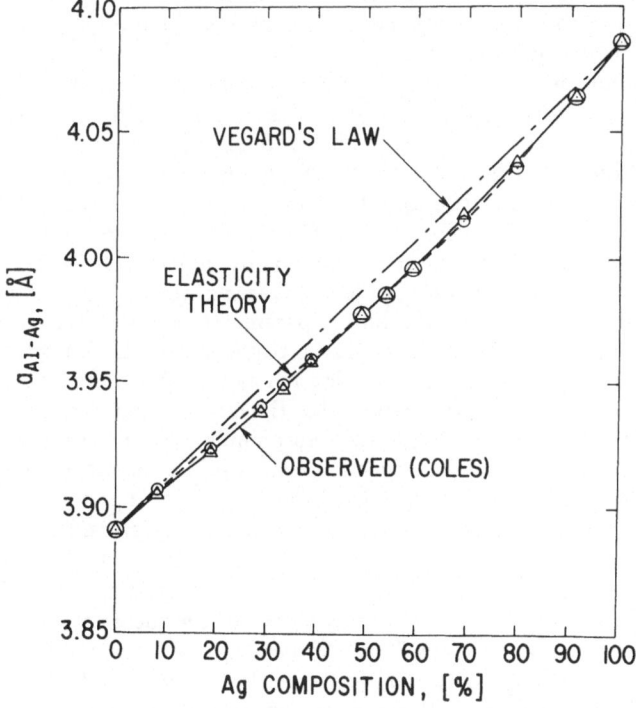

Figure 9. Lattice parameter-composition curve for Ag–Pd solid solutions.

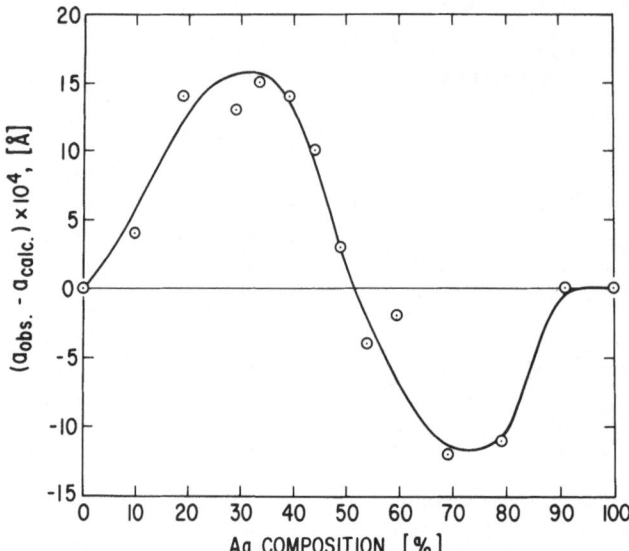

Figure 10. $a_{obs} - a_{calc}$ for Ag–Pd solid solutions.

is at the beginning of the long period and at the Ni–Ge end of the diagram shown, except when the simple volume scheme is employed. The reasons for this are not clear. Either the concept of a compressible shell surrounding a hard incompressible core merits further attention or, in the case of Ge, it may well be, as Raynor[10] has shown, that for Ge in the noble metals the valency effect is important. Yet the lattice spacing of Ge in Fe appears to be satisfactorily explained as we have demonstrated in the example of our atomic volume scheme. The larger discrepancy of Ge in Ni may be an experimental one and Pearson and Thompson[21] do not exclude this possibility when they imply some uncertainties in metallurgical and chemical analyses. Nevertheless, of all the methods employed here to evaluate the size effect, that of our simple scheme best fits the experimental data and this statement holds for the Ni–Ge system as well.

It can be seen from an inspection of Figure 6 that the increase of lattice constants is of the order Ti > Ge > V > Fe > Cr > Cu and this same general order is followed in the calculated values with the exception that the predicted increase for Fe in Ni is less than that of Cr in Ni when the compressible atoms are permitted, contrary to experimental fact. It is instructive to compare the lattice parameter–composition curves for the Fe solid solutions taken from Sutton and Hume-Rothery for the corresponding solutes shown in Figure 11. While a careful analysis such as that above for Ni solutions has not yet been carried out it is interesting to note that the lattice expansions of Cr and the corresponding Group VIII element (Ni) show a close similarity in both systems. The general scheme pointed out above is followed for the Fe solid solutions but there are some differences which require further attention beyond the scope of this paper.

It must be made clear that this paper considers only the size effect in lattice parameter relationships for solid solutions and that the other factors mentioned earlier may play a role here. To what extent these other factors are responsible for the experimental observations cannot be assessed at this stage of the theoretical developments and references such as Massalski and King[24] or Sivertsen and Nicholson[25] should be consulted for a discussion of electronegativity and of valence electron factors. It appears though that the predominant influence in the solutions studied here is the size effect and that further investigations should help in explaining the role of the electronic states. In particular,

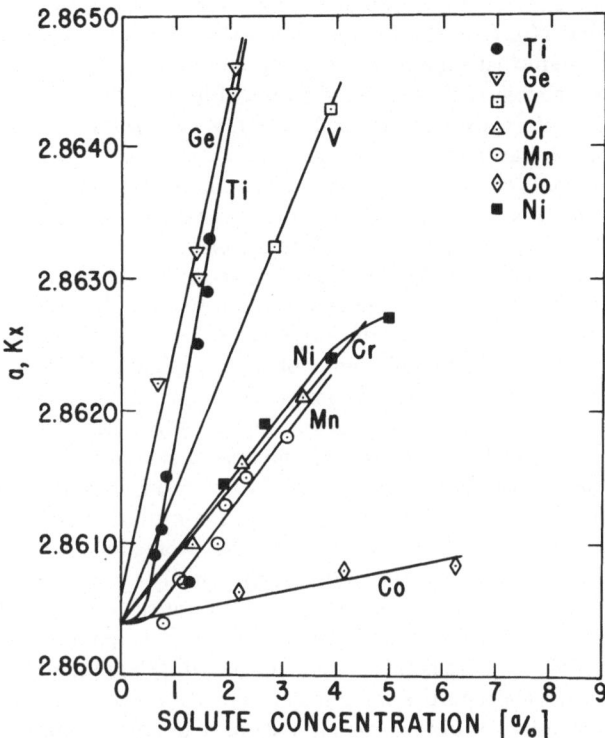

Figure 11. Iron solid solutions.

as Sutton and Hume-Rothery[27] point out, the sequence Cr to Cu in Fe and, to a lesser extent in Ni is due to an interplay of many factors including those which are electronic in nature. Whether or not this is due to some defects yet remaining in the theory of the size effects awaits further study and it is anticipated that this will be discussed in a future report. It should be mentioned in passing that the experimental uncertainties may be larger than most workers indicate. There are, for example, significant differences in the lattice parameters reported for V, Co, and Ni among others. Experimentalists in X-ray diffraction tend to exaggerate the accuracy of their lattice parameter measurements and it is hoped that the recent worldwide study on this subject carried out by the International Union of Crystallography[26] will have a restraining influence on these tendencies.

It is significant that Friedel's relationship predicts a much larger lattice distortion than those observed and this poor quantitative agreement has been noted by Friedel himself. Perhaps the most interesting aspect of this part of the paper is that the best quantitative agreement between calculated and observed lattice parameters is that due to the simple scheme outlined above to explain the Fe–Ge spacings. It suggests that Friedel's development may be profitably applied to only part of the atom which we have designated as smeared out and that an elastically incompressible core may be associated with each atom.

It will be recalled, as shown in Figures 1 and 2, that relative sizes of Mn were in doubt because of the different definitions of atomic size applied in these figures. If the distances of closest approach are used (Figure 2) then α–Mn appears to be considerably smaller than all the other elements while if the volume of the unit cell per atom is used (Figure 1), the Mn appears to be larger than the elements Cr to Cu. The experimental observation that Mn expands the lattice of Ni (Figure 5) and Fe (Figure 11) as well as

Cr and Co indicates that Figure 1 may be more realistic than Figure 2. However, Sutton and Hume-Rothery[27] imply from the way that they plot Vegard's law for the Fe–Mn system that they regard the Mn atom to be larger than the Fe atom although this is not consistent with the plot shown in Figure 2. The lattice parameter–composition curve of Fe–Mn shown in Figure 11 has a very small slope initially. Sutton and Hume-Rothery state that they do not understand this abnormality. If we consider only the linear range, there is close agreement with our scheme. We have found a similar effect in Fe–V alloys, but succeeded in eliminating it by a change in heat treatment.[3]

The Ag–Pd system offers the clearest example that the elastic continuum theory may be capable of predicting the size effect. The deviations of the observed spacings from those calculated occur at the composition where other physical properties show a change.[28,29] It is significant that the slope of the lattice spacings calculated is equivalent to that observed at the high–Pd ends of the composition axis.

There are several points which should be mentioned by way of clarifying several yet unsolved problems regarding the physical picture of the size effect presented here. The anisotropy inherent in the crystal structures and elastic properties, especially in the hexagonal close-packed alloys is not taken into acccount. Implicit in the derivations of Friedel and of Richmond is the assumption that the atom is isotropic and this is also averaged out in our Table I. Yet the atoms in some elemental structures such as Ti, Co, Zn, Mn, etc., are possibly not spherically symmetrical. This should be evident from the observation that these structures do not have true twelvefold coordination but have six nearest neighbors and six next-nearest neighbors. Another objection to the theories employed here is that it may not be valid to assume that the atoms retain their individual physical properties as elements when they form a solid solution. It may be that the present formulations of the problem may have to be recast in terms of a resultant compressibility* of the solid solution rather than, as is presently done, in terms of the compressibilities of the constituent elements and Poisson's ratio.[30]

The best agreement between observed and calculated data has so far been achieved when our volume scheme is employed. Even though the calculations based on elasticity theory show larger discrepancies, this theory is more attractive and is related to other physical properties of the crystalline solid. Therefore we feel that our volume scheme should serve as an indication of the boundary conditions to be applied to the theory. Further investigations are proceeding on this basis.

ACKNOWLEDGMENTS

The authors are grateful to Prof. C. S. Barrett and Dr. O. Richmond for many profitable discussions. They acknowledge with gratitude the technical assistance of Mr. G. R. Dunmyre, Mr. R. C. Peck, and Mr. J. M. Peck, the analytical work of Mr. G. W. Momeyer and Mr. B. W. Conroy, and the preparation of the high-purity alloys by Dr. B. F. Oliver and Mr. E. W. Troy; all staff members of this Laboratory.

REFERENCES

1. Sigurds Arajs, R. V. Colvin, Henry Chessin, and J. M. Peck, *J. Appl. Phys.* **32**: 857, 1961.
2. Henry Chessin, Sigurds Arajs, R. V. Colvin, and D. S. Miller, *J. of Phys. and Chem. of Solids* (submitted for publication).
3. Henry Chessin and Sigurds Arajs (to be published).

* As this paper went to press, we received an unpublished paper by K. A. Gschneidner, Jr. and G. H. Vineyard in which a size effect theory is proposed based on elasticity theory, avoiding the artificial assumption of individual compressibilities.

4. W. Hume-Rothery and B. R. Coles, *Advances in Physics, Vol. 3*, 1954, p. 150.
5. P. A. Flinn, B. L. Averbach, and M. Cohen, *Acta Met.* **1**: 664, 1953.
6. B. L. Averbach, P. A. Flinn, and M. Cohen, *Acta Met.* **2**: 92, 1954.
7. B. L. Averbach, "Theory of Alloy Phases," *Am. Soc. Metals*, Cleveland, 1956.
8. B. G. Warren and B. L. Averbach, "Modern Research Techniques in Physical Metallurgy," *Am. Soc. Metals*, Cleveland, 1953.
9. W. Hume-Rothery and G. V. Raynor, "The Structure of Metals and Alloys," *Inst. Metals* (*London*), 1954.
10. G. V. Raynor, *Trans. Faraday Soc.* **45**: 698, 1949.
11. V. M. Goldschmidt, *Z. Phys. Chem.* **133**: 397, 1928.
12. L. Pauling, *J. Am. Chem. Soc.* **69**: 542, 1947.
13. H. E. Swanson and E. Tatge, *Nat. Bur. Standards Circ. 539* **1**: 18, 1953.
14. M. E. Straumanis and E. Z. Aka, *J. Appl. Phys.* **23**: 330, 1952.
15. W. Köster and H. Franz, *Metals Rev.* **6**: 1, 1961.
16. J. D. Eshelby, in *Solid State Physics, Vol. 3*, Academic Press, Inc., New York, 1956, p. 79; *J. Appl. Phys.* **25**: 255, 1954.
17. P. H. Miller and B. R. Russell, *J. Appl. Phys.* **23**: 1163, 1952.
18. F. Seitz, *Rev. Mod. Phys.* **18**: 384, 1946.
19. J. Friedel, *Phil. Mag.* **46**: 514, 1955; *Advances in Physics, Vol. 3*, 1954, p. 446.
20. W. B. Pearson, *Can. J. Phys.* **35**: 358, 1957.
21. W. B. Pearson and L. T. Thompson, *Can. J. Phys.* **35**: 349, 1957.
22. W. B. Pearson, *Lattice Spacings and Structures of Metals and Alloys*, Pergamon Press, Inc., New York, 1958.
23. B. R. Coles, *J. Inst. Metals* **84**: 346, 1955–1956.
24. T. B. Massalski and H. W. King, *Progress in Materials Science* **10** (1): 1961.
25. J. M. Sivertsen and M. E. Nicholson, *Progress in Materials Science* **9** (5): 1961.
26. W. Parrish, *Acta Cryst.* **13**: 838, 1960.
27. A. L. Sutton and W. Hume-Rothery, *Phil. Mag.* **46**: 1295, 1955.
28. B. R. Coles, *Proc. Phys. Soc.* (*London*) B**65**: 221, 1952.
29. J. C. Taylor and B. R. Coles, *Phys. Rev.* **102**: 57, 1956.
30. Z. Hashin, *Bull. Research Council Israel, Sec. C***5**: 46, 1955.

DISCUSSION

H. M. Otte (RIAS): Did you try Eshelby's approach to this problem?

Henry Chessin: No. We are not certain that Eshelby's formulation of this problem is very different from Friedel's.

M. Semchyshen (Climax Molybdenum Company of Michigan): How do you take into account the fact that the material is not isotropic, but anisotropic? How do you take into account Poisson's ratio, etc?

Henry Chessin: We don't take anisotropy into account and, as far as we know, this has not been looked at in a formal way. This should be kept in mind. Poisson's ratio for the solvent shows up in the equation for atomic radius using elasticity theory.

X-RAY MEASUREMENT OF THE STATIC LATTICE DISTORTION IN THE SOLID SOLUTION OF OXYGEN IN TITANIUM*

F. R. L. Schoening and F. Witt

The Franklin Institute Laboratories
Philadelphia, Pennsylvania

ABSTRACT

Oxygen was introduced into a single crystal of titanium in successive stages. The intensities of the $h00$ and $00l$ reflections were measured with a single-crystal diffractometer. The observed variation of the intensities with oxygen concentration was attributed to three factors: (1) the additional scattering from the oxygen atoms, (2) a change in the Debye–Waller factor, and (3) an exponential factor originating from the distortion around the oxygen atom. The theory of X-ray scattering from crystals containing centers of distortion was applied to the hexagonal titanium containing interstitial oxygen atoms. Using the variation of the lattice constant with oxygen concentration, it was possible to predict the intensity reduction due to lattice strains. It was concluded that it would have been possible to obtain an estimate of the defect concentration from the X-ray measurements of lattice expansion and intensity reduction.

INTRODUCTION

It has been shown by Huang[1] that random static displacements of atoms in a crystal influence the diffraction of X-rays as follows: (1) the Laue–Bragg intensity is reduced by an exponential factor; (2) the reflections are shifted on the 2θ scale, indicating expansion or contraction of the average lattice; and (3) diffuse scattering appears in the vicinity of the reciprocal lattice points. This paper is concerned with effects (1) and (2) as they apply to the distortion of the titanium lattice by interstitial oxygen atoms. The experimentally obtained reduction of the intensities as a function of defect concentration (atomic percent oxygen) is compared to a theoretical prediction based on an application of existing treatments to the case of hexagonal titanium. If several assumptions are made, an expression for the intensity reduction can be obtained which uses the observed lattice expansion resulting from a known amount of interstitial oxygen atoms. The detailed calculations and a full description of the experiment will be published elsewhere.[2]

THEORY

The change of the Laue–Bragg intensity at a reciprocal lattice point
$$\mathbf{H}[|F_M(\mathbf{H})|^2 - |F(\mathbf{H})|^2]/|F(\mathbf{H})|^2,$$
is a function of the average structure factor $F_M(\mathbf{H})$, the average being taken over all unit

---

* This research was supported by the U.S. Atomic Energy Commission under Contract No. AT(30–1)–2585.
[1] Superscripts pertain to references at the end of the paper.

cells in the crystal. If the fact that the atoms surrounding the defect are displaced is taken into consideration, the structure factor for a particular unit cell is found to be

$$\sum_n f_n \exp\{2\pi i H \cdot (r_n + \delta_n)\}$$

where r_n is the position vector of the nth atom in the cell in the perfect crystal and δ_n is the displacement of that atom from the perfect lattice site. The average structure factor for the whole crystal

$$F_M(H) = \sum_n f_n \exp\{2\pi i H \cdot r_n\} \overline{\exp\{2\pi i H \cdot \delta_n\}} \tag{1}$$

can be simplified by assuming that the displacement of the atoms is the same for all atoms in the unit cell. This is strictly true for monatomic cells such as can be used to describe fcc metals but is an approximation for hcp metals which have two atoms in the cell. Because the displacement is centrosymmetric with respect to the interstitial atom, the average structure factor becomes $F(H) \exp\{- 2\pi^2 \overline{(\delta \cdot H)^2}\}$, where the average is taken over the symmetry-independent cells, i.e., half the number of cells in the crystal.

The displacement vector δ has been calculated for a model which is suggested by the observation that, when interstitial oxygen is introduced, the titanium lattice expands almost exclusively in the c direction. It is assumed that: (1) the crystal can be approximated by an elastic continuum having the elastic properties of titanium, (2) the interaction of the interstitial atom with the crystal lattice can be represented by a pair of equal and opposite forces (a doublet nucleus of force) acting in the elastic continuum along the direction equivalent to the hexagonal axis and normal to the surface of a small sphere cut out at the position where the interstitial atom is situated in the crystal, and (3) only the displacement in the direction of the hexagonal c axis is large enough to have an observable influence on the X-ray intensities.

For a force doublet acting at the origin in an infinite medium the z component, δ_z, of the displacement can be obtained from the solutions given by Elliott.[3] It is

$$\delta_z = - p[(v_1 S_{44} + \rho)(1 + v_1 R^2/z^2)^{-3/2} - (v_2 S_{44} + \rho)(1 + v_2 R^2/z^2)^{-3/2}]/4\pi z^2(v_1 - v_2) \tag{2}$$

where ρ has been substituted for the expression of elastic moduli

$$(S_{12} - S_{11})[2S_{13}^2 - S_{33}(S_{11} + S_{12})]/(S_{13}^2 - S_{11}S_{33})$$

The strength of the force doublet is expressed by p, which is equal to $2p'h$ where $\pm p'$ are two forces acting normal to the sphere of radius h cut out around the origin. If the sphere is made to shrink, $2p'h$ is kept constant and equal to p. The position vector in the plane perpendicular to the c axis and going through the origin ($z = 0$) is denoted by R. The anisotropy of the medium is reflected by the parameters v_1 and v_2, which are the roots of the equation

$$v^2(S_{13}^2 - S_{11}S_{33}) - v[2S_{13}(S_{12} - S_{11}) - S_{11}S_{44}] + (S_{12}^2 - S_{11}^2) = 0 \tag{3}$$

For an isotropic medium, in which case $v_1 = v_2 = 1$, the displacement reduces to

$$(\delta_z)_{\text{isotropic}} = - [(4\sigma - 1)z(z^2 + R^2)^{-3/2} - 3z^3(z^2 + R^2)^{-5/2}]p/16\pi G(\sigma - 1) \tag{4}$$

which is identical to that quoted by van Bueren[4]; G is the shear modulus and σ is Poisson's ratio.

The displacement δ_z can be calculated if p, which is a measure of the acting force, and the elastic moduli are known. The force p can be evaluated from the volume expansion, which in turn is related to the change of the lattice constant c. Ignoring image forces at present, we find the volume change

$$\Delta V_\infty = \int \delta_z \cdot n dA = (\Delta c/c) V_c/q \tag{5}$$

where n is the normal to the sphere over which the surface integral is taken, V_c is the volume of the unit cell, and q is the number of defects per unit cell. The integration leads to

$$p = \left(\frac{\Delta c}{c}\right)\left(\frac{V_c}{q}\right)(v_2 - v_1)\Big\{(v_1 S_{44} + \rho)[(v_1 - 1)^{-1} - (v_2 - 1)^{-3/2}\sin^{-1}\{(v_1 - 1)/v_1\}^{1/2}]$$

$$- (v_2 S_{44} + \rho)[(v_2 - 1)^{-1} - (v_2 - 1)^{-3/2}\sin^{-1}\{(v_2 - 1)/v_2\}^{1/2}]\Big\}^{-1} \tag{6}$$

which can be evaluated if $\Delta c/c$, q, and the elastic moduli are known.

If the usual assumption is made that the defects act independently of each other, the average structure factor becomes

$$F_M(H) = F(H) \exp\{ - 4\pi^2 q l^2 (\Sigma \delta_z^2)/c^2\} \tag{7}$$

where $\Sigma \delta_z^2$ is the sum over the displacement in half the number of cells.

Except for a specialization to the case of titanium, the above expression is identical to that derived by Cochran and Kartha,[5] who write $\exp\{ - 2\pi^2 p N \overline{(\epsilon \cdot S)^2}\}$ for the distortion factor; here p is the atomic fraction of defects, N is the total number of atoms, ϵ is the displacement of the atoms from an infinite perfect lattice which is distorted by one center of pressure, and S is the diffraction vector. For the 00l reflections $\overline{(\epsilon \cdot S)^2} = \overline{\epsilon_z^2} l^2/c^2$, the average being taken over all atoms in the crystal. Furthermore, $2q \Sigma \delta_z^2 = pN \overline{\epsilon_z^2}$ because the sum $\Sigma \delta_z^2$ is taken over half the number of cells.

NUMERICAL RESULTS

Using the elastic constants measured by Dr. G. Bradfield* (private communication), the elastic moduli for titanium are:

$$S_{11} = 0.949 \times 10^{-12} \text{ cm}^2/\text{dyne}$$
$$S_{33} = 0.667 \times 10^{-12} \text{ cm}^2/\text{dyne}$$
$$S_{44} = 2.146 \times 10^{-12} \text{ cm}^2/\text{dyne}$$
$$S_{12} = - 0.480 \times 10^{-12} \text{ cm}^2/\text{dyne}$$
$$S_{13} = - 0.162 \times 10^{-12} \text{ cm}^2/\text{dyne}$$

Hence, the roots $v_1 = 2.056$ and $v_2 = 0.537$ yield $p^2 = \Delta V_\infty^2\, 8.858 \times 10^{24}$ (dyne-cm)^2 and

$$F_M(H) = F(H) \exp\{ - 3.496(\Delta c/c)^2 l^2/q\} \tag{8}$$

The sum of the square of the displacements $\Sigma \delta_z^2$ was calculated with a UNIVAC Computer by summing an increasing number of cells. The partial sums were then plotted *versus* a parameter proportional to the number of cells included in the sum. By a graphical extrapolation, the limit for an infinite number of cells was estimated.

* National Physical Laboratory, Teddington, Middlesex, England.

These results will later be compared with experimental results obtained for 00l reflections from a single crystal. For this purpose it is convenient to write

$$F_M(00l) = F(00l) \exp\{- K_D(\sin^2\theta)/\lambda^2\} \tag{9}$$

where $K_D = 306(\Delta c/c)^2/q$.

If one does not assume an infinite medium, image forces have to be taken into account. As shown by Eshelby,[6] $\Delta V = \gamma\Delta V^\infty$, where the volume expansion ΔV is related to the volume expansion of an infinite crystal ΔV^∞ by the image force constant γ, which, for the special case of a homogeneous isotropic metal has the value 1.5. For titanium, a rough correction for image forces can be applied by letting γ equal 1.41 and dividing K_D by γ^2.

EXPERIMENTAL RESULTS

The experimental results given below were obtained from a single crystal of titanium approximately 0.18 mm × 0.47 mm × 0.78 mm, cut from a large grain in a zone-refined rod of iodide titanium. The orientation of the single crystal was such that the c axis and an a axis were in a plane perpendicular to the large dimension of the specimen. By rotating the crystal around this direction, the h00 and 00l reflections were measured in the equatorial plane of a conventional diffractometer.

Filtered molybdenum radiation and the "stationary counter, moving crystal" technique were used. Care was taken to include the whole spectral range of the reflections by using a sufficiently large aperture in front of the counter. The crystal was rotated through its reflection position with a speed of 5° per hour, and the number of counts were measured with a pulse-height analyzer and scaler. Instrumental fluctuations were reduced by recording the 100 reflection after every other reflection and working only with the ratio of measured intensity to 100 intensity. The intensities were corrected for background radiation, absorption, and Lorentz polarization effects. Every reflection pair, e.g., h00 and $\bar{h}$00, was measured at least once for every oxygen concentration.

Oxygen was introduced into the crystal by keeping it at 800°C for 16 hours in an atmosphere of oxygen at pressures ranging from 5 to 60 torr. After the oxidation run, the specimen tube was cooled in air to room temperature. At the higher oxygen concentrations, it became necessary to homogenize the specimen in vacuum at 800°C for several days. The oxygen concentration in the crystal was determined from the c parameters

Table I

Oxygen, at. %	K, h00 reflections	K, 00l reflections
0	0.761 ± 0.078	0.543 ± 0.044
2.2	0.842 ± 0.074	0.545 ± 0.037
4.2	0.739 ± 0.062	0.618 ± 0.026
8.6	0.717 ± 0.060	0.644 ± 0.034
10.7	0.656 ± 0.070	0.645 ± 0.042
17.1	0.752 ± 0.079	0.650 ± 0.048
20.0	0.749 ± 0.101	0.655 ± 0.048
24.5	0.705 ± 0.083	0.671 ± 0.054
27.2	0.690 ± 0.073	0.912 ± 0.055

Magneli *et al.*[7] obtained from the 008, 00$\bar{8}$ and 0010, 00$\bar{10}$ reflections measured at both sides of the direct beam.

The structure factors were calculated using the atomic scattering factors published by Qurashi[8] for titanium and by Freeman[9] for oxygen. The titanium values were corrected for dispersion. The left-hand side of the expression

$$\ln\left\{\frac{F_{\text{obs}}(hkl)/F_{\text{obs}}(100)}{F_{\text{calc}}(hkl)}\right\} = -K\frac{\sin^2\theta}{\lambda^2} + \text{const} \tag{10}$$

used for interpreting the experimental data is a linear function of $(\sin^2\theta)/\lambda^2$ with the slope $K = K_T + K_D$, where K_T and K_D refer to the temperature and distortion effects, respectively. The numerical values for K have been determined by least-square analysis and are given in Table I. They can be compared with the values calculated in the preceding section.

REMARKS

For 0 at. % oxygen the measured K values, $K(h00) = 0.761 \pm 0.078$ and $K(00l) = 0.543 \pm 0.044$, are related to the Debye–Waller factors $\exp\{-K_T(\sin^2\theta)/\lambda^2\}$, parallel and perpendicular to the c axis. For higher oxygen concentrations $K(h00)$ should reflect the change of the Debye temperature and $K(00l)$ should, in addition, increase if static lattice distortions of the kind calculated in earlier sections are present. Disregarding the value for 2.2 at. %, the $K(h00)$ values seem to indicate that the Debye–Waller factor does not change appreciably with oxygen concentration, although an increase between 0 and 10.7 at. % followed by a decrease would be consistent with the observations. The initial increase, suggesting an increase in the Debye temperature, could be compared to the increase in hardness observed for small oxygen concentrations (see McQuillan and McQuillan[10]).

The $K(00l)$ values increase between 0 and 27.2 at. % oxygen. If it is assumed that the Debye–Waller factor does not change in that region, the increase can be attributed to the presence of static lattice distortions. Should the Debye–Waller factor vary, it could either increase, as does the factor for the $h00$ reflections, or decrease. In the former case the contribution of the static lattice distortions would be even larger; in the latter case they would be smaller or absent. The following interpretation is based on the assumption that the observed increase of $K(00l)$ is due to static lattice distortions.

This interpretation can be checked by comparing the observed increase with the changes calculated in the sections on Theory and Numerical Results. Qualitatively, the calculations predict, because they have been based on a particular model, that only the $00l$ reflections would be affected by distortions. Figure 1 shows a quantitative comparison between the calculated values neglecting image forces, the calculated values using a crude estimate of the image forces, and the observed data. The agreement between the calculated and observed variation of K appears to be sufficient to substantiate the interpretation in terms of static lattice distortions.

In the calculation of the K_D values the experimentally determined dependence of the lattice parameter on oxygen concentration has been used. The nonlinear dependence of the lattice parameter on q leads to the nonlinear K_D curves in Figure 1. This non-linearity can be interpreted as a decrease of the strength p of the defect caused by an interaction between defects.

The large K observed for 27.2 at. % oxygen has to be considered separately because the presence of 001 and 003 reflections indicates that ordering of the oxygen atoms has taken place. Such ordering was not observed for the lower concentrations.

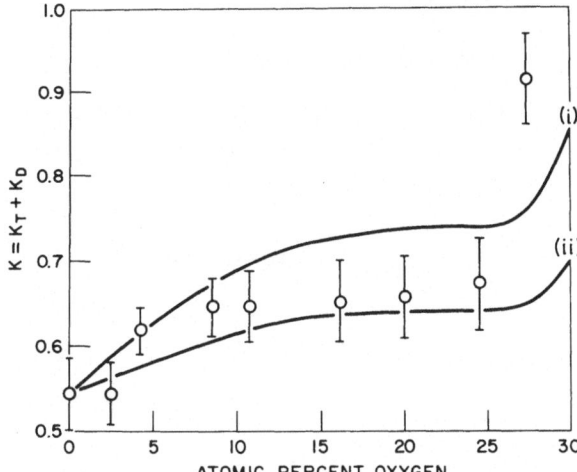

Figure 1. Observed values for K and their standard deviations are plotted against oxygen concentration. The straight lines are calculated assuming (i) no image forces and (ii) a crude estimate of image forces.

It may be concluded that the introduction of interstitial oxygen atoms into the titanium lattice leads predominantly to a displacement of titanium atoms in the c direction. These static distortions reduce the intensities of the $00l$ reflections by a factor which can be predicted from the observed lattice expansion and defect concentration by making rather simplified calculations. Therefore, it would have been possible to deduce the defect concentration from the experimental observations of volume expansion and intensity reduction. This is of interest in cases where the interstitial atoms are the same kind as those of the host material and direct analytical methods for the determination of the defect concentration are not possible.

REFERENCES

1. K. Huang, "X-ray Reflections from Dilute Solid Solutions," *Proc. Roy. Soc. (London)* *A***190**: 102, 1947.
2. Report for U.S. Atomic Energy Commission, Contract No. AT(30-1)-2585 (to be published).
3. H. A. Elliott, "Three-Dimensional Stress Distributions in Hexagonal Aeolotropic Crystals," *Proc. Cambridge Phil. Soc.* **44**: 522, 1948.
4. H. G. van Bueren, *Imperfections in Crystals*, North-Holland Publishing Company, Amsterdam, 1960, p. 92.
5. W. Cochran and G. Kartha, "Scattering of X-rays by Defect Structures. III. The Effect of Interstitial Atoms and Vacancies," *Acta Cryst.* **9**: 944, 1956.
6. J. D. Eshelby, "Distortion of a Crystal by Point Imperfections," *J. Appl. Phys.* **25**: 255, 1954.
7. S. Anderson, B. Collen, U. Kuylenstierna, and A. Magneli, "Phase Analysis Studies on the Titanium–Oxygen System," *Acta. Chem. Scand.* **11**: 1641, 1957.
8. M. M. Qurashi, "On the Completion and Extension of the Table of Atomic Scattering Factors Published by Viervoll and Ogrim," *Acta Cryst.* **7**: 310, 1954.
9. A. J. Freeman, "Atomic Scattering Factors of Spherical and Aspherical Charge Distributions," *Acta Cryst.* **12**: 261, 1959.
10. A. D. McQuillan and M. K. McQuillan, *Titanium*, Academic Press, Inc., New York, 1956.

PRECISION X-RAY DIFFRACTOMETRY USING POWDER SPECIMENS*

L. F. Vassamillet

Mellon Institute, Pittsburgh, Pennsylvania

and

H. W. King†

Imperial College, London, England

ABSTRACT

The counter tube diffractometer method for determining d spacings is often rejected for precision work because of lack of information concerning the nature and significance of the inherent errors. Errors concerned with the geometry of the method, the nature of the X-ray source, and the technique of collecting data have all been analyzed previously in some detail. The findings of these analyses, which are scattered throughout the literature, are reviewed briefly. Errors arising from imperfections in the instrument and misalignment of the X-ray source with respect to the diffractometer have been studied experimentally. The results are presented and discussed in terms of the resultant error in the determination of the lattice parameter of a cubic crystal. Errors determined both analytically and empirically are discussed in relation to the extrapolation procedures commonly used for diffractometers. It is shown that, depending on the construction of the instrument, the effect of imperfections in the gears may almost double the error in the final extrapolated value of a lattice parameter.

INTRODUCTION

While it has been shown that counter tube diffractometers are capable of measuring d spacings with an accuracy comparable to that of the Debye–Scherrer film method,[1] the facts indicate that relatively little use is as yet being made of diffractometers for this purpose.[2] Perhaps one of the major reasons for this state of affairs is that, while the various errors present in the film technique have been discussed collectively in several reviews, books, etc., treatments of the corresponding problems in diffractometry are either scattered throughout the literature or else are still in the stage of being analyzed. It is thus not surprising that the diffractometer method is often regarded with some suspicion for precision work. At the other extreme, an inadequate understanding of the errors and limitations of the diffractometer method often leads the less experienced diffractionist to place a blind faith in the numbers which are all too easily read off the instrument dials. The present aim is to dispel both of these false notions by drawing attention to the pertinent references in the literature and reporting further investigations of errors not previously considered, it being assumed throughout as a basis of discussion that the lattice parameter of a cubic crystal is required to an accuracy of 1 part in 20,000.

---

* This work was supported in part by the U.S. Atomic Energy Commission, Washington, D.C.
† Former Fellow of the Mellon Institute, Pittsburgh 13, Pennsylvania.
[1] Superscripts pertain to references at the end of the paper.

The various sources of error which enter into the determination of d spacings using the diffractometer method in accordance with the Bragg equation,

$$\lambda = 2d \sin \theta \tag{1}$$

may be conveniently grouped as follows:

1. Errors due to physical effects
    a. Errors in wavelength
    b. Dispersion, Lorentz and polarization
    c. Filters and monochromators
    d. Refraction
2. Errors arising from geometry
    a. Flat specimen and horizontal divergence
    b. Axial or vertical divergence
    c. Absorption
    d. Width of source and slit
3. Errors inherent in the instrument
    a. Errors in 2θ
    b. Errors in 2 : 1 following
    c. Backlash in gears
4. Errors associated with alignment
    a. Focal spot displaced from 180° 2θ
    b. Zero degrees 2θ off the beam path
    c. Specimen surface not coincident with diffractometer axis
    d. X-ray beam not in the diffractometer plane
    e. Mis-setting of $2\theta : \theta$ at 0° 2θ
5. Direct errors in the measurement of 2θ
    a. Accuracy required in $\Delta 2\theta$
    b. Peak position *vs.* center of gravity
    c. Electronic time constant
    d. Removal of backlash errors
6. Errors involved in extrapolation procedures
    a. The significance of errors when using $\cos^2\theta$ as extrapolation function
    b. Effect of dispersion, Lorenz and polarization
    c. Typical extrapolation plots against $\cos^2\theta$

ERRORS DUE TO PHYSICAL EFFECTS

Errors in the conversion of wavelengths from kX to Å units have been discussed by Lonsdale,[3] who has listed the "characteristic" wavelengths for various target materials. These so-called characteristic radiations are in fact distributed asymmetrically over a small band of wavelengths so that the peak value of the spectral distribution may differ significantly from its center of gravity (c.g.) or any other weighted mean of the distribution,[4] causing a problem as to which value of the wavelength should be used in the Bragg equation. The problem may be overcome by using wavelengths determined from peak values when Bragg reflections are measured using peak position and c.g. values of wavelength for Bragg angles measured from the c.g. of diffraction profiles, etc., for the various methods of measuring diffraction profiles. This principle is not so easy to apply in practice, however, since in most cases the methods used to measure the generally accepted values of the characteristic radiations are not described in sufficient detail. This uncertainty in the absolute value of λ is important when accuracies greater than 1 part in 20,000 are required.[4]

Since the relationship between λ and θ in the Bragg equation is not linear, the spread in θ resulting from an asymmetrical spread in λ will vary according to the Bragg angle θ, becoming most pronounced in the back-reflection region.[5] This dispersion effect causes the c.g. of the diffraction profile to be shifted to higher values of θ. The c.g. is also shifted toward higher values because of a variation of the Lorentz polarization factor across the diffraction profile. These three effects have been analyzed by Pike[6] who gives the following expression for the shift in the c.g. of a diffraction profile.

$$\Delta 2\theta = 3V/\bar{\lambda} \tan^3 \bar{\theta} \tag{2}$$

where V is the mean square of the breadth of the profile and $\bar{\lambda}$ and $\bar{\theta}$ the wavelength and Bragg angle, respectively, as indicated by c.g. measurements. This error $\Delta 2\theta$ may be related to the relative changes in the lattice parameter a of a cubic crystal by substituting in the differential form of the Bragg equation ($\Delta\theta$ being expressed in radians),

$$\Delta a/a = \Delta d/d = -\cot\theta\,\Delta\theta \tag{3}$$

i.e.,

$$\Delta a/a = -3V/2\bar{\lambda} \tan^2 \bar{\theta} \tag{4}$$

It is evident from equation (4) that these errors cannot be removed by conventional extrapolation procedures to $\theta = 90°$ (or $2\theta = 180°$) and that any such procedure will in fact enhance the errors (see section on extrapolation). The shift in the peak position is very much smaller than that for the c.g. in equation (2), but is not amenable to a simple analysis.[6]

The spread in wavelength of the X-ray source can be reduced, but not completely eliminated, by the judicious use of filters, curved crystal monochromators and, if the detector is a proportional or scintillation counter, by the use of pulse height discrimination. Although filters and monochromators are effective in cutting down the spread in wavelengths, these devices also modify the wavelength distribution of the characteristic radiation thereby introducing a further source of error in the Bragg angle θ.[7,8] The latter effect may be reduced by placing all filters, monochromators, etc., in the *diffracted beam*.

An additional complication, frequently overlooked, is the dependence of the wavelength distribution on the waveform of the voltage applied to the X-ray tube. This has been pointed out by Furnas (personal communication). Fortunately, there is a simple remedy for this error. The precision measurements of wavelength peaks and distributions may be obtained using a highly stabilized dc voltage. The constant dc source is a new feature now being included in commercial X-ray power supplies.

The influence of refraction has been discussed by Wilson.[9] For the case of small, roughly spherical crystals such as those used in powder specimens the change in the direction of the X-rays on entering and leaving the crystals may cause either an increase or a decrease in the Bragg angle, the net effect being that the diffraction profile is broadened symmetrically about 2θ but not shifted in angle. An error does occur with respect to λ, however, because the wavelength of the radiation inside the crystal differs slightly from that in air. This error may be corrected by increasing the calculated value of d of the a parameter of a cubic crystal by an amount given by [9]

$$+\Delta d = 2.70 \times 10^{-6} \lambda^2 \rho d\,\Sigma Z/\Sigma A \tag{5}$$

or

$$+\Delta a = 4.48 \times 10^{-6} (\lambda a)^2 \Sigma Z \tag{6}$$

where λ is measured in Å, ρ is the density in g/cm^3, ΣZ is the sum of the atomic numbers and ΣA, the sum of the atomic weights of the atoms in the unit cell.

ERRORS ARISING FROM GEOMETRY

The errors in Group 2, which arise because of the geometry of the method, are amenable to analysis in terms of the c.g. of the diffraction profile. They have been discussed in considerable detail by Wilson,[1] Pike,[10] Klug and Alexander,[11] and Parrish and Wilson.[12] When these analyses of $\Delta\theta$ are substituted in equation (3) the relative change in the lattice parameter of a cubic crystal takes the form,

$$\frac{\Delta a}{a} = \cos^2\theta \left[\frac{1}{2\mu R} + \frac{\alpha^2 + \delta_1{}^2}{12 \sin^2\theta}\right] - \frac{\delta_2{}^2}{12 \sin^2\theta} \tag{7}$$

where θ is the Bragg angle, α the horizontal divergence angle, $\delta_{1,2}$ the effective axial divergence angles, μ the linear absorption coefficient of the sample, and R the radius of the diffraction circle. The first term inside the bracket is the contribution of the absorption of the specimen.[1] It should be noted that, in contrast to the Debye–Scherrer method, a highly absorbing specimen causes less error than one with a low absorption coefficient. The other term inside the bracket refers to the contribution from the horizontal divergence angle α^1 and that part of axial divergence error which varies as $\cos^2\theta$. The results of Pike's analysis[10] of axial divergence are used here and the effective vertical divergence angles δ_1 and δ_2 are given by,

$$\delta_1 = (h/R)\sqrt{2Q_1} \qquad \delta_2 = (h/R)\sqrt{Q_1 - Q_2} \tag{8}$$

where Q_1 and Q_2 are constants for a given Soller slit system, and $2h$ is the height of the irradiated portion of the sample. Q_1 and Q_2 are rather complicated functions of the aperture q of the Soller slits. Typical examples of q, Q_1, Q_2, δ_1, and δ_2 are given in Table I below. Rows A and B are applicable to the Norelco and Siemens instruments where $q \sim 1.20$; C and D apply to the G. E. Diffractometer where $q = 0.85$ for the medium resolution slits. As indicated in equation (7), the δ_2 part of the shift in c.g., because of axial divergence, does not have $\cos^2\theta$ dependence and thus cannot be eliminated by extrapolation against $\cos^2\theta$ as will be discussed later.

The general error due to flat specimen absorption and the divergence effects cause a much larger change in the asymmetry of the diffraction line than a shift in the peak position. Thus even though shifts in peak position are more difficult to analyze than shifts

Table I. Effective Axial Divergence Angles for Some Conventional Diffractometers

No. Soller Slits		Q_1	Q_2	δ_1, deg	δ_2, deg	$\delta_2{}^2/12 \times 10^{-5}$
$q = 1.20$	$h = \frac{1}{2}$ cm	$R = 17$ cm*				
A	One set	1.24	0.29	2.65	1.65	6.92
B	Two sets	0.59	0.09	1.8	1.2	3.67
Med. Res.	$q = 0.85$	$h = \frac{1}{2}$ cm	$R = 14.6$ cm†			
C	One set	1.07	0.15	2.9	1.9	9.15
D	Two sets	0.37	0.026	1.55	1.05	2.8

* Siemens–Halske and Norelco $\qquad \delta_1 = (h/R)\sqrt{2Q_1} \qquad \delta_2 = (h/R)\sqrt{Q_1 - Q_2}$
† General Electric $\qquad\qquad\qquad q = (R/h)\Delta$, where Δ is the foil spacing/length

in c.g., the resultant error in $\Delta a/a$ when peak positions are used to measure θ will be less than that indicated in equation (7).

The effect of the finite width of the source and the receiving slits is to broaden the diffraction profile but not to change its position, except in the low-angle region where the $K_{\alpha_1 - \alpha_2}$ doublet is not resolved. Here a wider slit causes the *peak* to be shifted toward the c.g. of the two reflections.[12]

ERRORS INHERENT IN THE INSTRUMENT

The diffractometer is basically a calibrated gear wheel and worm drive combination. It should thus be borne in mind that a measurement of 2θ can be no more accurate than the hobbing of these gear teeth. Although the gear wheel/worm drive is common to all diffractometers, there are as many different methods of driving the specimen at half the speed of the main 2θ gear as there are makes of diffractometers. The situation is somewhat similar to that of an automobile engine where the camshaft turns at half the speed of the crankshaft, but each designer has his own method of linking the two drives together. As a consequence there are two sources of error inherent in the instrument: (1) the angular scale of 2θ and (2) the accuracy with which $\theta = \frac{1}{2}(2\theta)$. The latter is often referred to as the 2 : 1 following. Of course, θ can be set to equal $\frac{1}{2}(2\theta)$ at any desired 2θ value, since there is an adjustment available to the operator for alignment procedures. We are not concerned with this static 2 : 1 alignment at present, but that the 2 : 1 relationship once set will hold over all values of 2θ. In addition to such random errors arising from uneven wear of the teeth on the main gear and the variation of an oil film on the gear, systematic errors in 2θ may occur if the gear circle has some slight ellipticity or the center of the gear circle does not coincide exactly with the axis of rotation of the diffractometer. The accuracy of the 2 : 1 following will also depend on the precision to which the particular bisecting mechanism was machined.

Errors arising from the lack of precision of the diffractometer are not easy to measure by normal diffractometry since they cannot be separated from errors arising from the geometry of the method. Following the Pittsburgh Diffraction Conference last fall, at the urging of Mr. Karl Beu of the Apparatus Standards Committee of A.C.A. and with the generous assistance of Dr. Thomas Furnas, a study was made of several diffractometers in the Pittsburgh area using a standard twelve-sided optical polygon in conjunction with an optical autocollimating system.[13] The optical polygon was mounted on the axis of the diffractometer but rigidly connected to the detector arm so that it could be rotated through the angle 2θ. One of the faces of the polygon was aligned perpendicular to the collimator with the diffractometer scale set at $0°\ 2\theta$ and the error in 2θ was measured after the polygon was turned through successive intervals of $30°\ 2\theta$ (as indicated by the diffractometer dial) bringing successive faces of the polygon perpendicular to the auto-collimator. The form of the error was found to be a sinusoidal function of 2θ, the maximum deviation being of the order of 0.019–$0.045°\ 2\theta$ for the various instruments. The effectiveness of this error, however, depends on which particular point of the diffractometer circle is selected by the manufacturer to be the zero of the 2θ scale and whether or not the instrument is used to scan on both sides of the primary X-ray beam. Two extreme positions of the zero position are illustrated in Figure 1. In case A (solid lines) the effective $\Delta 2\theta$ error in the high-angle region is of the same order as the maximum $\Delta 2\theta$ (taken as $0.02°$) whereas in case B (broken lines) the effective $\Delta 2\theta$ is only half of the maximum $\Delta 2\theta$. In case A the error has the same sign on both sides of the primary beam, whereas in case B the error is positive on one side of zero and negative on the other. The latter type of error (case B) is important if the instrument is aligned by superimposing peaks scanned on

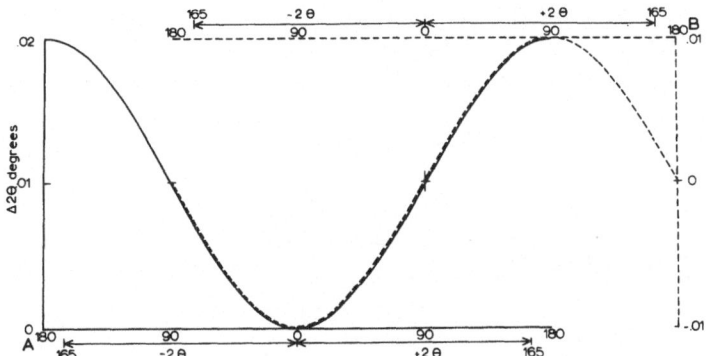

Figure 1. Typical instrumental errors. Solid lines, condition A; dashed lines, condition B.

both sides of the primary beam.[14] The position of the zero of the 2θ scale also affects the extrapolation plot of the error as discussed in a later section.

Backlash errors in 2θ were measured by taking the difference in readings of the auto-collimator when approaching a multiple of $30° 2\theta$ in both clockwise and counterclockwise directions. This error ranged from $0.003–0.002° 2\theta$ for the various diffractometers and for one any instrument showed a scatter of about $\pm 0.003°$ over the 2θ range examined. Errors in $2 : 1$ following were measured using two plane mirrors in conjunction with the autocollimator and were found to be of the order of $0.01–0.05°$. It is important to remember that these errors as well as those in Figure 1 reflect the current condition of the diffractometer, i.e., how much it had been used and in what manner.

Even though some resolution may be lost at various regions along the 2θ scale, the effect for powder diffractometry of an error in $2\theta : \theta$ following is not particularly serious because the intensities and peak positions will be affected only slightly. Backlash errors, on the other hand, may be eliminated by properly designed experiments as discussed later. If the backlash is too great ($> 0.02°$), the instrument probably needs reconditioning. In one case the backlash was reduced to a very low value of $0.003°$ merely by ensuring that the spring engaging the worm gear to the main 2θ gear was free to function.

If a diffractometer is to be used for precision work, it follows that it should be calibrated so that the measured values of 2θ can be corrected using curves such as those in Figure 1. Perhaps it is proper to suggest that the responsibility for this calibration be placed on the manufacturer, who should then provide a calibration curve for each instrument; but even so the instrument may still need recalibrating from time to time. While making suggestions to manufacturers, may we also make a plea that the main gears be cut as full throated gears so that the wear surfaces between gear and worm will be true surfaces and not just lines of contact. In fairness, it should be noted that some manufacturers are in fact ready to provide the service just suggested, as evidenced by the policies presented by the manufacturing spokesmen at the recent A.C.A. Meeting in Villanova (June 1962).

ERRORS ASSOCIATED WITH ALIGNMENT

The method of aligning a diffractometer has been discussed in some detail by Klug and Alexander.[11] In the general case this usually means following a set of directions and using special tools designed for the purpose. It is important, however, that the operator of the diffractometer understand which alignment procedures are critical and which

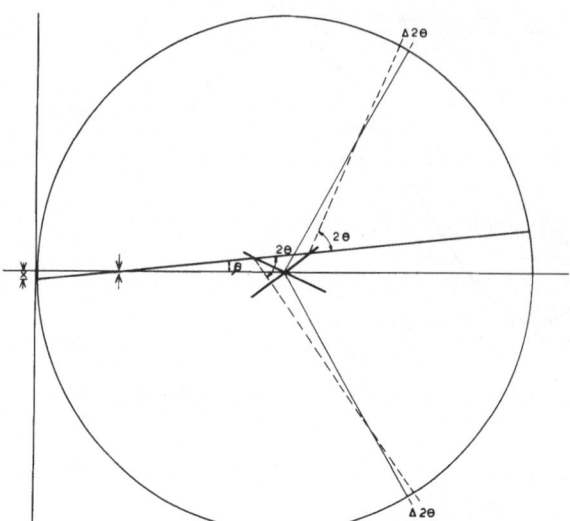

Figure 2. Diffractometer circle showing misalignment error X and β.

are insignificant in terms of the resultant error in $\Delta a/a$. A correctly aligned instrument should comply with the following conditions: (1) the source must lie directly over the 180° position; (2) the central ray of the direct beam must pass through both the axis of rotation of the specimen and directly over the 0° 2θ point of the diffractometer circle; (3) the central ray of the beam must lie in the same plane as that defined by the locus of points of the midpoint of the receiving slit as it is rotated through $\pm 180°\ 2\theta$; and (4) the surface of the specimen must be coincident with its axis of rotation and be parallel to the central ray of the beam when the detector is at $2\theta = 0$.

For convenience the various misalignments can be described in terms of four parameters: a displacement X of the focal spot along the tangent to the diffractometer circle; an angle β between the central ray of the direct beam and the true zero line in the diffractometer plane (these are shown schematically in Figure 2); an angle γ between the central ray and the plane of the diffractometer circle; and a displacement S of the sample surface from the rotation axis. According to the geometry in Figure 2, provided X is equal to zero, $\Delta 2\theta$ will also be zero even if β is not zero, i.e., a misalignment β is not at all critical. In practice, however, it is much easier to align the beam over 0° 2θ (i.e., make β zero) than to locate the focal spot over the 180° 2θ point. The effect of the latter misalignment (as measured by X) is very serious since, as shown in the following equation,

$$\Delta 2\theta = - X/R - 2S\cos\theta/R \tag{9}$$

it causes 2θ to be shifted to lower values by X/R radians, i.e., if $R = 17$ cm (Norelco, Siemens, Hilger, etc.), X must be > 0.015 mm for $\Delta 2\theta = 0.005°$ or > 0.003 mm for $\Delta 2\theta = 0.001°$ or if $R = 14.6$ cm (G. E., Picker, etc.), X must be > 0.013 mm or 0.002 mm, respectively. Also included in equation (9) is Wilson's[1] analysis of the error in 2θ due to a displacement of the specimen surface from its axis of rotation. This analysis applies to both the peak and the c.g. of the diffraction profile and the resultant error in 2θ is again quite significant particularly in the low-angle region. Since part of the displacement S is fixed by the manufacturer of the diffractometer, this error is frequently included among the geometrical errors of Group 2.

In Figure 2 the displacement X is assumed to be in the positive side of 180° 2θ, i.e., the same side as that scanned by the detector in normal operation. A displacement

S is positive if the specimen surface lies outside the focusing circle. If the sign of either of these two displacements is reversed, i.e. if X is on the negative side of 180° 2θ or the specimen surface lies inside the focusing circle, 2θ will be shifted in the opposite direction—toward higher values. Scanning on the opposite side (negative side in Figure 2) of the primary beam has the effect of changing the sign of X, but not that of S. Thus, provided the diffractometer is constructed to scan both sides of the beam, at least in the low-angle region, the critical alignment of the focal spot can be easily achieved to within the same accuracy as the measurement of 2θ (see the next section) by moving the 180° 2θ point of the diffractometer tangent to the diffractometer circle (along ± X) until the two diffraction profiles obtained by double scanning the same 2θ region on both sides of the direct beam are superimposed upon each other.[14] It has been demonstrated by the authors[14] that, provided the error in S is very small, the error due to any misalignment X can be measured from the separation between the two profiles of a low-angle Bragg reflection double scanned on either side of the direct beam to give a correction term that can be applied to all values of 2θ, since X/R is independent of 2θ [equation (9)]. The alignment of the focal spot can also be achieved with great precision if the X-ray tube is clamped to the diffractometer in such a manner that it is free to rotate about the focal spot in a plane parallel to that of the diffractometer.[15] The position of the focal spot can then be adjusted along ± X until no shift in 2θ is observed when the take-off angle of the X-rays is varied. If the errors in 2θ arising from X and S are substituted in equation (3), the resultant error in the lattice parameter of a cubic crystal is given by

$$\Delta a/a = X/2R \cot \theta + S \cos^2\theta / R \sin \theta \qquad (10)$$

Thus even though in many diffractometers these errors are difficult to remove by alignment, they can at least be minimized by extrapolation procedures.

The misalignment γ is not at all critical.[16] The resultant error in $\Delta a/a$ is given by (1−cos γ), which means that for an accuracy of 1 part in 20,000 the magnitude of γ must not exceed ± 0.57° (or ± 33′). This limit which represents the error due to the central ray not being in the plane of the diffractometer is quite easily achieved in practice. An additional error, similar in behavior and magnitude arises if there is a tilt of the specimen surface with respect to the axis of rotation. This, too, is easily kept below the prescribed limit.

The effect of an error in the static 2θ : θ setting, i.e., the specimen surface not being parallel to the central ray of the X-ray beam when the detector is set at 2θ = 0, was studied experimentally. It was found that if the diffractometer was otherwise well aligned, a change in 2θ : θ setting affected the breadth and maximum intensity of the diffraction profile but not its peak position or c.g. to within the experimental reproducibility of 0.001° 2θ. If, however, there was also a misalignment X, a change in 2θ : θ setting resulted in a shift in both the peak and the c.g. of the profile. Since the maximum intensity of the profile is sensitive to the 2θ : θ setting, any errors are easily detected when aligning the diffractometer by superimposing peaks on either side of the primary beam.

DIRECT ERRORS IN THE MEASUREMENT OF 2θ

A direct error of Δ2θ in the measurement of the position of a diffraction profile influences the lattice parameter of a cubic crystal according to equation (3). This error is of course superimposed on the other contributions to Δθ from physical, geometrical, instrumental, and alignment errors. Using the double scanning method and measuring peak positions to ± 0.005° 2θ it has been found that the total error in $\Delta a/a$ is approximately constant over the range 45° < θ < 80°, being of the order of 6.5×10^{-5}.[14]

Thus, if the limit of accuracy in $\Delta a/a$ is set at 1 part in 20,000 or 5×10^{-5}, the measurement of several Bragg reflections occurring at angles greater than $45°$ θ can provide the accuracy desired. For example, when studying Al or Ag using Cu K_α radiation, five reflections are obtained in the high-angle region and the mean value of the a_0s corrected for systematic errors will have an error of $1/\sqrt{5}$ of that of an individual measurement, i.e., approximately 3×10^{-5}, which is somewhat better than can be justified on the basis of the physical arguments given above. To reduce the product $\cot\theta\ \Delta\theta$ appreciably below 6.5×10^{-5} would require 2θ to $0.001°$ by point counting across the lines and using c.g. measurements with much greater time spent obtaining the data and a correspondingly greater time in analyzing it.[17,18]

As discussed by Parrish and Wilson[12] and others[17,18] the justification for using the c.g. as the measure of the position of a diffraction profile is that this property of the profile can be determined more accurately and is more amenable to analysis than the peak position. On the other hand, although the peak position is often harder to identify (particularly at large angles), it is somewhat more easily measured and is a less strongly varying property for the several errors. Thus, if 2θ is measured using peak positions, the equations for the correction of errors analyzed in terms of a shift in c.g. may be regarded as a guide to the variation in $\Delta a/a$ with θ and in a quantitative manner as an upper limit to the degree of error in the determination.

The error due to the time constant of the recording circuits can be eliminated by scanning across the diffraction profile in both clockwise (cw) and counterclockwise (ccw) direction and taking the average of the two measurements of 2θ[14] or alternatively by the more time-consuming step-scanning techniques. The backlash error is not as easily handled since the appropriate method of eliminating it depends on the manner in which $0°$ 2θ has been determined. Where possible $0°$ 2θ should be established by scanning both edges of the direct beam in directions away from zero toward higher angles, taking suitable precaution to protect the detector and adjusting the diffractometer so that traces of the profiles of the two edges are superimposed. This is in effect applying the double-scanning technique to the direct beam instead of a diffracted beam as in the alignment of $180°$ 2θ (X in Figure 2). When using this method to align $0°$ 2θ, half the backlash error is assigned to the cw direction and half to the ccw direction, and the total error can thus be conveniently eliminated at the same time as the electronic time constant error by scanning all diffraction profiles in both cw and ccw directions and taking the mean peak position (or c.g.).[14]

On the other hand, if, as in the case of the Norelco instrument, the zero position is obtained using the "O" alignment tool provided with it, the zero will then be defined with respect to the cw direction alone and no backlash errors will occur providing all scans are made in this same direction. This condition with respect to scanning direction means that the electronic time constant error can only be removed by step-scanning—again only in the cw direction. The time-constant error can of course be reduced to within the accuracy set for the investigation by selecting a very slow scanning speed but this can make the experiment very long.

ERRORS INVOLVED IN EXTRAPOLATION PROCEDURES

The various contributions to the error in the lattice parameter of a cubic crystal in equations (4), (6), (7), and (10) may be summarized in the following general equation in terms of θ,

$$\frac{\Delta a}{a} = \cos^2\theta \left[\frac{A}{\sin^2\theta} + \frac{B}{\sin\theta} + C\right] - \frac{D}{\sin^2\theta} + E\cot\theta - F\tan^2\theta + G \qquad (11)$$

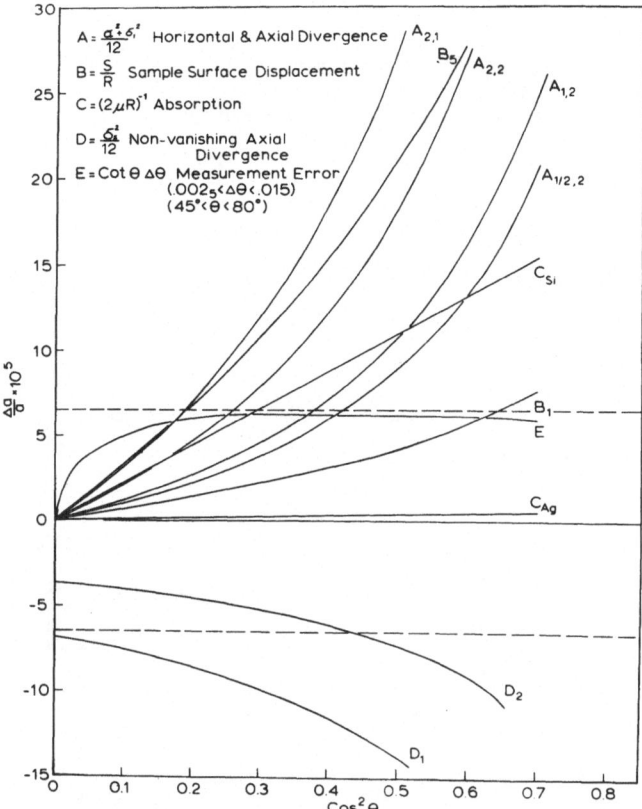

Figure 3. Extrapolation curves *vs.* $\cos^2\theta$.

It is immediately obvious that in contrast to the Debye–Scherrer film method no single extrapolation function can be selected such that all the errors will extrapolate nearly to zero as $\theta \to 90°$ (or $2\theta \to 180°$). Nevertheless, the dominant term $\cos^2\theta$ is often used as an extrapolation function. It is therefore instructive to assess the additional errors involved in this procedure by plotting the various terms in equation (11) against $\cos^2\theta$ as shown in Figures 3 and 4. The dotted lines at $\pm 6.5 \times 10^{-5}$ represent the total experimental error observed when peak positions are used to measure 2θ and have been included to give a significance limit to the various sources of error which will be considered in turn.

The curves labeled A in Figure 3 refer to the combined horizontal and axial divergence, equation (7), resulting from the following experimental conditions: $A_{2,1}$—2° div slit, single Soller slit, $A_{2,2}$—2° div slit, double Soller slit, $A_{1,2}$—1° div slit, double Soller slit, $A_{1/2,2}$—$\frac{1}{2}$° div slit, double Soller slit.

It is found that little improvement in error is gained by reducing the divergent angle below 1° because the axial divergence effect is clearly dominant even when two Soller slits are used.

The B curves correspond to small displacements of the sample surface normal to the diffractometer axis, equation (10): B_1 corresponds to 0.01 mm or 0.0004 in. and B_5 corresponds to 0.05 mm or 0.002 in. The plots show that in order for this error to be less than the significance limit over the range of $\cos^2\theta$ to 0.4 the specimen surface should

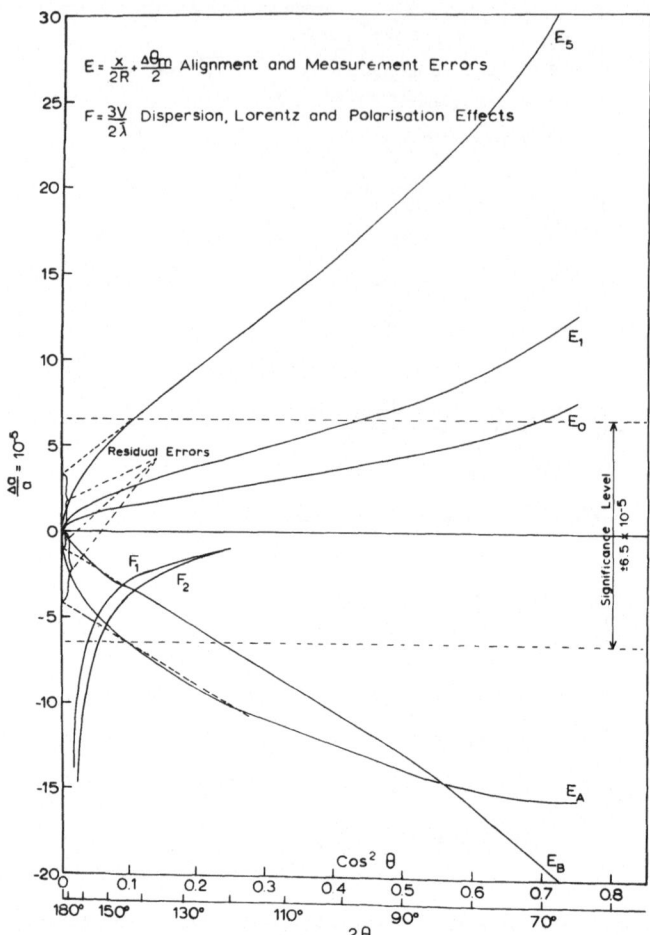

Figure 4. Extrapolation curves *vs.* cos²θ.

be located to within 0.01–0.02 mm of the diffractometer axis. As mentioned earlier the sign of this error can be either positive or negative depending on whether the surface lies inside or outside the focusing circle which explains why extrapolation plots of diffractometer results may have either positive or negative slopes.

The slopes of the C curves are determined by the linear absorption coefficient μ of the sample, equation (7). Shown in Figure 3, on the same scale with other errors, are the results for silver and silicon (copper radiation), where again the contrast between the Debye–Scherrer and the diffractometer methods is evident. Although the results for silicon lie just outside the significance limit, the extrapolated value of a_0 will not be affected since this error is truly a linear function of cos²θ.

The D curves come from the nonvanishing axial divergence term, equation (7), and are shown (Figure 3) for the case of single Soller slits D_1, and double Soller slits D_2. These plots tend to offset the positive deviations of A, B, and C and have significant residual errors at $\theta = 90°$ which must be added on to the extrapolated value of a_0.

The curve labeled E shows the resultant product of $\Delta\theta \cot \theta$ where from experience we have found that the measurement error $\Delta\theta$ is increasing more or less linearly with θ

while at the same time $\cot \theta$ is decreasing. The net effect is a nearly constant error in $\Delta a/a$ due to measurement. The E curves shown in Figure 4 refer to the three different sources of error which have $\cot \theta$ dependence. These errors are: a misalignment X (Figure 2) of the focal spot with respect to the $180° \ 2\theta$ point, an error $\Delta 2\theta_m$ in the measurement of 2θ and an error $\Delta 2\theta_I$ in the value of 2θ indicated on the diffractometer scale (Figure 1). It should be noted that all three of these errors may be either positive or negative in sign and may therefore either reinforce or oppose each other or any of the other errors in curves A to D. To avoid too many plots overlapping in Figure 4 it is assumed that the error in $\Delta a/a$ is positive for the E_0, E_1, and E_5 curves and negative for the E_A and E_B curves. If all the E errors have the same sign as curves A, B, and C, the resulting plot will lie well outside the significance level and would have a very steep slope. On the other hand, if the E curves have the opposite sign curves A to C, the resultant plot will be almost horizontal. The experimental conditions for the E curves in Figure 4 are listed in Table II.

The E_0 curve lies below the significance level for $\cos^2\theta$ up to 0.7 (i.e., 2θ from 60–180°) confirming that the limit of $\pm 0.005°$ set for the measurement of 2θ is appropriate for the required accuracy in $\Delta a/a$. The E_1 curve indicates that a misalignment of 0.01 mm in X (Figure 2) can be tolerated, provided a_0 is calculated from 2θ values in the range 100–180°. It is false security, however, to assume that an error caused by a misalignment of ± 0.05 mm in X (which in any case lies outside the significance level) can be removed by extrapolating against $\cos^2\theta$ since, as shown by the E_5 curve, a linear extrapolation drawn through the data for $\cos^2\theta = 0.1$–0.5 will leave a considerable residual error in $\Delta a/a$. This illustrates one of the drawbacks of using $\cos^2\theta$ as an extrapolation function, particularly as the alignment of the focal spot is the most difficult alignment procedure to accomplish in practice.

The curve E_A shows the error in $\Delta a/a$ when the instrumental errors in Figure 1A are superimposed on an error of $0.005°$ in the measurement of 2θ. The resultant error in $\Delta a/a$ is clearly outside the tolerated limit of accuracy and leaves a very significant residual error on extrapolating a linear plot to $180° \ 2\theta$. In curve E_B, based on the $\Delta 2\theta_I$ errors in Figure 1B, a considerable error in $\Delta a/a$ occurs at low angles but the data conform more closely to a linear plot with a relatively small residual error left at $180° \ 2\theta$. Thus a given total error $\Delta 2\theta_{max}$ is less harmful to a lattice parameter determination if the zero of the 2θ scale lies near a point of inflection on the sinusoidal plot as in Figure 1B, than is the case when the zero of 2θ lies near a maximum or minimum as in Figure 1A. Since the error $\Delta 2\theta_{max}$ in Figure 1 is toward the lower limit of that found in commercial diffractometers and in any case the form of the error (type A or B) cannot be ascertained without doing a calibration, the magnitude of the errors in curve E_A is sufficient warrant to have a diffractometer accurately calibrated before applying it to precision parameter determinations.

One of the strongest objections to the use of an extrapolation procedure to $180° \ 2\theta$

Table II

E_0:	$X = 0$ (or eliminated by double scanning)	$\Delta 2\theta_m = -0.005°, \Delta 2\theta_I = 0$ ⎫ (or corrected from
E_1:	$X = 0.01$ mm or 0.0004 in.	$\Delta 2\theta_m = -0.005°, \Delta 2\theta_I = 0$ ⎬ a calibration curve)
E_5:	$X = 0.05$ mm or 0.002 in.	$\Delta 2\theta_m = -0.005°, \Delta 2\theta_I = 0$ ⎭
E_A:	$X = 0$	$\Delta 2\theta_m = +0.005°, \Delta 2\theta_I = $ as in Figure 1A for $+2\theta$
E_B:	$X = 0$	$\Delta 2\theta_m = +0.005°, \Delta 2\theta_I = $ as in Figure 1B for $+2\theta$

in conjunction with the diffractometer method is that it *enhances* the errors due to dispersion, Lorentz, and polarization factors, equation (4). This is demonstrated by the F curves in Figure 4, which are based on the $\Delta 2\theta$ shifts in the c.g. profiles calculated by Pike[6] for Cu K_α radiation. In curves F_1 and F_2 the c.g. of the Cu K_α doublet was calculated using $10\times$ and $30\times$ the half width of the diffraction profiles, respectively. The F plot shows that the error in a_0 resulting from this c.g. shift is below the significance level up to $\cos^2\theta = 0.0047$ (or 2θ up to 155°), but thereafter the negative error increases exponentially. Thus, when 2θ is measured using the c.g. of diffraction data, it is important that no a_0 values calculated from Bragg reflections at $2\theta > 155°$ be included in the extrapolation plots.

The constant G in equation (11) contains the terms which are independent of the Bragg angle, i.e., the refraction correction, equation (6), and any correction necessary to relate the result to a reference temperature. These terms, together with the residual part of the D curves which remain at $2\theta = 180°$, must be added algebraically to the extrapolated value of a_0 if the results are to have any absolute significance.

Since only the curves C and E in Figures 3 and 4 actually refer to shifts in peak position (curves A, B, D, and F being for c.g. shifts which may be much greater than peak shifts), the deviations of the remaining curves, A, B, D, and F from a $\cos^2\theta$ plot are in fact greater than would be observed when measuring Bragg reflections by peak positions. It is thus of interest to compare these analyses with some typical extrapolation plots using data obtained by the double-scanning method.

Figure 5 shows typical results for measurements on silver and silicon. In the case of silver there seems to be no detectable slope in the plot for $\cos^2\theta < 0.43$. The appropriate A, C, and D curves for these experimental conditions in Figure 3 are $A_{1,2}$, C_{Si} or C_{Ag}, and D_2 respectively. Since S is determined both by the manufacturer and the technique

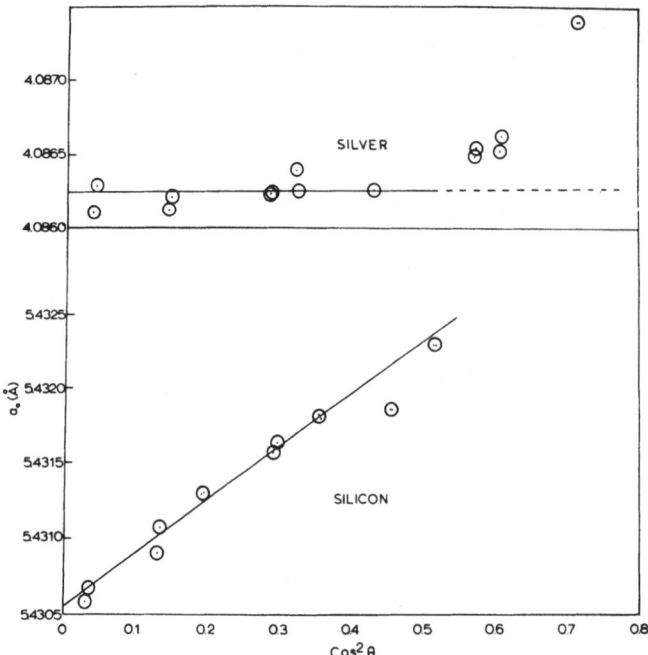

Figure 5. Extrapolation plots for silver and silicon.

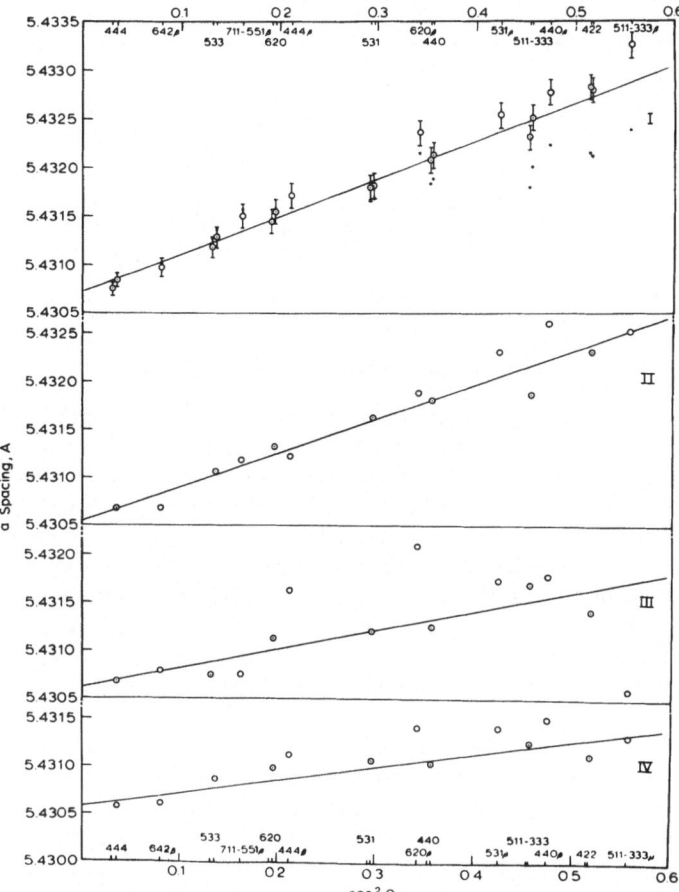

Figure 6. Extrapolation plots for silicon.
I. Point counting
II. Double scan, standard conditions
III. Double scan, sample surface displaced ($S \neq 0$)
IV. Double scan, $S \neq 0$, one set Soller slits removed

of sample preparation, B is unknown. It can be calculated that if S were no greater than 0.01 mm and negative (Figure 2) the plot of the sum of the errors A–D would be quite flat for values of $\cos^2\theta$ up to about 0.5. In fact, even if it were necessary to use very wide divergence slits and/or only one Soller slit, a judicious choice of the value of S could nevertheless result in a nearly horizontal plot of a_0 vs. $\cos^2\theta$. This is shown in Figure 6 where the pattern for silicon was scanned using the four experimental conditions listed in Table III.

In Case IV the rear Soller slits have been removed which should about double the error due to nonvanishing axial divergence and change the extrapolated value of a_0 (assuming the Bragg angles were measures using the c.g. of the lines) by an amount $\Delta a/a = 3 \times 10^{-5}$. However, this is just equal to the stated accuracy of measurement of a_0 itself, based on five lines. Pending more accurate measurements involving the axial divergence effect, these results justify a final correction of $+\frac{1}{2}D$ to the extrapolated value of a_0.

Table III

Case	Type of Measurement	A	B	C	D	a_0 (Least Squares)
I	Point counting	$A_{1/2,\,2}$	$S = 0$	20×10^{-5}	D_2	5.4307_2
II	Double scan	$A_{1/2,\,2}$	$S = 0$	20×10^{-5}	D_2	5.4305_5
III	Double scan	$A_{2,\,2}$	$S = 0.06$ mm	20×10^{-5}	D_2	5.4306_2
IV	Double scan	$A_{2,\,1}$	$S = 0.06$ mm	20×10^{-5}	D_1	5.4305_8

SUMMARY AND CONCLUSIONS

The present review of the errors involved in the application of the diffractometer to precision lattice parameter measurements has shown that while the instrument may provide a very accurate measure of the diffraction angle, it is also very sensitive to quite a variety of errors, each of which must be controlled if the final result is to be of any absolute significance. At the present time the upper limit of accuracy in $\Delta a/a$ is limited by the uncertainty in the wavelength of the characteristic radiation to 1 part in 20,000 and so it is not worthwhile measuring 2θ to a precision greater than 0.005° by the more accurate but very time-consuming c.g. methods. It has been demonstrated that the various geometrical, instrumental, alignment, and operational errors can all be kept to within this limit when measuring peak positions to 0.005° 2θ provided: (1) the instrument is properly calibrated and accurately aligned with respect to the X-ray source; (2) the powdered specimen is carefully prepared to have a truly flat surface which is aligned with respect to the diffractometer axis; (3) a suitable combination of Soller slits and divergence slits is selected; (4) the peaks are measured in such a way as to cancel out errors due to backlash and electronic time constant; and (5) the extrapolation procedures are not abused, particularly with respect to data from diffraction peaks at 2θ greater than 155°.

ACKNOWLEDGMENTS

The authors are grateful to Dr. T. B. Massalski and Dr. L. E. Alexander for helpful discussions. They are also grateful to Mr. K. Beu and Dr. T. Furnas for their helpful cooperation with respect to the instrument evaluation program.

REFERENCES

1. A. J. C. Wilson, "Geiger-Counter X-ray Spectrometer—Influence of Size and Absorption Coefficient of Specimen on Position and Shape of Powder Diffraction Maxima," *J. Sci. Instr.* **27**(12): 321, December, 1950.
2. W. Parrish, "Results of the I.U.Cr. Precision Lattice-Parameter Project," *Acta Cryst.* **13**(10): 838, October, 1960.
3. K. Lonsdale, *International Tables for X-ray Crystallography,* Vol. *III*, 1962, Kynach Press, Birmingham, England; *Acta Cryst.* **3**: 400, 1950.
4. A. J. C. Wilson, "Some Problems in the Definition of Wavelengths in X-ray Crystallography," *Zeit. Krist.* **3**: 471, 1959.
5. A. R. Lang, "Effect of Dispersion and Geometrical Intensity Factors on X-ray Back-Reflection Line Profiles," *J. Appl. Phys.* **27**: 485, 1956.
6. E. R. Pike, "Counter Diffractometer—The Effect of Dispersion, Lorentz and Polarization Factors on the Position of X-ray Powder Diffraction Lines in Terms of the Center of Gravity of the Lines," *Acta Cryst.* **12**: 87, 1959.

7. A. J. C. Wilson, "Effect of Absorption on Mean Wave-Length of X-ray Emission Lines," *Proc. Phys. Soc. (London)* **72**: 924, 1958.

8. J. Cermak, "The Intensity Distribution in the Faces of Curved Crystal Monochromators and an Estimate of Its Influence on Precision Measurements of Lattice Parameters," *Acta Cryst.* **13**: 832, 1960.

9. A. J. C. Wilson, "Correction of Lattice Spacings for Refraction," *Proc. Comp. Phil. Soc.* **36**: 485, 1940.

10. E. R. Pike, "Counter Diffractometer—The Effect of Vertical Divergence on the Displacement and Breadth of Powder Diffraction Lines," *J. Sci. Instr.* **34**: 355, September, 1957 and **36**: 52, January, 1959.

11. H. P. Klug and L. E. Alexander, *X-ray Diffraction Procedures*, first edition, John Wiley & Sons, Inc., New York, 1954.

12. W. Parrish and A. J. C. Wilson, *International Tables for X-ray Crystallography, Vol. II*: 216, 1959, Kynach Press, Birmingham, England.

13. J. C. Evans and C. O. Taylerson, "Measurement of Angle in Engineering," *Nat. Phys. Lab. Notes on Appl. Sci.* No. 26, 1961, H.M. Stationery Office, London.

14. H. W. King and L. F. Vassamillet, "Precision Lattice Parameter Determination by Double Scanning Diffractometry," *Advances in X-ray Analysis, Vol. 5*, University of Denver, Plenum Press, New York, 1962, pp. 78–85.

15. T. C. Furnas, Jr. and E. W. White, "New Instruments for X-ray Analysis," *Advances in X-ray Analysis, Vol. 4*, University of Denver, Plenum Press, New York, 1960, p. 521.

16. W. L. Bond, "Precision Lattice Constant Determination," *Acta Cryst.* **13**: 814, 1960.

17. E. R. Pike and A. J. C. Wilson, "Counter Diffractometer—The Theory of the Use of Centroids of Diffraction Profiles for High Accuracy in the Measurement of Diffraction Angles," *Brit. J. Appl. Phys.* **10**: 57, 1959.

18. J. Ladell, W. Parrish, and J. Taylor, "Center-of-Gravity Method of Precision Lattice Parameter Determination," *Acta Cryst.* **12**: 253, 1959 and "Interpretation of Diffractometer Line Profiles," *Acta Cryst.* **12**: 561, 1959.

DISCUSSION

M. Semchyshen (Climax Molybdenum Co. of Michigan): In view of this big scare that you have given us, would you say that the old-fashioned Straumanis camera and some back-reflection camera would help in comparison to the diffractometry method?

L. F. Vassamillet: If you want precision greater than what we have been talking about, I think you are justified in looking at it very carefully. You can do very well at routine work of this order of accuracy. By the way, much of the data in Pearson has an accuracy listed which is considerably less than one part in 20,000. I think you would be justified in using the diffractometry method routinely (it's fast) to get data of this accuracy.

THE CHARACTERIZATION OF LARGE SINGLE CRYSTALS BY HIGH-VOLTAGE X-RAY LAUE PHOTOGRAPHS

H. S. Peiser and E. P. Levine*

*National Bureau of Standards
Washington, D.C.*

ABSTRACT

Large single crystals can be examined by conventional X-ray diffraction procedures only at their surface or by destructive sectioning. Within the limitations inherent in polychromatic X-ray photography, high-voltage Laue pictures are shown to give some information on the internal quality of large crystals.

Asterism in conventional Laue photographs is contrasted with streaks due to geometric effects in Laue patterns of large crystals. Detail within the streaks reveals subgrain structure. A primary extinction effect can be used as striking proof of good crystals being capable of scattering coherently over large distances.

INTRODUCTION

The rapid advance in the extent and diversity of technological and scientific uses for single crystals probably has barely started. The directional dependence of properties in crystals and their possession of properties not observable in polycrystalline or amorphous substances place crystals among the most promising materials of the present and the future.

The National Bureau of Standards tries to provide reference standards and a capability of precise measurement; so in this field especially it must prepare for future demands for critical measurements that characterize single crystals.

Only one small aspect of this wide program forms the subject of this paper, namely, crystals that are dimensionally large compared with the penetration of conventional electron or X-ray beams used in diffraction or diffraction microscopy. Naturally such large crystals can be subjected to "surface" diffraction examination or destructive studies by sectioning, but these procedures are always tedious and usually unsatisfactory. It is fair therefore to ask to what extent one could use more penetrating radiations. One possibility relies on neutron-diffraction techniques. These have already been used for such purposes,[1] and we hope they will soon be attempted at the National Bureau of Standards. The work here to be reported employs high-voltage X-rays, up to about 220 kv, for which the penetration is more than an order of magnitude greater than that of conventional X-rays used in diffraction experiments. It is to be stressed that this work is preliminary and exploratory. It has been our intention only to show that the method is promising and deserving of more careful attention.

---

* Student Trainee from the University of California (Berkeley).
[1] Superscripts pertain to references at the end of the paper.

THE HIGH-VOLTAGE LAUE METHOD

At kilovoltages above those corresponding to the K line of the heaviest elements we lack convenient monochromatic radiation sources of adequate brilliance for diffraction. Thus it is necessary to revert to the Laue technique, in which a stationary crystal is exposed to collimated "white" radiation. The Laue method, although quite popular especially in this country in the 1920s, has gone almost out of use in modern crystallography. It is, however, well described in some textbooks, such as that by Henry, Lipson, and Wooster,[2] and we will therefore confine ourselves to specific comments relating to high-voltage Laue photographs. Even these are not entirely new, having been described previously by Peiser and Rait[3,4].

Bragg's law states:

$$\frac{\lambda}{2d_{hkl}} = \sin\theta$$

where λ is the wavelength of the coherently diffracted beam; d_{hkl} is the distance between adjacent planes of the family of equidistant parallel planes, designated hkl, which symbols denote the number of planes traversed in any unit translations parallel, respectively, to the three unit-cell edge directions; and θ is the glancing angle of the incident and diffracted beams, which are coplanar with the normal to the planes.

In the Laue method, the θ values are given by the orientation of the crystal to the incident beam. The λ values range from the short-wave limit to long wavelengths for which there is a fall-off of X-ray tube output and increased absorption in the crystal as well as in the X-ray generator itself. Typically we can expect a wavelength range of from about 0.05 to 0.5 Å. The d_{hkl} values are limited at the upper end by the geometry of the atomic pattern in the crystal. Usually there are few spacings with d_{hkl} values greater than 2.5 Å. The lower d_{hkl} limit is effectively set by simple well-known intensity considerations. When the d_{hkl} values are small compared with the dimensions of the electronic orbits in the atom, destructive interference within atoms lowers the relevant diffracting power of the crystal. At below about 0.5 Å this becomes such a dominant effect that diffractions can no longer be observed experimentally. The useful range of $\sin\theta$ is thus given by

$$\frac{0.05}{2 \times 2.5} \leqslant \sin\theta \leqslant \frac{0.5}{2 \times 0.5}$$

or, in terms of the deviation angle (2θ), $1° < 2\theta < 60°$ approximately.

SOME EXPERIMENTAL DETAILS

It can thus be deduced by the simplest theory that a good collimation system is needed that, owing to the greater penetrating power of the X-ray to be used, is more cumbersome and more critical than for conventional X-ray diffraction wavelengths. Our design is shown schematically in Figure 1.

This slit system is placed as close as possible to the X-ray tube focus. It is preferable to employ a fine-focus X-ray tube, but such has not been available to us so far. Provisions are made for mounting the crystal specimen and for placing the photographic film. However, for voltages lower than the maximum, additional useful magnification can be obtained by placing the film a greater distance from the crystal specimen. Exposures are of the order of half an hour. The most important feature of the system is that the crystal penetrated by the X-ray beam is not negligible in linear size compared with its

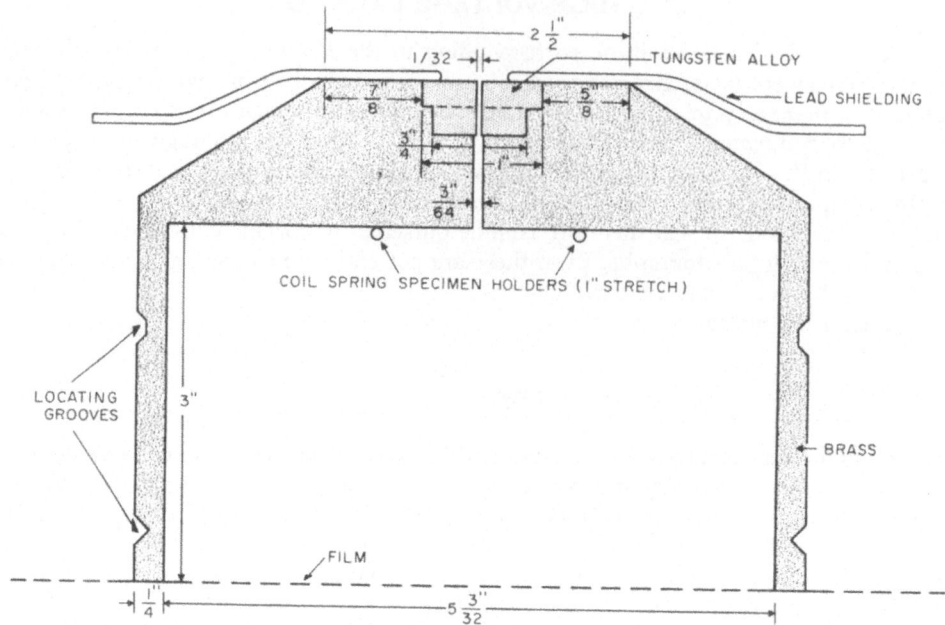

Figure 1. High-voltage X-ray diffraction slit.

distance from the photographic film. In this respect the method here described differs from conventional Laue photographs. The diffractions are therefore produced as streaks of length $t \cdot \tan 2\theta$, where t is the crystal thickness, instead of individual spots. Streaking of conventional Laue spots, "asterism" as it is usually called, is caused by crystal imperfections, such as strain, polygonization, and so forth.[5] However, examination of Figure

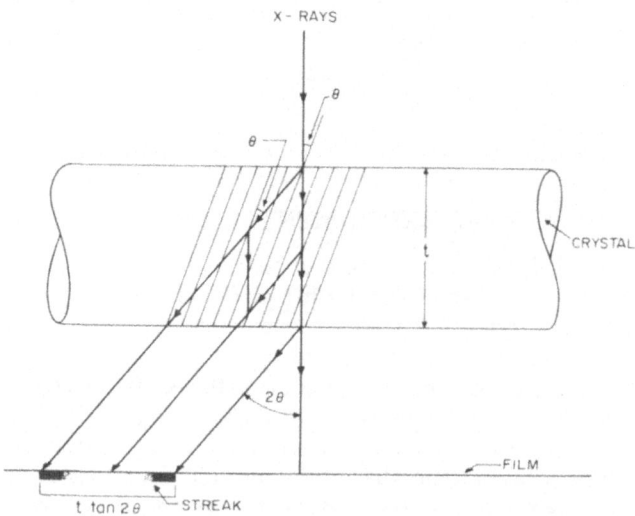

Figure 2. Production of Laue streak and primary extinction effect.

2 will readily show that in our experiments for any single diffraction each position along the streak corresponds to a particular portion of the crystal. Asterism, if present, results in additional radial elongations of the streaks.

SOME RESULTS

A glance at the X-ray photograph of a good aluminum single crystal of about 2 cm thickness (Figure 3) shows that the streaks are far from uniform. The wealth of detail along the streaks reveals a form of mosaic subcrystals. The periodic fine-structure along the streaks may be visualized as due to slightly misset crystallites failing to produce streak intensity in one place and enhancing it in another.

We conclude with the demonstration of a primary extinction effect. In a rather perfect crystal a diffracted ray (see Figure 2) is still inclined at the Bragg angle θ to the selfsame diffracting planes. A second diffraction will reduce the diffracted beam and would also interfere with the primary beam, with which it is out of phase by an angle π (there is a phase shift of $\pi/2$ at each diffraction), if it were not for the fact that in our experimental arrangement these rays are not collinear. A third diffraction, however, can be collinear and therefore can interfere with a once-diffracted ray. It is now clear from Figure 2 that this process will reduce the intensities of the middle of the streaks much more than those of the ends, for which there are no or at least fewer multiple paths. The streaks will thus show what one might call a "saddle effect," exhibited rather strikingly by the diffraction picture of another aluminum single crystal (Figure 4). The observation of primary extinction implies coherence, that is, high perfection over distances comparable with the penetration length of the crystal. This is the more remarkable in this specimen as at the same time it shows clear indications of subgrain structure as discussed above. This subgrain structure evidently does not eliminate a high degree of precise long-range order.

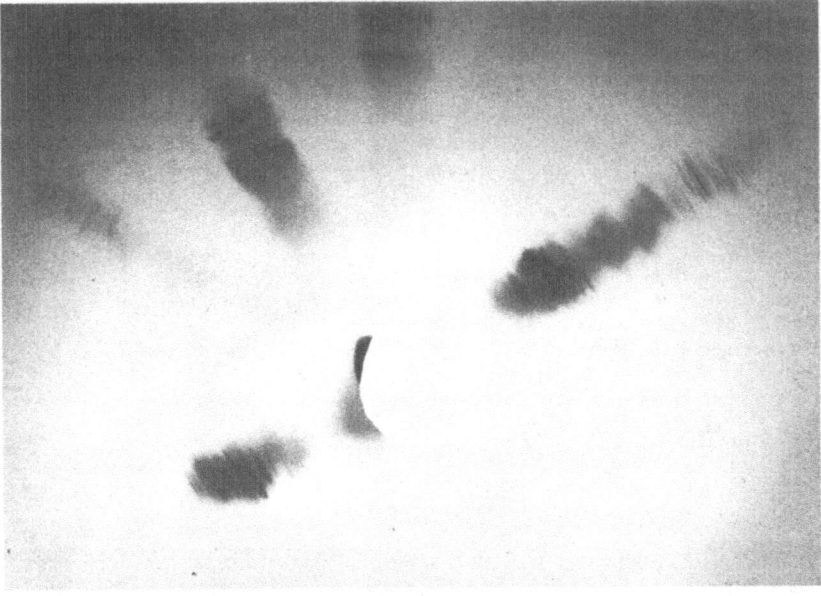

Figure 3. Laue photograph of aluminum single crystal (detail). Film was 5 in. from crystal, exposed for 10 min to a beam of 250 kv at 10 ma.

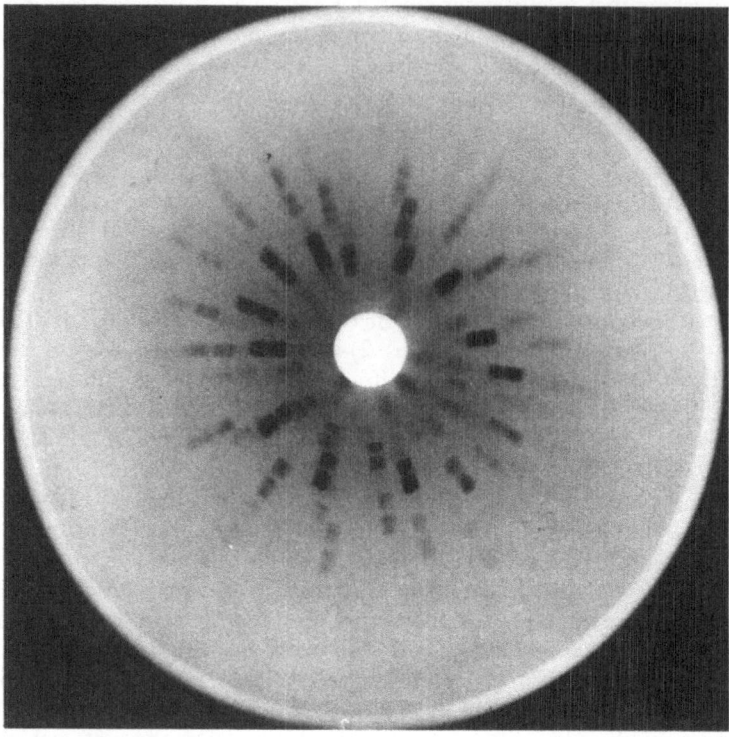

Figure 4. Laue photograph of nearly-perfect large single crystal of aluminum, showing primary extinction effect. Film was 3 in. from crystal, exposed for 30 min to a beam of 190 kv at 10 ma.

ACKNOWLEDGMENTS

The authors are indebted to Dr. S. W. Smith and Mr. L. A. Dobak of the Radiological Equipment Section of the National Bureau of Standards for advice and loan of some equipment. Mr. Dobak also took most of the X-ray pictures. In addition we thank Mr. Theodore Orem of the Metal Reactions Section of the National Bureau of Standards for providing us with suitable single crystals of aluminum.

REFERENCES

1. G. E. Bacon, *Neutron Diffraction*, Clarendon Press, Oxford, 1962.
2. N. F. M. Henry, H. Lipson, and W. A. Wooster, *The Interpretation of X-ray Diffraction Photographs*, Macmillan and Co., London, 1960.
3. H. S. Peiser and J. R. Rait, "A New Type of X-ray Diffraction Picture (High Voltage Laue Photographs)," *Acta Cryst.* **8**: 738, 1955.
4. J. R. Rait and H. S. Peiser, "The Potentialities of X-ray Diffraction Studies in Nondestructive Testing," *Brit. J. Appl. Phys.*, *Suppl.* **6**: 13, 1957.
5. H. G. Van Bueren, *Imperfection of Crystals*, North-Holland Publishing Company, Amsterdam, 1960.

DISCUSSION

F. Laves (Zurich, Switzerland): Why is there such surprising sharpness between the light and the dark portions of the streaks instead of a gradient?

H. S. Peiser: I am afraid this may in part be due to the slide reproduction but, to a great extent, it arises from a real periodicity of the slight misalignment of neighboring crystallites.

H. E. Kissinger (General Electric Co.): It is my impression that pronounced extinction has as an effect an anomalous increase in the absorption coefficient. How do you account for the beam being strong enough to cause a strong diffraction spot at the point where it leaves the crystal?

H. S. Peiser: Going back to Figure 2, I think I can make this fairly clear. Primary extinction arises essentially from the trebly-diffracted beam interfering through collinearity with the once-diffracted ray. By our geometry, in which the specimen diameter is by no means negligible compared with the specimen-to-film distance, we physically separate the beams in a very definite way. Whereas the once-diffracted ray originating from the specimen center has a trebly-diffracted beam with which it is collinear and is consequently attenuated, the ray diffracted from the bottom of the crystal, just before the primary beam leaves the crystal, has no possible trebly-diffracted ray that could travel collinearly with it.

H. E. Kissinger: I would like to supplement this question a little bit. Is there not a doubly-diffracted beam traveling in the same direction as the incident beam that would cause a decrease in the intensity of the direct beam?

H. S. Peiser: It is perfectly true and I am glad Homer Kissinger points out that primary extinction is also caused by a doubly-diffracted beam coinciding with the undiffracted primary beam. Now to answer the point. Our beam is a collimated beam that travels in a straight line perpendicular to the specimen axis. Multiple diffraction by the inclined family of planes shown necessarily displaces the diffracted rays to the left, away from the path of the primary beam. No further diffraction by these planes can bring a portion of the beam back into collinearity with the primary beam, which therefore traverses the specimen unattenuated by extinction. Does this answer the question?

H. E. Kissinger: Not completely, because this incident beam does have a finite dimension—it is not an infinitely small beam—and a double diffraction could occur in this very small space. In fact, it seems to me that it would, rather than travel any great distance.

H. S. Peiser: The spread of the beam due to finite size and divergence is small compared with the geometric effects discussed. The probability of diffraction occurring in a given length of crystal is of course far lower for high-voltage X-rays than for conventional radiations used for diffraction experiments.

Robert Kelsey (Pratt-Whitney Aircraft): What is used for a high-voltage source?

H. S. Peiser: All these photographs were taken with conventional radiographic X-ray equipment. That is, one that has a focal spot size hundreds of times bigger than we would like it to be. Yet it was adequate to see whether such a method would work at all. Now we know that it gives useful results, we thought it was reasonable to obtain a fine-focus tube, which of course makes the slit system very much easier to build and will greatly enhance the useful intensity. Any party who would like to try this method I think would be well advised to obtain one of the several fine-focus equipments which are available. This is the source which we will use in future.

DIFFRACTION EFFECTS FROM IRRADIATED ALUMINUM SINGLE CRYSTALS*

H. E. Kissinger

General Electric Company
Richland, Washington

ABSTRACT

The effect of substructural perfection on the accumulation of irradiation damage in aluminum was examined. Large single crystals with extensive substructure and crystals essentially free of substructure, all with faces cut parallel to crystal planes, were subjected to neutron irradiation. Subsequent examination by X-ray diffraction revealed pronounced changes in integrated intensity and Debye–Waller temperature factor for the substructure-free crystals; these effects disappeared upon re-etching of the surface. Laue photographs showed that the normal single-crystal pattern was partially obscured by a polycrystalline effect which also disappeared upon etching. Crystals with extensive substructure showed no such effects.

This diffraction evidence supports the view that irradiation-induced defects in aluminum migrate to and collect at the crystal surface if no internal trapping sites exist.

INTRODUCTION

The irradiation of a crystalline material by high-energy neutrons creates a disorder in the crystal lattice. This irradiation-induced disorder can be removed in many cases by annealing at a sufficiently high temperature. The nature of the defects generated by irradiation and the mechanism of the annealing are not too well understood at present. Elementary defects probably generated are unoccupied lattice sites, or vacancies, and atoms not in normal sites but in interstitial positions.

In a metal, vacancies and interstitials may exist as isolated defects. They may be annihilated through combination of unlike defects, or like defects may combine to form clusters. At temperatures high enough to make these defects sufficiently mobile, they may move to sinks at dislocations, grain boundaries, or other trapping sites.

The concentrations of vacancies and interstitials during irradiation will be governed by a balance between the rate of generation and the rate of annealing. In a crystal with a high density of structural imperfections serving as sinks for vacancies and interstitials, irradiated at a temperature high enough that both types of defect are mobile, very little observable damage would be expected. On the other hand, if sinks for irradiation-induced defects were fewer, there would be more opportunity for coalescence of single defects into clusters. These clusters, if they remained stable after irradiation, might cause changes in the X-ray diffraction results and might be detected by this method.

The availability of single crystals of aluminum having very pronounced differences in crystalline perfection suggested the experiments to be described. By simultaneously

* Work performed by the General Electric Company for the Atomic Energy Commission under Contract No. AT(45–1)–1350.

irradiating crystals having highly perfect lattices and those much less perfect, it should be possible to demonstrate the effect of substructure on the annealing behavior. Warren[1] irradiated an aluminum crystal at reactor-ambient temperature and observed no effects by X-ray diffraction. However, resistivity measurements by McReynolds *et al.* of aluminum samples irradiated at very low temperatures[2] showed that the irradiation effects annealed at temperatures far below room temperature. The annealing process in nearly perfect crystals would be more likely to leave stable clusters, or other evidence from which the annealing mechanisms might be deduced, since the lack of nearby sinks would make the interaction of defects a more probable process.

MATERIALS

Single crystals of 99.996% pure aluminum were grown by L. J. Friesen of Hanford Laboratories for use as monochromators for neutron diffraction. Crystals were grown by the Bridgman technique[3] and by the modified Bridgman method described by Noggle.[4] The Bridgman method produces single crystals with a higher concentration of structural imperfections than does the Noggle method. With care, crystals can be grown by the Noggle method which are essentially free of substructure.[5]

Transmission neutron diffraction patterns of the crystals chosen for this experiment are shown in Figure 1. These patterns were made by D. A. Kottwitz on a double-crystal neutron spectrometer. The samples were in the form of ingots approximately 3 in. in diameter and 8 in. long. It is apparent that only a few subgrains exist in the Noggle-method crystal, despite the large volume irradiate. The Bridgman-grown crystal is quite imperfect by comparison.

EXPERIMENTAL PROCEDURE

Specimens approximately $1 \times 1 \times 3$ cm were cut with a jeweler's saw from larger aluminum crystals. The (111) planes were oriented by back-reflection Laue photographs after which the samples were mounted in a self-curing resin in the desired orientation. Faces were ground parallel to the chosen planes with an abrasive wheel. A heavy etch in 20% NaOH solution then removed about 2 mm from the ground surface. The resulting uneven surface was smoothed by conventional metallographic techniques, each polishing stage followed by a light etch.

The crystal surfaces were examined by X-ray diffraction at intervals during the grinding of the faces, with the intensities of the first- and second-order reflections serving as an indication of surface perfection. When the surfaces were perfectly plane and parallel to the desired crystal planes within 1°, and when additional etching produced no change in integrated diffracted intensity, the polishing was terminated. It was found that a highly polished surface was unnecessary, equivalent results being obtained from a satin-smooth surface. Surface roughness effects were small, as an unetched ground surface yielded diffracted intensities approaching the theoretical values.

Integrated diffracted intensities were measured on a G.E. XRD-5 diffractometer equipped with a single crystal orienter. The open-counter scan method was used,[6] with a scintillation counter detector. The counter opening was sufficiently wide ($\frac{1}{2}$ in.) that no appreciable scanning error existed. Molybdenum K_α and copper K_α wavelengths were used, monochromatized with a flat lithium fluoride crystal. The incident beam intensity was measured by calibrated foils and by diffraction from a crystal of sodium chloride.

---

[1] Superscripts pertain to references at the end of the paper.

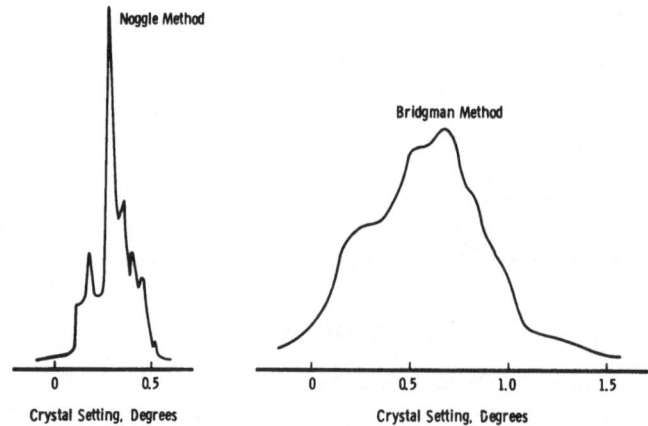

Figure 1. Transmission neutron diffraction patterns of aluminum single crystals grown by the Noggle method (left) and the Bridgman method (right).

Counting errors were minimized by use of absorber foils to reduce the counting rate The antiparallel orientation of monochromating and sample crystals was used. Beam dimensions at the monochromator were approximately 0.2 × 0.2 mm, with a vertical divergence of about 0.4°.

Lattice parameters were measured by determining the differences in angular position of the crystal for successive orders. Such measurements are free of positioning errors except for the uncertainty in the zero of the angle scale. There is, however, a small correction for refraction and for the Lorentz-polarization factor,[7] both of which were neglected here. The zero error is constant for all orders and may be eliminated analytically or by trial and error. A small correction for vertical divergence (less than 0.001°) has also been neglected.

For irradiation the crystals were placed in small cylindrical capsules. The cut and polished faces were placed against the inner capsule wall so that contact was made only at the face edges. The irradiations were performed with the outside of the capsule in contact with the cooling water at about 70°C; however, the poor thermal contact of the crystals with the capsule made any reliable estimate of the sample temperature during irradiation impossible. No visible change in the polished surfaces could be observed after irradiation except for a slight yellowish discoloration near the edges.

RESULTS AND OBSERVATIONS

Each irradiated crystal was examined by X-ray diffraction at least three times: before irradiation, after irradiation, and again after a light etching of the surface. The crystals grown by the Bridgman technique were found to exhibit considerable subgrain structure, as shown in Figure 2. The overlapping peaks made both intensity measurements and lattice parameter measurements subject to considerable error. The crystals grown by the Noggle technique produced patterns with sharp single peaks.

The measured diffraction intensities are given in Table I. Intensities calculated for the "ideally imperfect" crystal at 0°K and at room temperature, 21°C, are also given. The measured intensities are in absolute units and have not been normalized.

Figure 3 shows the data of Table I for the (111) orders of the Noggle-type crystals as the logarithm of the ratio of observed intensity to that calculated without inclusion of a temperature factor, plotted against $\sin^2 \theta/\lambda^2$. This type of plot should produce a straight line having a slope equal to the Debye–Waller temperature factor. It may be seen in the plot that the three highest-order points do indeed define a straight line. The lower-order points show considerable deviation due to extinction in the case of the unirradiated material, much less deviation for the irradiated sample and, again, a greater deviation after a light surface etch. In addition, the slope of the line defined by the points is considerably different for the irradiated material.

The same data are shown in Figure 4 for the Bridgman-grown crystal. Changes in the intensities and temperature factors were in the same direction as for the Noggle-grown crystals, but were much smaller in magnitude.

Table I. Diffraction Intensities—Aluminum Single Crystals

A. Bridgman-Method Crystals

	hkl	Unirradiated	Irradiated 10^{18}	Irradiated, etched
Mo K_α	111	448×10^{-6}	450×10^{-6}	358×10^{-6}
	222	129	121	124
	333	25.9	25.5	26.3
	444	4.5	4.28	4.80
	555	1.12	1.06	1.26
Cu K_α	111	242	320	256
	222	72	80	75

B. Noggle-Method Crystals

	hkl	Unirradiated	Irradiated 10^{18}	Irradiated 10^{19}	Irradiated, etched
Mo K_α	111	381×10^{-6}	520×10^{-6}	430×10^{-6}	326×10^{-6}
	222	107	138	109	129
	333	25.6	22.0	21.6	26.5
	444	4.90	3.50	3.92	4.74
	555	1.30	0.93	0.92	1.30
Cu K_α	111	238	294	320	240
	222	75	77	55	74

C. Calculated Intensities

	hk	$0°K$	$21°C$
Mo K_α	111	851×10^{-6}	830×10^{-6}
	222	210	164
	333	52.6	26.5
	444	15.8	5.10
	555	8.76	1.26
Cu K_α	111	425	393
	222	106	78

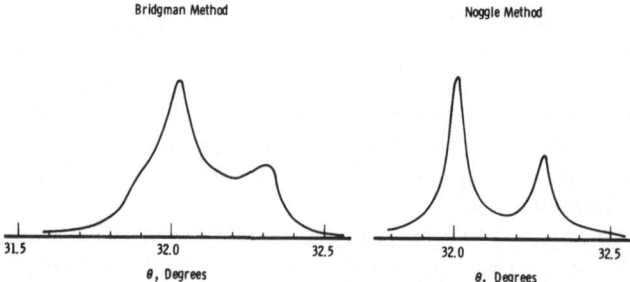

Figure 2. Single-crystal rocking curves for aluminum single crystals, showing substructure differences between Bridgman method crystals (left) and Noggle-method crystals (right). Molybdenum K_α radiation, LiF monochromator.

Lattice parameters decreased slightly upon irradiation, from 4.051 Å to 4.050 Å, for the Noggle-grown crystals. Postirradiation etching produced no significant change. Because of the substructure in the peaks of the Bridgman-grown crystals, the lattice parameters could not be measured accurately.

Line profiles were normal for the unirradiated Noggle-grown crystals, consisting of sharp, symmetric peaks. After irradiation, the peaks tended to be slightly asymmetric (Figure 5), with the steeper slope on the high-angle side. The postirradiation etch treatment restored the symmetry.

Laue patterns of the unirradiated surfaces gave moderately sharp to very sharp patterns. After irradiation, the single-crystal pattern became weakened and somewhat diffuse, with traces of an overlaying, weak polycrystalline pattern. Etching restored the sharp Laue pattern in every case.

The experimental data for the Noggle-method crystals are summarized in Table II.

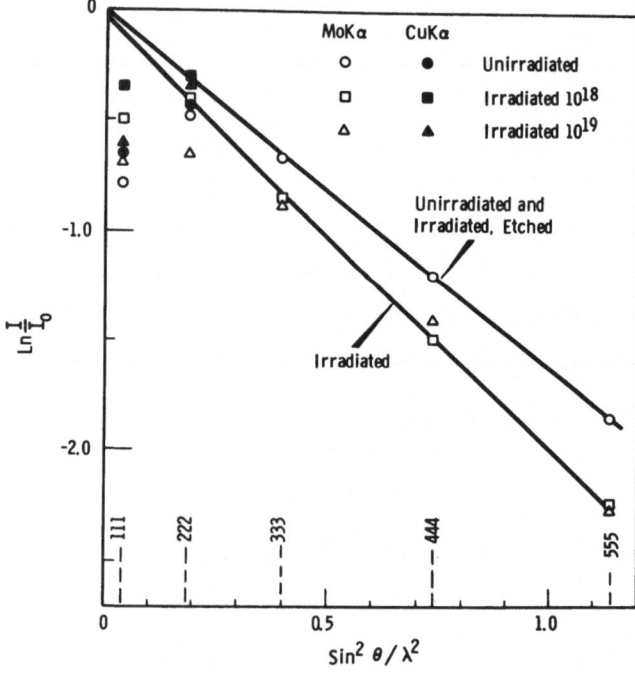

Figure 3. Ratios of observed intensity to intensity calculated omitting temperature factor plotted against $\sin^2\theta/\lambda^2$, for crystals grown by the Noggle method.

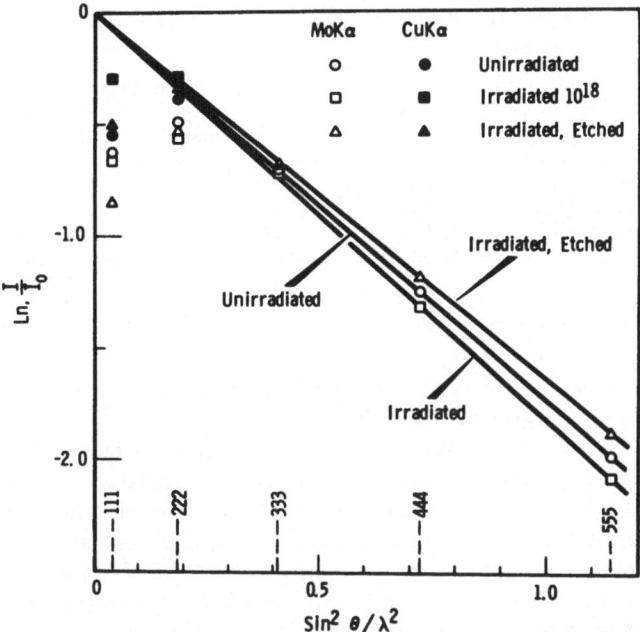

Figure 4. Intensity ratios for Bridgman-method crystals, plotted against $\sin^2\theta/\lambda^2$.

The effects observed after irradiation are evidently the result of a surface layer of relatively imperfect material. The low-order reflections are enhanced, indicating less extinction. The peak asymmetry also suggests a surface effect. The temperature factor increase may result from a static displacement of individual atoms due to disorder, possibly an effect such as that described by Huang.[8] All these effects were very much reduced in the

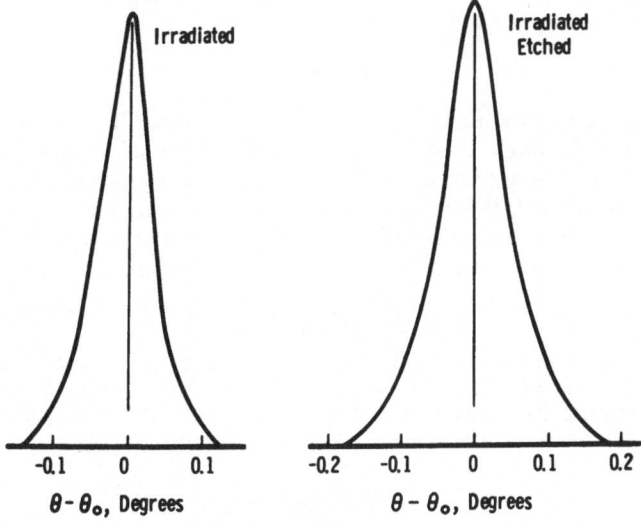

Figure 5. Profiles of 440 reflection from an irradiated Noggle-method crystal, before and after etching.

Table II. Summary of Experimental Results

Property	Sample condition		
	Unirradiated	Irradiated	Irradiated, etched
X-ray extinction	Pronounced	Moderate	Pronounced
Temperature factor	Normal	Abnormally large	Normal
Lattice parameter	4.051 Å	4.050 Å	4.050 Å
Rocking curves	Normal	Slightly asymmetric	Normal
Laue photos	Sharp	Weak, diffuse	Very sharp

irradiated Bridgman-method crystals.. No visible abrasion or other mechanical distortion could be detected on the ground faces after irradiation.

All these effects were removed by etching, leaving a surface as free, if not more free, of crystalline imperfections than the unirradiated surface. Apparently no stable cluster configurations remain within the volume of the material. The disordered surface layer probably forms through the accumulation of point defects—and possibly clustered defects as well—at the crystal surface. The energy barrier opposing the complete annihilation of defects in this thin layer is unknown, possibly involving interactions between potential fields at the surface itself and around the moving defects.

A second possibility is that the disordered layer is not directly a result of irradiation, but of residual cold work in the aluminum crystal. This view is supported by the slightly higher lattice parameter of the unirradiated crystal. Irradiation may, by some unknown mechanism, cause a concentration of dislocation lines near the surface and a corresponding dearth of imperfections in the interior.

Warren[1] reported a surface effect in irradiation of a copper–silicon alloy single crystal. However, he found the diffraction peaks to be relatively unaffected in the surface layer, becoming broadened and weakened only after etching. This effect and that observed in the present work are possibly related, the movement of and interaction of the mobile defects being modified by the alloying atoms.

SUMMARY

Diffraction experiments indicate that a disordered layer is formed on the surface of irradiated aluminum single crystals having a high degree of crystal perfection. More imperfect crystals do not appear to form this surface layer. Irradiation-induced point defects moving toward the free surface are suggested as a possible cause of the disordered layer. The interior of the crystal retains its highly perfect state after irradiation, from which it may be concluded that stable defect configurations are not present.

ACKNOWLEDGMENT

Grateful appreciation is expressed to W. J. Friesen for making the sample crystals available and for furnishing the neutron diffraction patterns, and to M. L. Sorick and J. M. Leahy for their patient work in preparing the crystal surfaces. This work was performed as a part of a program for investigation of the effects of irradiation on metals, under contract with the Atomic Energy Commission.

REFERENCES

1. B. E. Warren, *USAEC Reports*, NYO-3732, 1952; NYO-6511, 1954.
2. A. W. McReynolds, W. Augustyniak, M. McKeown, and D. B. Rosenblatt, *Phys. Rev.* 98: 418, 1955.

3. P. W. Bridgman, *Proc. Nat. Acad. Sci.* **60**: 306, 1925.
4. T. S. Noggle, *Rev. Sci. Instr.* **24**: 184, 1953.
5. T. S. Noggle and J. S. Koehler, *Acta Met.* **3**: 260, 1955.
6. T. C. Furnas, *Single Crystal Orienter Manual*, General Electric Co., Milwaukee, Wisc., 1956, p. 96.
7. W. L. Bond, *Acta Cryst.* **13**: 814, 1960.
8. K. Huang, *Proc. Roy. Soc. (London)* *A***190**: 102, 1947.

DISCUSSION

F. R. L. Schoening (Franklin Institute): What is the purity of your material?

H. E. Kissinger: Kaiser Aluminum, 99.996% pure.

David T. Keating (Brookhaven National Laboratory): Was not Warren's work, as quoted in your paper, in the opposite direction to your work?

H. E. Kissinger: You are absolutely right, it was in the opposite direction.

ORIENTED SINGLE CRYSTALS OF ALUMINUM FOR X-RAY ANALYSIS*

Joseph M. Dhosi, Charles P. Gazzara, and
Raymond M. Middleton

Watertown Arsenal Laboratories
Watertown, Massachusetts

ABSTRACT

A simple gradient heating apparatus was developed and used successfully in growing oriented single crystals of high-purity aluminum suitable for X-ray diffraction analysis.

INTRODUCTION

Considerable effort has been devoted to the growing of aluminum single crystals. Since aluminum does not undergo an allotropic transformation, it would appear that the Czochralski technique† could be used to grow rather large, oriented single crystals of aluminum free of low-angle grain boundaries. Upon investigation of X-ray Laue back-reflection spots from several aluminum single crystals pulled from the melt, small-angle grain boundaries of approximately 1 to 2° angular separation were revealed. In order to grow oriented aluminum single crystals having dislocation densities below 10^6 lines/cm^2 with no apparent small-angle boundaries, Elbaum[1] found that the crystal had to be grown below 2 mm in diameter and at a rate of growth not greater than 10^{-3} cm/sec. This work supports the theory that collapsing vacancy disks constitute a very important source of dislocation lines in crystals grown from the melt. Howe and Elbaum[2] were able to produce virtually dislocation-free aluminum crystals less than 0.5 mm in diameter. In order to grow larger oriented crystals of aluminum for X-ray diffraction examination, a strain anneal technique is recommended by Honeycombe[3] in an excellent review of methods for growing single crystals of metals. In growing oriented single crystals of aluminum by the traveling gradient strain anneal method, the furnace design is critical and the specimens must be handled with care. This report describes a method which afforded the greatest success in growing oriented aluminum single crystals by a strain anneal technique.

EXPERIMENTAL PROCEDURE

The starting material used in this investigation was 99.99 + % aluminum rod $\frac{1}{4}$ in. in diameter. Cylindrical rod was selected because the desired crystal orientation could be more closely selected initially by rotating the rod about its longitudinal axis, thereby minimizing the degree of bending for further growth of the selected crystal through

---

* The statements and opinions expressed in this article are those of the Authors and do not necessarily indicate the views or policy of the Army Ordnance Corps.

† This method of growing single crystals, as well as many others, is described in Reference 3.

[1] Superscripts pertain to references at the end of the paper.

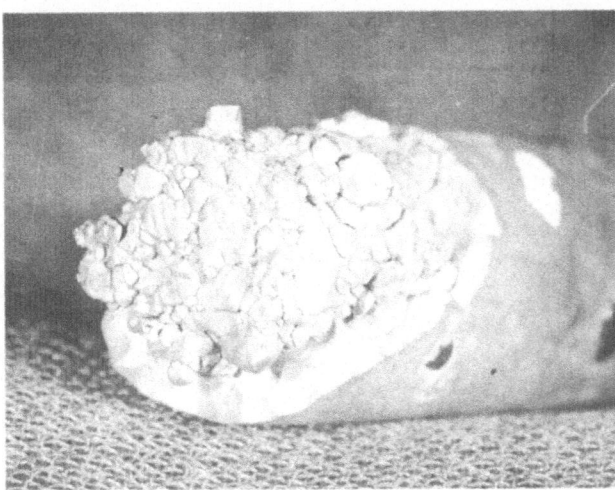

Figure 1. Aluminum single-
crystal cylinder.

the bent portion of the rod. The limiting amount of twist and bend that a crystal may be subjected to has been determined by Hagg and Kärlsson[4] to be 30 to 45° and 15 to 20° respectively. One of the main requirements for growing single crystals of high-purity aluminum is to provide a maximum temperature gradient across the aluminum rod. Initially, attempts were made to grow single crystals in these rods by immersing the cold-worked rods in a salt bath at approximately 650°C. With this procedure, it may appear that a large grain has been obtained, but if the apparent single crystalline section is cut and examined, a polycrystalline substructure may be seen by etching in NaOH solution or by "selective embrittlement cutting" using liquid metal. Liquid gallium metal was deposited on a grain boundary of the aluminum specimen and the outer

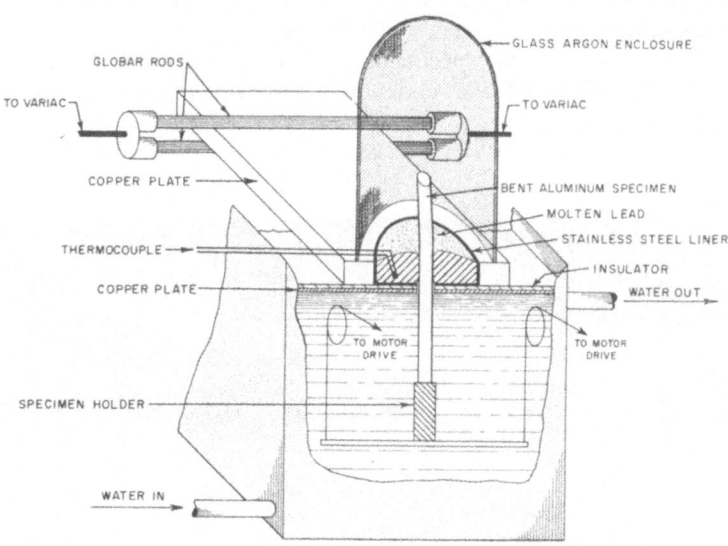

Figure 2. Gradient furnace. Molten metal—water arrangement.

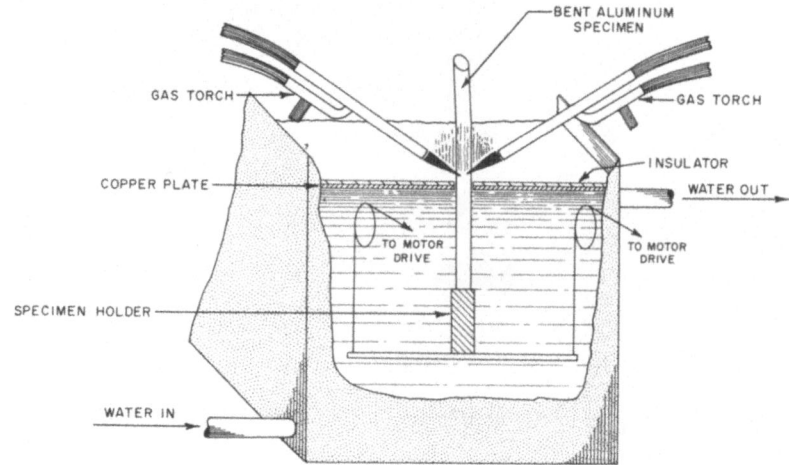

Figure 3. Gradient furnace. Gas flame—water arrangement.

aluminum crystal shell was detached from the specimen, revealing the polycrystalline substructure shown in Figure 1.

The aluminum rods were prepared as follows:

1. The $\frac{1}{4}$-in.-diameter rod was swaged and straightened to 0.156 in. diameter.
2. Polished tensile specimens having gage lengths of approximately 6 in. were sealed in quartz tubes and evacuated to 0.1 μ.
3. Annealing was performed for 1 hr at 380°C, which was 30° above the minimum temperature necessary for complete recrystallization of these specimens. At 425°C or above a duplex grain size resulted.
4. The specimens were marked every $\frac{1}{2}$ in. and strained from 1 to 2% elongation, making certain that the strain was uniform.
5. The gripped ends of the specimens were cut with a jeweler's saw and etched in 10% NaOH solution for several minutes to remove any cold work introduced by the cutting.
6. The aluminum rod was then placed in either of the gradient furnaces shown in

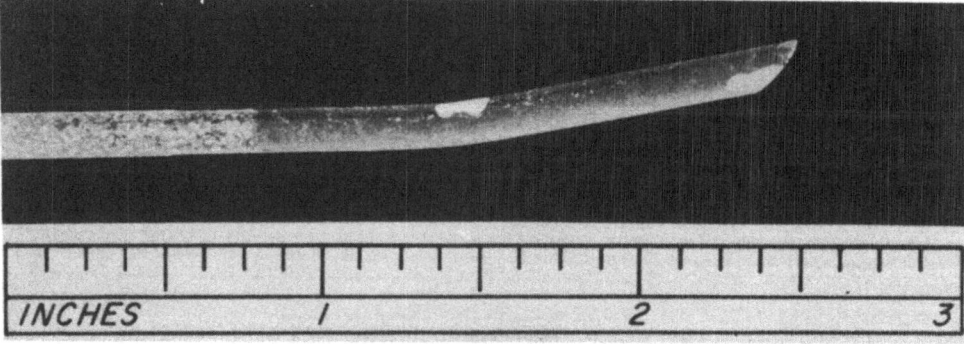

Figure 4. Bent aluminum single crystal (2×).

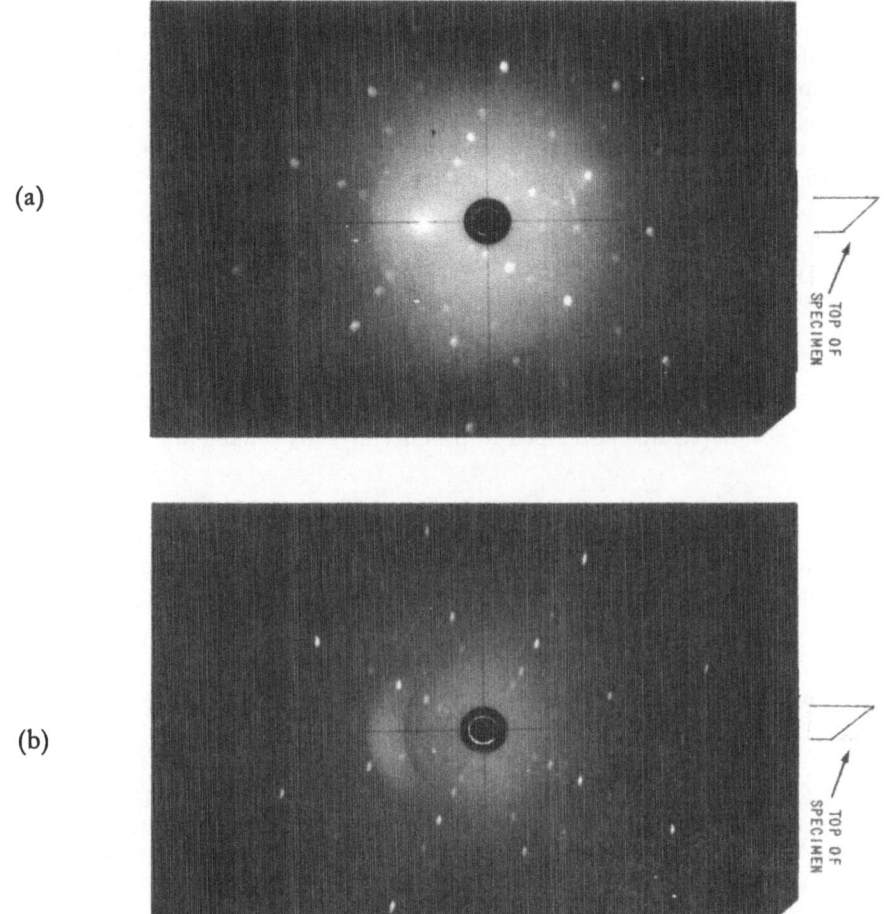

Figure 5. (a) Single crystal of aluminum with 200 plane approximately 10° to specimen surface. (b) Single crystal of aluminum oriented on 200 plane.

Figures 2 and 3 and raised at a rate of 1 in./hr through the hot zone until the top of the rod was approximately 1 in. above the hot zone.

7. The rod was then removed, etched in a 10% NaOH solution for 15 sec, and mounted on a goniometer, and back-reflection X-ray patterns were taken to select a particular plane, within 15° of the specimen face perpendicular. Since the rod is cylindrical, it may be rotated to eliminate twisting of the specimen. The specimen is bent in the polycrystallized region the required number of degrees and placed back in the gradient furnace to continue growing the selected grain. An example of a grown single crystal of aluminum is shown in Figure 4; the Laue back-reflection patterns of the unoriented and oriented crystal are shown in Figure 5.

The gradient furnace shown in Figure 2 utilizes a molten lead pot to make intimate contact with the aluminum rod as it emerges from the cold-water bath. The temperature of the lead was controlled at 660°C, and oxidation was prevented by the argon enclosure. The clearance between the aluminum rod and the stainless steel liner was 0.005 in., the molten lead being retained by surface tension.

A simpler arrangement, shown in Figure 3, replaces the lead pot with a torch at a specified distance above the tank. The flame temperature is maintained at just below the temperature at which a dummy aluminum rod would melt. This method is recommended over that employing a lead pot since the construction of a heater and temperature control apparatus is avoided and the likelihood of damage to the aluminum crystal is minimized.

The crystals produced to date have given a half-width of 8 min of arc, as determined with a double-crystal spectrometer.

CONCLUSION

A simple, dependable method of growing oriented aluminum single crystals by a strain anneal technique was developed to produce crystals suitable for X-ray diffraction analysis.

ACKNOWLEDGMENT

The authors should like to thank Mr. David Tracy for his help in preparing and growing the aluminum single crystals and Mr. John A. Alexander for helpful suggestions.

REFERENCES

1. C. Elbaum, "Dislocation in Metal Crystals Grown from the Melt," *J. Appl. Phys.* **31**: 1413, 1960.
2. S. Howe and C. Elbaum, "Dislocation Free Aluminum Crystals," *J. Appl. Phys.* **32**: 742, 1961.
3. R. W. Honeycombe, "The Growth of Metal Single Crystals," *Metal. Rev.* **4** (13): 1, 1959.
4. G. Hagg and N. Kärlsson, "Aluminum Monochromator with Double Curvature for High Intensity X-ray Powder Photographs," *Acta Cryst.* **5**: 728, 1952.

AN X-RAY DETERMINATION OF THE DEBYE– WALLER FACTORS FOR Cu₂O AND UO₂ AND THE ATOMIC SCATTERING FACTOR FOR Cu IN Cu₂O

C. J. Sparks, Jr. and B. S. Borie

Oak Ridge National Laboratory*
Oak Ridge, Tennessee

ABSTRACT

The Debye–Waller factor for Cu_2O and the atomic scattering factor for copper were determined independently of each other at one temperature from measurements of both the diffuse scattering and the integrated intensities of the Bragg maxima. The scattering factor agreed well with that computed from theory. The characteristic temperature computed from the Debye–Waller factor at 298°K was found to be 182°K. At 88°K its value is 152°K.

Similar measurements of the Bragg maxima with a powder sample of UO_2 were not useful because of preferred orientation. However, the diffuse intensity is not so sensitive to preferred orientation and may be used to determine the characteristic temperature. Its value for UO_2 was found to be 188°K at room temperature.

INTRODUCTION

Because the atoms in a crystal are not at rest, the intensities of the Bragg reflections of its X-ray diffraction pattern are reduced by the factor e^{-2M}, the Debye–Waller temperature factor. This intensity loss from the Bragg reflections is distributed as temperature diffuse scattering (TDS) throughout reciprocal space. For a simple material, one may obtain its characteristic temperature from a measurement of $2M$. This quantity is the simplest one-parameter description of the frequency spectrum of the solid, and in many cases it is adequate to compute the temperature-sensitive properties of the material. A review of the methods of measuring characteristic temperatures has been written by Herbstein.[1]

Usually $2M$ is determined by measuring the integrated intensities of the Bragg maxima and then reducing the data to values of $f^2 e^{-2M}$. One then either assumes that f may be computed theoretically to obtain $2M$, or one makes measurements at two temperatures and assumes that the temperature dependence of $2M$ is known from theory.

An alternative[2] is to measure not only the integrated intensities but the diffuse scattering as well, which also depends on f and $2M$. These two sets of measurements may then be used to determine both f and $2M$ without changing the temperature. Described here are the results of such measurements for Cu_2O.

In the case of UO_2, this method failed because preferred orientation made the integrated intensity measurements useless. However, the TDS determinations, which

---

* Operated by the Union Carbide Corporation for the U.S. Atomic Energy Commission.
[1] Superscripts pertain to references at the end of the paper.

are very much less sensitive to preferred orientation, were combined with Thomas–Fermi values of f_U to obtain $2M$ for this material.

Values of the characteristic temperatures for both of these oxides were obtained from the experimentally determined Debye–Waller factors.

THEORY

It is desirable to treat the TDS in terms of Warren's[3] theory developed for close-packed cubic powders. He showed that

$$I_{TD} = f^2(1 - e^{-2M})C_1 \tag{1}$$

where I_{TD} is the intensity in electron units per atom, f the atomic scattering factor, and C_1 a modulating function of average value unity. C_1 arises from the coupling between atoms and is calculated by Warren for a close-packed cubic powder pattern.

To use this expression, intensity measurements are made and converted to absolute units. Then, each intensity measurement far enough away not to include any part of the Bragg reflection is divided by the value C_1 at that position in reciprocal space. This reduces Warren's TDS expression to $f^2(1 - e^{-2M})$, which is further reduced to $(e^{2M} - 1)$ when divided by f^2e^{-2M} as determined from the integrated intensity of the Bragg reflections. In using this value of C_1, one assumes that Cu_2O and UO_2 behave as close-packed cubic materials.

In both Cu_2O and UO_2, the metal atoms occupy a face-centered cubic lattice, and the TDS may be treated as from close-packed cubic powders if the oxygen contribution is negligible. This is very nearly true for Cu_2O because the ratio of $(4f_{Cu})^2/(2f_O)^2$ is always greater than forty for the 2θ range investigated. Of those Brillouin zones for which the structure factor contains both copper and oxygen contributions, half have a structure factor of $(4f_{Cu} + 2f_O)$ and the other half $(4f_{Cu} - 2f_O)$. Hence, the contributions to the powder pattern TDS from the cross terms of F^2 for these zones will tend to cancel each other, leaving $[(4f_{Cu})^2 + (2f_O)^2]$. The remaining Bragg maxima have either $4f_{Cu}$ or $2f_O$ as their structure factor. Under these circumstances, it is adequate to account for the very small contribution of oxygen to the TDS by writing equation (1) as

$$I_{TD} = [(4f_{Cu})^2 + (2f_O)^2](1 - e^{-2M})C_1 \tag{2}$$

This small correction for oxygen in equation (2) further assumes that both atoms have the same Debye–Waller factor, which in this case is very nearly so. The fact that the oxygen atoms lie on a body-centered cubic lattice is not expected to alter C_1 appreciably since oxygen does make such a small contribution to the TDS.

Similar arguments for neglecting the oxygen contribution are valid for UO_2. The ratio of $(4f_U)^2/(8f_O)^2$ is always greater than forty for the 2θ range over which intensity measurements were made. In addition, one-fourth of the Bragg maxima have a structure factor $(4f_U + 8f_O)$ and one-fourth have $(4f_U - 8f_O)$. On averaging their contribution to the TDS for a powder, the oxygen contribution is again negligible. The remaining Bragg maxima have a structure factor $4f_U$. Thus, we write equation (1) for UO_2 as

$$I_{TD} = (4f_U)^2(1 - e^{-2M})C_1 \tag{3}$$

DETERMINATION OF $2M$ AND f_{Cu} FOR Cu_2O

Reagent-grade Cu_2O powder was treated at $900°C$ for 12 hr in 2 mm Hg of air to convert any free Cu to Cu_2O. The material was then lightly ground with a mortar and pestle to achieve small particle size and to minimize extinction effects, but not enough

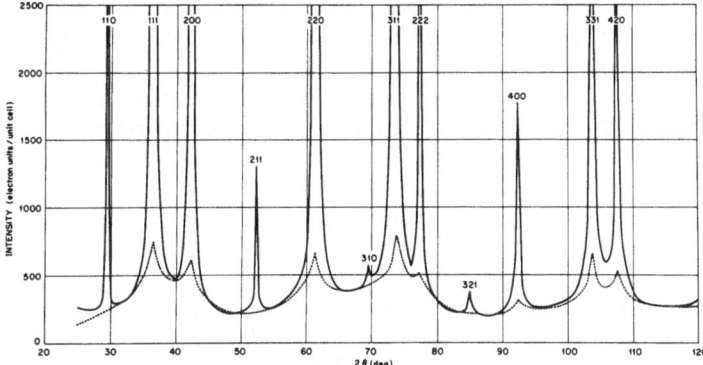

Figure 1. Measured X-ray coherent diffuse intensity for Cu_2O at 298°K. Dashed curve is Warren's I_{TD} expression fitted to the measured curve.

to broaden significantly the Bragg peaks. A flat sample holder was carefully back-loaded with <400 mesh powder to minimize preferential alignment of the particles. That a smooth curve could be drawn through the values of fe^{-M} determined from the integrated intensity measurements attests to the fact that preferred orientation was negligible.

The powder samples were mounted on a GE XRD-5 diffractometer in an evacuated cryostat to eliminate air scattering. Monochromatic Cu K_α radiation from a doubly bent LiF crystal, with Ni–Co filters balanced to eliminate the half-wavelength component of the beam, resulted in a high signal-to-noise ratio by reducing fluorescing and other unwanted radiation.

The intensity was measured at every degree 2θ between the Bragg reflections and as often as every 0.05° across the Bragg peaks. The intensity measurements were converted to absolute units (electron units per unit cell) by measuring the power of the incident beam by scattering it from polystyrene $(C_8H_8)_x$ at 100° 2θ. At this angle polystyrene

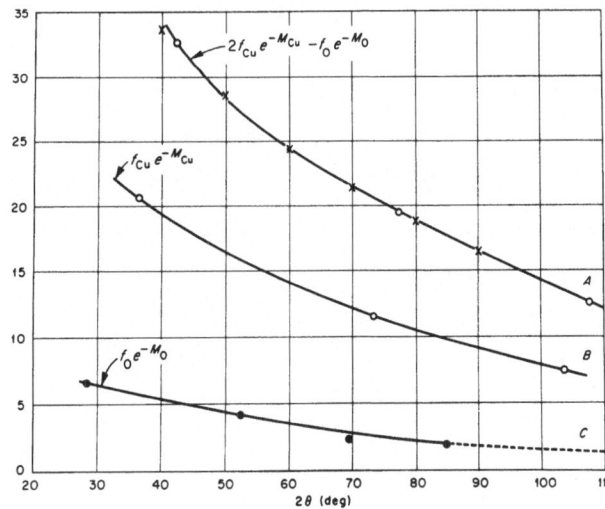

Figure 2. Form factors with their Debye–Waller factors for Cu_2O as determined from integrated intensity measurements. Crosses on Curve A are computed from Curves B and C.

may be assumed structureless. The Compton scattering was computed and subtracted from the measurements with the use of values given by Freeman[4] for copper and Compton and Allison[5] for oxygen.

Figure 1 shows the measured coherent diffuse intensity in electron units per unit cell for Cu_2O at room temperature. The dashed curve represents the I_{TD} function $[(4f_{Cu})^2 + (2f_O)^2](1 - \exp(-2M_{Cu}))C_1$. This was obtained by fitting Warren's expression to the valleys and then calculating it to extend under the Bragg peaks. One may now make accurate integrated intensity measurements of the Bragg maxima.

Figure 2 shows the atomic scattering factors with Debye–Waller factors determined from these integrated intensity measurements. The internal self-consistency of these data is shown by computing the crosses on Curve A from Curves B and C.

The curves of Figure 2 were combined to obtain values of $[(4f_{Cu})^2 + (2f_O)^2]$ $\exp(-2M_{Cu})$. These values were used with the TDS data of Figure 1 to obtain values of $2M_{Cu}$ independently of f_{Cu} and f_O. The result is shown in Figure 3. Because the diffuse intensity is so insensitive to the thermal motion of the oxygen atoms, $2M$ determined here was interpreted to be that of copper. $2M_O$ is probably not too different from $2M_{Cu}$, as can be estimated by comparing f_O from theory with the experimental $f_O \exp(-M_O)$.

Since

$$2M = 2B \frac{\sin^2\theta}{\lambda^2} = \frac{16}{3}\pi^2\overline{u^2}\frac{\sin^2\theta}{\lambda^2} \tag{4}$$

the slope of the straight line in Figure 3 gives the value of $2B$ from which the root-mean-square displacement $(\sqrt{\overline{u^2}})$ of the copper atoms from their lattice sites is calculated to be 0.25 Å at 298°K. Using the experimentally determined value of $2M_{Cu} = 3.3 \sin^2\theta/\lambda^2$ and the values of $f_{Cu} \exp(-M_{Cu})$ of Curve B, Figure 2, one may finally reduce the data to f_{Cu}. This quantity is compared in Figure 4 with the theoretical values given by Freeman and Watson[6] with a dispersion correction as tabulated by Dauben and Templeton.[7]

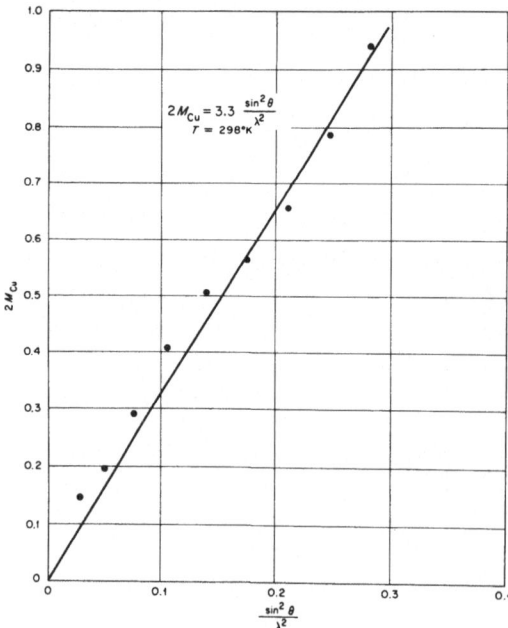

Figure 3. The Debye–Waller factor for copper in Cu_2O as determined from the data of Figures 1 and 2.

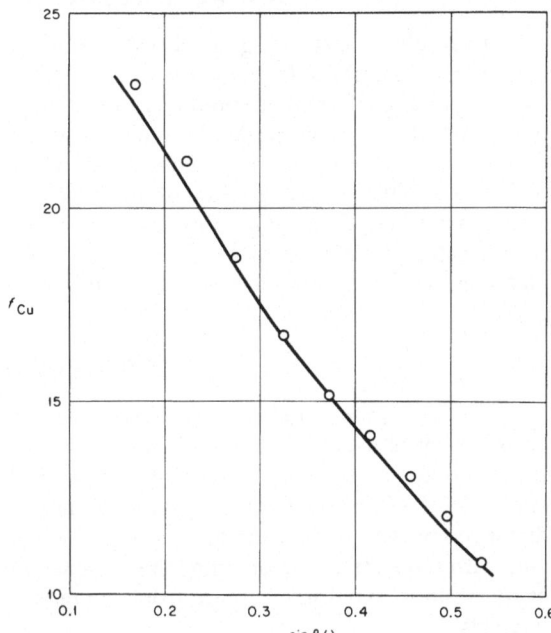

Figure 4. Comparison of observed atomic scattering factor for copper with that computed by Freeman and Watson. Observed factors are shown as circles, computed factors as a solid line.

The I_{TDS} from Cu_2O was also determined at a temperature of 88°K. The ratio of the diffuse intensity away from the Bragg reflections gives

$$\frac{I_{TDS\ 88°K}}{I_{TDS\ 298°K}} = \frac{1 - \exp\left(-2M_{88}\right)}{1 - \exp\left(-2M_{298}\right)} \tag{5}$$

but $1 - \exp\left(-2M_{298}\right)$ is known from the room temperature data and the values of $2M$ at 88°K are thus determined. A plot of $2M_{88°K}$ as a function of $\sin^2\theta/\lambda^2$ shown in Figure 5 gives a value of $2B = 1.50$ Å² from the slope of the line. The corresponding root-mean-square displacement of the copper atoms from their lattice sites is 0.17 Å at 88°K.

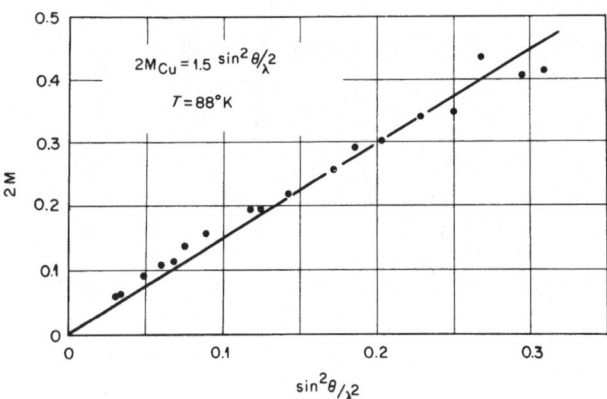

Figure 5. The Debye–Waller factor for copper in Cu_2O as a function of $\sin^2\theta/\lambda^2$ at 88°K.

DETERMINATION OF $2M$ FOR UO_2

A sample of very fine-grained uranium dioxide, which chemical analysis showed to have the composition $UO_{2.029}$, was prepared in a manner similar to that for Cu_2O. Nevertheless, preferred orientation made it impossible to measure the true integrated intensities of the Bragg peaks, but this is not expected to affect the TDS appreciably. The diffuse scattering measurements made with this sample at 300°K are shown as circles in Figure 6. They are compared with the solid curve calculated using the Thomas–Fermi atomic scattering factor for uranium[8] with a dispersion correction of -5 units[7] and with a value of $2B$ of 0.88 Å^2. The corresponding value of the root-mean-square displacement of the uranium atoms from their sites is 0.12 Å at room temperature.

OBSERVATIONS

In general, the fit between experiment and theory shown in the figures is reasonably good. Strain is usually introduced into the samples during preparation and this causes the tails of the Bragg reflections to interfere with the diffuse intensity. To avoid extinction effects on the integrated intensity measurements, however, it is desirable to have some distortion present in the sample. Possibly, the best solution would be to use two samples: One with some strain for the integrated intensity measurements, and the other with little or no strain to avoid interference from the tails of the Bragg reflections with the diffuse intensity.

An assignment of precision to the values of $2M$ obtained is not easy and at best subjective. Probably the values obtained for Cu_2O are more accurate than that for UO_2. The dispersion corrections and the atomic form factor for uranium are computed from theoretical considerations and have not been tested experimentally. The absorption coefficient for uranium is large and probably not accurately known. There exists the possibility of some multiple scattering contributing to the diffuse intensity measurements. Warren's[9] investigation of the magnitude of this contribution shows it to be very small except possibly at very low angles. Surface roughness effects, though probably small, may contribute to the inaccuracy of the final results.

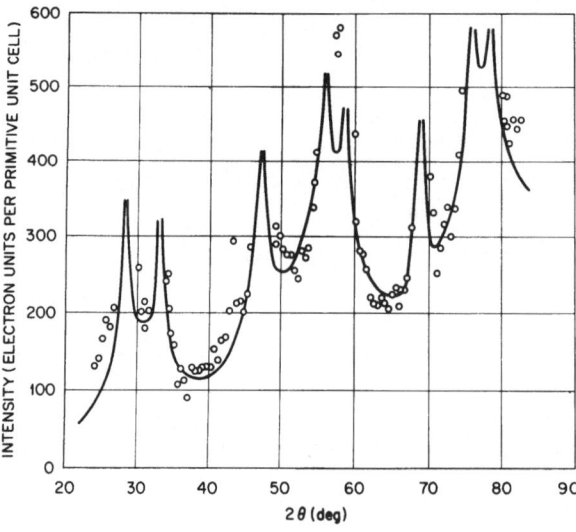

Figure 6. Temperature diffuse X-ray scattering for UO_2 at 300°K. Solid curve computed for $2M = 0.80 \sin^2\theta/\lambda^2$.

The only diffraction data available for comparison with our results are neutron diffraction measurements of the integrated intensities of UO_2 by Willis et al.[10] Their result for the root-mean-square displacement of uranium atoms from their sites at 293°K is 0.09 Å compared with our value of 0.12 Å.

Having determined the parameter $2M$, one may calculate the characteristic temperature Θ_M which is related to $2M$ by the expression

$$M = \frac{6h^2T}{mk\Theta_M{}^2}\left(\Phi(x) + \frac{x}{4}\right)\frac{\sin^2\theta}{\lambda^2} \tag{6}$$

where h and k are Planck's and Boltzmann's constants, T the absolute temperature, m the mass of the harmonic oscillator in grams, and $\Phi(x) + x/4$ is a quantity near unity if the temperature is not too low.

If there is more than one kind of atom in the crystal, the meaning of the characteristic temperature is no longer precise. The frequency spectrum is not a continuous distribution with a single maximum frequency ν_M which is related to Θ_D by $h\nu_M = k\Theta_D$. It is divided into acoustic and optic branches, the two being separated by a band of forbidden frequencies.

Nevertheless, it is common practice to associate with a measurement of $2M$ for such crystals a value of Θ_M as given by equation (6). It is usual to follow a suggestion made by Lonsdale[11] to use for m in this expression simply the average of the masses of the kinds of atoms.

This procedure is practical if the masses are not very different. The band of forbidden frequencies is then narrow and unimportant, and one wishes Θ_M to correspond to the maximum frequency of the optic branch.

If, however, the mass difference is very large, the optic modes have very high frequencies distributed over a very narrow range, and the band of forbidden frequencies is quite wide. Under these circumstances, it would seem better to associate Θ_M with the maximum frequency of the acoustic branch and neglect the optic modes. In effect, instead of interpreting equation (6) in terms of the limiting case that the masses approach equality, we propose to use the equation in the limit that the smaller of the masses approaches zero.

If this is done, m of equation (6) is the mass of the heavier atom. The value of Θ_M for Cu_2O at 298°K computed this way is 182°K. At 88°K, it is 152°K. For UO_2, $\Theta_M = 188$°K at 298°K.

A recent article[12] on the heat capacity of cuprous oxide in the low-temperature range of 2.8 to 21°K gave the result of 125 to 140°K for the Debye characteristic temperature. This result compares favorably with the value determined by the present authors of 152°K and 88°K.

There are other arguments for interpreting $2M$ for heavy-metal oxides in this way. Perhaps the best one is that it tends to give values of the characteristic temperature in agreement with low-temperature specific heat measurements. For example, Jones et al.,[13] find a value of 160°K for UO_2 from low-temperature specific heat data, while Willis et al. interpret their diffraction data in terms of Lonsdale's suggestion and obtain a value of 395°K. Roof,[14] using the average mass for PuO_2, finds from diffraction measurements a value of $\Theta_M = 415$°K. If only the mass of plutonium is used, there results a value of the characteristic temperature of 256°K, in good agreement with Sandenaw's[15] value of 253°K determined from heat capacity measurements at low temperatures.

In any case, the concept of a characteristic temperature for the materials here considered is indistinct, and its value should not be overemphasized.

ACKNOWLEDGMENTS

It is a pleasure to acknowledge the help given by Dave Welch in obtaining the data and by Dr. D. L. McElroy, who furnished the UO_2 samples.

REFERENCES

1. F. H. Herbstein, *Advances in Physics*, *Vol. 10*, 1961, p. 313.
2. B. Borie, *Acta Cryst.* **9**: 617, 1956.
3. B. E. Warren, *Acta Cryst.* **6**: 803, 1953.
4. A. J. Freeman, *Acta Cryst.* **12**: 274, 1959.
5. A. H. Compton and S. K. Allison, *X-rays in Theory and Experiment*, D. Van Nostrand Co., Inc., Princeton, N.J., 1935.
6. A. J. Freeman and R. E. Watson, *Acta Cryst.* **14**: 231, 1961.
7. C. H. Dauben and D. H. Templeton, *Acta Cryst.* **8**: 841, 1955.
8. *International Tables for the Determination of Crystal Structures*, *Vol. 2*, Gebruder-Borntraeger, Berlin, 1935.
9. B. E. Warren, *J. Appl. Phys.* **30**: 1111, 1959.
10. T. M. Willis, K. A. D. Lambe, and T. M. Valentine, U.K. A.E.A. Research Group Report, AERE-R 4001, 1962.
11. K. Lonsdale, *Acta Cryst.* **1**: 142, 1948.
12. L. V. Gregor, *J. Phys. Chem.* **66**: 1645, 1962.
13. W. M. Jones, J. Gordon, and E. A. Long, *J. Chem. Phys.* **20**: 695, 1952.
14. R. B. Roof, Jr., *J. Nuclear Materials* **2**: 39, 1960.
15. T. A. Sandenaw, to be published, 1962.

DISCUSSION

H. M. Otte (RIAS): I did not quite understand the curves where you showed $f_{Cu} \exp(-M_{Cu})$ and $f_O \exp(-M_O)$. How did you develop them?

C. J. Sparks: The curves to which you refer are shown in Figure 2. These curves were derived from the integrated intensity of the Bragg reflections. With reference to Figure 1, the areas under the Bragg peaks are planimetered, with the entire peaks plotted, of course. Now, knowing the power of the incident X-ray beam and all the constants in the intensity equation, one has determined $F^2 (\exp - 2M)$. We then take the square root to get the structure factor, which is plotted in Figure 2. In our case, with Cu_2O, there are three or four points on each curve from the different Bragg peaks, so we could plot the structure factors as a continuous function with respect to 2θ.

H. S. Peiser (National Bureau of Standards): Do anharmonic terms contribute to the temperature diffuse scattering here in such a way as to make it a factor in your consideration?

C. J. Sparks: No, I do not believe so. There has been some work on this problem which leads one to expect that the effect of anharmonic waves would be negligible except at temperatures approaching the melting point. Inelastic neutron scattering from single crystals of lead suggested that anharmonic effects are small.

Since UO_2 has more than one atom per unit cell, there are three acoustical and six optical modes associated with the transverse and longitudinal waves. This makes the system quite complicated, and one cannot properly expect to correct for higher-order effects such as the anharmonic waves.

AN X-RAY STUDY OF THE STRUCTURE OF
THE ALKALINE EARTH OXIDE CATHODE

Paul Lublin

General Telephone and Electronics Laboratories, Inc.
Bayside, New York

ABSTRACT

An X-ray diffraction study has been conducted on the emissive coating (Ba, Sr, CaO) of experimental diodes from which thermionic emission data were taken before X-ray analysis. The tubes were then opened and the oxide protected by special techniques in order to prevent the formation of the hydroxide. In addition to X-ray diffraction, other techniques were used to give a complete description of the structures present.

For many years there has been interest in determining the crystal structure of the oxide cathode. It was hoped that some relationship between structure and emission could be revealed by careful X-ray techniques and other supporting methods. One of the difficulties encountered when a tube was opened to the air was the reaction of the coating with the moisture and oxygen of the air. A preliminary paper on work done at our laboratory was given in 1959.[1]

Most of the work reported in the literature was aimed at determining the crystallite size of the oxide and relating it to the emission. Eisenstein[2] investigated the double oxide (BaSr)O system using film techniques and protective coatings such as polystyrene and waxes while opening the tubes in dry nitrogen. He examined the crystallite size in both the carbonate and the oxide, and also the lattice parameter as a function of composition for the double system. Yamaka[3] constructed a special tube with mica windows that he used to study crystal growth and size, which he then tried to relate to emission. Ostapchenko[4] used primarily dry nitrogen and protective waxes to examine his cathodes. He found a single phase on samples of triple oxide which had been decomposed at high temperatures, and related this to good emission. Terada[5] also used dry nitrogen and protective coatings for his X-ray techniques. He discussed the triple oxide as a system consisting of three separate phases or compositions. Rooksby,[6] again using similar protective techniques, measured the crystallite sizes of the double oxide using microbeam techniques. Up to now everyone had noticed the broadness of the reflections, but had attributed this to different causes. He found the crystallite size to be 2 to 5 μ; therefore this should not contribute to the broadening of the diffraction lines.

Efforts at our laboratory have been devoted to the study of the double and the triple oxide systems on some experimental diodes. A picture of the diode is shown in Figure 1. The insert shows the oxide-coated nickel cap in greater detail. Early attempts involved the use of dry boxes and protective coatings similar to those reported on in the literature. These generally proved unsatisfactory, due to the formation of some hydroxide in many

---

[1] Superscripts pertain to references at the end of the paper.

Figure 1. Experimental diode with nickel cap.

cases, and also due to a few strong reflections produced by the waxes. The technique reported earlier,[1] which was employed at this time, involved the use of an oil bath under which the tube could be opened. Cargill's oil, which is nondrying and anhydrous, has low fluorescence and critically controlled optical properties, was selected as the immersion medium. The bath was heated to about 70°C in order to drive off any moisture or air bubbles surrounding the tube. The tube was opened while completely immersed, and the oil immediately impregnated the oxide coating. The cap was removed and mounted on a special adjustable specimen post on a conventional X-ray goniometer, as shown in Figure 2. Excess oil was removed and a special box containing a $\frac{1}{4}$-mil mylar window was mounted on the goniometer surrounding the cathode cap as shown in Figure 3. Dry nitrogen was passed through the box continuously while the diffraction pattern was being recorded.

The samples had only 1.1 mg of coating, and it became necessary to increase the sensitivity of the goniometer as much as possible. The length of the slits was reduced in order to lower the background. The scintillation counter and pulse height analyzer were used, and the nickel filter was transferred from the receiving slit to the divergence slit. All of these changes increased the sensitivity of detection for small peaks. Subsequent to the recording of the diffraction pattern, the cap was placed in a bottle of oil for future microscopic examination.

Initially, the lattice constants were measured using the low-angle region, but later

Figure 2. Specimen post mounted on the diffractometer.

procedures involved the back-reflection lines for greater precision. The line breadth measurements were also evaluated more accurately in the latter part of the program. Figure 4 illustrates the type of patterns obtained.

Emission data were taken after 250 hours of life testing. Tubes in which any unusual condition occurred during life were not used. Two types of heating schedules were used: the E-1, which has a maximum of 1200°C; and the A, which has a maximum of 1100°C. Pulsed emission was measured at 1000°K and is expressed in amperes per square centimeter.

A number of tubes were prepared with the double compositions of 80%Ba–20%Sr, 50%Ba–50%Sr, and 20%Ba–80%Sr. The lattice constants were averaged for each set and are plotted against composition in Figure 5. The lattice constants for the pure oxides (from the literature) are 5.539 Å for BaO, 5.160 Å for SrO, and 4.811 Å for CaO. As can be seen from the graph, there is 100% solid solubility of BaO in SrO. Deviations

Figure 3. Atmosphere box on the diffractometer.

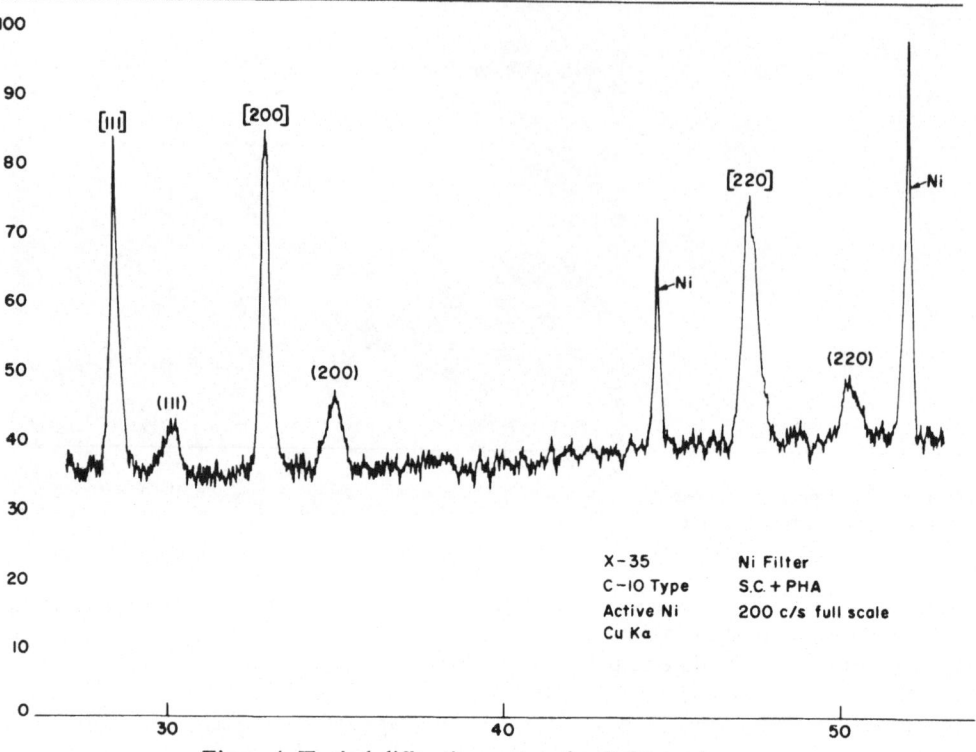

Figure 4. Typical diffraction pattern for C–10 coating.

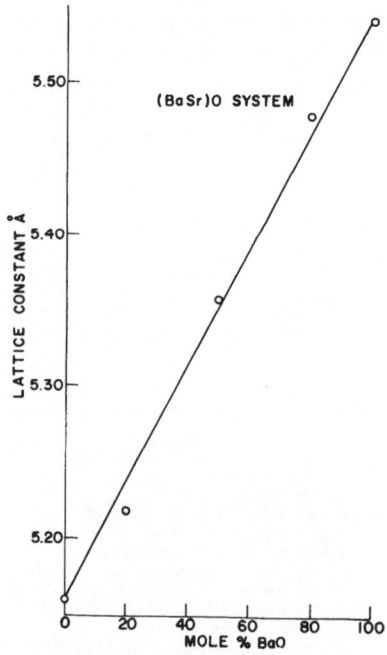

Figure 5. Plot of lattice constants against composition for the double oxide systems.

Table I. Composition of C-10 and C-4

Tube number	Emission, amp/cm^2	Schedule	a_0, main phase	Relative amount second phase	a_0, second phase	Half-width, relative
C-10 (57.2% Ba, 38.2% Sr, 4.0% Ca)						
148-1	4.6	A	5.365	M	5.10	47
148-2	4.1	A	5.401	W	5.11	52
148-4	4.0	A	5.368	W	5.11	35
148-5	4.2	E-1	5.370	N.D.	—	43
148-6	5.8	E-1	5.365	W	5.11	44
C-4 (48.0% Ba, 42.5% Sr, 9.5% Ca						
149-1	1.1	A	5.354	S	5.11	33
149-2	2.5	A	5.386	S	5.09	35
149-3	5.4	E-1	5.347	S	5.10	41
149-4	5.4	E-1	5.347	M	5.11	33
149-5	1.2	A	5.389	S	5.11	30
149-6	4.9	E-1	5.348	M	5.07	36

from linearity are probably due to the fact that the compositions listed are nominal and are not determined on the heated cathode.

The triple oxide systems studied were the commercial C-10 and C-4 carbonates and variations of these. The base metal in all cases was N-4 nickel alloy. The composition of C-10 and C-4 and the results obtained are listed in Table I. After studying this table some tentative conclusions can be drawn:

1. The double oxide cathodes contain only a single face-centered cubic solid solution.
2. The triple oxide system generally results in two face-centered cubic solid solutions with the minor phase probably consisting of an SrO–CaO solid solution. The major phase is probably a BaO–SrO solid solution containing a small amount of CaO.
3. A second phase is still apparent in the good emitter of the triple system, which is contrary to what Ostapchenko[4] and others have stated. This is attributed to the greater sensitivity of the present X-ray method.
4. C-4 cathodes require the higher temperature schedule to produce good emission, while C-10 cathodes generally give good emission on both schedules. This applies only for the N-4 base alloy used.
5. The second or minor phase is of lower concentration in the C-10 tubes than in the C-4. This is probably due to the lower calcium content of C-10.
6. The lattice constants of the major phase are generally lower in the C-4 than in the C-10 for the same schedule. Again, this may be due to the higher calcium content of the C-4.
7. The line breadths of the C-10 are generally greater than those of the C-4. The half-widths of the original carbonate powders were also measured but no correlation with the resultant oxide was found.

In conclusion, it can be said that correlation of the emission and structure of the oxide coating is not easily resolved. There are some trends indicated, but a more statistical approach may be required to establish more positive relationships.

REFERENCES

1. G. F. Tufts, P. Lublin, and B. Wolk, "A Technique for the Analysis of Activated Oxide-Coated Cathodes Using X-ray Diffraction and Light Microscopy," presented at the Electrochemical Society Meeting, May 1959.
2. A. Eisenstein, "A Study of Oxide Cathodes by X-ray Diffraction Methods," *J. Appl. Phys.* **17**: 436, 654, 1946.
3. E. Yamaka, "A Study of the Oxide Coated Cathode by X-ray Diffraction," *J. Appl. Phys.* **23**: 937, 1952.
4. E. P. Ostapchenko, "X-ray Investigation of the System of Double and Triple Carbonates and Oxides of Alkaline Earth Metal," *Izvest. Akad. Nauk. S.S.S.R. Ser. Fiz.* **20**: 1105, 1956.
5. J. Terada, "An X-ray Study of Barium–Strontium–Calcium Triple Oxide," *J. Phys. Soc. Japan* **10**: 555, 1955.
6. H. P. Rooksby, "Changes in the Structure of Oxide Cathodes at High Temperatures," *Brit. J. Appl. Phys.* **6**: 272, 1955.

DISCUSSION

Z. Wilchinsky (Esso Research and Engineering Co.): Can you give some indication of how much improvement you got by changing the nickel filter from the receiving slit to the divergence slit?

P. Lublin: As I recall, for a particular chart recording the improvement must have been close to half, or 50%. In general the background was lowered by a factor of one-third to one-half from what it was previously.

Initially we ran the samples using the diffractometer and Geiger counter. When the background seemed unusually high we transferred the nickel filter to reduce the background. Subsequently, we restricted the X-ray beam to further reduce the background, and then switched to the scintillation counter and PHA to improve the sensitivity.

APPLICABILITY OF ROUTINE METHODS OF CRYSTALLITE SIZE ANALYSIS*

Robert C. Rau

General Electric Company
Cincinnati, Ohio

ABSTRACT

Several methods for the routine determination of crystallite size by means of X-ray diffraction line-broadening have previously been reported. Although these techniques have proven useful and reliable when utilized with the single X-ray diffractometer and instrumental geometry used to originally develop the methods, it was not known whether other instruments would provide similar reliability. Therefore a study was performed to evaluate the applicability of routine methods of crystallite size analysis to other X-ray diffraction units. A series of six beryllium oxide powder specimens, whose average crystallite sizes ranged stepwise from about 35 to nearly 3000 Å, were used to test a number of X-ray diffractometers. By using a predetermined diffraction geometry for each instrument tested, measured crystallite sizes were found to be quite reproducible and well within the limits of experimental error. The testing procedure, instrumental conditions, and individual performance results are presented in this paper.

INTRODUCTION

At a previous Denver X-ray Conference, several techniques for routine crystallite size analysis by means of X-ray diffraction line-broadening were presented.[1] These techniques were all derived using a single X-ray diffractometer in the author's laboratory, and although accurate and reproducible results could be obtained with that instrument, it was not known whether similar agreement could be attained using other instruments. Therefore, a study was undertaken to determine the validity of routine crystallite size determinations performed with other X-ray diffractometers. This paper presents the results of that study and illustrates the general applicability of routine methods of crystallite size analysis.

ROUTINE METHODS FOR CRYSTALLITE SIZE ANALYSIS

Four modifications of the usual method for routinely determining crystallite size by X-ray line-broadening have been developed, including the use of (1) a graded set of powder photographs, (2) a computer program, (3) sets of crystallite size curves for given materials, and (4) a standard set of curves for any strain-free material. Since these modifications were discussed at the 1961 Denver Conference,[1] they will not be described here. It will be sufficient at this time to point out that the latter two techniques, involving crystallite size curves, have proven to be by far the most useful and readily applied, and

---

* Work performed under U.S. Atomic Energy Commission Contract AT(40–1)–2847.
[1] Superscripts pertain to references at the end of the paper.

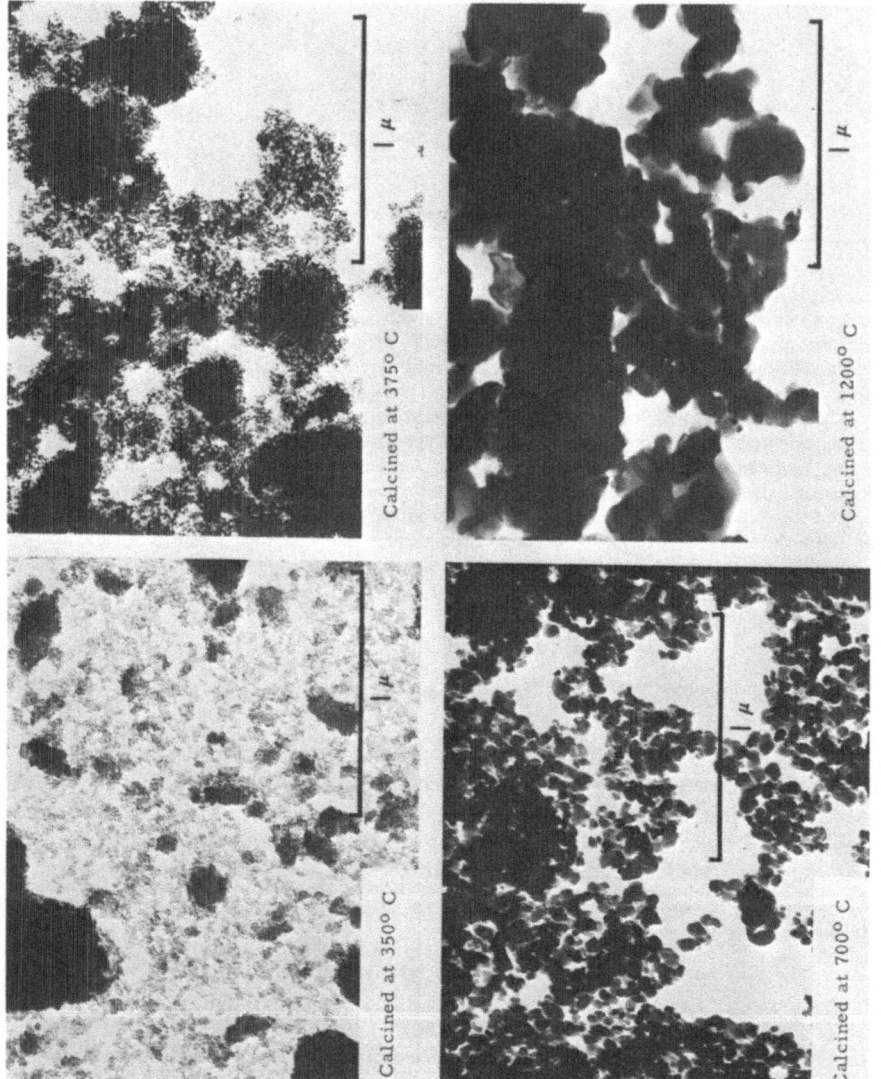

Figure 1. Electron micrographs of calcined BeO.

were therefore used in studying the applicability of routine methods with other X-ray diffractometers. However, since all four modifications were derived with the same instrumental setup in the author's laboratory, the results of the present study are applicable to all four techniques.

The four routine crystallite size techniques are all based on the Scherrer equation,* which relates diffraction peak breadths to crystal size; and the graphical aids to solving that equation, as described by Klug and Alexander.[2] In deriving the routine methods, standardized instrumental conditions were chosen and certain line-breadth correction curves were determined for those conditions. Although these correction curves, together with the routine crystallite size methods based upon them, were felt to be reliable as long as the original X-ray diffraction instrument was used, it was not known whether the reliability could be duplicated on other instruments.

Of particular concern was the validity of a derived curve showing the inherent instrumental broadening as a function of diffraction angle. Although the instrumental conditions used in deriving this curve could essentially be duplicated with a diffractometer of the same manufacture, it was feared that this duplication would not be exact enough to reproduce that curve. Furthermore, it was not known whether X-ray instruments of different manufacture would duplicate the original experimental conditions since different instrumental geometries are employed. Thus, if the instrumental broadening curve was not duplicated by other diffractometers, routine crystallite size methods based on that curve would not be valid for other diffractometers.

METHOD OF TESTING APPLICABILITY OF ROUTINE METHODS

Rather than determine an instrumental broadening curve for each diffraction instrument to be tested, it was decided that an easier course would be to assume general validity of the crystallite size curves and use them in conjunction with those diffractometers to determine crystallite sizes of a known series of specimens. This approach would test not only the validity of the original instrumental broadening curve, but would test the reliability of all the relations used in deriving the routine methods. Agreement in crystallite sizes measured by the different diffractometers would then establish the methods as generally applicable.

For use as the standard set of samples, a series of beryllium oxide powders calcined at different temperatures was chosen. The crystallite sizes of these powders had been checked previously by electron microscopy and X-ray line-broadening.[3] Six specimens were used which displayed crystallite sizes ranging stepwise from about 35 up to about 3000 Å. Four of the standard specimens are shown in the electron micrographs of Figure 1. These micrographs, all at the same magnification, illustrate the great range in crystallite sizes covered by this study.

For line-broadening work, slow diffractometer scanning speeds and fairly fast chart speeds are employed. This combination assures recorded diffraction patterns whose peak breadths can be easily and accurately read. In Figure 2 recorded patterns from the four specimens of Figure 1 are shown. This figure shows the dependence of peak breadth on crystallite size, and also illustrates some of the difficulties encountered in analyzing crystal sizes near the lower and upper size limits which can be handled. In particular, very tiny crystallites produce an almost amorphous type of pattern in which discrete diffraction peaks tend to become indistinguishable from background, while large (> 3000 Å)

* $D = (K\lambda/\beta \cos \theta)$, where D is crystallite size, K is a shape constant, λ is the X-ray wavelength, β is the corrected peak breadth, and θ is the Bragg angle.

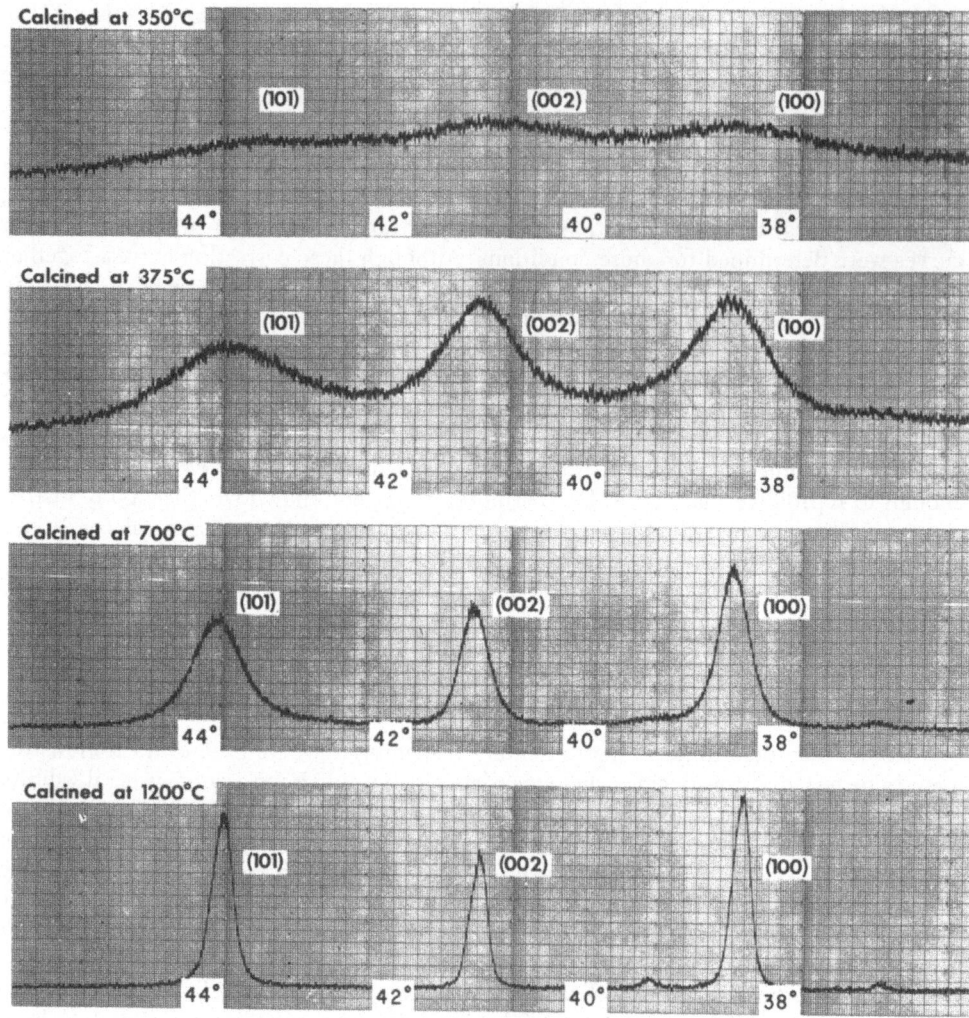

Figure 2. X-ray diffraction patterns of calcined BeO.

crystallites produce sharp peaks whose widths approach that due to instrumental factors.

For routine crystallite size analysis, recorded diffraction peak breadths are measured, in degrees 2θ, at half-maximum intensity, as shown in Figure 3. In this figure, which shows the (002) peak of beryllia calcined at 700°C, lines designating background and maximum peak intensities have been drawn. These lines are located at the mean of the recorder pen fluctuations. Also shown in Figure 3 is a line representing the half-maximum peak height. The peak breadth observed at the half-maximum height is then located on the proper crystallite size curve to determine average crystal thickness perpendicular to the diffracting crystallographic planes.[1] Figure 4 shows the set of curves used for the beryllium oxide specimens of this study. By locating the diffraction peak breadth, 0.320°, observed in Figure 3, on the (002) curve of Figure 4, an average crystallite dimension of 338 Å is obtained.

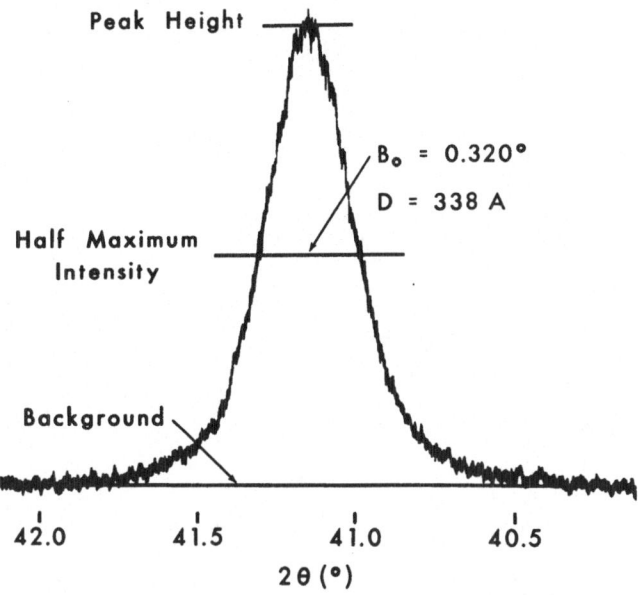

Figure 3. Diffraction peak (002) of BeO calcined at 700°C.

INSTRUMENTS AND EXPERIMENTAL CONDITIONS

For this study, six different X-ray diffractometers were used, including the General Electric XRD-3 unit used in the original work,[1] three General Electric XRD-5 units, and two Norelco units. All six instruments were operated with X-rays from the line source of copper target X-ray tubes, and utilized detector circuits and recorders with linear response. The six diffractometers, together with instrumental conditions used, are listed

Table I. Instrumental Conditions

Instrument number	Instrument[a] location	Instrument type	Slit[b] system	Goniometer speed, deg/min	Chart speed, in./hr	Detector
1	GE-NMPO	GE XRD-3	1	0.1	24	Krypton Proportional
2	UC-Geol	GE XRD-5	1	0.2	30	Xenon Proportional
3	GE-FPLD	GE XRD-5	1	0.2	60	Scintillation
4	USPHS	GE XRD-5	1	0.1	24	Xenon Proportional
5	GE-NMPO	Norelco	2	0.125	30	Xenon Proportional
6	USIC	Norelco	2	0.125	30	Scintillation

[a] GE-NMPO: General Electric Company, Nuclear Materials and Propulsion Operation, Cincinnati, Ohio.

UC-Geol: University of Cincinnati, Department of Geology, Cincinnati, Ohio.

GE-FPLD: General Electric Company, Flight Propulsion Laboratory Department, Cincinnati, Ohio.

USIC: United States Industrial Chemicals Company, Cincinnati, Ohio.

USPHS: United States Public Health Service, Occupational Health Facility, Cincinnati, Ohio.

[b] Slit System 1:1° beam slit, MR Soller slit, 0.1° detector slit.

Slit System 2:4° divergence slit, 0.006-in. receiving slit, 1° scatter slit.

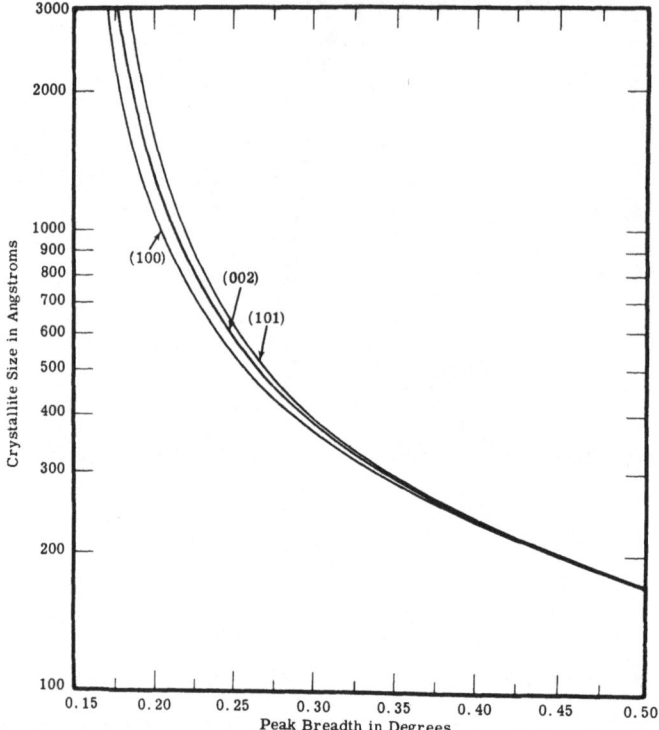

Figure 4. Crystallite size *vs.* half-peak breadth for BeO.

in Table I. No special effort was made to realign the diffractometers, but rather it was assumed that the instruments were in satisfactory alignment for normal X-ray diffraction work. Rough checks of this alignment were made before beginning work on each instrument.

As seen in Table I, two different slit systems were used for the General Electric and Norelco instruments, respectively. Slit System 1 was originally chosen for the derivation of the routine crystallite size methods due to its general usefulness for routine diffractometry. That system sufficiently limits the X-ray beam to prevent excessive instrumental line-broadening but yet allows sufficient intensity to reach the detector so that weak, diffuse diffraction peaks from very small crystallite size material can be measured. Slit System 2 was subsequently chosen for use with the Norelco diffractometers since that combination produced diffraction peak profiles most nearly duplicating those of the original work.

Table I indicates that various goniometer scanning speeds and recorder chart speeds were used. This variety of speeds caused no discrepancies in observed peak breadths since all measurements were in terms of degrees 2θ, not chart divisions. Likewise, the several kinds of detectors used also seemed to have no effect on recorded peak breadths.

In addition to the instrumental constants listed in Table I, recorder ranges and time constants were chosen which provided recorded peaks with smooth profiles and nearly full-scale heights. Short time constants were used to prevent sluggishness of the recorder pen response. A recent study has shown that time constants of 4 sec or less have little or no effect on recorded diffraction peak breadths when those peaks are scanned at the slow speeds normally used for line-broadening work.[4]

The beryllium oxide specimens were in the form of powders which were mounted in thin layers on glass microscope slides. Each powder was prepared by first mixing it with a solution of ethyl and amyl acetate containing about 10 vol.% Duco cement. This mixture was then coated uniformly on a glass slide and allowed to dry. The type of sample thus produced has been found to be quite useful for X-ray line-broadening work, especially when dealing with materials of low X-ray absorbing power. Thicker layers of such materials produce asymmetric line-broadening due to diffraction occurring from depths within the specimen which are displaced from the center of focus of the diffractometer.[5]

RESULTS

The individual performance results of the six X-ray diffractometers are shown in the following tables. Table II lists the measured diffraction peak breadths obtained with each instrument, while Table III lists the crystallite sizes determined from those peak breadths. As shown by these tables, the agreement among the various instruments is quite good, especially over the smaller crystallite size range. In general, the variation seen in all of the measurements of Tables II and III are felt to be within the normal limits of experimental error.

Comparison of the peak breadths of Table II with crystallite sizes of Table III

Table II. Observed Peak Breadths for BeO

Peak Breadths Measured in Degrees 2θ at Half-Maximum Intensity

Instrument number	Calcination temperature					
	350°C	375°C	700°C	1000°C	1200°C	1500°C
(100) Line						
1	2.03	0.965	0.335	0.243	0.213	0.1675
2	2.19	0.952	0.340	0.240	0.208	0.172
3	2.02	0.936	0.336	0.242	0.212	0.172
4	2.14	1.005	0.342	0.248	0.215	0.1725
5	2.02	0.932	0.340	0.235	0.210	0.1625
6	1.95	0.945	0.342	0.243	0.213	0.1625
(002) Line						
1	2.32	1.015	0.312	0.230	0.210	0.180
2	2.05	0.996	0.316	0.228	0.204	0.180
3	2.05	0.956	0.316	0.234	0.208	0.180
4	2.13	1.040	0.320	0.235	0.210	0.180
5	2.41	1.022	0.312	0.225	0.207	0.1675
6	2.25	1.042	0.335	0.232	0.207	0.1675
(101) Line						
1	2.44	1.515	0.578	0.340	0.260	0.190
2	2.46	1.600	0.588	0.336	0.260	0.188
3	2.22	1.546	0.596	0.338	0.258	0.192
4	2.59	1.780	0.592	0.337	0.257	0.195
5	2.58	1.758	0.595	0.337	0.252	0.1825
6	2.40	1.738	0.590	0.342	0.255	0.180

Table III. Crystallite Dimensions of BeO

Dimensions in Angstroms, Determined from Observed Peak Breadths.

Instrument number	Calcination temperature					
	350°C	375°C	700°C	1000°C	1200°C	1500°C
(100) Line						
1	40	84	304	580	850	3400
2	37	85	297	595	910	2600
3	40	87	303	583	860	2600
4	38	81	293	553	815	2530
5	40	87	297	627	885	5150
6	41	86	280	580	850	5150
(002) Line						
1	34	80	354	728	1010	2570
2	39	81	347	747	1140	2570
3	39	85	347	692	1040	2570
4	38	78	338	685	1010	2570
5	33	79	354	785	1060	7400
6	36	78	312	705	1060	7400
(101) Line						
1	32	52	147	312	557	2200
2	32	50	143	320	557	2390
3	37	51	142	316	567	2020
4	31	45	142	317	570	1800
5	31	45	142	317	600	3350
6	33	46	143	309	587	4050

shows that small variations in peak width represent much greater size variations in large crystallite material than in small crystallite material. Size measurements of the large crystallite BeO calcined at 1500°C show considerable variation in results. In particular, the two Norelco instruments, numbers 5 and 6 in the tables, tend to give narrower diffraction peaks and larger crystallite dimensions than the four General Electric X-ray units. This is probably due to the different instrumental geometries and slit systems used. The large variations in crystallite measurements of the large-size material are to be expected since those sizes are near the upper limit of applicability of the line-broadening method.

The variations in crystallite size results listed in Table III are shown graphically in Figure 5, which shows crystallite size range plotted as a function of average crystallite size. Vertical bars represent the range of sizes determined with the six diffractometers from each of the three diffraction peaks of the standard samples. The average of the six determinations from each peak is used as the location of each bar along the abscissa. From Figure 5 it can be seen that the relative variation in measured sizes is quite small, especially when the average crystallite size of the sample falls between approximately 80 and 300 Å. A slight increase in the relative variation is seen to occur when crystallites smaller and larger than these sizes are studied. Wide deviations in measured sizes are shown when the average crystallite size exceeds 2000 Å.

A further analysis of the data obtained by this study consisted of determining the

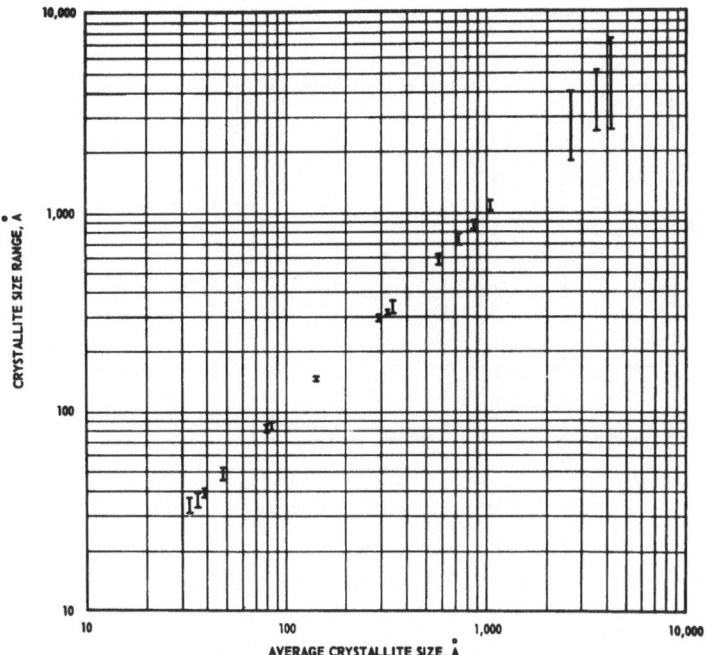

Figure 5. Variations in measured crystallite sizes.

percent deviation from the mean for each group of six crystallite size determinations. This analysis was performed by applying the following relation to each determined crystallite size value listed in Table III:

$$\text{Percent deviation} = \frac{|D_{\text{avg}} - D_{\text{obs}}|}{D_{\text{avg}}} \times 100$$

where D_{avg} is the average value of each group of six size determinations, and D_{obs} is an individual observed size within that group. Percent deviation from the mean was then obtained by averaging the individual deviations in each group of determinations.

A plot of mean deviation as a function of average crystallite size is shown in Figure 6. From the plot it can be seen that the average deviation was quite low, remaining well below 10% for all but the largest crystallite dimensions. For sizes above 2500 Å the deviation increased considerably. As shown by Figure 6, a minimum in percent deviation exists near the 200 Å crystallite size region. At smaller crystallite sizes, the mean deviation increases due to the difficulty in defining the true profiles of very broad diffraction peaks, while the increased deviation at larger sizes is caused by the large changes in crystallite size corresponding to small changes in peak breadths. The exact point at which mean deviation begins to increase rapidly for large crystallite sizes could not be determined since no specimens with average sizes between 1000 and 2500 Å were available.

CONCLUSIONS

Previously, the feasibility of performing crystallite size analyses routinely by means of X-ray diffraction line-broadening methods was shown. It has now been demonstrated that the techniques, originally derived with a single X-ray diffractometer, can be applied

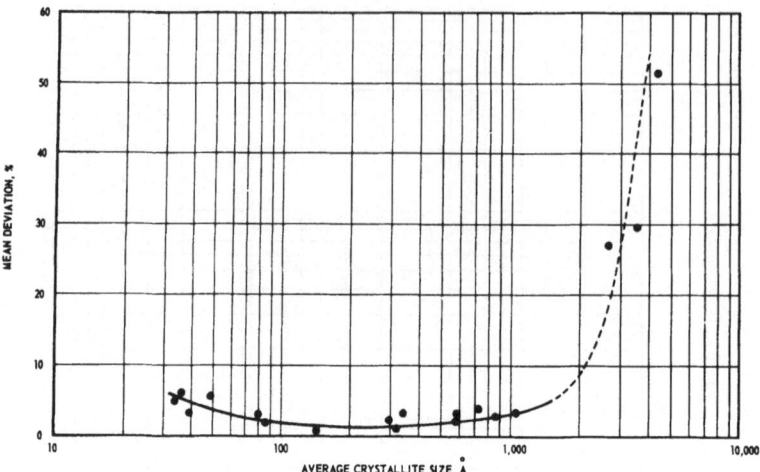

Figure 6. Mean deviation of measured crystallite sizes.

to other diffractometers using a standard set of instrumental conditions. Tests of several X-ray units were made using a standard series of specimens of known crystallite size. Size values obtained were found to agree quite well with the originally determined values, especially when instruments of the same manufacture were used. Deviations from the mean were less than 10% except when measuring crystallite dimensions near the upper limit of applicability usually associated with the line-broadening method.

ACKNOWLEDGMENTS

The author is indebted to the following persons for their cooperation and help in performing this study: Dr. Frank L. Koucky, University of Cincinnati, Department of Geology; Mr. Ottis L. Isaacs, General Electric Company, Flight Propulsion Laboratory Department; Mr. Richard E. Kupel, United States Public Health Service, Occupational Health Facility; and Mr. Roger Laib, United States Industrial Chemicals Company.

REFERENCES

1. R. C. Rau, "Routine Crystallite-Size Determination by X-ray Diffraction Line Broadening," *Advances in X-ray Analysis, Vol. 5*, University of Denver, Plenum Press, New York, 1962, pp. 104–115.
2. H. Klug and L. Alexander, *X-ray Diffraction Procedures*, John Wiley & Sons, Inc., New York, 1954, pp. 491–538.
3. R. C. Rau, "Calcination of BeO from Basic Acetate Derived Be(OH)$_2$," presented at the Fall Meeting, Basic Science Division, American Ceramic Society, Columbus, Ohio, October 8–9, 1962.
4. E. A. Rosauer and R. L. Handy, "Crystallite-Size Determination of MgO by X-ray Diffraction Line Broadening," Iowa Academy of Science Meeting, Simpson College, Indianola, Iowa, April 14, 1961.
5. S. F. Bartram, "Crystallite-Size and Particle-Size Measurements on BeO Powders by X-ray Methods," *Advances in X-ray Analysis, Vol. 4*, University of Denver, Plenum Press, New York, 1961, pp. 40–62.

DISCUSSION

R. B. Scott (Parke, Davis and Company): Did you use the same amount of specimen in each analysis?

R. C. Rau: A single set of specimens was made up and was used in each diffractometer.

H. D. Orcutt (Aerojet-General Corp.): What precautions were taken in preparing and running the specimen?

R. C. Rau: The sample was spread out on a glass microscope slide to cover a large enough area so that the beam area would be entirely covered by sample.

H. D. Orcutt: Did you cover the sample or was it left open?

R. C. Rau: No, it was open, but the binder holds it tight.

G. Brown (Siemens, Inc.): Would the measurement with the monochromatic peak give better results than with the K_α doublet used?

R. C. Rau: It probably would. In materials where we have large crystallites and where we get diffraction at higher angles and better resolution, we just used the K_α.

LOW-TEMPERATURE TRANSITIONS OF SOME AMMONIUM SALTS

M. Stammler, D. Orcutt, and P. C. Colodny

Aerojet-General Corporation
Sacramento, California

ABSTRACT

X-ray diffraction studies were performed on ammonium fluoroborate [NH_4BF_4], ammonium perchlorate [NH_4ClO_4], ammonium chloride [NH_4Cl] and ammonium phosphate [$(NH_4)H_2PO_4$] within a temperature range of ambient to $-190°C$. For these experiments a specimen holder was designed to effect rapid cooling of the sample. Results indicate that the fluoroborate and perchlorate salts undergo a polymorphic transition at $-190°C$. The decreasing intensities of the (011) and (112) reflections of the orthorhombic modification indicates a first-order reaction. No transition was observed when these salts were cooled to $-40°C$. The ammonium chloride shows a transition at about $-30°C$ which confirms the results obtained by G. H. Goldschmidt and D. G. Hurst. The ammonium phosphate $(NH_4)H_2PO_4$, which is tetragonal at ambient temperature also undergoes a transition if cooled to $-190°C$. The kinetics of this reaction are discussed.

INTRODUCTION

The polymorphism of ammonium salts is well known and has been the subject of several investigations. For example, studies of simple compounds of the type AB, such as the ammonium halides, have been reported.[1-4] Here, the polymorphic transformations at low temperatures appear to be analogous to the case of methane, inasmuch as the NH_4^+ ion has the same structure as CH_4. The transitions of low-temperature (α) modifications to the high-temperature (β) modifications are accompanied by considerable changes in density as shown in Table I.

The denser α modifications crystallize in the CsI lattice, the less dense high-temperature β modification in the NaCl lattice. In both lattices, the NH_4^+ ion is to be regarded as rotationally symmetrical, and their interconversion is not connected with a change in the rotational state. The converse is true in the case of transformations which the α modifications undergo at temperatures below $-30°C$. Below this temperature, the ammonium halides exhibit an abnormal behavior in their coefficients of expansion and their specific heats, although the crystal structure changes but slightly. Ammonium bromide is transformed at $-58°C$ and ammonium iodide at $-39°C$ from the α modification with the CsI lattice into a tetragonal γ modification with a structure of the type also observed with phosphonium iodide. At temperatures below $-110°C$ with deuterated ammonium bromide, another regular modification identical with the α form, which is stable above $-58°C$, has been observed.[1-3]

At these low transition temperatures, the rotation of the ammonium ion, which at higher temperatures leads to a spherical symmetry, changes to an oscillation. The

---

[1] Superscripts pertain to references at the end of the paper.

Table I

Compound	Transition temperature	Density	
		α	β
NH_4Cl	184.3	1.526[20]*	1.266[250]
NH_4Br	137.8	2.436[20]	1.97[250]
NH_4I	− 17.6	2.86[−20]	2.513[20]

* Represents the temperature at which the density was measured.

tetrahedral arrangement of the hydrogen atoms causes two of the four halogen ions, forming a square above and below each cation at opposite corners, to move slightly downward and the other two slightly upward. With ammonium chloride, however, the cubic symmetry is preserved during the transformation, which takes place at −30.5°C, but it no longer corresponds to holohedrism, as shown by the appearance of piezoelectricity in γ-NH_4Cl.[1,2,4]

Another ammonium compound which has been studied very extensively is ammonium nitrate. Here, both cation and anion can create an increase in symmetry by rotation. This is probably the cause for the numerous modifications of this salt. The following reversible transformations have been observed:

Modification:	V		IV		III		II		I
Symmetry:	Hexagonal	$\rightleftharpoons$	Orthorhombic II	$\rightleftharpoons$	Orthorhombic I	$\rightleftharpoons$	Tetragonal	$\rightleftharpoons$	Cubic
Transition Temperature:	−18°C		+32.3°C		84.2°C		125°C		

Ammonium nitrate shows the general trend that with increasing temperature an increase in symmetry can be observed. R. N. Brown and A. C. McLaren[5] reported electrical measurements and NMR studies on NH_4NO_3 at temperatures as low as −190°C. These investigations also indicate that at all temperatures above −190°C the NH_4^+ ions are rotating about their centers of mass at a frequency of about 10^5 cps.

Work directly concerned with studies of the transition mechanism was performed by Erofeev et al.[6] and Brown et al.[5] For ammonium nitrate, Erofeev et al. found that the rate of transition and the temperature do not obey the Arrhenius law. Brown et al. showed that in NH_4NO_3 the transitions II → III and IV $\rightleftharpoons$ III do not occur in the dry solid state but require the presence of some moisture. The transition takes place by the dissolution and recrystallization of the solid.

Thus far no investigations at very low temperatures have been made on ammonium salts with complex anions except ammonium nitrate. Therefore, one of the objectives of this work was to ascertain that in these salts a change in symmetry occurs at low temperatures due to a change in the energy state of the NH_4^+ ion or caused by orientation of hydrogen bonds at low temperatures.

Another objective was to study the mechanism of these transitions at low temperature, since comparatively little work has been devoted to the mechanism of such changes.

DESCRIPTION OF THE APPARATUS

All X-ray data were obtained with a General Electric XRD-5 diffractometer and a CA-7 copper target tube with a nickel foil filter.

To study the transition of ammonium salts at low temperatures, a redesigned

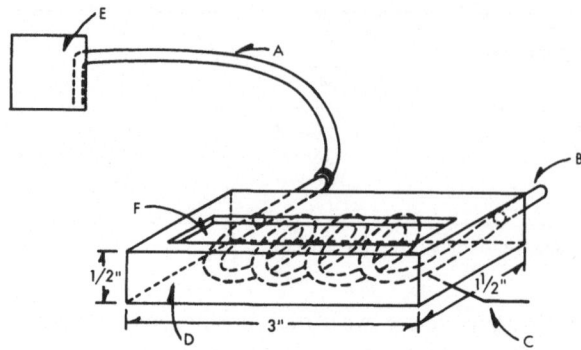

Figure 1. Specimen holder for low-temperature diffraction studies. A—flexible stainless steel tubing; B—copper tubing imbedded in holder; C—thermocouple ; D—Cerrobend metal specimen holder ($3 \times 1\frac{1}{2} \times \frac{1}{2}$in.); E—liquid-nitrogen dewar, pressurized with helium; F—$\frac{1}{16}$-in.-deep trough for sample.

specimen holder was used to achieve effective cooling of the sample. This specimen holder was similar in design to the one reported by Weltman.[7] Figure 1 shows a schematic drawing of the sample holder. The thermocouple junction was placed in the sample. In the sample holder, a copper coil was imbedded in Cerrobend and liquid nitrogen was circulated through the coil to maintain temperature of the sample at $-190 \pm 5°C$. An iron–constantan thermocouple was utilized with a Leeds and Northrup Millivolt Potentiometer to determine the temperature. The Cerrobend block had a $\frac{1}{16} \times \frac{3}{4} \times 2\frac{1}{2}$in. sample trough and the thermocouple was imbedded so as to be in good contact with the sample.

An alternative method of cooling was tried wherein nitrogen gas was made to flow through copper coils immersed in liquid nitrogen and then through the Cerrobend sample holder. In this case, the temperature can be varied by changing the rate of the nitrogen flow. This method, however, was not applied in the present case since no transition was observed at temperatures above $-100°C$. Therefore, it was desirable to achieve temperatures as low as possible.

The applied procedure was to scan first at ambient temperature and then cool the sample to $-190°C$. The sample was allowed to remain at this temperature for at least 1 hr and was scanned again. A jet of dry nitrogen gas was directed onto the sample to prevent ice formation.

Investigations on ammonium perchlorate were conducted in a slightly different manner. After the scan at ambient temperature and the attainment of $-190°C$, successive scans of $2\theta = 14$ to $52°$ were commenced so as to record the growth and disappearance rates of the phases present.

EXPERIMENTAL RESULTS

With the setup described above, X-ray diffraction patterns were taken of ammonium perchlorate, ammonium fluoroborate, NH_4Cl and $(NH_4)_2HPO_4$ at temperatures down to $-190°C$.

By cooling the samples with liquid nitrogen to $-190°C$, new reflections were observed while some of the original peaks decreased considerably in intensity. However, none of the original peaks vanished completely. These changes were most obvious at $2\theta = 20$ to $50°$. Sections of the diffraction patterns of ammonium perchlorate, ammonium fluoroborate, and $(NH_4)_2HPO_4$ are reproduced in Figure 2. The ambient temperature patterns are compared with those obtained at $-190°C$.

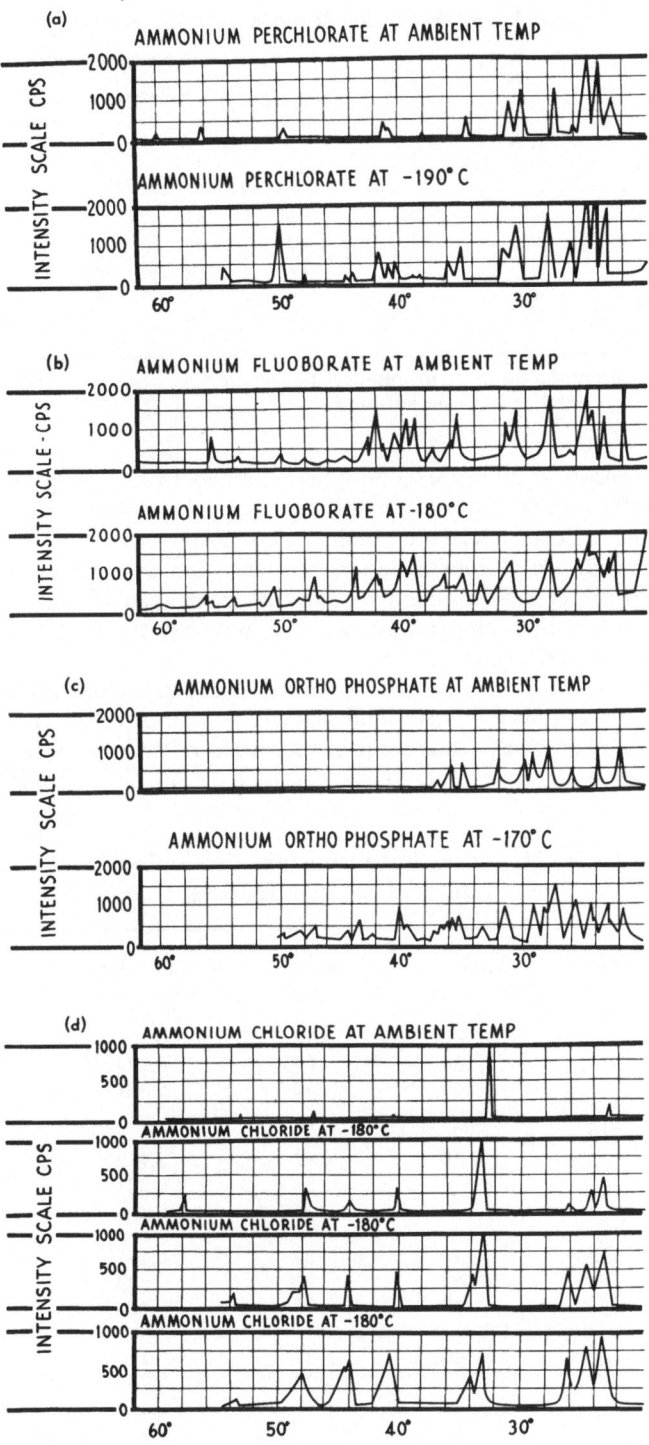

Figure 2. X-ray diffraction patterns at ambient and −190°C.

Table II. Reflections Decreasing in Intensity

No.	d, Å	d, Å $-190°C$	hkl	Relative intensity I^*				
				I_0	I_{20}	I_{40}	I_{60}	I_{80}
1	4.58	4.53	011	91	83	75	64	50
2	3.61	3.56	210	60	77	62	55	52
5	2.59	2.57	121	29	36	27	20	19
4	2.89	2.95	202	25	35	25	21	—
3	2.97	2.94	112	40	40	27	20	16

* Subscripts pertain to the time in minutes the sample was kept at $-190°C$. The intensity was measured as peak height minus background.

In order to ensure equilibrium conditions for all cold runs, samples were stored for about 2 hr at $-190°C$ in the specimen holder and were then scanned at this temperature. After the low-temperature pattern was obtained, the coolant was turned off and the sample was allowed to warm up to room temperature when it was scanned again. The time between the cold run and the ambient run was not more than 15 min. In all cases, the reflections observed at low temperature vanished and only the well-known pattern for ambient temperatures remained.

An attempt was made to study the kinetics of the low-temperature transition of ammonium perchlorate since it was found that the new lines which appear at $-190°C$ gradually increase in intensity while some of the strong reflections of the ambient modification show a decreasing tendency. These data are listed in Tables II, III, and IV.

Table III. First-Order Reaction Rate Constant for Reflections with Decreasing Intensities of NH_4ClO_4

$d_{-190°C}$	hkl	Time, min	$a - x^*$	$\log [a^*/(a-x)]$	K, min^{-1}
4.53	011	20	83	0.80	0.920×10^{-2}
		40	75	0.125	0.720×10^{-2}
		60	64	0.193	0.730×10^{-2}
		80	50	0.301	0.806×10^{-2}
3.56	210	20	77	0.077	0.885×10^{-2}
		40	62	0.172	0.991×10^{-2}
		60	55	0.223	0.860×10^{-2}
		80	52	0.248	0.716×10^{-2}
2.94	112	40	27	0.193	1.11×10^{-2}
		60	20	0.322	1.24×10^{-2}
		80	16	0.420	1.21×10^{-2}
2.59	121	40	27	0.160	0.922
		60	20	0.289	1.110
		80	19	0.312	0.900

* a represents the extrapolated intensity of a reflection at the time zero, while x represents the increased or decreased intensity of this reflection after the time t in min.

Prior to the appearance of new lines, e.g., the transition at temperatures close to $-190°C$, the unit cell of the orthorhombic form shrinks. The contraction of the d-values was found to be of the order of 5%. The smallest contraction appears to occur in c direction.

Interpretation of these data in terms of a first-order reaction yield the constants given in Table III.

The relatively large deviation of the rate constant for the 20-min values (Table III) may be explained by the fact that the first 20 min include the time required to cool the sample from room temperature to the desired $-190°C$. This effect is particularly pronounced for reflections of medium or low intensities. Therefore, no consideration was given to these values. The average rate constant, based on the first two strong lines was determined to be $K_d = 0.837 \times 10^{-2}$ min^{-1}. Graphical representation of log $[a/(a-x)]$ vs. time is shown in Figure 3. Using the K value determined above the time required to complete one-half of the reaction is given by $t_{0.5} = (\ln 2/K) = 83$ min.

If the same treatment were applied to the intensities of the new growing lines a rate constant which is of nearly the same value can be calculated. This is demonstrated in Table IV. Eliminating the K values calculated from the first scan after 20 min for the same reason as mentioned above, we determined the average rate constant to be: $K_i = 0.920 \times 10^{-2}$ min^{-1}. The time to complete one-half of the reaction using K_i was determined to be $t_{0.5} = 75.3$ min.

The agreement between K_d and K_i is considered to be good since the accuracy with which the intensity can be reproduced is of the order of 10%. In addition, small temperature changes during the scans may increase this error further.

The ammonium fluoroborate which is isomorphous with NH_4ClO_4 due to the same

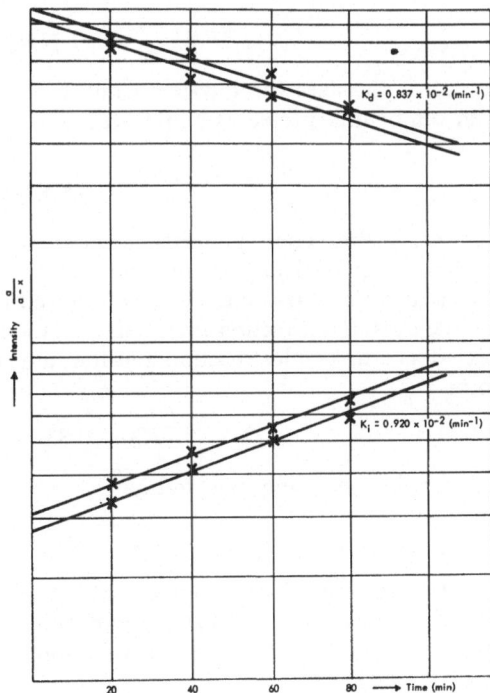

Figure 3. Transition of orthorhombic NH_4ClO_4 at $-190°C$.

$$\alpha - NH_4ClO_4 \underset{-190°C}{\overset{K}{\rightleftarrows}} \gamma - NH_4ClO_4$$

Table IV. First-Order Rate Constants for Reflections with Increasing Intensities

$d_{-190°C}$	Time, min	$a - x$	$\log [a/(a - x)]$	K, min^{-1}
2.66	20	33	0.056	0.644×10^{-2}
	40	42	0.160	0.921×10^{-2}
	60	51	0.245	0.944×10^{-2}
	80	59	0.308	0.890×10^{-2}
2.04	20	38	0.061	0.700×10^{-2}
	40	47	0.153	0.881×10^{-2}
	60	55	0.222	0.856×10^{-2}
	80	66	0.301	0.871×10^{-2}
1.91	20	24	0.068	0.782×10^{-2}
	40	30	0.166	0.955×10^{-2}
	60	36	0.245	0.944×10^{-2}
	80	44	0.332	0.960×10^{-2}
1.88	20	12	0.079	0.908×10^{-2}
	40	15	0.176	1.011×10^{-2}
	60	17	0.231	0.890×10^{-2}
	80	21	0.322	0.930×10^{-2}

symmetry of the BF_4^- and the ClO_4^- shows the same changes in the X-ray diffraction pattern as the ammonium perchlorate.

$$\gamma\text{-}NH_4ClO_4 \overset{-190°C}{\rightleftharpoons} \alpha\text{-}NH_4ClO_4 \overset{240°C}{\rightleftharpoons} \beta\text{-}NH_4ClO_4$$

$$\gamma\text{-}NH_4BF_4 \overset{-190°C}{\rightleftharpoons} \alpha\text{-}NH_4BF_4 \overset{236°C}{\rightleftharpoons} \beta\text{-}NH_4BF_4$$

In both cases the low-temperature transition measured over a time period of 80 min can be interpreted as a first-order reaction.

SUMMARY

X-ray diffraction patterns for ammonium perchlorate, ammonium fluoroborate, ammonium phosphate and ammonium chloride showed a transition at about $-190°C$. The kinetics of this transition was followed by the change of the intensity of increasing and decreasing diffraction lines using ammonium perchlorate. The results can be interpreted as a first-order reaction and the rate constant was calculated for decreasing lines to be

$$K_d = 0.837 \times 10^{-2} \text{ min}^{-1}$$

and for the increasing lines to be

$$K_d = 0.920 \times 10^{-2} \text{ min}^{-1}$$

It is believed that the agreement between the two rate constants can be considered good in view of the fact that the intensities can be reproduced only with a variation of about 10% and the temperature variations during the time of the experiment within $\pm 5°C$.

For the ammonium salts with complex anions, the low-temperature transition may be due to the onset of disorder resulting in a random orientation of the ammonium ions at liquid nitrogen temperature. Neutron diffraction studies of NH_4Cl and ND_4Cl phase III have been reported by G. H. Goldschmidt and D. G. Hurst[8] and have confirmed the order–disorder hypothesis for these compounds.

The diffraction pattern of ammonium chloride which was obtained at $-190°C$ could not be indexed as a body-centered cubic lattice and is thus different from Phase III stable below $-23.8°C$ which has a CsCl-type structure.

REFERENCES

1. J. A. Keteclaar, *Nature* **134**: 250, 1934.
2. J. Weigle and H. Saini, *Helv. Phys. Acta* **9**: 515, 1936.
3. K. Clusius, A. Kruis, and W. Schanzer, *Z. anorg. Chem.* **236**: 26, 1938.
4. A. Smits, J. A. Keteclaar, and G. J. Muller, *Z. phys. Chem.* **A75**: 359, 1936.
5. R. N. Brown and A. C. McLaren, *Proc. Roy. Soc. (London) A*: 239, 1962.
6. B. V. Erofeev and N. I. Mitskevich, *J. Phys. Chem. USSR* **24**: 1235, 1950; **26**: 848, 1631, 1952; **27**: 118, 1953.
7. H. J. Weltman, "Low Temperature X-ray Diffraction of Frozen Electrolytes," *Advances in X-ray Analysis, Vol. 5*, University of Denver, Plenum Press, New York, 1962, pp. 48–56.
8. G. H. Goldschmidt and D. G. Hurst, *Phys. Rev.* **83**: 88, 1951.

DISCUSSION

H. Steffen Peiser (National Bureau of Standards): Have you seen any evidence of nonuniformity of temperature in the sample?

D. Orcutt: We have not, but we assume there is. In cooling down and remaining at low temperature there was a tendency for ice to form. We used helium or dry nitrogen to remove the ice. We noticed there was some tendency for the sample to cool down. It probably did result in the cooling of the surface of the sample.

N. O. Smith (Fordham University): Did you change the particle size and did you find that it transformed at a different rate? Did the constituents form solid solutions with each other?

D. Orcutt: No, we did not. (To the second question)—I do not know.

N. O. Smith: Is the fluoroborate isomorphous?

D. Orcutt: Yes, it is.

X-RAY DIFFRACTION ANALYSIS OF AEROSOLS FROM EXPLODING WIRES

A. G. Barkow, F. G. Karioris, and J. J. Stoffels*

Physics Department, Marquette University
Milwaukee, Wisconsin

ABSTRACT

In this study, X-ray diffraction analysis is used to investigate the composition of aerosols produced by exploding wires with the current surge from a 4000-joule capacitor. Qualitative analyses of aerosols from 15 different metals exploded in air or inert atmosphere indicate that the particles are crystalline with the normal crystal structure and that explosion of noble metals in air and of base metals in argon produces aerosols consisting of metallic particles. Base metals exploded in air produce aerosols consisting primarily of oxides. Nitrides were not observed. An analytical scheme is described for the $Cu–Cu_2O–CuO$ mixtures collected on membrane filters from explosions of copper wires in air. The composition of aerosols is determined for various initial voltages (2–18 kv) on the 20-μf capacitor bank for two series of wires. In one, the weight fraction of CuO increases rapidly with voltage until it accounts for almost the entire sample while the Cu_2O and Cu content decrease smoothly. In the other, CuO and Cu_2O are about equal for explosions above 6 kv while Cu decreases. Differences are attributed to a change in the circuit and mass of wire used.

INTRODUCTION

The exploding wire phenomenon occurs when a high-energy capacitor is discharged very rapidly through a small wire. Most research pertaining to exploding wires has been concerned with basic mechanisms involved in the explosion itself or with associated applications of the light output, shock wave, and other phenomena.[1–3] Little attention, however, has been given to the nature of aerosols produced by such explosions. Abrams and others[4] have used the technique to produce aerosols of uranium and plutonium for animal inhalation studies by exploding aluminum foil folded around small pieces of uranium or plutonium.

Recently, an aerosol generator[5] based on the exploding wire was constructed at the Oak Ridge National Laboratory for use by the Health Physics Division. In this device a capacitor bank capable of storing 4000 joules is discharged through a wire by a sphere-gap switch. The resulting explosion, characterized by a brilliant flash of light and a loud noise, generates and disperses smokes which can be removed from the explosion chamber and collected on membrane filters† for study.

Aerosols from 15 different metals were prepared with this apparatus using an

---

* Present address, Hanford Atomic Products Operation, General Electric Company, Richland, Washington.

[1] Superscripts pertain to references at the end of the paper.

† Type HA, Millipore Filter Corporation, Redford, Massachusetts.

atmosphere of air at room temperature and barometric pressure. Copper wires were exploded in argon. Electron micrographs of the aerosol particles, collected by a thermal precipitator, show that the exploding wire phenomenon produces typical smokes. The particles are composed of small primary spheres which join to form long chains and complex aggregates (Figure 1).

From X-ray diffraction patterns of deposits collected on membrane filters, it is evident that the particulate matter is crystalline with the usual crystal structure for metals or oxides which can be expected from the materials used.

A series of copper wires was exploded with various initial voltages (1 to 18 kv) on the 20-μf capacitor bank. The preliminary X-ray studies of these samples suggested a variation in composition with energy input to the explosion. The determination of the quantitative composition of the aerosols from exploded copper wires as a function of the voltage used for explosion is the core of this investigation. The aerosol from a copper wire exploded in air presents a three-component system including copper (Cu), cuprous oxide (Cu_2O), and cupric oxide (CuO). There is no reaction of copper with nitrogen at temperatures up to 900°C,[6] but since temperatures of the order of 25,000°C[7] may be attained in wire explosions, the reaction might be expected. However, no copper nitride (Cu_3N) was detected.

Chemical methods for the quantitative analysis of mixtures of copper and its oxides have been developed.[8–12] Pollard *et al.*[13] report a chromatographic method. The analysis has been done by a statistical method[15] of counting the various particles under a microscope. None of these methods has been applied to the analysis of aerosols from exploded copper wires.

The X-ray method of analysis described in this paper is believed to be accurate and fairly rapid, and it can be reduced to routine once the necessary working curves have been obtained. The diffraction technique is advantageous since it can be applied directly to the small amounts of material collected on membrane filters by standard procedures. Because the technique is nondestructive, it is possible to repeat analyses and use the unaffected samples for other studies.

COMPOSITION OF AEROSOLS FROM EXPLODED WIRES

Specimens of 15 different metals in 2.5-cm lengths of wire were exploded electrically in an atmosphere of air with 10 kv on the 20-μf capacitor bank. Copper wires were exploded in argon. For each case, the aerosol was drawn from the explosion chamber through a membrane filter to collect the solid particulate matter for X-ray diffraction analysis with a General Electric XRD-5 automatic recording diffractometer.

The aerosol deposit, of the order of 10 mg, occupies the center 3.5-cm section of a 47-mm-diameter membrane filter. From this, a strip 1.5 cm wide is cut for use with the diffractometer. The strip was originally mounted on a glass slide with a full backing of double-faced transparent tape, but since X-rays reflected from the tape produced a very high background, it was subsequently used only along the edges of the strip. This kept the tape out of the X-ray beam and still produced the flat sample required for the diffractometer.

Qualitative results from X-ray diffraction analyses of these samples, shown in Table I, indicate that aerosols consisting of metallic particles result from the explosion of noble metals in air or from exploding base metals in argon. Base metal wires exploded in air produce aerosols consisting primarily of oxides. Oxides of noble metals were not detected even though they were specifically sought. Nitrides were not detected. Some line broadening was observed in practically all cases. In most respects, the smokes from

Figure 1. Electron micrographs of particles from exploded wires. a. Aerosol particles collected with thermal precipitator from Fe wire explosion in air with 10 kv. b. Aerosol particles collected with thermal precipitator from Cu wire explosion in air with 10 kv.

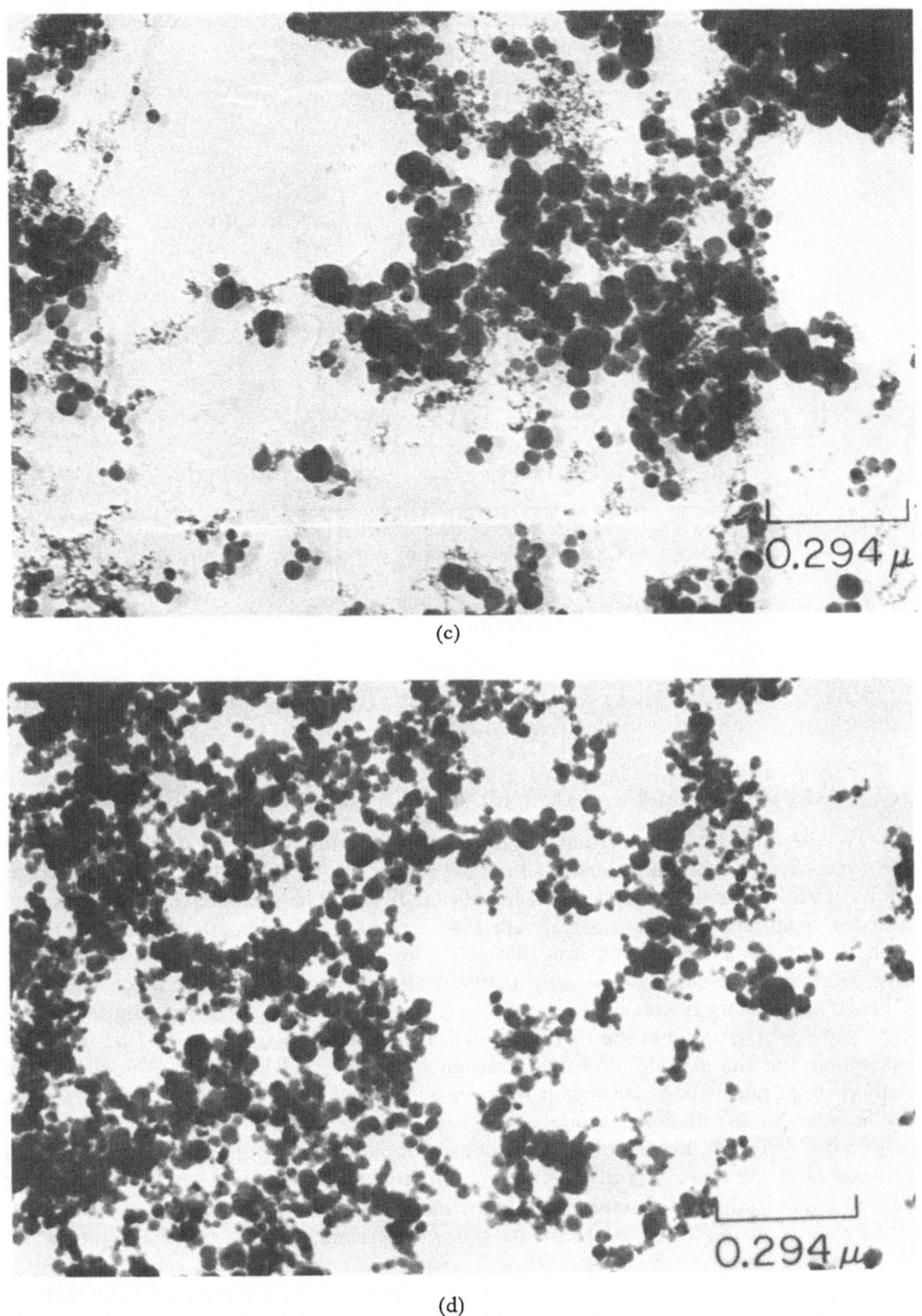

(c)

(d)

Figure 1. c. Primary particles from Cu wire explosion in air with 2 kv redispersed from membrane filter. d. Primary particles from Cu wire explosion in air with 10 kv redispersed from membrane filter.

Table I. Composition of Aerosols from Exploded Wires[a]

Metal	Color of sample	Composition
Ag	Black	Ag, Ag_2S[b]
Al	Light gray	Al_2O_3 (many phases)
Au	Brown	Au
Cu[c]	Black	Cu
Cu	Brown	Cu, Cu_2O, CuO
Fe	Brown	Fe_3O_4
Mg	White	MgO
Mo	Sky blue	MoO_2, MoO_3[d]
Ni	Gray	Ni, NiO
Pb	White	$2PbCO_3 \cdot Pb(OH)_2$[d]
Pt	Black	Pt
Sn	Beige	SnO_2
Ta	Gray	Ta, Ta_2O_5
Th	Lavender	ThO_2
U	Dark gray-green	U_3O_8, β-UO_2
W	Purple	W, WO_3[d]

[a] 2.5-cm lengths of wire, all exploded with 10 kv on 20-μf capacitor.
[b] Trace amount, possibly formed after explosion by exposure to open air.
[c] Exploded in argon atmosphere, all others in air.
[d] Incomplete or doubtful results.

exploding wires are similar to those from electric arcs[14] except that regular polyhedral particles are not observed in electron micrographs.

ANALYSIS OF Cu–Cu_2O–CuO MIXTURES ON MEMBRANE FILTERS

To determine the composition of aerosols as a function of voltage used for explosion of copper wires, a method of analysis for mixtures of Cu–Cu_2O–CuO was devised which can be applied directly to deposits on membrane filters. The methods[16] most commonly used for quantitative X-ray analysis are the internal-standard method, the single-line method, and the direct-comparison method. The internal-standard technique was not used because of the small mass to be analyzed and a desire to keep the aerosol samples intact. The aerosol, as collected on the filter, presents a flat specimen completely free from any preferred orientation of the crystallites. This characteristic could be impaired by remounting the sample after the addition of an internal standard. The single-line method could not be used because it requires a specimen of effectively infinite thickness, that is, a thickness sufficient for total absorption of the beam. The samples to be analyzed here do not approach this criterion. They consist of only 5 to 10 mg of aerosol deposited in a thickness of the order of 5 μ. The direct comparison method is based on the intensity ratio of a line from the substance of unknown amount to a line from another component in the same mixture. The use of this method is necessitated by the rather unique nature of the aerosol samples.

The intensity of the characteristic diffraction pattern of a material in a mixture depends upon the concentration of the material, except for an absorption correction.[17] No absorption correction need be made for a mixture of copper oxides since their linear absorption coefficients for Cu K_α radiation are practically identical. Thus, the line

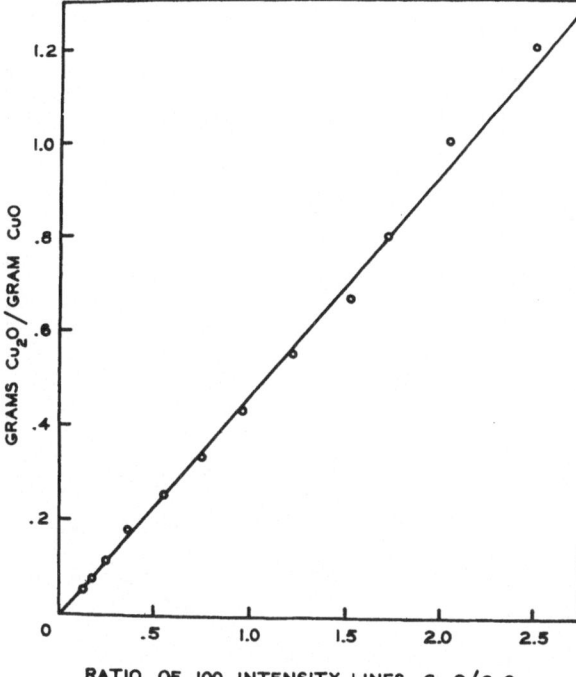

Figure 2. Intensity ratio–concentration plot from prepared mixtures of cuprous and cupric oxides. Integrated intensities were used.

intensities of cuprous and cupric oxides vary directly with their concentration in a mixture of the two for infinitely thick specimens. For the aerosol samples, line intensity varies not only with relative concentration but also with sample thickness. However, since the intensity remains directly proportional to the mass of material in the volume irradiated,[18] the intensity ratio between lines from each of the oxides is directly related to the mass ratio of the oxides, no matter how thick or thin the sample may be. Figure 2 shows this linear relationship as obtained from synthetic mixtures of the oxides prepared as standards. The most intense line from each of the oxides is used.

The presence of the third component, copper, requires another intensity–concentration curve. A series of synthetic mixtures was prepared consisting of varying amounts of copper added to a matrix composed of cupric and cuprous oxides in the constant ratio by weight $CuO/Cu_2O = 3$. The integrated intensity ratios of the copper with respect to each of its oxides were determined and plotted as functions of the mass ratios. Because the absorption coefficient of copper is different from that of each of its oxides, such curves are not linear over the entire range of copper concentration, 0–100%. However, for concentrations of copper up to 25%, linear approximations (Figure 3) can be made which fall within the statistical precision of the experimental data. Percentages of copper greater than this are not encountered in the aerosol samples.

The standard working curves make it possible to determine, from integrated intensity data, the mass ratios between cuprous and cupric oxide and between copper and either of its oxides. The Cu/Cu_2O ratio was used for aerosol samples exploded at low kv since Cu_2O content was found to be high in these. For samples exploded at high kv, the Cu/CuO ratio was used. The one other relation needed to determine the composition completely is that for weight fractions, $Cu + Cu_2O + CuO = 1$. Each component is then given as a weight fraction of the total sample.

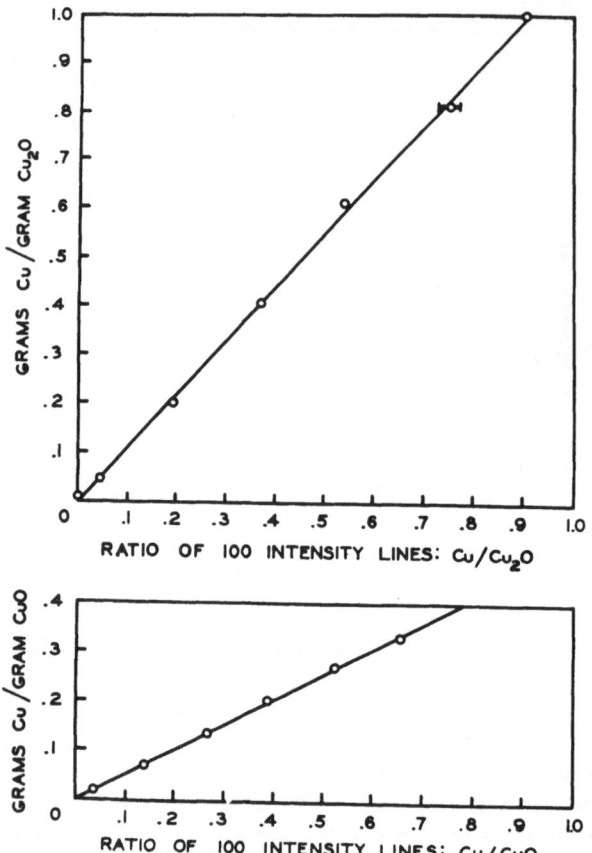

Figure 3. Intensity ratio–concentration curves from prepared mixtures of copper in a matrix consisting of copper oxides in the constant ratio by weight: $CuO/Cu_2O = 3$. Integrated intensities were used.

EQUIPMENT AND EXPERIMENTAL PROCEDURE

The General Electric XRD-5 automatic recording X-ray spectrometer was used in this investigation and the following instrumental parameters were kept constant during the entire project: radiation—Cu K_α; filter—two layers of 0.00035-in. nickel foil; target viewed at 4°; beam slit—3°; detector slit—0.1°, medium-resolution Soller slits. The detector used was a xenon-filled proportional counter tube (G-E No. 6 SPG). This tube[19] has an efficiency of 75% for detecting Cu K_α radiation. The efficiency remains constant for all counting rates since it is only wavelength-dependent. Calibration by the multiple foil method showed the counter to be linear in response up to at least 10,000 cps, and, in practice, counting rates did not exceed 5000 cps. Thus, it was unnecessary to correct observed count rates for either the dead time of the counter or the pulsed nature of the X-ray beam.

Automatic recording of the diffraction peaks was used exclusively in this investigation. An undistorted record for the diffraction pattern is obtained by employing both a slow scanning speed and a time constant less than one-half the time width of the detector slit.[20] A scanning speed of 0.2° 2θ/min was used in conjunction with a chart speed of 12 in./hr. The time constants used ranged from 2 to 8 sec and the time width was 30 sec.

As a check on the accuracy of recorded intensities, a calibration curve was prepared comparing count rates obtained from recorder deflections with those obtained by direct

Table II. Prominent Diffraction Lines of Copper and Copper Oxides[a]

Material	hkl	d, Å	2θ°	I/I₀
CuO	$\begin{Bmatrix}002 \\ \bar{1}11\end{Bmatrix}$	$\begin{Bmatrix}2.530 \\ 2.523\end{Bmatrix}$	$\begin{Bmatrix}35.48 \\ 35.58\end{Bmatrix}$	$\begin{Bmatrix}49 \\ 100\end{Bmatrix}$
Cu₂O	111	2.465	36.45	100
CuO	$\begin{Bmatrix}111 \\ 200\end{Bmatrix}$	$\begin{Bmatrix}2.323 \\ 2.312\end{Bmatrix}$	$\begin{Bmatrix}38.75 \\ 38.95\end{Bmatrix}$	$\begin{Bmatrix}96 \\ 30\end{Bmatrix}$
Cu₂O	200	2.135	42.33	37
Cu	111	2.088	43.33	100

[a] Data obtained from the ASTM file.

counting. This calibration curve was used to correct all intensity data. It is not difficult to make the correction even for integrated intensity measurements. By determining the proper line height and background level, the correct peak and base of the diffraction line can be sketched. There is no measurable change in the sides of the line because of the steep slope.

The diffraction lines best suited to the direct-comparison method of analysis are the most intense ones. Unfortunately, overlapping occurs between the 100 intensity lines* of the two copper oxides (Table II). These lines were delineated by superimposing the line profiles obtained from pure samples of each oxide. Since no overlapping occurs with the oxide lines second in intensity, they might be considered more suitable for quantitative analysis in this case. However, the 37 intensity cuprous oxide line cannot be measured with sufficient precision at low concentrations. The 96 intensity cupric oxide line was not used since the delineation of the 100 intensity lines provided all the data necessary for the analysis.

Peak area rather than peak height had to be used as the measure of line intensity because particle-size broadening was encountered in the aerosol samples. The area of each peak was measured as one-third of three consecutive traces with a planimeter. The limited resolving power of crystallites smaller than about 1000 Å causes the energy in a diffraction line to be distributed over a wider angular range. This broadening reduces maximum intensity but not integrated intensity since the latter is directly proportional to the diffracted energy.[21] In cases where the intensity distribution for lines is the same in all samples, the areas of corresponding lines are proportional to the line heights,[22] and maximum intensity measurements can be used directly. With the aerosol samples, however, broadening is slight in those exploded at low kv and becomes quite pronounced in those exploded at high kv.

The average aerosol particle size was determined from line broadening by the standard procedure outlined by Barrett.[23] The width of a line at half-height, measured in radians (2θ), is compared with the same line for a standard sample. The width of the standard line incorporates all broadening due to instrumental factors and any increase in width is attributed to size broadening. If W_a is the width of an aerosol line and W_s is the width of the standard line, then the broadening due to particle size B_{ps} is given by

$$B_{ps}^2 = W_a^2 - W_s^2$$

---

* The unresolved (002) ($\bar{1}$11) and (111) (200) doublets of CuO are here referred to respectively as the 100 intensity and the 96 intensity lines.

The particle thickness L_{hkl} in the direction normal to the reflecting plane (hkl) is given in angstroms by the relation

$$L_{hkl} = 0.89\lambda/B_{ps} \cos\theta$$

where λ is the wavelength in angstroms and θ is the Bragg angle. This technique was applied to the 96 intensity CuO line because of its freedom from overlapping. The average particle size was determined for samples exploded at various initial voltages on the capacitor bank. The results for samples of the first series (Figure 4) show that the aerosol particle size decreases rapidly as kv increases and has a fairly constant value for explosions above 10 kv. This agrees favorably with results obtained by Karioris et al.[24] from electron photomicrographs of aerosols from uranium wires exploded with the Oak Ridge National Laboratory device. They report a continuous distribution of particle sizes between 100 and 1000 Å with broad peaks occurring in the distribution curves at 130 Å for an explosion at 15 kv and at 250 Å for explosions at 5 kv and 2 kv.

Other difficulties found in practice in quantitative X-ray analysis are caused by preferred orientation and extinction. That the aerosol samples are completely free from any preferred orientation of the crystallites was determined from Laue transmission photographs of the samples. In the preparation of the standards, preferred orientation was minimized through a method of sample mounting devised by McCreery.[25] Two slight modifications were made to the sample holder described. A rectangular cell rather than a circular one was used to hold the powder samples. The cell was made long enough to intercept the entire X-ray beam at the lowest diffraction angle ($2\theta = 34°$) used. In addition, the edges of the cell were beveled to hold the powder sample more securely.

Two types of extinction, primary and secondary, are both due to the shielding of lower layers of crystallites by parallel upper layers.[26] Primary extinction is appreciable only in crystals larger than about $0.1\ \mu$ and secondary extinction in crystals larger than 10 to $100\ \mu$.[27] Both effects can be neglected for the aerosol samples. The particle size of the standards is not well known. The Cu_2O powder, as-received, could be packed into the sample holder without binder. The CuO and Cu were fine enough to pass through a

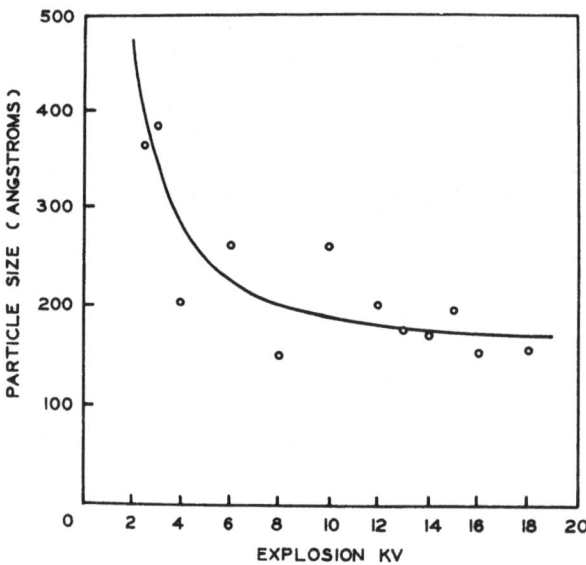

Figure 4. Average size of crystallites from exploded Cu wires as determined by line-broadening measurements.

325-mesh sieve, but would not pack into the sample holder. Grinding the CuO for 10 min in an agate mortar reduced it to a size which would pack without binder. The Cu could not be ground because of its ductility. The particle size of the standards was certainly less than 44 μ. A maximum size of 5 μ is needed for good reproducibility of intensity data.[28] The reverse argument indicates that the particle size of the standards was less than 5 μ since very good reproducibility was obtained with them. Secondary extinction is, therefore, negligible for the standards. If the assumption is made that all three materials in the standards have the same particle size, the effects of primary extinction may also be neglected.[29]

RESULTS AND DISCUSSION

The direct-comparison method of quantitative X-ray diffraction analysis was applied to two series of aerosols from exploded copper wires. In the first series, wires 2 cm long and weighing 8.5 mg were exploded in air with voltages between 2 and 18 kv applied to the capacitor bank. In the second series, 12.4-mg wires 3 cm long were exploded in dry air at voltages between 3 and 13 kv and with a rearrangement of the connections of the capacitor bank. The quantitative composition of the aerosols in Series 1 is shown as a function of the voltage in Figure 5. Below 1.5 kv explosion did not occur. At 1.5 kv there was not enough particulate matter collected on the filter to give a diffraction pattern and at 2 kv, the amount collected was insufficient for accurate quantitative analysis. Between 2.25 and 5 kv the relative amount of CuO increases rapidly, while the amount of Cu_2O decreases. The relative amount of unoxidized Cu also decreases in this voltage range. Above 5 kv the composition remains fairly constant, with CuO accounting for almost the entire sample.

Quantitative analysis of the second series of aerosol samples gave the results shown in Figure 6. The increase of CuO and decrease of Cu_2O between 3 and 6 kv are more gradual than in Series 1. Above 6 kv, the CuO and Cu_2O data points do not lie on a smooth curve but both fluctuate widely about a mean weight fraction of approximately 0.5. The Cu content, however, decreases smoothly throughout the entire range as it did in Series 1.

Series 2 differed from Series 1 in that: (1) the explosions took place in dry air; (2)

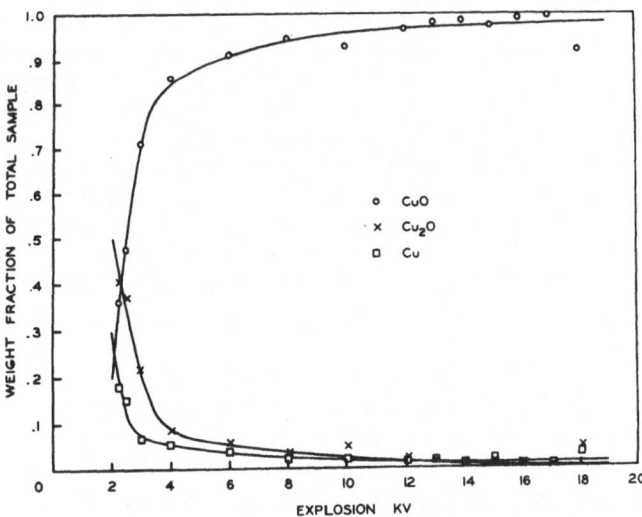

Figure 5. Composition of aerosols from exploded copper wires (Series 1).

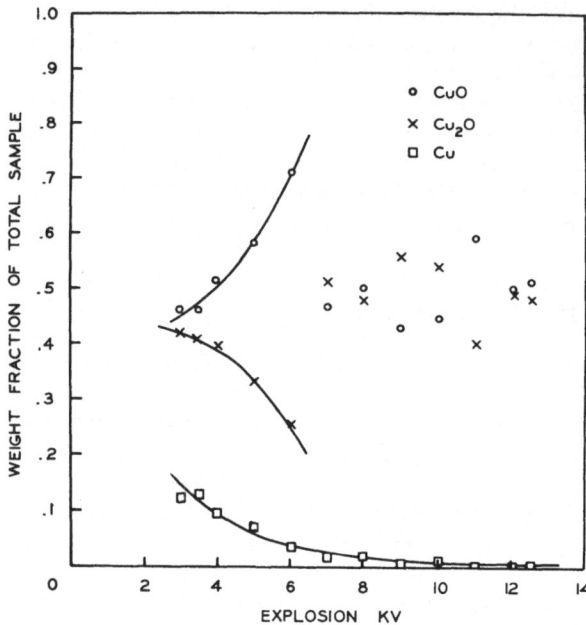

Figure 6. Composition of aerosols from exploded copper wires (Series 2).

a larger mass of wire was exploded; and (3) the arrangement of the capacitor bank was changed before the explosion of the second series. Drying of the air in the explosion chamber seems to be least responsible for the differences in composition observed. A wire was exploded at 12.5 kv in humid air under conditions which were otherwise identical to those for Series 2. The composition of this sample was not appreciably different from the dry-air sample. The rearrangement of the capacitor bank is a probable cause of the differences between the two series. The twenty 1-μf capacitors were originally arranged in line so that they could be connected by minimum lengths of straight copper bars. During the interval between the preparation of the two series of aerosol samples, the capacitors were placed in a cabinet which necessitated stacking them on shelves and connecting them with lengths of copper strap. It is believed that this arrangement greatly increased the inductance of the circuit, thereby causing a slower discharge of the capacitor bank. The explosion of a larger wire adds to this effect, since more energy is required to vaporize the increased mass. Both of these factors could account for the more gradual variation in the composition of the Series 2 aerosol samples between 3 and 6 kv.

The variation in oxide content above 6 kv for Series 2 is quite another matter. It is very difficult to attribute these erratic results to one or several of the many variables which influence the explosion process. It is sufficient here to say that the deviation of the data from a smooth curve must be attributed to the explosion process rather than to the analysis of the samples. The results of analyses are very reproducible. Each of the samples prepared as a standard was remounted and scanned three times. The standard deviation for the three runs is contained in the size of data points on the working curves. Each of the slide-mounted aerosol samples was scanned twice, with the sample rotated 180° between runs. Here again, the standard deviation for each of the three components was within 1 or 2% of the amount present, except at low concentrations.

No analytical relation between energy input and product formation has been derived. Tucker[30] has found, in studies of exploding gold wires, that the total energy input may exceed the theoretical vaporization energy by factors as large as three. Recovery of metal from oxide aerosols varies with the mass of wire and with voltage used for explosion approaching 100% for the explosion of small wires with high voltage.[24] That the composition of smokes from wires exploded in air varies with the voltage used for explosion indicates that mechanisms other than simple ignition and burning are involved in the production and dispersal of these aerosols.

ACKNOWLEDGMENTS

The authors would like to express their appreciation to Mr. B. R. Fish, Health Physics Division, Oak Ridge National Laboratory, for his cooperation in producing these aerosols and for many helpful discussions. This work was supported, in part, by the Milwaukee-Pere Marquette Foundation, Inc.

REFERENCES

1. W. G. Chace, AFCRC-TN-58-457, Air Force Cambridge Research Center, Bedford, Mass., 1958.
2. W. H. Richardson, SCR-53, Sandia Corp., Albuquerque, N.M., 1958.
3. W. G. Chace and H. K. Moore, eds., *Exploding Wires*, Plenum Press, New York, 1959.
4. R. Abrams *et al.*, CH-3629, Argonne National Laboratory, Chicago, 1946.
5. F. G. Karioris and B. R. Fish, *J. Colloid Sci.* **17**: 155, 1962.
6. M. Hansen and K. Anderko, *Constitution of Binary Alloys*, McGraw-Hill Book Co., Inc., New York, 1958, p. 600.
7. F. D. Bennett, *Sci. American* **206**: 102, 1962.
8. S. M. Mehta and R. M. Bharacha, *Proc. Indian Acad. Sci. A* **37**: 29, 1953.
9. V. N. Podchainova, *J. Anal. Chem. USSR* **7**: 343, 1952.
10. I. Ubaldini and F. Guerrieri, *Ann. chim. appl.* **38**: 695, 1948.
11. A. K. Lavrukhina, *Zhur. Anal. Khim.* **1**: 73, 1946.
12. D. Nishida and K. Hirabayashi, *J. Chem. Ind. (Japan)* **26**: 1123, 1923.
13. F. H. Pollard, J. F. W. McOmie, and A. J. Banister, *Chem. & Ind. (London)* 1955, p. 1598.
14. J. Harvey, H. I. Mathews, and H. Wilman, *Discussions Faraday Soc.* **30**: 113–123, 1960.
15. S. Zerfoss and M. L. Willard, *Ind. Eng. Chem., Anal. Ed.* **8**: 303, 1936.
16. B. D. Cullity, *Elements of X-ray Diffraction*, Addison-Wesley Publishing Co., Inc., Reading, Mass., 1956, p. 388.
17. G. M. Faulring and R. D. Carpenter, *Advances in X-ray Analysis, Vol. 1*, University of Denver, Plenum Press, New York, 1960, p. 60.
18. S. T. Gross and D. E. Martin, *Ind. Eng. Chem., Anal. Ed.* **16**: 95, 1944.
19. General Electric Company, X-ray Department, Direction 12593A.
20. H. L. Klug and L. E. Alexander, *X-ray Diffraction Procedures*, John Wiley & Sons, Inc., New York, 1954, p. 310.
21. B. L. Averbach and M. Cohen, *Metals Technology* 15, T.P. No. 2342, 1948.
22. J. C. M. Brentano, *J. Appl. Phys.* **20**: 1215, 1949.
23. C. S. Barrett, *Structure of Metals*, second edition, McGraw-Hill Book Co., Inc., New York, 1952, pp. 156–158.
24. F. G. Karioris, B. R. Fish, and G. W. Royster, Jr., "Aerosols from Exploding Wires." (To be published in Proceedings of the Second Conference on the Exploding Wire Phenomenon, Boston, 1961.)
25. G. L. McCreery, *J. Am. Ceram. Soc.* **32**: 141, 1949.
26. W. L. Bragg, C. G. Darwin, and R. W. James, *Phil. Mag.* **1**: 897, 1926.
27. R. J. Havighurst, *Proc. Natl. Acad. Sci.* **12**: 375, 1926.
28. H. P. Klug, L. Alexander, and E. Kummer, *Anal. Chem.* **20**: 607, 1948.
29. B. D. Cullity, *Elements of X-ray Diffraction*, Addison-Wesley Publishing Co., Inc., Reading, Mass., 1956, p. 400.
30. T. J. Tucker, *J. Appl. Phys.* **32**: 1894, 1961.

DISCUSSION

D. Rodier (Grumman Aircraft Engineering Corporation): The particles show agglomeration. Are there any that do not show agglomeration?

F. G. Karioris: The aerosol particles that were collected by the thermal precipitator showed a typical chain formation, rather than an agglomeration. The others were redispersed from a millipore filter by taking a small amount of material, shaking it with a supersonic vibrator, and then putting it on an electron microscope grid. These also show agglomeration, but not the typical chain formation. We have produced, we think, free spheres by producing an aerosol, putting it into a plenum chamber and fanning for perhaps 20 min to 1 hr. This drives the larger aggregates to the wall and leaves a much more dilute aerosol consisting essentially of these very, very small spheres of the order 0.025 or 0.05 μ. We think we can produce isolated spheres by diluting the aerosol very, very quickly. When the aerosol is left in its concentrated state, it coagulates very rapidly and we have seen chains that are 2 or 3 cm long and can be seen in a Tyndall beam.

D. Rodier: Do they form free and then agglomerate?

F. G. Karioris: The spheres are formed by condensation. Our work now consists of the explosion of metals in inert atmospheres, so we know a little bit more about the physical properties and the situation is not complicated by reaction with oxygen. We have been exploding all of these metals and have been including indium, cadmium, and zirconium in nitrogen and argon.

D. Rodier: How large a sphere can you produce by this method?

F. G. Karioris: We have an idea that the metal vapor will nucleate into larger particles or spheres if the ambient pressure of the indifferent gas is low. But one cannot get too low or one has a vacuum vaporization process where everything ends up on the walls. We have done explosion with vacuums and we find all sorts of interesting things. For example, wires that will not explode at a 100-μ pressure will explode when you get down to 0.1-μ pressure because at 100 μ the Paschen curve for the gas is such that the discharge takes place in the gas rather than through the metal, and the wire just stays there. So we have tried many things; however, one has to end the paper somewhere, and we ended it when the electron diffraction work on the $Cu–Cu_2O–CuO$ system was completed.

R. E. LaLonde (Owens-Illinois Glass Co.): The particle size varies with kv. Have you noticed any variance in the three phases?

F. G. Karioris: It has been piquing my curiosity whether each phase is a separate particle or whether each particle consists of all three phases. I don't know. The particles are too small to do electron diffraction on an individual particle. One needs a size of about 1 μ. Our particles are of the order of 0.05 μ. The fact that the particle size as determined by X-ray line broadening is in the same ball park as the particle size determined by electron micrographs suggests that each particle has one phase. Also, there are nice spheres, and the fact that they end up giving the diffraction pattern of crystallites is very intriguing. We thought we might have glassy spheres rather than crystallites. People have published electron micrographs of smokes which do show polyhedral particles, and there are some classic photographs of magnesium oxide smokes, where one can see the very, very tiny cubes, but we have not been able to observe them. The net answer is, it seems probable that each particle is a single phase of the mixture. We have been trying to think of a way of finding this out, because it is very difficult to separate these particles by any convenient method.

THE PREPARATION OF POLE FIGURES
FOR POLYMERS BY COMPUTER TECHNIQUES

J. W. Jones

E. I. du Pont de Nemours and Co., Inc.
Wilmington, Delaware

ABSTRACT

Preferred orientation in polymers has been studied by preparation of pole figures. The diffraction data are collected on punch tape under the control of a programmer for the orienter which permits stepwise or continuous scanning. A computer is programmed to prepare the data for an off-line automatic curve plotter. Several projections are available.

INTRODUCTION

The value of pole figures in the study of preferred orientations has long been recognized. In practice, the tedious problem of data reduction has limited the application of the method. The direct recording of pole figures has been proposed by Geisler[1] but such devices are specific and are not easily modified for such changes as the investigator may desire. Holland, Engler, and Powers[2] brought to bear the advantages of computer techniques in the preparation of pole figures from reflection studies with metals. However, the data had to be transcribed to punched cards before the information could be processed.

In this study the orienter was programmed in a simple manner and the data were taken directly in punched-tape form. The data were processed by computer and the pole figures plotted by an automatic plotter. The technique has been applied to polymers and polymer films using transmission methods.

EQUIPMENT AND PROCEDURE

The X-ray equipment consisted of a General Electric XRD-5 with No. 2SPG detector and single-crystal orienter.[3] The motions of the orienter were motorized with synchronous motors on ϕ (rotation) and χ (tilt) motion, and the samples prepared in the manner described by Heffelfinger and Burton[4] (Figure 1). Cams with indentations were attached to the drive gears and these cams actuate microswitches which provide pulses to monitor or control the position of the orienter. Cams cut for 1° change in χ and 2° change in ϕ have been satisfactory for most of the work, but it is possible to decrease the intervals to one-fifth or perhaps more without additional gears.

The control of the orienter and data reading is in a separate panel, Figure 2. The upper panel has switches for control of the orienter position, program, and a readout of orienter position. The next panel is a scanner/coupler which reads the counter and timer and translates these readings to a form suitable for punching on paper tape. There is a

---

[1] Superscripts pertain to references at the end of the paper.

Figure 1. Motorized single-crystal orienter.

parameter board for entering the information on sample description and necessary instructions for the computer. And, finally, the tape punch is in the base.

The program control of the orienter consists of two accounting systems, which operate in tandem and each has three modes of operation. The accounting systems totalize the number of steps made by the orienter from pip count. At preset intervals the first system can stop the orienter and advise the XRD-5 to take a data point. When the

Figure 2. Control panel for orienter and
punched-tape preparation.

datum is obtained the system then proceeds to advance the orienter, repeating its operations until the preset limit is attained. The first accounting system is reset and control given to the second accounting system which advances the second variable a preset interval. Control is then returned to the first system, which proceeds as described above. The whole process can be repeated until the preset limit of the second system is reached, at which time the operation is terminated. The limits may be set as high as 999 and the interval from 0 to 10. Mode 1 provides the step scanning described above. Mode 2 is a positioning operation or order to go to the set limit. In Mode 3 the orienter is continuously moved while counts are being taken. When both ϕ and χ are continuously changed we obtain a spiral scan. We have used this mode for the preparation of rapid scans in transmission similar to what Holden[5] has described for reflection.

The scan time will depend upon the detail desired and upon the counting rates. For most polymer samples we find that data for a single diffracting plane can be collected over a hemisphere in 90 min or less.

THE COMPUTATION PROCEDURE

The computer was a Bendix G-15D, which is a small computer with a magnetic drum memory. There are approximately 2000 words of storage available for data and program. The speed of the computer is 30 ms/drum revolution (108 words). The programs are in machine language for fixed point calculation. Up to 600 count points are processed as a unit and several units can be processed successively. The output from the computer is a punched tape for an off-line XY automatic Moseley Autograph curve plotter.

The information for the computer consists of a punched-paper data tape and a set of parameters. The parameters describe the initial position of the sample ($\chi_0\phi_0$), the manner in which the sample is moved (stepwise or continuous), the rate of motion in the two directions (k for ϕ and m for χ), the maximum value (χ_m or ϕ_m), the type of pole figure to be prepared, and the selected level values. The parameters may preface the data tape and be read by the computer or entered by the operator from a typewriter. The data tape contains a series of counts and the counting times. Normally, we maintain the counting time as a constant and the computer sorts out the counting data and arranges it in blocks in the computer memory.

The data, after being read into the computer, may be corrected for absorption or other effects, but the absorption correction is small for many polymers and generally we use the data without these corrections. The programs make available the plotting of sections of the reflection hemisphere or various projections of the hemisphere of reflection. We have used the following terms to describe the plots—plane orientation diagrams, equiangular projections, equiarea projections, and normal projections.

The plane orientation diagram is a plot of intensity as radius vs. angular position (ϕ or χ). The orientation angle may be read from these plots. These diagrams are plane sections of the reflection hemisphere. Normally they are made with one angle (ϕ or χ) held constant and the other varied, but it is possible to make variations in the section.

The normal projection is an orthographic projection on a base plane. The concept is simple but because of the distortion of the information, we have made little use of this form.

The equiangular projection is the projection of the hemisphere on normal polar coordinate paper. Equal angles in χ are equal radius changes and equal angles in ϕ correspond to equal radial sweeps. This type of plot is convenient for describing the location of diffraction peaks and orientations in angle rotation.

The equiarea projection is closely related to the equiangular projection. The

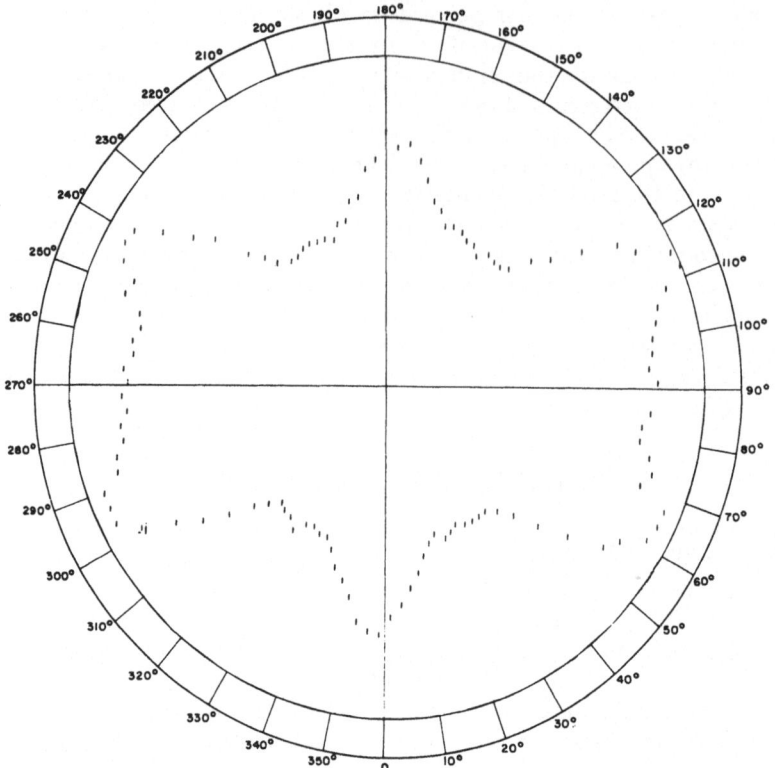

Figure 3. Plane orientation diagram. Polyethylene terephthalate (100), plane-of-film to edge ($\chi = 0°$).

projection area is directly proportional to the area of the reflection hemisphere. It is most suitable for integrated area measurements.[6]

The plots are available with all-points or level change only. All points are plotted in the plane orientation diagram and can be plotted on the projections. The number of levels determined by the plotter print wheels is six. The programs are written for this number, but it is possible to make this less by choice of level, or more by rescanning the data stored in the computer. A level or contour in a plot may be a multiple of the random counting rate or any other arbitrary counting rate selected. When the all-points program is specified, the level and location of each datum is indicated, and in level change, only those points where changes occur are indicated. It is a relatively simple alteration to the program to provide stereographic projections if this is desired.

The running time on the computer will depend upon the number of points which must be calculated. When level changes only are indicated, a projection can be calculated and punched out in 5 to 8 min. The plotter is an off-line device so that one set of data can be plotted while a second set is being calculated.

The position calculations for the various projections for use by the XY plotter are:

R_0 = radius of pole figure	N = number of datum
R = radius of point to be plotted	$X = R_0 + R \cos \phi$
k = degrees per second for ϕ	$Y = R_0 + R \sin \phi$
m = degrees per second for χ	$\phi = \phi_0 + kNt$
t = seconds per datum	$\chi = \chi_0 + mNt$

The Plane Orientation Diagram

$$I = \text{measured intensity}$$
$$I_m = \text{maximum intensity}$$
$$R = R_0(I/I_m)$$

Equiangular Projection

$$R = R_0(360 - 4\chi)/360$$

Equiarea Projection

$$R = R_0(1 - \sin \chi)^{1/2}$$

Normal Projection

$$A = (k + m)Nt + \phi_0 + \chi_0$$
$$B = (k - m)Nt + \phi_0 - \chi_0$$
$$X = R_0[1 + \tfrac{1}{2}(\sin A + \sin B)]$$
$$Y = R_0[1 + \tfrac{1}{2}(\cos A + \cos B)]$$

EXAMPLES

Pole figures for a sample of biaxially oriented $3X$ and heat-set polyethylene terephthalate film were prepared. The sample, which consisted of a stack of 1-mil films

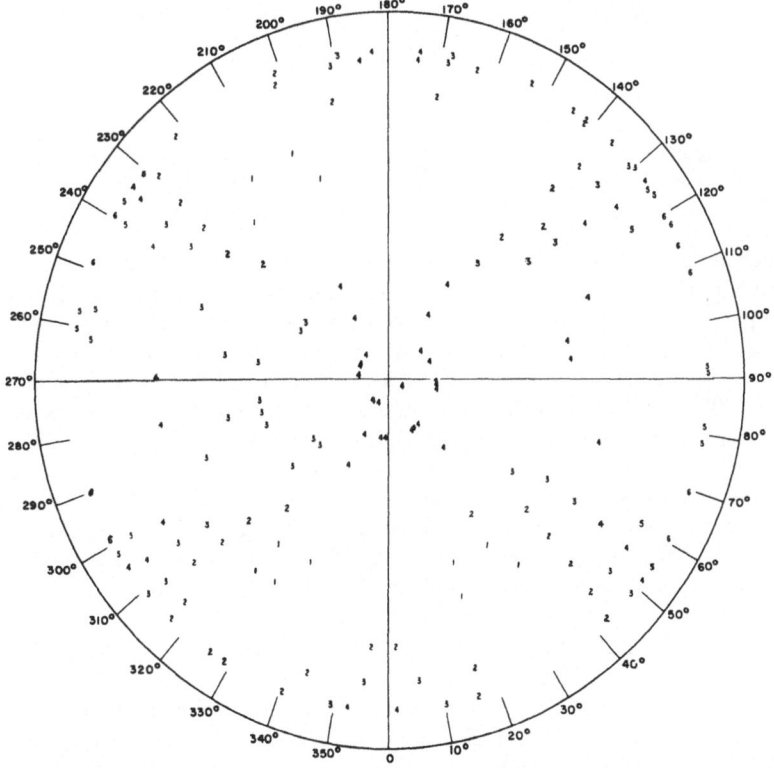

Figure 4. Equiangular pole figure. Polyethylene terephthalate (100), prepared with level change selection: Level 1 is 380 counts/sec. The level difference is 88 counts/sec.

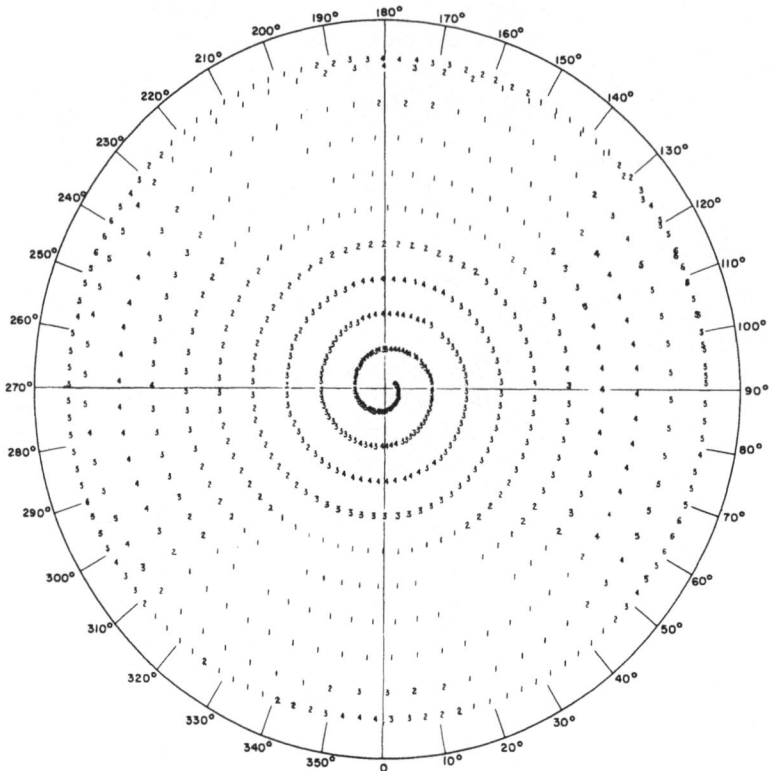

Figure 5. Equiangular pole figure (all-points). This figure is identical to Figure 4 except that the level of each experimental point is indicated.

50 mils thick and 50 mils wide mounted on a pin, was placed in the single-crystal orienter and aligned with the MD (Machine Direction) parallel with the beam and TD (Transverse Direction) perpendicular to the plane of diffraction at $2\theta = 0°$. The goniometer was set at the 2θ angle corresponding to the peak of the selected diffracting plane. In this arrangement, diffracting planes in the plane of the film are at $\chi = 0°$, $\phi = 0°$, and those in the edge at $\chi = 90°$. For the normals to the planes, the plane-of-film normal is at $\phi = 90°$, $\chi = 0°$ the edge normal is at $\chi = 90°$ and the end normal is at $\phi = 0°$, $\chi = 0°$.

In Mode 3, the scan time per plane was 90 min. There were two parts to a complete scan. In the first part ϕ was scanned from 0 to 360° at χ equal to zero. Data were taken at 3° intervals. In the second part, both ϕ and χ were varied. Data were taken every 5.6° in ϕ with a 10° change in χ for each 360° in ϕ. The data for the first part were used to prepare a plane orientation diagram and combined with part two to prepare the full pole figure.

The plane orientation diagram for the (100) plane ($2\theta = 17.35°$ Cu K_α) of this polyethylene terephthalate film is shown in Figure 3. The equiangular pole figure in Figure 4 was prepared with level selection, and in Figure 5 the all-points program was used. The selection of levels is usually done by inspection of the strip chart from the X-ray recorder. If the data are to be normalized in terms of the random level we usually proceed to plot a figure with arbitrary levels, planimeter the areas to obtain the total or integrated intensity, and then rescan the data with the computer using a new series of

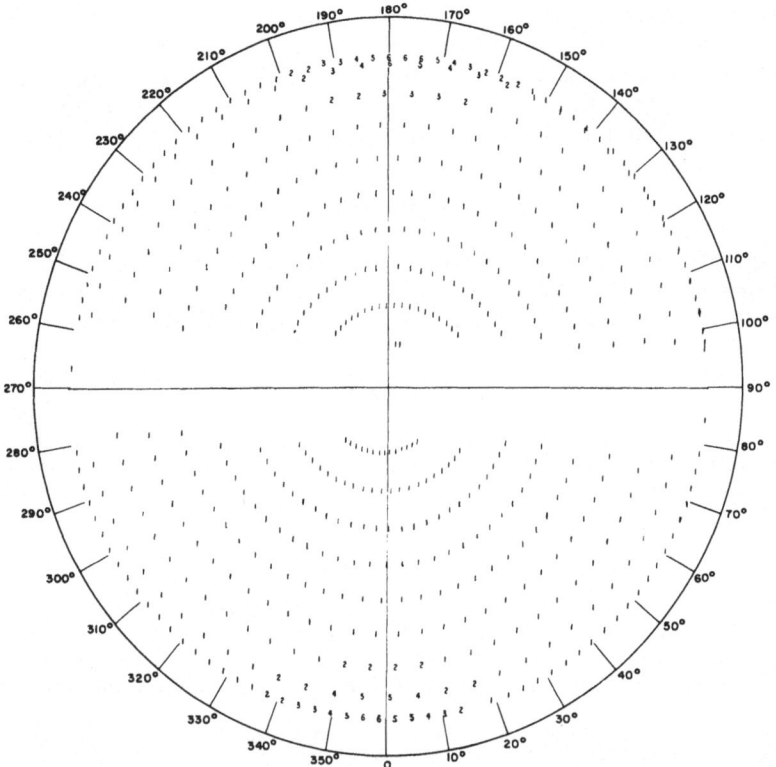

Figure 6. Equiarea pole figure. Polyethylene terephthalate (010). Level 1 is 300 counts/sec. The level difference is 1500 counts/sec.

levels. This approach is to be preferred to a so-called random sample because the degree of crystallinity of a sample can vary depending upon the method of preparation. Each sample is thus normalized to its own random level.

More than one diffracting plane in a sample must be studied if the texture is to be interpreted. In Figure 6, the pole figure for the (100) plane ($2\theta = 25.75°$) is presented. This figure is an example of an equiarea projection. The predominant orientation for the 100 planes is the plane of the film. The double 010 planes center about the edge and end. There is a smaller component of 010 planes in the plane of the film.

The data presented here were taken by rapid scanning. When detail or precision is necessary the counting data can be taken under the control of the programmer for periods such as overnight or over a weekend. Where precision is required, step scanning (Mode 1) is to be preferred. In this mode, counts are made at fixed ϕ and χ, and no slurring of counts over an angular interval occurs. The scan rates can be changed either by gears or motor changes. The motors are relatively inexpensive and simple to change. We have preferred the motor change method.

Six levels of counts can be specified to the computer. Counts below the lowest level (zero level) are not indicated and all points above the sixth level are in level six. When we wish to indicate more than six levels, the data in the computer are reprocessed. The sixth level in the first processing is discarded and it in turn becomes the first level for the second processing. In order to clearly indicate the levels on the graph, either the print wheel or ink ribbon color is changed.

The intervals chosen for ϕ and χ, 2° and 1°, respectively, may appear large to those who are accustomed to work with single crystals. However, most polymer samples have orientation angles (width at half-maximum of a peak) of 15° or more. Further, when film stacks are prepared, the alignment of a stack is within the range of $\pm 2°$ in ϕ and $\pm 1°$ in χ. Definition within these limits is provided.

CONCLUSIONS

The preparation of pole figures for polymer films has been simplified by collection of the diffraction data on punched tape under the control of a program device. The computer rapidly reduces these data so that they can be handled by a plotter.

ACKNOWLEDGMENT

The author wishes to express his appreciation to the du Pont Central Research Department Instrument Group for electronics design and assembly of the X-ray programmer, and to R. E. Clark of the Film Department for taking the data.

REFERENCES

1. A. H. Geisler, "Crystal Orientation and Pole Figure Determination," Modern Research Techniques in Physical Metallurgy, ASM Cleveland, 1953, pp. 131–153.
2. J. R. Holland, N. Engler, and W. Powers, "The Use of Computer Techniques to Plot Pole Figures," *Advances in X-ray Analysis, Vol. 4*, University of Denver, Plenum Press, New York, 1961, pp. 74–84.
3. T. C. Furnas, Jr., *Single-Crystal Orienter Instruction Manual*, General Electric Company, Milwaukee, Wisc., 1957.
4. C. J. Heffelfinger and R. L. Burton, "X-ray Determination of the Crystallite Orientation Distribution of Polyethylene Terephthalate Films," *J. Polymer Sci.* **48**: 289, 1960.
5. A. N. Holden, "A Spiral-Scanning X-ray Reflection Goniometer for the Rapid Determination of Preferred Orientations," *Rev. Sci. Instr.* **24**: 10, 1953.
6. ASTM E-81-54T, "Tentative Method for Preparing Quantitative Pole Figures of Metals," *ASTM Specifications, Vol. 3*, 1961, pp. 780–795.

DISCUSSION

Z. Wilchinsky (Esso Research and Engineering Co.): Would it have been possible to put in a correction for sample geometry if you wanted to improve the precision still further?

J. S. Jones: We have not provided for these extra corrections but it can be done. Our major problem here was just getting things arranged so that the information we got from the point of the X-ray into the computer mechanical and programming problems of handling long strips of continuous information. But these changes are relatively simply made.

RECENT DEVELOPMENTS IN THE MEASUREMENT OF ORIENTATION IN POLYMERS BY X-RAY DIFFRACTION

Zigmond W. Wilchinsky

Esso Research and Engineering Company
Linden, N. J.

ABSTRACT

From pole figure data, the orientation of a crystallographic direction $\mathbf{R}$ in any reference direction $\mathbf{Q}$ in a sample can be quantitatively expressed by the average $\langle \cos^2 \sigma \rangle$, where σ is the angle between $\mathbf{Q}$ and $\mathbf{R}$ in a crystal. In general, $\langle \cos^2 \sigma \rangle$ can be evaluated in terms of the experimental averages $\langle \cos^2 \phi_{hkl} \rangle$ where ϕ_{hkl} is the angle between $\mathbf{Q}$ and the normal to the (hkl) planes. The number of independent $\langle \cos^2 \phi_{hkl} \rangle$ measurements needed varies from one to five depending on the crystal system, hkl indices, and other factors discussed. Experimental techniques found useful for measuring $\langle \cos^2 \phi \rangle$ are described.

INTRODUCTION

Preferred orientation in polymers is of practical interest because the structural anisotropy thus produced affects many of the polymer's physical properties.

In polymers, a convenient manner of expressing preferred orientation of the crystallographic direction $\mathbf{R}$ is in terms of an orientation parameter[1,2] $\langle \cos^2 \sigma \rangle$, where σ is the angle between the direction $\mathbf{R}$ in the crystal and some chosen reference direction $\mathbf{Q}$ in the sample*; the angular bracket denotes a weight average. Until rather recently, the X-ray diffraction methods used for the rigorous evaluation of $\langle \cos^2 \sigma \rangle$ were limited to directions that are perpendicular to convenient reflecting planes. This limitation was recently removed when it was shown that $\langle \cos^2 \sigma \rangle$, can, in general, be evaluated even if there are no planes perpendicular to the crystallographic direction of interest.[3] The parameter $\langle \cos^2 \sigma \rangle$ can be expressed in terms of one or more similar parameters $\langle \cos^2 \phi_{hkl} \rangle$ where ϕ_{hkl} is the angle that the normal to the hkl plane makes with the reference direction $\mathbf{Q}$. This discussion will be concerned with procedures for evaluating $\langle \cos^2 \sigma \rangle$, their limitations and their application, particularly to polymers.

EXPRESSION FOR THE INDIRECT EVALUATION OF $\langle \cos^2 \sigma \rangle$

A general expression will first be presented for $\langle \cos^2 \sigma \rangle$ in terms of the orientation parameters $\langle \cos^2 \phi_{hkl} \rangle$. From this, the explicit solution for $\langle \cos^2 \sigma \rangle$ for a particular case can be obtained by a straightforward procedure.

---

[1] Superscripts pertain to references at the end of the paper.
* A complete list of notation appears at the end of the paper.

General Relationship for $\langle \cos^2 \sigma \rangle$

With reference to Figure 1, let the desired crystallographic direction **R** coincide with the z axis of a Cartesian coordinate system. Except for this restriction, the crystal may be fixed in any arbitrary position with respect to the x and y axes. The position of the crystal with respect to the direction **Q** will then be specified by the angles δ, ϵ, and σ, i.e., by the angles the xyz axes make with **Q**.

Let the normal **P** to a set of (hkl) planes make angles E, F, and G with the xyz axes, respectively. For a given set of indices in a crystal these angles are fixed; however, the angles δ, ϵ, σ, and ϕ (the angle **Q** makes with the x, y, z, and **P** directions) depend on the crystal orientation. Let **Q** and **P** be represented as unit vectors and let $e = \cos E$, $f = \cos F$, and $g = \cos G$. The vectors then can be written

$$\mathbf{Q} = (\cos\delta)\mathbf{i} + (\cos\epsilon)\mathbf{j} + (\cos\sigma)\mathbf{k} \tag{1}$$

and

$$\mathbf{P} = e\mathbf{i} + f\mathbf{j} + g\mathbf{k} \tag{2}$$

and the scalar product is

$$\mathbf{P} \cdot \mathbf{Q} = \cos\phi = e\cos\delta + f\cos\epsilon + g\cos\sigma$$

A general relationship between $\langle \cos^2\phi \rangle$ and the angles δ, ϵ, and σ is then

$$\langle \cos^2\phi \rangle = e^2 \langle \cos^2\delta \rangle + f^2 \langle \cos^2\epsilon \rangle + g^2 \langle \cos^2\sigma \rangle + 2ef \langle \cos\delta \ \cos\epsilon \rangle$$

$$+ 2fg \langle \cos\epsilon \ \cos\sigma \rangle + 2ge \langle \cos\sigma \cos\delta \rangle \tag{3}$$

This can be regarded as an equation in six unknowns, i.e., the averages, since e, f, and g are assumed known from unit-cell geometry. The desired quantity $\langle \cos^2\sigma \rangle$ can in the general case be evaluated by solving six simultaneous equations in these six unknowns. Five of the equations would be provided by equation (3) in which $\langle \cos^2\phi_{hkl} \rangle$ is evaluated for five $\{hkl\}$ planes having normals in different directions from each other. The sixth equation is the orthogonality relationship

$$\langle \cos^2\delta \rangle + \langle \cos^2\epsilon \rangle + \langle \cos^2\sigma \rangle = 1 \tag{4}$$

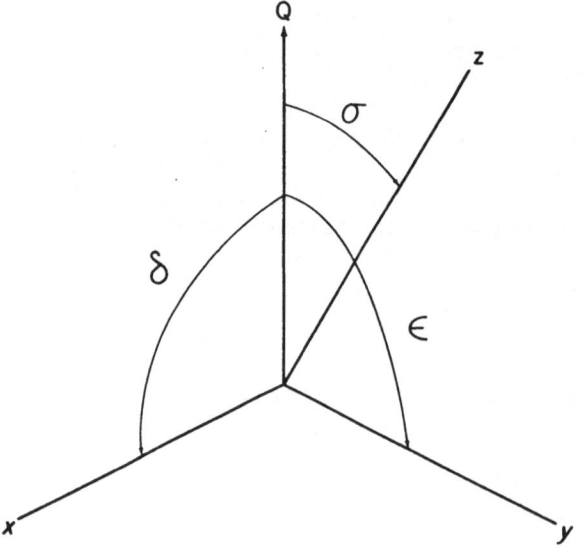

Figure 1. Angular relationships between the reference direction **Q** in the sample and the coordinate system xyz in a crystal.

Although the formal explicit solution for $\langle\cos^2\sigma\rangle$ for the general case can be written it is very cumbersome and perhaps not too useful in that form. A more practical procedure is to obtain the solution for a specific case after introducing simplifications resulting from the choice of crystallographic direction $\mathbf{R}$, choice of $\{hkl\}$ planes, and symmetry relationships. It will be shown that the number of $\{hkl\}$ planes required for obtaining $\langle\cos^2\sigma\rangle$ will usually be fewer than five and often only one or two.

Choice of Crystallographic Direction R

If, by symmetry considerations, there are no other directions in the crystal equivalent to $\mathbf{R}$ except possibly $-\mathbf{R}$, then $\mathbf{R}$ will be termed unique. The unique directions for the various crystal systems are listed in Table I. For the case where $\mathbf{R}$ is unique, the value of $\langle\cos^2\sigma\rangle$ is unity for complete orientation parallel to $\mathbf{Q}$, zero for complete orientation perpendicular to $\mathbf{Q}$, and $\frac{1}{3}$ for randomness. If $\mathbf{R}$ is not unique, the value of $\langle\cos^2\sigma\rangle$ for randomness is still $\frac{1}{3}$. However, since all the equivalent directions cannot be simultaneously either parallel to $\mathbf{Q}$ or perpendicular to $\mathbf{Q}$, the value of $\langle\cos^2\sigma\rangle$ for nonunique $\mathbf{R}$ cannot be unity or zero. In the case of polymers, the crystallographic direction for which $\langle\cos^2\sigma\rangle$ is desired is usually unique. For this reason and also to avoid excessive complexity, only unique directions will be considered in the subsequent discussion. Furthermore, the unique $\mathbf{R}$ direction in polymers will be taken as the crystal c axis unless otherwise specified.

Simplification Due to Symmetry

For polycrystalline materials, the diffraction from the (hkl) planes cannot be distinguished from that from the $(\bar{h}\bar{k}\bar{l})$ planes. For the $(\bar{h}\bar{k}\bar{l})$ planes, the direction cosines are $-e$, $-f$, $-g$, and $-\cos\phi$. If this substitution is made in equation (3), nothing is changed. Furthermore, a transformation through a center of symmetry likewise will not affect equation (3) or equation (4).

Twofold Rotation Axis or Mirror Plane. If z is a twofold rotation axis or if there is a mirror plane perpendicular to z, then for every set of (hkl) planes there are also the equivalent planes $(\bar{h}\bar{k}l)$, $(\bar{h}k\bar{l})$, and $(h\bar{k}\bar{l})$. In evaluating $\cos^2\phi$ for each of these cases and averaging the four values one finds that

$$fg\langle\cos\epsilon\ \cos\sigma\rangle = ge\langle\cos\sigma\ \cos\delta\rangle = 0 \tag{5}$$

As far as the simplification of equation (3) is concerned this is equivalent to keeping

Table I. Unique Directions in the Various Crystal Systems

Crystal system	Unique directions
Triclinic	All directions
Monoclinic	
(1) $b \perp ac$ plane	a, b, and c axes; directions in ac plane
(2) $c \perp ab$ plane	a, b, and c axes; directions in ab plane
Orthorhombic	a, b, and c axes.
Tetragonal	c axis
Hexagonal	c axis
Cubic	None

e, f, and g invariant and rotating the xy axes through 0 and 180°. In this case, the coordinates of the reference direction will change and $\mathbf{P}$ will be the following vectors:

and

$$\mathbf{P}_1 = (\cos \delta)\mathbf{i} + (\cos \epsilon)\mathbf{j} + (\cos \sigma)\mathbf{k}$$

$$\mathbf{P}_2 = - (\cos \delta)\mathbf{i} - (\cos \epsilon)\mathbf{j} + (\cos \sigma)\mathbf{k}$$

If $\cos^2\phi$ for these two values is carried out and the results averaged, then one will find that

$$\langle \cos \epsilon \cos \sigma \rangle = \langle \cos \sigma \cos \delta \rangle = 0 \tag{6}$$

The simplification produced in equation (3) by equation (6) is, of course, identical with that produced by equation (5).

Threefold, Fourfold, or Sixfold Rotation Axis. For a crystal having threefold rotation symmetry about the c axis, all the equivalent reflections can be represented by rotation of the xy axes through 0, 120, and 240°, while e, f, and g remain invariant. When the calculations are carried through as in the case of the twofold axis one obtains

and also

$$\langle \cos \delta \cos \epsilon \rangle = \langle \cos \epsilon \cos \sigma \rangle = \langle \cos \sigma \cos \delta \rangle = 0$$

$$\langle \cos^2\delta \rangle = \langle \cos^2\epsilon \rangle$$

The same results are obtained for the case of a fourfold or sixfold axis of rotation.

Mechanically Introduced Symmetry. If the distribution of orientations about the z axis (i.e., crystal c axis) is random, then the values of $\cos \delta$ and $\cos \epsilon$ are equally likely to be positive as negative and both go through the same sequence of orientations. Therefore, the same results obtain as for the threefold rotation axis.

It should be pointed out that a rotation about the reference direction $\mathbf{Q}$ cannot produce any simplifications to equation (3) since the direction cosines along $\mathbf{Q}$ are unchanged by this operation.

**Table II. Simplifications to Equation (1) Due to Symmetry When
σ is the Angle Between the c Axis and the Reference Direction
Q in the Sample**

Symmetry condition	Resultant simplification
Monoclinic system	
(1) $b \perp ac$ plane	$\langle \cos \delta \cos \epsilon \rangle = \langle \cos \epsilon \cos \sigma \rangle = 0$
(2) $c \perp ab$ plane	$\langle \cos \epsilon \cos \sigma \rangle = \langle \cos \sigma \cos \delta \rangle = 0$
Orthorhombic system	All cross-product averages are zero
Tetragonal or hexagonal systems	All cross-product averages are zero, $\langle \cos^2\delta \rangle = \langle \cos^2\epsilon \rangle$
$\{hk0\}$ planes	$g = 0$
$\{001\}$ planes and $c \perp a$ and b	$e = f = 0$, $g = 1$
Randomness about c axis	All cross-product averages are zero, $\langle \cos^2\delta \rangle = \langle \cos^2\epsilon \rangle$

Table III. Number of Independent {hkl} Planes Needed to Determine $\langle \cos^2\sigma \rangle$

Crystal system	hkl, general	hk0	h0k	001	Randomness about the c axis
Triclinic	5	3	5	—	1
Monoclinic					
(1) $b \perp ac$ plane	3	2	3	—	1
(2) $c \perp ab$ plane	3	3	2	1	1
Orthorhombic	2	2	2	1	1
Hexagonal	1	1	1	1	1
Tetragonal	1	1	1	1	1

Choice of {hkl} Planes. In polymers there are usually strongly reflecting {hk0} planes present. Therefore, $g = 0$, leading to a possible further simplification of equation (3). Similarly e or f may be zero, leading to further possible simplifications.

Number of Independent {hkl} Planes Required

The simplifying relationships are listed in Table II for the various crystal systems, and the number of independent {hkl} planes necessary for the evaluation of $\langle \cos^2\sigma \rangle$ are listed in Table III. Results equivalent to most of those in Table II and III have been reported by Sack.[4]

EXPERIMENTAL DETERMINATIONS OF $\langle \cos^2\phi \rangle$

From the relationships in the preceding discussions one may, for a specific case, obtain an explicit solution for $\langle \cos^2\sigma \rangle$ in terms of $\langle \cos^2\phi_{hkl} \rangle$ evaluated for one or more {hkl} planes. In this final solution all cross-product averages $\langle \cos^2\delta \rangle$ and $\langle \cos^2\epsilon \rangle$ have been eliminated. Thus, it only remains to measure $\langle \cos^2\phi_{hkl} \rangle$ experimentally. This can of course, be done from quantitative measurement of diffracted intensities.

With reference to Figure 2, consider the pole P of an hkl plane, i.e., the intersection of the plane normal with the surface of a unit sphere circumscribed about the sample. The coordinates of the pole are the spherical coordinates ϕ and ψ. Let $I(\phi, \psi)$ be the pole concentration, representing the relative amount of material having {hkl} plane normals in the direction ϕ, ψ. Then $\langle \cos^2\phi \rangle$ can be evaluated from the distribution of $I(\phi, \psi)$ over half of the unit sphere by the relationship

$$\langle \cos^2\phi \rangle = \frac{\int_0^{2\pi} \int_0^{\pi/2} I(\phi, \psi) \cos^2\phi \sin\phi \, d\phi \, d\psi}{\int_0^{2\pi} \int_0^{\pi/2} I(\phi, \psi) \sin\phi \, d\phi d\psi} = \frac{\int_0^{\pi/2} I(\phi) \cos^2\phi \, d\phi}{\int_0^{\pi/2} I(\phi) \sin\phi \, d\phi} \quad (7)$$

where

$$I(\phi) = \int_0^{2\pi} I(\phi, \psi) \, d\psi$$

The distribution function $I(\phi, \psi)$ is proportional to the integrated intensity diffracted from the {hkl} planes with orientation ϕ, ψ. In practice, the integrated intensity can usually be obtained by multiplying the peak intensity by the integral line width, after the peak

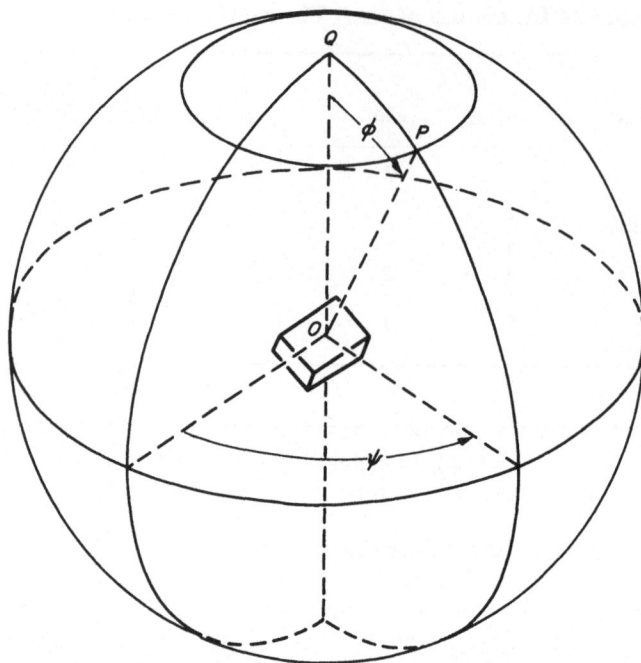

Figure 2. Position of plane normal **P** specified by the spherical coordinates ϕ and ψ.

intensity had been corrected for background intensity. If necessary, additional corrections to the intensity are made for the absorption of X-radiation in the sample and for the relationship of beam geometry to sample geometry.

At present, the intensity as a function of ϕ and ψ can be most conveniently measured by some form of pole figure device in conjunction with an X-ray quantum counter. The techniques that have been found particularly applicable to polymers will be briefly described.

Complete Pole Figures

The most versatile data are those obtained in a complete pole figure. Among the simplest of the pole figure devices to use are those for either plane or cylindrical samples.

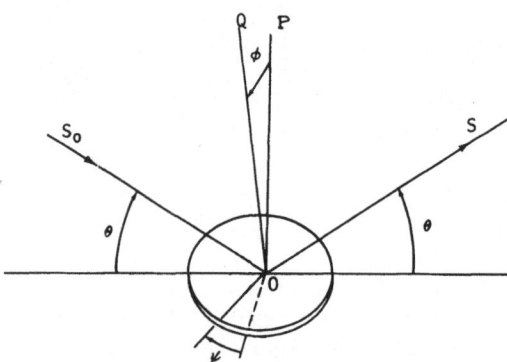

Figure 3. The angles ϕ and ψ obtained with a Schulz type pole figure device.

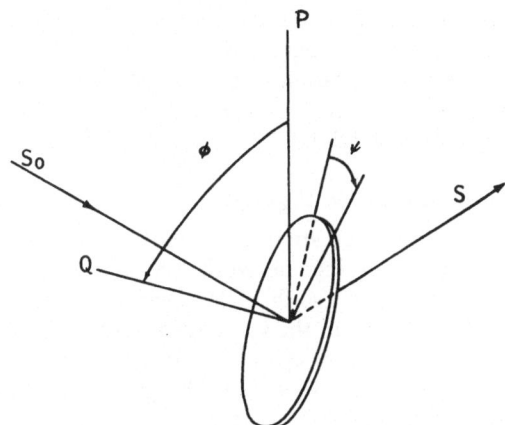

Figure 4. The angles ϕ and ψ with a Decker type of pole figure device.

A device for obtaining pole figure data by reflection from a flat sample has been described by Schulz.[5] The values of ϕ and ψ are obtained directly from the position of the sample, as indicated in Figure 3. In this operation the reference direction **Q** is normal to the plane of the sample. Initially, let **Q** be parallel to the bisector of the angle formed by the incident beam S_0 and the diffracted beam S entering the detector. Throughout the measurements, the detector is held fixed at the Bragg angle for the selected (hkl) planes. For any other position of the sample, ϕ is the angle through which **Q** has been tilted from its initial direction. The angle ψ is determined from the projection of **P** onto the plane of the sample; the value of ψ is the angle of rotation of the sample in its own plane, measured from a reference meridian. In operation, the sample is tilted so that **Q** is confined to a plane perpendicular to S_0OS. With this type of pole figure device, the range of ϕ between 0 and somewhat over 50° can be conveniently covered.

The range of ϕ between 50° and 90° can usually be covered for the same sample by transmission diffraction in a device described by Decker *et al.*[6] A similar relationship of sample position to ϕ and ψ holds, as indicated in Figure 4.

A pole figure device for cylindrical samples has been described by Cullity and Freda.[7] The measurements of ϕ and ψ are indicated in Figure 5; they are similar to those in the Schulz type. However, since the diffraction is by transmission, the entire range of ϕ from 0 to 90° can be covered.

With these devices $I(\phi, \psi)$ can be obtained for the entire range of ϕ, and ψ, i.e., over

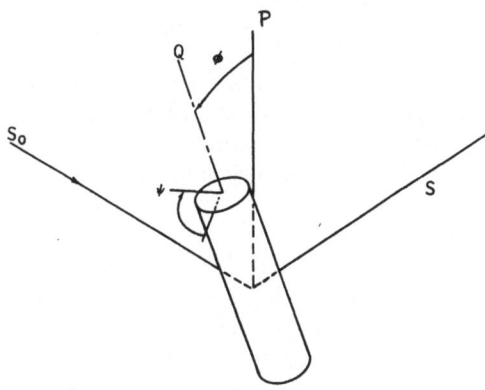

Figure 5. The angles ϕ and ψ with a Cullity and Freda type of pole figure device.

the entire surface of the coordinate sphere, Figure 2. From these data $\langle \cos^2\phi \rangle$ could then be calculated by equation (7). Also, the data may be plotted in the conventional pole figure representation to give additional information on the orientation. Although these data are obtained for a specific reference direction they can be transformed to any other reference direction.[8]

Restricted Pole Figures

If the orientation is desired only for the **Q** directions indicated in Figures 3 or 5, then the experimental work can be considerably simplified. The sample may be spun about the Q axis so that the integration with respect to ψ in equation (7) is automatically performed. From the intensities, $I(\phi)$ is obtained directly.

In the case of cylindrical samples (such as most fibers) having a random distribution of crystal orientations about the Q axis, it is not necessary to spin the sample in using the above method.

APPLICATIONS

The application of the above methods to polymers will be illustrated with selected examples from the writer's investigations.

Fiber

An experimental filament of poly-4-methylpentene-1 was selected for measurement of c-axis orientation.[9] This polymer crystallizes in the tetragonal system, hence, from Table III, only one set of $\{hkl\}$ planes is needed. The strongly reflecting $\{200\}$ planes were used.

For the hexagonal or tetragonal case in general, the solution for $\langle \cos^2\sigma \rangle$ is

$$\langle \cos^2\sigma \rangle = \frac{1 - g^2 - 2\langle \cos^2\phi_{hkl} \rangle}{1 - 3g^2} \qquad (9)$$

For the case of $(hk0)$ planes, $g = 0$, hence,

$$\langle \cos^2\sigma \rangle = 1 - 2\langle \cos^2\phi_{hk0} \rangle \qquad (10)$$

Because of symmetry about the filament axis the evaluation of c-axis orientation along the filament axis could be carried out with the restricted pole figure method without spinning. The diffraction sample was made up of a sheet of parallel fibers, and a constant area of this sample was intercepted by the incident beam. This arrangement eliminated the necessity for correcting the intensity for factors other than the background scattering.

The curve of $I(\phi)$ for $hk0$ planes in arbitrary units and the derived curves $I(\phi) \sin \phi$ and $I(\phi) \cos^2\phi \sin \phi$ are shown plotted vs. ϕ in Figure 6. From the latter two curves, the value of $\langle \cos^2\phi \rangle$ by equation (7) is 0.232, and hence from equation (10), the quantitative measure of the c-axis orientation is $\langle \cos^2\sigma \rangle = 0.536$.

Film

The preferred orientation of the b and c axes of a polypropylene film was measured.[8] Convenient reflection planes for this purpose were the $\{040\}$ and $\{110\}$ planes. Since the crystalline form is monoclinic with the b axis perpendicular to the ac plane, the above two planes were sufficient for the determinations (see Table III). Also, since the (040) planes are perpendicular to the b axis, $\langle \cos^2\phi_{040} \rangle$ is the orientation parameter for the b axis. The orientation parameters for the b and c axes were determined in the following three

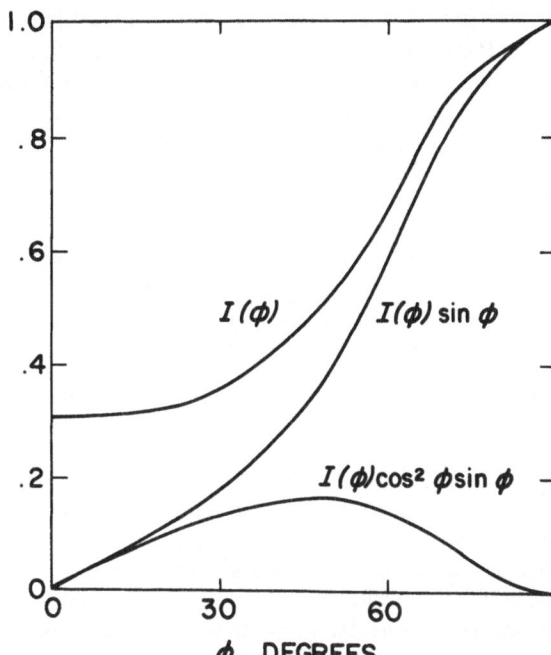

Figure 6. Experimental curves used for evaluating $\langle \cos^2\sigma \rangle$ in a filament sample of poly-4-methylpentene-1.

principal directions: A, direction of maximum c-axis orientation in the plane of the film (as determined by birefringement measurements); B, direction in plane of sample transverse to A; and N, normal to the plane of the film.

For this case, the b axis was taken along the y axis of the xyz coordinate system, hence, for the (040) planes $e = 0, f = 1, g = 0$. Since the normal to the (110) planes makes an angle of $72.5°$ with the b axis, $e = \sin 72.5°$ $f = \cos 72.5°$, and $g = 0$ for these planes. The final equation for $\langle \cos^2\sigma \rangle$ then becomes

$$\langle \cos^2\sigma \rangle = 1 - 0.901 \langle \cos^2\phi_{040} \rangle - 1.099 \langle \cos^2\phi_{110} \rangle \tag{11}$$

For measuring the orientation in the A direction, a strip of film was rolled into a tight cylinder whose axis was parallel to the A direction. $I(\phi_{040})$ and $I(\phi_{110})$ were then measured by the restricted pole technique with no spinning. From these values, $\langle \cos^2\phi_{040} \rangle$ and $\langle \cos^2\phi_{110} \rangle$, and hence $\langle \cos^2\sigma \rangle$ were determined. Similar measurements were made for the orientation parameter in the transverse direction, B, from a cylinder rolled so that its axis was parallel to the B direction in the film. To obtain the orientation in the direction normal to the plane of the film, the orthogonality relationship was used:

$$\langle \cos^2\sigma \rangle_A + \langle \cos^2\sigma \rangle_B + \langle \cos^2\sigma \rangle_N = 1$$

The orientation parameters were also measured by the complete pole figure method. Results obtained from the two types of measurements, summarized in Table IV are in satisfactory agreement. They both show a high c-axis orientation in the A direction in the film, and a high b-axis orientation normal to the plane of the film.

CONCLUDING REMARKS

To add to the completeness of the discussion several items pertaining to measuring orientation will be briefly mentioned.

Table IV. Orientation Parameters for Polypropylene Film in Three Principal Directions: A and B in Plane of Film, and N Normal to Plane of Film

Reference direction in sample	Restricted pole figure method		Complete pole figure method	
	c axis $\langle\cos^2\sigma\rangle$	b axis, $\langle\cos^2\phi_{040}\rangle$	c axis, $\langle\cos^2\sigma\rangle$	b axis, $\langle\cos^2\phi_{040}\rangle$
A	0.87	0.02	0.83	0.06
B	0.10	0.19	0.09	0.23
N	0.03	0.79	0.09	0.71

The degree of accuracy of the results depends significantly on the choice of $\{hkl\}$ planes. A poor choice may sometimes lead to an indeterminate solution. For example, in equation (9) if g, the direction cosine of the plane normal on the z axis is $\frac{1}{3}\sqrt{3}$, then $\langle\cos^2\phi\rangle$ is $\frac{1}{3}$, and the solution for $\langle\cos^2\sigma\rangle$ is indeterminate. This subject is discussed more fully by Sack.[4] Another point he discusses is using more than the minimum number of $\{hkl\}$ reflections. He shows that these can be used in a least-square method for improving the accuracy of the results. He also indicates how one could handle the problem of using diffraction lines that cannot be resolved from each other but the unresolved group can be accurately measured as a unit.

Another consideration is that of reducing the amount of work involved in the measurements. If the orientation is required for three principal directions in a film, it has been shown that the X-ray measurements made for any one of these directions can be combined with birefringence measurements (which can be made quickly) to give the desired information.[8]

Finally, the application of these methods to nonpolymeric materials might be desired. The methods are quite general and should be applicable to other polycrystalline materials especially if the direction $\mathbf{R}$ for which $\langle\cos^2\sigma\rangle$ is desired is unique. As has been pointed out, the interpretation of the orientation parameter $\langle\cos^2\sigma\rangle$ depends on whether or not $\mathbf{R}$ is unique. In the extreme case, e.g., cubic symmetry, the evaluation is trivial since all orientation parameters have a value of $\frac{1}{3}$. However, in less extreme cases, where $\mathbf{R}$ is not unique, but an equivalent group of $\mathbf{R}$ directions make only small angles among themselves, then $\langle\cos^2\sigma\rangle$ evaluated by the method discussed may still have a useful interpretation. Another special case of interest arises for materials of high crystallographic symmetry when the orientation is of a sufficiently high degree so that the poles are restricted to rather well-defined regions. In this case, it may be possible to work with only a selected portion of the diffraction data to obtain the desired information. For example, if the cubic $\{001\}$ direction is preferentially oriented parallel to the axis of a cylinder but cannot be satisfactorily measured directly from the (001) planes for certain reasons, then the equatorial reflection from the (100), ($\bar{1}$00), (010), and (0$\bar{1}$0) planes can be used. The value of $\langle\cos^2\phi\rangle$ for these planes is evaluated by equation (7) over a limited range of ϕ, and $\langle\cos^2\sigma_{001}\rangle$ can then be evaluated by equation (10).

NOTATION

x, y, z = Coordinate axes fixed in a crystal
$\mathbf{i, j, k}$ = Unit vectors along x, y, and z, respectively

$\mathbf{Q}$ = Unit vector along a reference direction in the sample
$\mathbf{P}$ = Unit vector along the normal to a crystal plane
$\mathbf{R}$ = Vector defining a crystallographic direction
ϕ = Spherical coordinate, angle of colatitude
ψ = Spherical coordinate, angle of longitude
E = Angle between $\mathbf{P}$ and x axis
F = Angle between $\mathbf{P}$ and y axis
G = Angle between $\mathbf{P}$ and z axis
e = cos E
f = cos F
g = cos G
δ = Angle between $\mathbf{Q}$ and x axis
ϵ = Angle between $\mathbf{Q}$ and y axis
σ = Angle between $\mathbf{Q}$ and z axis
$I(\phi, \psi)$ = Concentration of plane normals having coordinates ϕ, ψ

REFERENCES

1. W. Kuhn and F. Grun, *Kolloid-Z.* **101**: 248, 1942.
2. R. S. Stein, *J. Polymer Sci.* **34**: 709, 1959.
3. Z. W. Wilchinsky, *J. Appl. Phys.* **30**: 792, 1959.
4. R. A. Sack, *J. Polymer Sci.* **54**: 543, 1961.
5. L. G. Schulz, *J. Appl. Phys.* **20**: 1030, 1949.
6. B. F. Decker, E. T. Asp, and D. Harker, *J. Appl. Phys.* **19**: 388, 1948.
7. B. D. Cullity and A. Freda, *J. Appl. Phys.* **29**: 25, 1958.
8. Z. W. Wilchinsky, *J. Appl. Phys.* **31**: 1969, 1960.
9. G. Schael (Unpublished work in collaboration with the writer).

DISCUSSION

H. Steffen Peiser (National Bureau of Standards): Does the speaker assume a Gaussian distribution about the most probable axis?

Z. W. Wilchinsky: I have not assumed anything. The equation for the orientations for particular crystal directions is evaluated from the pole figure data regardless of how it comes out. I have some slides indicating some peculiar types of orientation behavior, such as cold-rolled polymers, which are definitely not a Gaussian distribution.

H. Steffen Peiser: Has an account been taken that most fibers of polymers do not have a crystallographic orientation?

Z. W. Wilchinsky: I may have failed to mention that this direction could be any direction whatsoever. It could be a crystallographic direction in the strict sense. When we have a direction in the crystal we wish to determine the orientation of that direction with some reference direction in the sample.

MEASUREMENT OF THE LATTICE CONSTANTS OF NEON ISOTOPES IN THE TEMPERATURE RANGE 4–24°K

L. H. Bolz and F. A. Mauer

National Bureau of Standards
Washington, D.C.

ABSTRACT

Using an X-ray diffractometer cryostat, the lattice constants of the isotopes ^{20}Ne and ^{22}Ne as well as the naturally occurring mixture have been measured throughout most of the temperature range in which they exist as solids (0 to approximately 24°K). The heavier isotope has the smaller lattice constant, the values obtained at 4.2°K being: ^{20}Ne—4.462$_4$ Å, ^{22}Ne—4.454$_0$ Å, and ^{n}Ne—4.462$_2$ Å.* The absolute error in these values is believed to be no greater than 0.001 Å. The volume expansion coefficient, which does not appear to differ significantly in the three cases, increases to a value of $6.3 \times 10^{-3}/°K$ at 24°K.

INTRODUCTION

The inert-element solids xenon, krypton, argon, and neon have often served as models in studies of the theory of simple solids but, because of their low melting points, they have not found wide favor in experimental work. Few accurate measurements of their physical properties have been made and theoretical studies have often been based on inadequate experimental data.

Of the properties of interest, the interatomic distance, density and expansivity can best be measured by low-temperature X-ray diffraction techniques. Such measurements, supplemented with bulk density measurements, have been carried out during the last six or seven years on argon, krypton, and xenon by various investigators under the leadership of E. R. Dobbs and G. O. Jones[1-4] at Queen Mary College, London. However, the measurements made by their group have not extended below 20°K, and solid neon, which melts at $\sim 24.5°K$, was not included in the study.

Values of the lattice constant of neon at 4.2°K have been reported by several investigators. Agreement among these values is rather poor, the range being more than 100 times the experimental error expected in present-day X-ray measurements. Values reported by Kogan, Lazarev, and Bulatova[5] for the separate isotopes ^{20}Ne and ^{22}Ne show a surprisingly large difference. When these were compared with a preliminary value for natural neon obtained in our laboratory, the unexpected result was that the lattice constant of the isotope that constitutes only 8.8% of natural neon agreed with our value for natural neon, while that for the more abundant isotope did not. The need for additional measurements was obvious.

Using a liquid helium cryostat and a commercial diffractometer we have measured

---

* ^{n}Ne is used to designate the naturally occurring mixture, 90.92% ^{20}Ne–0.26% ^{21}Ne–8.82% ^{22}Ne.
[1] Superscripts pertain to references at the end of the paper.

the lattice constants of the isotopes[20]Ne and [22]Ne as well as the naturally occurring mixture (90.02% [20]Ne–0.26% [21]Ne–8.82% [22]Ne)[6] throughout most of the temperature range in which they exist as solids (0 to 24.5°K). From the values obtained, the interatomic distance, molecular volume, density, and expansivity can easily be calculated.

Of the remaining physical properties of interest, data on the atomic heat capacity and heat of fusion of natural neon[7] and the isotopes [20]Ne and [22]Ne are available.[8] However, compressibility data are limited to a single value of the isothermal compressibility of natural neon at 4.2°K.[9] It is to be hoped that additional measurements will be forthcoming.

APPARATUS

Diffractometer Cryostat

The liquid–helium cryostat used with a General Electric XRD-5 diffractometer in making these measurements was described earlier.[10,11] Most of its features are shown in a vertical section that includes the axis of the double dewar vessel (Figure 1). The inner vessel (21) has a capacity of $2\frac{1}{2}$ liters and is filled with liquid helium, hydrogen, or nitrogen. Surrounding this vessel and separated from it by an insulating vacuum space is a hollow cylindrical tank (22) which is normally filled with liquid nitrogen. Radiation shields (3, 19) attached to this tank extend around the inner vessel at the top and bottom, shielding it from thermal radiation. The nitrogen tank, in turn is separated from the outer vacuum wall by an insulating vacuum space.

Attached to the bottom of the inner vessel is a cylindrical copper block (4) having one side cut out to leave a plane sample-deposition surface (30) with semicircular flanges above and below. Mylar* film 0.0007 in. thick cemented to these flanges with an epoxy cement confines the vapor in equilibrium with the specimen but permits the X-ray beam to pass with negligible attenuation. Measurements can be made at temperatures corresponding to vapor pressures as high as 1 atm without loss of the insulating vacuum.

An inlet tube (27) with an internal heater (29) is required for introducing samples into the chamber behind the mylar window. Circumferential windows of beryllium, in the outer wall, and of aluminum foil in the radiation shield permit the passage of the incident and diffracted beams throughout the angular range 0–165° 2θ. Rotational and translational adjustments (33, 34) are used to align the specimen surface so that it is tangent to the focusing circle.

The copper block (4) is joined to the helium vessel by means of a thin-walled tube (5) of 304 stainless steel selected because of its very low thermal conductivity at low temperatures. When the plug (35) is inserted in the top of this tube, as shown, to exclude liquid from the region below, the copper block can be heated electrically to maintain a desired temperature. The heater is wound on a copper bobbin (2) which is pressed into a hole in the copper block. A carbon resistor embedded in the bobbin serves as the sensing element for the automatic temperature controller.

A separate carbon resistor and a Au + 2.11 at. % Co vs. Cu thermocouple are provided for measuring the temperature of the copper block. Both of these are calibrated during each experiment at the following fixed points.[12]

Boiling point of He .... 4.216°K
Triple point of n-H$_2$ .... 13.957°K
Boiling point of n-H$_2$ ...... 20.39°K

The equation of Clement and Quinell[13] was used to interpolate readings of the carbon

---

* Polyethyleneterephthalate. Mylar is a registered trademark of the Film Department, E. I. du Pont de Nemours & Co. Inc., Wilmington, Delaware.

(1) thermocouple,
(2) heater bobbin,
(3) radiation shield,
(4) copper block,
(5) thin-walled support tube (stainless steel),
(6) copper ring,
(7) soft-solder joint,
(8) ring with centering pins,
(9) flange with O-ring seal,
(10) inner vacuum wall,
(11) heater leads,
(12) vacuum valve
(13) radiation baffle,
(14) helium-vessel neck stopper,
(15) block vent tube,
(16) liquid-helium depth probe,
(17) helium vent,
(18) nitrogen fill-tube and vent,
(19) radiation shield,
(20) spacing spring,
(21) helium vessel,
(22) nitrogen vessel,
(23) inlet-tube-heater terminal,
(24) by-pass valve,
(25) sample-inlet valve,
(26) manifold block,
(27) sample-inlet tube (stainless steel),
(28) mylar plastic window,
(29) sample-inlet-tube heater,
(30) sample-deposition surface,
(31) aluminum foil window (0.00035 in. thick),
(32) opening for 0.025-in.-thick beryllium X-ray window,
(33) translational-adjustment screw,
(34) rotational-alignment gear sector,
(35) plug,
(36) internal siphon.

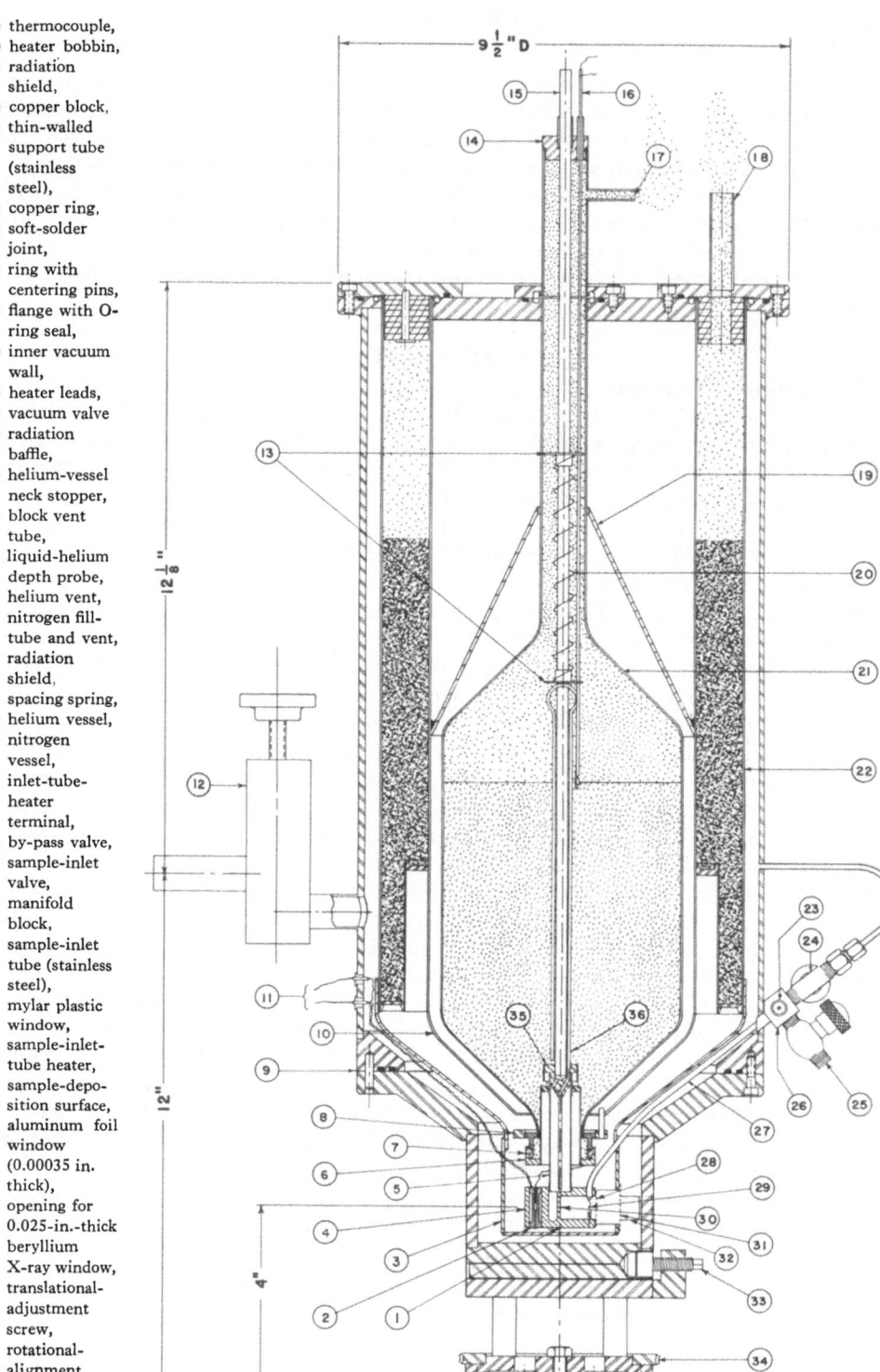

Figure 1. Vertical section through X-ray diffractometer cryostat.

resistance thermometer. Temperature measurements in the region 4.2–24.5°K are believed to be accurate to $\pm 0.2°$K.

Temperature Controller

The controller is shown schematically in Figure 2. The carbon resistance thermometer serves as one arm of a Wheatstone bridge which is balanced for the resistance corresponding to the desired temperature. The output of the bridge is amplified by the servo amplifier from a recording potentiometer and used to drive a reversible motor geared to the variable autotransformer that supplies power to the heater. Any fluctuation in temperature is counteracted by a change in the setting of the variable transformer.

Stability of the system is improved by incorporating a miniature generator which is geared to the motor shaft. Whenever the motor is turning, the generator produces a dc voltage, the polarity and magnitude of which depends on the direction and speed of rotation. A small portion of this voltage is fed back to the amplifier phased in such a way that it tends to drive the motor in the reverse direction. When the magnitude of this signal is suitably chosen it is very effective in eliminating cycling of the controller about the control point.

Samples

Natural neon (90.92% ^{20}Ne–0.26% ^{21}Ne–8.82% ^{22}Ne)[6] was obtained from a commercial source. The producer's specifications for the maximum limits of possible impurities (mole %) were: He—0.03%; N_2—0.01%; H_2—0.0005%; O_2—0.0005%. Semiquantitative analysis of the gas as received and as recovered from our apparatus did not show any significant departures from the above specifications.

Samples of the separate isotopes were supplied by the Mound Laboratories, Miamisburg, Ohio, and the following analyses (mole %) were given:

Sample	Neon Isotopes	Impurities
^{20}Ne	^{20}Ne–98.6%	N_2–0.28%
	^{21}Ne– 0.1%	O_2–0.12%
	^{22}Ne– 0.9%	
^{22}Ne	^{20}Ne– 0.85%	Ar–0.08%
	^{21}Ne– 0.18%	N_2–0.06%
	^{22}Ne–98.83%	

Method

Nickel-filtered copper radiation ($\lambda = 1.54050$ Å for K_{α_1}) was used throughout this work. A preliminary check of the alignment and calibration of the apparatus at low temperatures was carried out using high-purity diamond powder ($a_0 = 3.56685 \pm 0.0001$ Å[14]) as a standard and making a correction for its thermal expansion, which is very small.[15] Subsequently, the gold-plated deposition surface was used as a standard for checking the alignment at the beginning of each experiment. When samples were deposited on this surface it was necessary to realign the cryostat to allow for the finite thickness of the deposit. After alignment, measurements of lattice constants are believed to be accurate to ± 0.001 Å.

As a further check on the accuracy of our method, it is worth noting that measurements of the lattice constants of argon, krypton, and xenon at 20.3°K give values which do not differ by more than 0.002 Å from values obtained by the group at Queen Mary College.

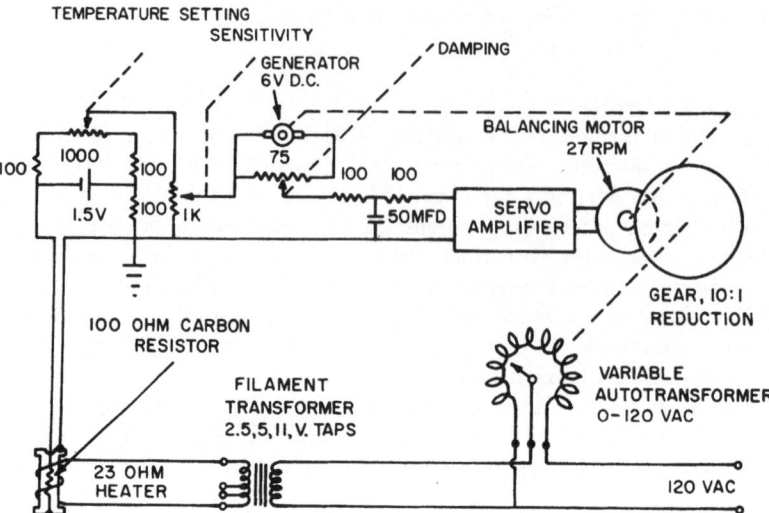

Figure 2. Block diagram of temperature controller.

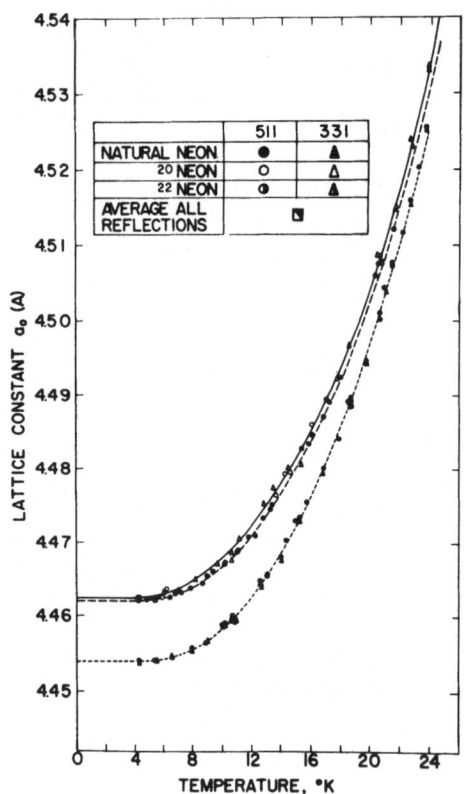

Figure 3. The lattice constant as a function of temperature for ^{20}Ne, ^{22}Ne, and the naturally occurring mixture.

Table I. The Volume Coefficient of Thermal Expansion of Neon at Several Temperatures

No significant difference in the values for the separate isotopes was detected and the values given are averages for the three samples.

T, °K	β
0	0.00×10^{-3}/°K
4	0.05*
8	0.74
12	1.7
16	2.7
20	4.0
24	6.3

* Approximate value obtained by interpolation.

Measurements in each case at 4.2°K have confirmed the results obtained at Queen Mary College by extrapolation from 20°K.

Neon samples were formed by introducing the gas into the sample chamber, which was kept at a temperature of approximately 20°K. The film thickness was monitored during deposition by measuring the attenuation of the diffraction pattern of the gold substrate. In most cases, deposition was continued until a thickness of approximately 0.1 mm was reached, but in one case, the final thickness of the deposit was 0.5 mm.

Measurements on the 111 reflection at approximately 34.8° 2θ were used only in checking the alignment of the cryostat. Values of the lattice constant are based on measurements of the 511 and 331 reflections at approximately 127.5° and 97.6° 2θ, respectively. Whenever the temperature of the sample was changed, the effect on the lattice constant was immediate. There was no indication that it was necessary to wait for thermal equilibrium to be established. In general, the sample was kept at each new temperature for at least 5 min before the reflection was scanned.

RESULTS

The lattice constants of [20]Ne, [22]Ne, and [n]Ne are shown as a function of temperature in Figure 3. At each temperature, the difference between the values for the two isotopes is about eight times the estimated error for each of the two values. The heavier isotope has the smaller lattice constant. Values for the naturally occurring mixture fall in between but closer to the more abundant isotope, as expected from Vegard's rule. The difference in the molar volumes of the two isotopes at 0°K is 0.564% that of the lighter isotope. In each case, the volume increases with increasing temperature by approximately 5.3% between absolute zero and the triple point. The volume expansion coefficients $[\beta = (1/V_0)(dV/dT)]$ do not appear to differ significantly in the three cases, and mean values at several temperatures are given in Table I.

Table II. Summary of Published Values for the Lattice Constant of Neon

Isotope	a_0, Å	T, °K	Method	Source
^{n}Ne	4.53	~ 4.5	X-ray	de Smedt, Keesom, and Mooy (1930)[16]
^{n}Ne	4.429 ± 0.010	4.2	neutron	Henshaw (1958)[17]
^{20}Ne	4.480 ± 0.004	4.2	X-ray	Kogan, Lazarev, and Bulatova (1961)[5]
^{22}Ne	4.464 ± 0.004			
^{n}Ne	$4.462_2 \pm 0.001$	4.2	X-ray	Bolz and Mauer (1962) In this paper
^{20}Ne	$4.462_4 \pm 0.001$			
^{22}Ne	$4.454_0 \pm 0.001$			

OBSERVATIONS

Values of the lattice constant of neon reported by various investigators are summarized in Table II. All previous measurements were made at 4.2°K and there are no thermal expansion data with which to compare our results.

The first value in Table II is mainly of historical interest. That obtained by neutron diffraction is smaller than ours by three times the sum of the estimated errors for the two sets of measurements, while the values reported by Kogan, Lazarev, and Bulatova for the isotopes are larger than our corresponding values by from two to four times the sum of the estimated errors. It should also be pointed out that the difference in the lattice constants of ^{20}Ne and ^{22}Ne at 4.2°K reported here is only half as great as that reported by the Russian group. In spite of these discrepancies, it is considered unlikely that the accuracy of ± 0.001 Å claimed for the present results is unduly optimistic.

Theoretical values of the lattice constants of ^{20}Ne and ^{22}Ne at several temperatures have been calculated by Johns.[18] At absolute zero these are about 1.7% larger in each case than the values obtained by extrapolation of the experimental curve. In the temperature range 0 to 24°K the calculated values increase by about 1%, whereas the experimental values increase by 1.6%. However, the difference in the calculated values of the lattice constants of the two isotopes at absolute zero (0.0085 Å) agrees exactly with our experimental result. At 24°K, the calculated difference (0.006 Å) is smaller than that at absolute zero. The calculated change, which is only two and one-half times the estimated error in the experimental results, cannot be detected with certainty.

The rather poor agreement between the calculated and experimental values of the lattice constants at absolute zero can probably be traced to the data used by Johns in evaluating his potential function. The interatomic separation was no doubt chosen to make the calculated lattice constant consistent with the value reported by de Smedt, Keesom, and Mooy,[16] which was the only one available prior to 1958. It is hoped that the values reported here will be useful in refining the theory.

REFERENCES

1. E. R. Dobbs and G. O. Jones, *Repts. Prog. Phys.* **20**: 516–564, 1957.
2. E. R. Dobbs, B. F. Figgins, G. O. Jones, D. C. Piercey, and D. P. Riley, *Nature* **178**: 483, 1956.
3. B. F. Figgins and B. L. Smith, *Phil. Mag.* **5**: 186, 1960.

4. A. J. Eatwell and B. L. Smith, *Phil. Mag.* **6**: 461, 1961.

5. V. S. Kogan, B. G. Lazarev, and R. F. Bulatova, *Soviet Physics JETP* **13**: 19, 1961; Original **40**: 29–31, 1961.

6. *American Institute of Physics Handbook*, D. E. Gray, coordinating editor, McGraw-Hill Book Co., Inc., New York, 1957, pp. 8–154.

7. K. Clusius, Z. *Phys. Chem.* **B31**: 459, 1936.

8. K. Clusius, P. Flubacher, U. Piesbergen, K. Schleich, and A. Sperandio, *Z. Naturforschung* **15a**: 1, 1960.

9. J. W. Stewart, *Phys. Rev.* **97**: 578, 1955.

10. I. A. Black, L. H. Bolz, F. P. Brooks, F. A. Mauer, and H. S. Peiser, *J. Research N. Bur. Standards* **61**: 367, 1958.

11. F. A. Mauer and L. H. Bolz. *J. Research N. Bur. Standards* **65C**: 225, 1961.

12. R. B. Scott, *Cryogenic Engineering*, D. Van Nostrand Co., Inc., Princeton, N.J., 1959, Tables 9.19 and 9.35.

13. J. R. Clement and E. H. Quinell, *Rev. Sci. Instr.* **23**: 213, 1952.

14. B. J. Skinner, *Am. Mineralogist* **42**: 39, 1957.

15. J. Thewlis and A. R. Davey, *Phil. Mag.* **1**: 409, 1956.

16. J. de Smedt, W. H. Keesom, and H. H. Mooy, *Commun. Phys. Lab. Univ. Leiden* **203e**: 1930.

17. D. G. Henshaw, *Phys. Rev.* **111**: 1470, 1958.

18. T. F. Johns, *Phil. Mag.* **3**: 229, 1958.

A HIGH-TEMPERATURE X-RAY DIFFRACTOMETER FURNACE UTILIZING HIGH-FREQUENCY HEATING*

E. W. Franklin and S. M. Lang

Owens-Illinois Technical Center
Toledo, Ohio

ABSTRACT

The adaptation of high-frequency heating techniques to a vertical diffractometer will be discussed. The heating system functions as a portion of an integrated system that provides a wide range of atmospheric and temperature control. Some of the design problems and their solutions and operating characteristics of the system will be described. The useful temperature range is from less than 200°C to greater than 1600°C, depending upon the furnace atmosphere and susceptors used. Gaseous pressures may be from vacuo of about 10^{-6} mm to about 30 psia; and, the sample may be heated in oxidizing, neutral, or reducing atmospheres.

INTRODUCTION

Two recent papers[1,2] have surveyed the chronological development of high-temperature X-ray diffraction equipment, with particular emphasis upon applications. As will be noted from a study of these papers, the most frequently used methods of heating have been variations of electrical resistance heating. Resistance elements embedded in refractory materials or supported in close proximity to the sample that heat the sample by conduction and/or by direct radiation have been a popular approach. The sample itself, in the form of a foil or strip, has been used as the element or it is common to place the sample in a dispersion, float it, and dry it on the surface of such a resistance element. Information on the use of induction heating for X-ray diffraction equipment is relatively scarce. The only equipment for which known design information has been published is that described by Edwards, Speiser, and Johnston[3] for a film camera apparatus.

It is the intent of this paper to describe only the problems, and some of the solutions, involved in the use of an RF induction heating system for a high-temperature X-ray diffractometer. Only the briefest description is given of the entire diffractometer system, which is intended merely to orient the reader to both the general and the particular problems.

THE PROBLEM

For the present approach to the design of an inductively heated high-temperature X-ray diffractometer furnace, one of the self-imposed limitations was that the instrument be readily acceptable to at least one type of commercially available equipment. In the design by the authors, described in ARL 62-315,[2] the conversion from the normal to the

---

* Partial support of the research effort was provided by the Metallurgy and Ceramics Branch, Aeronautical Research Laboratory, Office of Aerospace Research, U.S. Air Force.
[1] Superscripts pertain to references at the end of the paper.

diffractometer furnace arrangement requires only the removal of the normal specimen post and the insertion of the diffractometer and its shaft assembly; there is no relocation of the goniometer. Once the alignment of the furnace has been established,[4] the furnace assembly and the standard sample post may be interchanged without disturbing the base alignments of either the diffractometer furnace or the goniometer.

The other main design criteria for the entire system included the following: (1) operation to at least 1600°C; (2) operation with selected atmospheres from a vacuo of about 5×10^{-6} mm Hg to 30 psia; and (3) the ability to "scan" over the range from less than 0 to about 160° 2θ with an unobstructed X-ray window opening for the range from just under 0 to greater than 180° 2θ. These criteria and their solutions will not be discussed in this paper.

Because the 1600°C requirement approaches the usual resistance-heated upper operating temperature limits in oxidizing atmospheres, consideration was given to the use of RF induction heating of relatively massive pieces, in comparison to the resistance heating of thin foils, of rhodium, iridium, etc., for oxidizing atmospheres and for molybdenum, tantalum, tungsten, etc., for neutral, reducing, and evacuated furnace chamber atmospheres. In addition, the method offers the possibility of utilizing conducting oxides[5,6] as the susceptors for oxidizing atmosphere temperatures in excess of 2200°C. Some of the other factors influencing the selection of the induction heating method were: (1) the ability to concentrate the available power into a relatively small volume, thereby permitting the very rapid attainment of the desired test temperatures; (2) the heating element (either the sample itself or the sample holder-susceptor) is not attached to electrodes, thereby eliminating some of the difficult thermal expansion and distortion problems normally associated with the resistance heating of thin strips or foils so that the problem of placing the sample surface on the focusing circle is somewhat simplified; (3) although induction heating generators offer continuously variable levels of power input into the system being heated, the precision of temperature control usually achieved with normal control devices is only of the order of $\frac{1}{2}$ to 1%, but the high temperatures desired are only a function of the generator capability, materials, etc., and not of the feasibility of the method; and, (4) when furnaces using resistance wire heating elements are overpowered, the usual consequence is a major rewinding job but, if an inductively heated furnace is overpowered, it is anticipated that only the sample and/or its support will be deformed or melted and a simple replacement should be easily accomplished.

In order to maintain a perspective of all the various design considerations, it must be pointed out that the prime objective of this portion of the research effort was to design a system that would operate at very high temperatures—it is anticipated that temperatures of at least 2200 to 2500°C can be achieved—with highly accurate and precise diffraction measurements and with as good thermal control as possible, even if the accuracy of the thermal measurement is poor. As a point of fact, no design features were considered that would, in any manner, compromise the possibility of attaining high sample temperatures with precision and accuracy of the diffraction measurements. The problems of sample preparation, handling, loading, etc., and those concerned with accuracy of the thermal measurement (but not the control) were deferred for later study.

A number of facets (materials, geometry, etc.) of inductively heated systems are so interrelated[7-9] that it would have been both difficult and time-consuming to study each of them independently. It was considered to be more desirable to prove (or disprove) the feasibility of such a heating method for high-temperature X-ray diffractometry than to restudy the elements of the method for the particular case. Therefore, a number of the features were "fixed" initially and the studies reported were those considered to be of immediate importance.

The size of the sample holder-susceptor was fixed at $\frac{1}{2}$-in. OD with a $\frac{3}{8}$-in.-diameter sample depression. Years of previous experience had shown that perfectly satisfactory X-ray diffraction examinations could be made using that (and smaller) surface area samples of well-crystallized materials. Also, $\frac{1}{4}$-in.-thick buttons of the same (and smaller) diameter are relatively easily heated inductively to very high temperatures. A $2\frac{1}{2}$ kw output RF generator, about 440-kc frequency, was selected as the power source because it had been used to inductively heat samples of the sizes described. Further, once a system has been assembled, the generator capacity (assuming no changes in the mode or frequency of operation) merely limits the maximum temperature attainable and, if that temperature is not sufficient for the proposed experimentation, larger-capacity generators can be substituted with little or no physical modification to the system. The susceptor material selected was either Pt or the 60 Pt/40 Rh alloy because their susceptibility is approximately the average of that of almost all of the susceptor materials that are considered to be useful for the system. It was decided that for the present a susceptor-work coil gap of $\frac{1}{16}$ in. on a side would be convenient for the assembly of the system inside of the limited volume of the furnace chamber, particularly because the assembly would be made "blind." It seemed apparent that some type of an electromagnetic transformer would be required to efficiently concentrate sufficient inductive power into the small holder-susceptor using the $2\frac{1}{2}$-kw generator, although both a one- and two-turn work coil had been successfully used in early prototype studies.

Having "fixed" these features of the general design, it appeared that there were four aspects of the conceptual design that required study before the feasibility of the design could be established: (1) the work coil–concentrator design; (2) the method of providing power to the work coil from the RF generator; (3) the ionization and discharge phenomena of the gases in the chamber due to the inductive field of the work coil, etc.; and (4) the effects of the RF fields emitted by the generator, transformer–concentrator, work coil, etc., on all of the associated electronic circuits of the X-ray generating and detecting systems and the various components of the supporting subsystems for the entire apparatus.

These are the topics that are discussed in this report.

THE SOLUTION

Figure 1 shows the sample holder-susceptor used in the preliminary investigations and Figure 2 is a drawing of the support system for that type of susceptor configuration. The simple configuration was chosen to provide a means of evaluating the feasibility of using the induction heating technique without becoming involved with the side effects of sample size and shape. The assumption was made that anyone using high-temperature X-ray diffraction techniques would have some appreciation of the high-temperature behavior of the material being investigated. Therefore, the method of sample preparation should be selected according to the nature of the problem. This sample holder can be used for the same type of specimens normally examined at room temperature: (1) powder samples; (2) bulk samples formed (as shown in Figure 1) without the depression in the top of the holder, provided that the material is capable of being heated inductively; and, (3) wafer-type samples that can be fitted into the sample holder depression. Ease of fabrication of the sample holder from a variety of materials was also considered. No difficulty has been encountered in forming sample holders from platinum-type metals and alloys, tantalum, tungsten, molybdenum, and graphite, etc. The sample holder is easily replaceable on its support and, if desired, can carry the control and temperature measuring thermocouple.

An essentially empirical approach to the work coil design was undertaken. Since

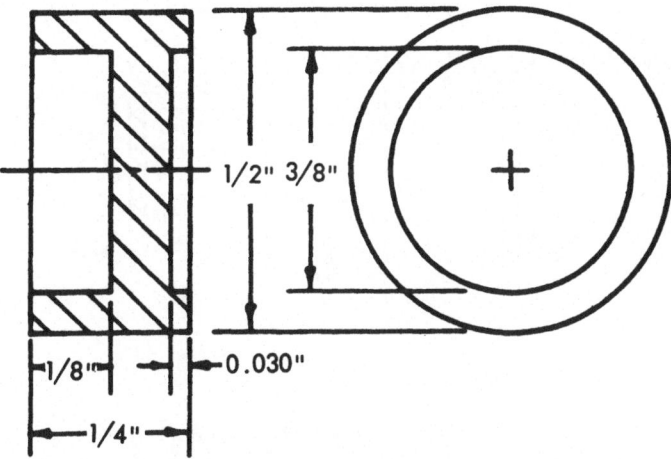

Figure 1. Sample holder-susceptor.

the restrictions imposed by the X-ray diffractometer geometry could not be by-passed, the heating system was adapted to the furnace design at some undetermined sacrifice in efficiency in order to obtain an operating unit. Then, further studies could be undertaken to improve the efficiency of the unit. At the same time, "good" induction heating techniques were used wherever possible. Copper was used in the fabrication of all work coils with silver brazing as required. All work coils incorporated water cooling and the normal air gap between the susceptor outside diameter and coil inside diameter is $\frac{1}{16}$ in. The spacing ensures rather effective coupling of the work load with the coil without posing problems in centering the susceptor in the coil; other gap designs can be considered later.

Figure 3 shows two concentrator-type work coils that were investigated early in the work. At that time, it was planned to use flexible leads to carry the high-frequency alternating current from the generator and the cooling water for the coil simultaneously.

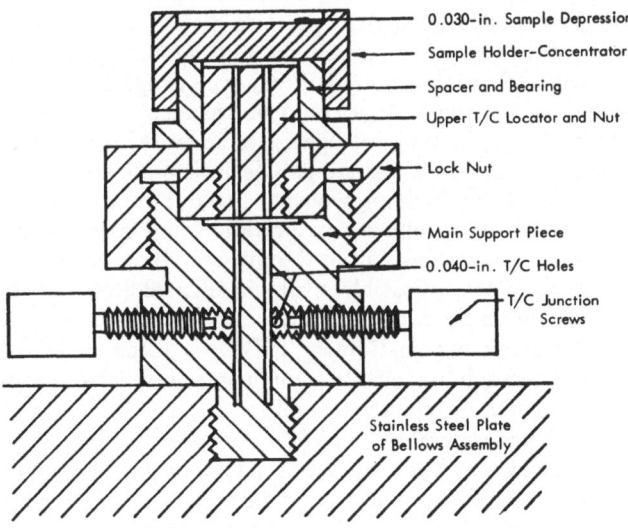

Figure 2. Sample support system.

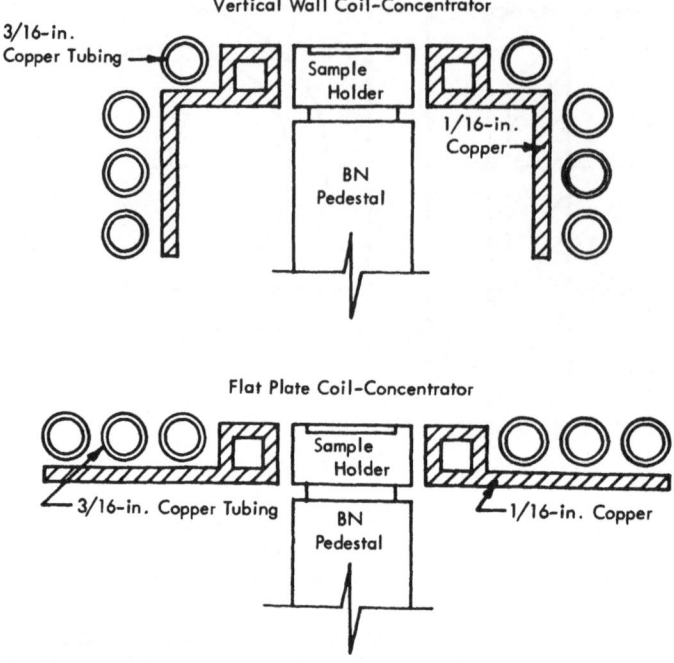

Figure 3. Two types of work coils.

It should be noted that the uppermost portions of all work coil assemblies lie approximately $\frac{1}{32}$ in. below the upper surface of the sample holder to allow a full 180° passage of the X-ray beam. Experimental trials using these designs were made under a bell jar and with a simulated pedestal and support system for the sample holder-susceptor. The high-frequency leads were brought into the bell jar through teflon glands in the base plate. The system was so arranged that the sample holder could be continuously inductively heated at various gas pressures, ranging from less than 1×10^{-5} mm Hg to pressures above atmospheric, with the convenience of continuous monitoring of those pressures in the vacuum range with either ionization or Pirani-type gauges.

The atmosphere within the bell jar could be changed as desired by admitting any gas chosen for study. Argon, helium, nitrogen, and air were investigated to indicate both the degree and type of problem that is existing. The decision for placement of the primary coil of the generator had been made previously. Placing the coil outside of the furnace chamber should minimize the problems of ionization and gaseous discharge that may occur when the work coil is located within the furnace. Since with a work coil outside the furnace, it is difficult, if not impossible, to obtain efficient heating, the choice was made to place the work coil within the furnace chamber and to supply the power to that work coil through an electromagnetic transformer "fed" by a primary coil from the generator. For these preliminary tests, however, all components were placed within the bell jar. In the actual diffractometer installation, therefore, the conditions would not be worse, and were expected to be considerably better.

The problem of ionization was very real: the type of gas, the pressure of the gas in the chamber, and the coil potential all played important roles. A number of studies[10] have been made relating the potential which causes ionization as a function of pressure and the distance between plane parallel electrodes with a uniform field between the electrodes. Since the ionization potential depends not only on the pressure and the gas composition but also upon the frequency and the electrode size and shape, the data provided by the experiments performed under standard conditions serve best only to place the various gases in some relative position with respect to each other.

The sample holder for these trials was fabricated from a 60Pt/40Rh alloy and it had a flat top surface. The temperatures of this surface were measured with an optical pyrometer sighted through the glass of the bell jar. The temperatures measured were not corrected because only comparative results were desired. After evacuating the system to a pressure of less than 1×10^{-3} mm Hg, the sample was heated to an indicated temperature of 1550°C. The actual or real sample temperature was higher than that because both the sample emissivity and radiation absorption through the glass of the bell jar are additive corrections. No temperature gradient could be observed across the surface of the "sample." The system was blanked-off from the vacuum pumps and the chamber pressure was increased by the introduction of the selected gas. The pressure was continuously monitored and the pressure at which a gaseous discharge started was noted. The pressure was maintained constant at that level and the sample temperature then lowered until the discharge stopped. A number of experiments of this type were made and the results indicated that in no case for each gas used was there a discharge at a pressure less than 20μ. Figure 4 is a graphical representation of these results and shows the critical

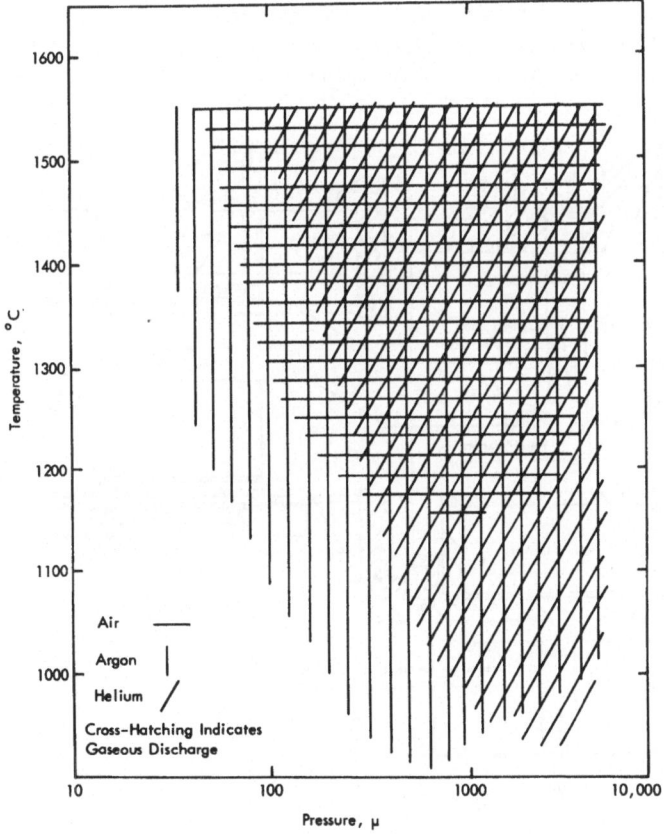

Figure 4. Gaseous discharge phenomena.

temperature–pressure region in which some discharge ionization occurs. The area is essentially independent of the type of gas used. By raising or lowering the pressure outside of the critical range, the glow discharge disappears. Since the glow-discharge ranges appear at partial pressures not normally expected to be used in the proposed research investigations, the discharge problem was not considered serious enough to cause rejection of the use of the induction-heating method. Only one limitation has been encountered during these studies (and operation of the actual system): When helium gas of over 500 mm Hg pressure is used in the diffractometer chamber, the full output of the RF $2\frac{1}{2}$-kw generator is not sufficient to heat the sample holder-susceptor at temperatures above about 1350 to 1400°C because of the high thermal conductivity and heat capacity of that gas. When air, nitrogen, or argon gases are used or the chamber is evacuated, temperatures of 1600°C are attained using only somewhat less than $\frac{1}{2}$ to $\frac{2}{3}$ of the power available from the same RF generator. This is not a serious condition because the substitution of a larger-capacity generator, say a 5-kw unit, should provide ample power for much higher-temperature operation when helium atmospheres are required.

The method of bringing the power to the work coil (within the furnace shell) was of some concern. One of the requirements is that the work coil retain a fixed relationship to the sample surface; therefore, a means had to be devised to permit a rotational movement of greater than 80° between the work coil and the generator. Normally, this type of movement can be obtained through the use of flexible leads. Because the sample was to be placed into the furnace through an opening in the bottom and the chamber was to be evacuated through an opening in the furnace front (the shaft being considered too small for an efficient evacuation port), little room remained for attaching flexible power lines to the work coil. Using either side of the furnace could cause interference difficulties with

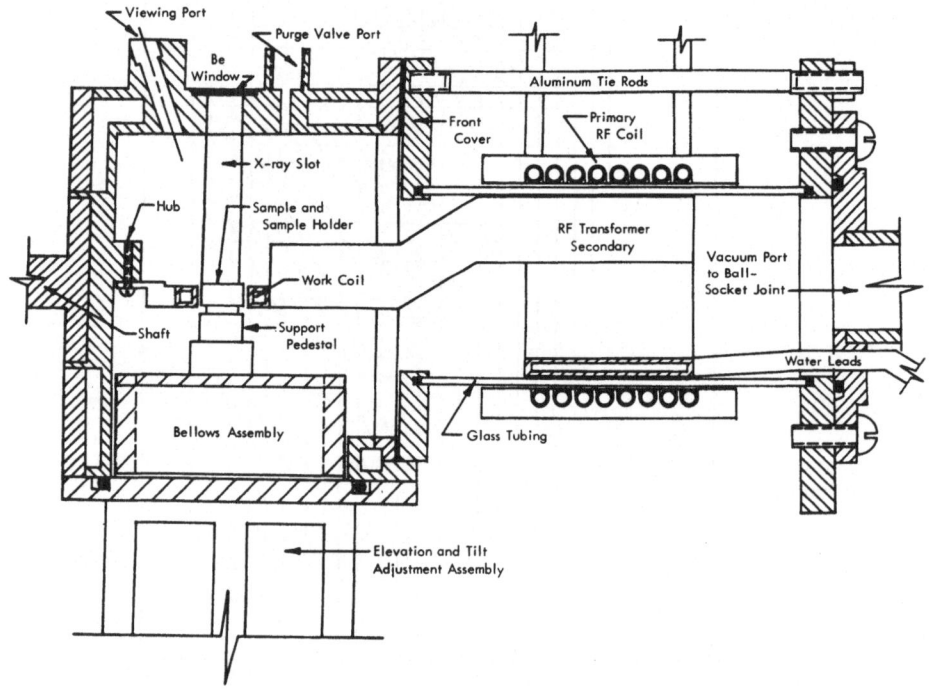

Figure 5. Sectional view of furnace and coil assembly.

the tower or detector arm and terminating flexible lines is something of a problem because of the massive size of the fittings required even for a $2\frac{1}{2}$-kw-output generator.

It was at this time that consideration was given to eliminating the flexible high-frequency lines design and substituting a rigid coaxial line that would terminate at the multiturn primary coil (fitted around a glass tube) of an electromagnetic transformer. Inside of the glass tube a water-cooled secondary is fitted with two copper "wings" that terminate at a one-turn work coil that could be located, and rotated, with respect to the sample holder-susceptor. A line drawing of the arrangement is shown in Figure 5.

The primary advantage of this arrangement is that it provides a method of permitting movement of the diffractometer through the electromagnetic coupling and, therefore, does not load or restrict the movement of the diffractometer. Note the small tip at the left of the work coil; this is an integral portion of the coil assembly. It is attached to a hub protruding from the rear wall of the furnace shell to prevent any movement of the work coil that might cause it to interfere either with the X-ray beam or with the sample holder. Extending to the right of the work coil are the "wings" of the transformer secondary. These are formed from $\frac{1}{16}$-in. copper sheet and carry $\frac{1}{4}$-in. diameter copper-cooling water tubes to their points of attachment at the top portion of the transformer secondary. The inner surfaces of the "wings" are parallel and closely spaced to provide a low inductance and increased efficiency; it is expected that a smaller separation may be more effective. The transformer secondary is a hollow-walled tube that is split at the top portion to accommodate the wings. At the bottom right of that section are the water leads for cooling the entire work coil–transformer assembly. Figure 6 is a photograph of the assembly in an exploded view. The vacuum sealing of the parts is done by means of O-rings. Where possible, the metal parts are made from copper or aluminum to minimize inductive heating from any stray radio-frequency fields. The generator coil (the primary of the

Figure 6. Exploded view of primary coil.

Figure 7. Assembly of primary coil.

transformer) is shown in place in its internally grooved plastic housing. This form serves to evenly space the turns and, also, to protect the thin-walled copper tubing from damage. At the left is shown another typical coil. The offset tubes at the top are spaced to provide a standard gap for attachment to the rigid coaxial leads and to permit easy substitution of other multiturn primary coils for testing purposes. The three aluminum tie rods and the glass tube have been designed so that when the system is assembled, as shown in Figure 7, the components are metal-to-metal and the O-rings are properly compressed.

Below the male portion of the ball-socket joint are shown the water lines carrying the water to the work coil–transformer unit. Where necessary, the joints are silver-brazed and vacuum-tight. Although the transformer secondary is effectively grounded through the water lines and metal parts of the assembly, no problems are encountered by this design because this is electrically neutral. In fact, the previously mentioned fastening of the work coil tip to the furnace hub may be done with a metal fitting because that portion of the work coil is electrically neutral also. During trials with a mock-up of the design, a sample susceptor was heated at various temperatures up to about 1600°C and grounding wires connected to the water lines and coil tip without any noticeable effects. Although it has been done successfully in the laboratory, it is not recommended that the work coil be plated because of the possibility of heating the thin metal coating by the high-frequency currents carried by the copper portions of the assembly.

The coaxial lead for connecting the generator to the primary coil is connected at both ends by standard brass tubing connectors. Two concentric copper tubes separated by a teflon insulating sleeve form the basic coaxial assembly. Another copper tube, silver brazed to the exterior tube before assembly, forms the ground side of the coil leads. Standard-wall copper refrigeration tubing, $\frac{3}{8}$-in. and $\frac{1}{4}$-in. OD, has been used to make a number of such rigid coaxial leads for other laboratory applications. After assembly the tubing may be bent easily as required for placement. This simple type of coaxial lead can carry power levels up to about 5 kw. A more elaborate type, with a water-cooled outer jacket, would be required at higher power levels to prevent overheating of the coaxial leads. When using such an outer jacket, some flexibility for positioning the line is lost because of one's inability to bend the assembly without collapsing the water passage.

In practice, some difficulty is experienced in adapting this type of heating to a diffractometer when the proportional counting circuit is used. When a Geiger–Muller tube is used, however, no difficulties have ever been experienced under any operating conditions. Even when the high-voltage current is not applied to the proportional counter, large spurious counts are registered when the counter arm is rotated; the high count positions are not reproducible. Movement of the interconnecting cable from the counter to the circuit panel also produces large numbers of spurious counts. At this time, these problems are being studied through several approaches: (1) shielding and grounding of the interconnecting cable; (2) shielding of the proportional counter and the primary coil; and (3) redesign of the proportional counter assembly to provide shielding and physical isolation of the preamplifier circuit. As an example of the insidious nature of the radiation effects, a 20-kw induction generator, of the same frequency as the X-ray furnace generator, is located only 18 in. from the detector arm of the goniometer, but there has never been any difficulty or interference with the proportional counter circuitry. The interfering $2\frac{1}{2}$-kw generator is about 3 ft away from the detector arm, with the primary transformer coil always in the same position relative to the arm.

Rather precise thermal control was achieved using thermocouple sensing; it appears advisable to describe that system briefly. Two newly developed systems became available just after the prototype feasibility trials had begun: (1) a precision-type thermocouple control unit that electronically delivers a control signal of a magnitude directly related to the difference between a standard signal (the control) and that of the measurement thermocouple; and, (2) an electronic RF generator oscillator output control (by phase shifting the thyraton firing) whose "level" is directly related to the magnitude of an input milliampere signal. Some "electrical matching" was required before the two units could be incorporated in the same system. The suppliers of these units have been asked to consider some minor circuitry changes that would permit the use of the two systems without the "in field" modifications.

With this system, extremely fast heating and thermal equilibration are achieved. For example, a temperature increment of 1000 to 1200°C can be obtained and the control system equilibrated in appreciably less than 2 min. Precision of the system, as monitored by a fast, AZAR-type recorder, is less than $\frac{1}{2}$°C for sample-holder temperatures from about 120 to about 1200°C and less than 1°C from about 1200 to 1600°C. Because the accuracy of the temperature measurement is dependent upon the physical and thermal properties of the sample being examined and the size, type, and location of the thermocouple, no values can be given for the inherent accuracy. However, it is believed that at least 1 to 2% accuracy should be attainable without undue effort. Presently, some effort is being directed toward increasing the thermal measurement accuracy using thermocouple and/or radiation-type sensing elements while simultaneously studying various powder sample preparation techniques.

One high-temperature X-ray diffractometer apparatus, incorporating the various features described, has been in operation for almost a year and another has been used for about three months; these demonstrate the feasibility of using the RF induction-heating method. Although all of the features have not been optimized as yet, the apparatus and the procedures developed for its use permit the operator to achieve both accuracy and precision of the diffraction measurement and precision of temperature control permitting

Figure 8. Glass cross.

accurate differential but, perhaps, not absolute thermal accuracy measurement for high-temperature experimentation.

Even though the units that have been operating are proving to be most satisfactory in practically all respects, the program is being continued to investigate and evaluate design modifications, the operational and procedural techniques, and the technology of high-temperature X-ray diffractometry. To perform the necessary tests to carry out the program, while using the present apparatus for research experimentation, a separate apparatus has been assembled.

In Figure 8 is shown a glass cross, fabricated from a standard piece of glassware and fitted with a baseplate and flanges made from aluminum or brass, that simulates the X-ray diffractometer furnace. At the left is a flange that accommodates a vacuum line and the pressure gauge; at the top, the flange has a fused quartz window to permit temperature measurement by radiation devices; the right side has a flange and appropriate fittings for the induction work coil and associated components; and, the bottom flange rests upon a baseplate–sample support assembly. Among the items that are to be evaluated are: (1) gaseous discharge phenomena; (2) work-coil secondary–primary modifications designed to improve efficiency; (3) sample support systems for use above 2000°C; (4) modifications to permit preheating a conducting ceramic susceptor so that temperatures above 2000°C in oxidizing atmospheres may be reached; (5) various thermocouple and radiation-type thermal measurement and control systems and materials for use at sample temperatures above 1600°C; and, (6) a variety of sample preparation techniques. When these and other contemplated features have been evaluated and proved useful or advantageous, they will be incorporated in the present apparatus and/or methodology.

REFERENCES

1. W. J. Campbell, S. Stecura, and C. Grain, "High-Temperature Furnaces for X-ray Diffractometers," *Bureau of Mines Rept. of Investigations*, 1960, p. 5738.
2. S. M. Lang and E. W. Franklin, Research in High-Temperature X-ray Diffraction Technology, Aeronautical Research Laboratory, ARL 62-315.
3. J. W. Edwards, R. Speiser, and H. L. Johnston, "A High-Temperature X-ray Diffraction Camera," *Rev. Sci. Instr.* **20**: 343, 1949.
4. S. M. Lang, to be published.
5. S. M. Lang and R. F. Geller, "The Construction and Operation of Thoria Resistor-Type Furnaces," *J. Am. Ceram. Soc.* **34**: 193, 1951.
6. M. H. Leipold and J. L. Taylor, Ultra-High-Frequency Oxide Induction-Heating Furnace, Technical Report No. 32-32, March 17, 1961. Jet Propulsion Laboratory, California Institute of Technology, Pasadena, Calif.
7. C. A. Tudbury, *Basics of Induction Heating*, John F. Rider, Publisher, Inc., New York, 1960.
8. D. Warburton-Brown, *Induction Heating Practice*, Philosophical Library, Inc., New York, 1956.
9. R. M. Baker, "Design and Calculation of Induction-Heating Coils," *Trans. AIEE*, March 1957, pp. 31–40.
10. M. J. Druyvesteyn and F. M. Perming, "The Mechanism of Electrical Discharges in Gases of Low Pressure," *Rev. Mod. Phys.* **12**(2): 87–174, 1940.

DISCUSSION

D. Rodier (Grumman Aircraft Engineering Corp.): Can the Franklin apparatus be used in the vertical as well as the horizontal position?

E. W. Franklin: One would have a problem in maintaining the sample in a vertical position as in any high-temperature application.

SOME X-RAY GENERATOR CHARACTERISTICS TO CONSIDER IN ORDER TO REALIZE THE OPTIMUM STABILITY AND REPRODUCIBILITY OF INTENSITY MEASUREMENTS

R. Torkildsen

General Electric Company
Milwaukee, Wisconsin

ABSTRACT

This paper presents some of the physical considerations which must be considered in operating an X-ray generator to obtain maximum stability and reproducibility of intensity measurements. In addition, a typical X-ray generator will be discussed from the standpoint of enabling the user to realize all of the stability and reproducibility capabilities which were designed into the equipment.

INTRODUCTION

Increasing emphasis is presently being placed on stability and reproducibility of measured X-ray intensities in both emission X-ray analysis and diffraction analysis. While it is a fairly simple matter to make frequent use of a standard sample, and in some cases an absolute necessity, as with elements low in the atomic chart, it still proves very desirable to first of all understand the functional limitations of the equipment with which we are working and, secondly, apply this understanding in a way which enables us to take maximum advantage of the characteristics which are available.

This paper has as its purpose the presentation of a basic X-ray generator in such a manner as to reveal its inherent limitations and what can be done to optimize the results which we are depending on this equipment to give us.

The material presented here is not new; however, these are significant factors which should be kept in mind by the equipment operator desiring to optimize the stability and reproducibility of his X-ray generation equipment.

X-RAY GENERATOR CHARACTERISTICS

An X-ray generator as used for emission or diffraction work is shown functionally in Figure 1. It consists of five principal parts: An X-ray tube, and X-ray tube voltage control, an X-ray tube current control, a heat exchanger, and related power control and protective circuitry.

All of these component parts are subject to various influences which affect the stability of the X-ray tube output. These include: environment, power line variations, tap water supply variations, physical limitations of X-ray generation and excitation, operational history, and component degradation or failure.

In order to understand the stability problem it is necessary to look closely at the X-ray tube and its inherent characteristics and limitations.

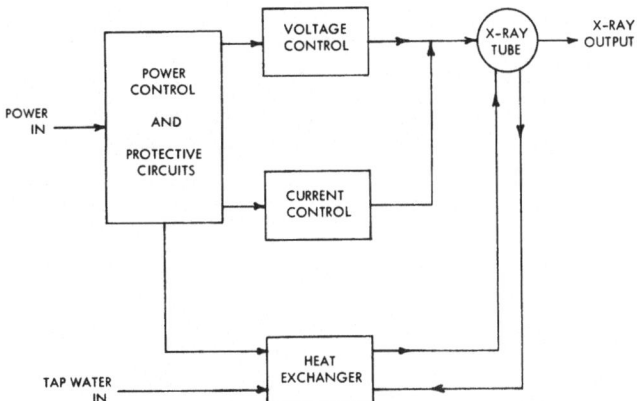

Figure 1. Schematic diagram of the X-ray generator.

The factors which influence an X-ray tube's stability can be divided as follows:

Group I: 1. Limitations of X-ray physics
             2. X-ray tube degradation and operational history
             3. Environment

Group II: 1. X-ray tube voltage (KVP)
             2. X-ray tube current (MA)
             3. Anode and tube temperature.

Group I lists those factors which cannot or do not readily lend themselves to control by any direct function of the equipment itself, while Group II lists other factors which are normally controlled by an X-ray generator.

Let us consider first, from Group I, the X-ray physics or physical limitations of the X-ray generation and excitation processes. Ideally, we want to generate a beam of X-rays which is invariant with respect to intensity and photon energy distribution. While it would be significant from an actual measurement standpoint, to know what this intensity and distribution might be, from a stability consideration invariance is the only important consideration. Were this the case, we would only have to deal with normal statistical behavior in our measurements. Typically, the X-ray beam photon energy distribution will start, at the low-energy end, with the photons which are sufficiently penetrating to get through the X-ray tube window. The beam will contain photons of all energies up to a maximum corresponding to the KVP at which the tube is being operated.

The intensity or number of photons at any given energy will depend on both KVP and MA at which the X-ray tube is being operated and, in addition, on the target material in the particular X-ray tube. Where the KVP is sufficient, the characteristic lines of the target element will be excited and the number of photons will be greatly increased in this particular energy region.

A characteristic excitation line of the X-ray tube target material, which occurs slightly below the KVP at which the X-ray tube is operating, will very definitely help to increase intensity, due to the unusually rapid increase of photons of this characteristic energy, with a small increase in voltage. However, any instability of intensity caused by variations in the KVP will be augmented to the same degree.

What may be called sample excitation efficiency is another characteristic which must be considered when evaluating overall stability and reproducibility. Classically, in emission work, the intensity of output is approximately proportional to the difference squared

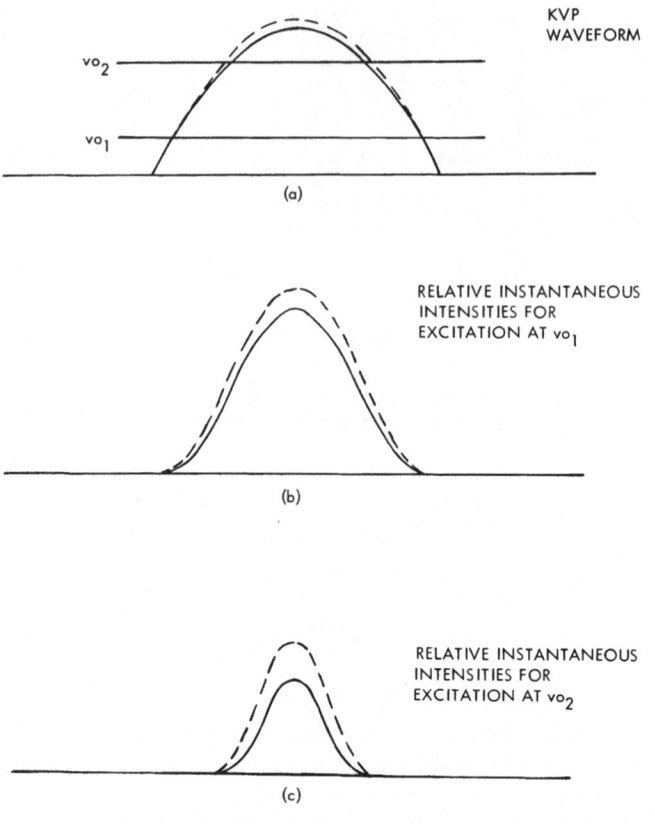

Figure 2

between the voltage on the X-ray tube and the excitation energy of the element under observation. Thus

$$\text{Intensity} = K(v - vo)^2$$

where v is instantaneous voltage on X-ray tube, vo is characteristic excitation voltage of fluorescing element, and K is a constant in a given situation. A typical situation which demonstrates this effect is shown in Figures 2 and 3.

Figure 2a shows an ideal voltage waveform for a full-wave rectified X-ray tube high-voltage supply. Figure 2b shows the instantaneous numerical intensity distribution for an element which excites at potential vo_1. Figure 2c shows distribution for an element exciting at higher potential vo_2. The dashed curve in each case shows what happens for a small change in voltage.

However, if the high voltage contains harmonics as shown in Figure 3a the intensity distribution will be as shown in Figure 3b for the element exciting at vo_1 and Figure 3c for the element exciting at the higher potential vo_2. Here the dashed line shows what could happen to intensity for a small change in harmonic content.

The overall effect which can be observed here is twofold. First, in the case of an element with a higher characteristic excitation potential, the intensity will normally be less for a given KVP and MA. In addition, the harmonics (or voltage waveform variations) will represent a larger portion of the difference $(v - vo_2)$. Where this difference is smaller, the resultant intensity will change with voltage to a much greater extent for the element with a higher excitation potential.

This effect can be further augmented when a characteristic excitation of the X-ray tube target material occurs near the KVP voltage value at which the X-ray tube is being operated.

Other factors such as increased background may dictate an optimum operating voltage as being somewhat less than the maximum obtainable in a particular case.

The second factor affecting X-ray tube output is degradation. This factor depends generally on operating history, and may occur in any one or a combination of several ways. One of these would be normal end of life, where the filament has evaporated to the point where the increased filament power, required to maintain emission, causes increased evaporation, etc. in an exponential cycle to the point of rupture.

A second and related type of degradation occurs when this evaporated filament material is deposited on the target anode. With a target material other than tungsten, which is also a prevalent filament material, a tungsten contamination will result. This will, of course, introduce a variable tungsten interference into the analysis, which may directly or indirectly change measured intensities.

Any material which is deposited on the inside of the X-ray tube window will, of course, be extremely detrimental to intensity measurement. A very small amount of tungsten, for example, will reduce intensity very sharply. A reduction of intensity will generally also come about due to any foreign materials or depositions on the outer surface of the window.

Corrosion caused by condensation of moisture on the outer surface of the rather active beryllium X-ray tube window will, if permitted to occur too often, reduce the X-ray output and eventually cause the tube to puncture.

In addition to the more direct types of X-ray tube degradation given above, degradation resulting from such things as operation of X-ray tubes outside of ratings may produce some mild to catastrophic effects.

Among these are premature filament failure or target contamination, target melting, tube gassing, puncture of glass insulation, and window burnout.

The checks for any of the above must be made by either checking intensities on standard samples or running scans with a permaquartz or equivalent sample. In other cases where stability is suspect, comparison with a new tube may be necessary.

Environment will generally have little or no direct effect on stability of the X-ray beam. However, the effects may be quite significant on the components used to control the three factors listed above in Group II.

The factors in Group II above are generally recognized as those requiring control since they are all dependent on a varying source. The degree of control necessary varies depending on which factor is involved.

In the case of voltage we have, perhaps, the most critical factor. Since the change in intensity for a given voltage change can be anything from approximately a one-to-one relationship to something much larger, it is necessary to control this factor more closely than the others. Control must be maintained in the presence of power-line variations, temperature variations, barometric pressure variations, and changes in humidity.

In the case of constant potential operation, where the excitation potential of the element being observed is very small compared with the KVP on the X-ray tube, the intensity change of the sample being observed will be approximately the same as the voltage change experienced by the X-ray tube. Thus for a $\pm 0.1\%$ peak voltage change, the intensity could be expected to change by at least $\pm 0.1\%$ from this cause. If the excitation potential of the element were to be of the order of $\frac{2}{3}$ of the KVP, however, changes of at

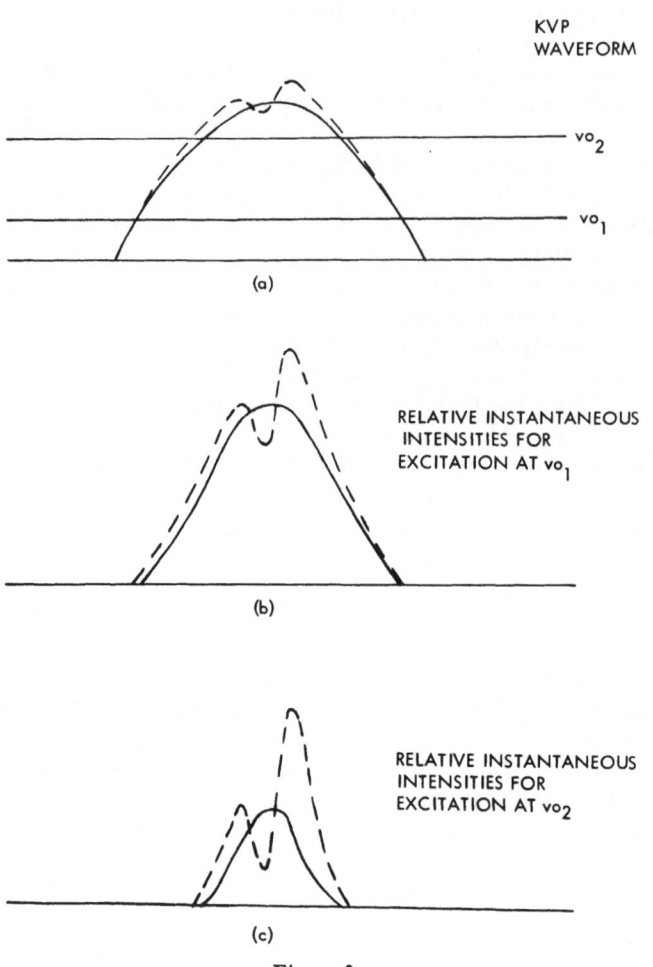

Figure 3

least $\pm 0.6\%$ could be anticipated. This number will increase rapidly as vo approaches v.

When full-wave or half-wave voltage is applied to the X-ray tube, however, these calculations will represent a much smaller magnitude than changes actually observed, since the voltage waveform factor will magnify the instability.

Current control is generally thought of as a straight one-to-one relationship so that for a $\pm 0.1\%$ control on MA in the X-ray tube a $\pm 0.1\%$ change in intensity should result. This is true except in cases where changes in voltage caused by the high-voltage regulation present in the X-ray tube power supply come into play.

Here again the problem is most significant when working with elements whose excitation potential is close to the KVP of operation.

In a case of this kind, a very small voltage regulation change brought about by the 0.1% current change can result in a very undesirable intensity change.

Anode and tube temperature changes will only affect stability insofar as the relative geometry of the X-ray tube anode structure, the X-ray tube window, and the sample presentation is affected. Where the temperature range is small the stability will be good.

SUMMARY

There are several recommendations from an operational and/or equipment standpoint which will improve stability of the X-ray generation process. These are:

1. Operate with a sufficiently high KVP for the element being measured so as not to magnify instability effects.
2. Select a line-voltage stabilizer with a minimum of harmonic or waveform distortion.
3. Use a constant potential accessory.
4. Select a high-voltage transformer with good regulation.

SPECIFICATIONS AND PERFORMANCE DATA FOR THE ARL ELECTRON MICROPROBE X-RAY ANALYZER

L. P. OBrien

Applied Research Laboratories, Inc.
Glendale, Calif.

ABSTRACT

Detailed EMX specifications are presented. Performance data obtained on several production instruments are discussed and some future applications of microprobe analysis are suggested.

INTRODUCTION

For many years scientists have had need for a rapid, nondestructive, accurate method of analyzing sample volumes of a few cubic microns. With the development of the ARL Electron Microprobe X-ray Analyzer (EMX) it is now possible to obtain quantitative analyses of these minute areas. Figure 1 illustrates an EMX laboratory consisting of four units: The spectrometer in the center, the power supply unit to the right, the electron beam scanning system on the left, and the recording console at the extreme left. The EMX Spectrometer utilizes three types of optical systems—electron optics, light optics, and X-ray optics.

ELECTRON OPTICAL SYSTEM

The triode, self-biased electron gun, consisting of a 0.004-in. tungsten hair pin filament, grid cap, and anode plate, is of special design which permits centering of the filament with respect to the grid aperture while the gun is in operation. The high-voltage supply is variable from 3 to 50 kv with 0.005% regulation and residual ripple of less than 0.2% peak to trough and a long-term stability of 0.02%. The electron beam is focused at the specimen surface (from less than 0.3 to 300 μ in diameter) by means of two magnetic lenses. The current to the magnetic lenses is adjustable from 1 to 300 ma, with coarse and fine adjustments, regulation is 0.01%, and ripple is less than 0.2% peak to trough. The special inverted design of the lower or objective lens provides maximum space for the viewing system, a high emergence of angle 52.5° between the emitted X-rays and the specimen surface, and a large Bragg angle range for the focusing X-ray monochromators. This high take-off angle and the normal incidence of the electron beam on the specimen enhance X-ray sensitivities and reduce interelemental effects and the number of correction systems required. Sample preparation is less critical than with instruments of lower take-off angle. Figure 2 illustrates the components and design of the EMX Spectrometer.

ELECTRON BEAM SCANNING SYSTEM

Perhaps the most significant improvement in the EMX has been the development of the Electron Beam Scanning system (EBS). This addition has greatly increased the

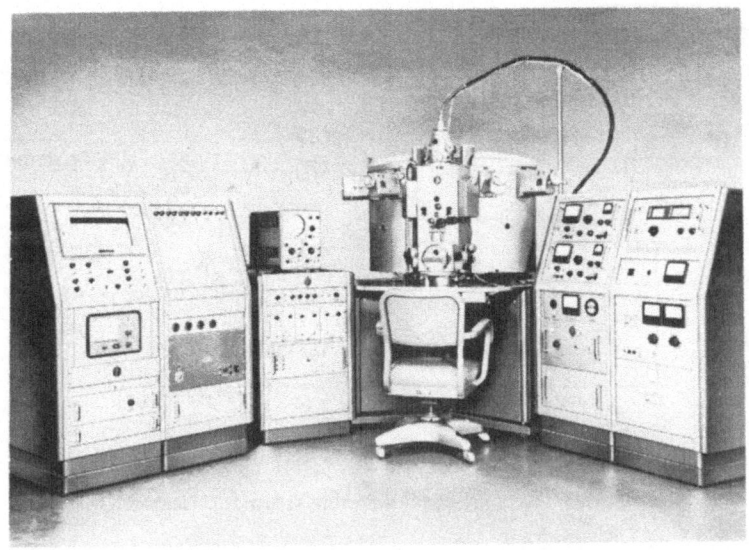

Figure 1. Electron microprobe laboratory.

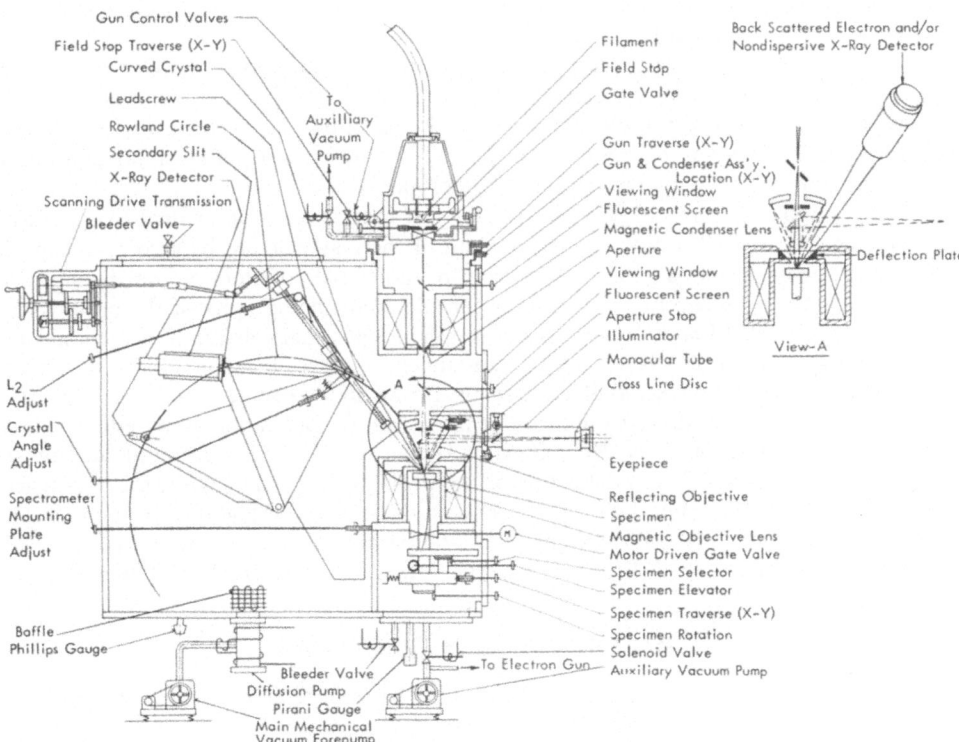

Figure 2. EMX schematic diagram.

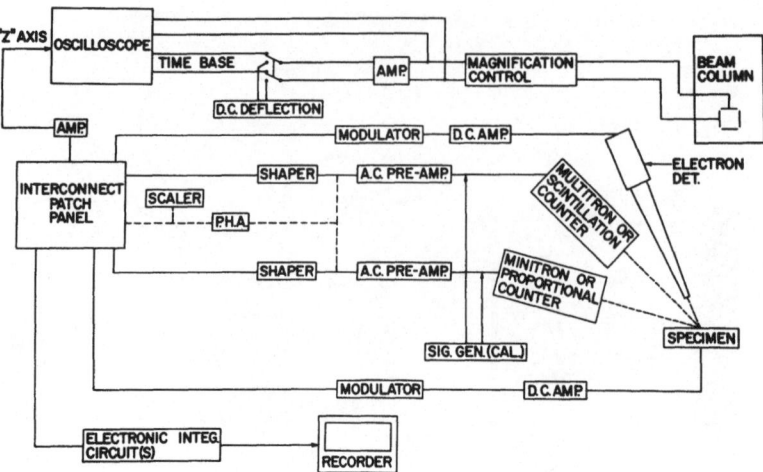

Figure 3. Electron beam scanning schematic diagram.

versatility and application of microprobe analysis since it is now possible to present oscilloscope pictures of the analysis. This is especially valuable in the metallurgical field since the EBS system provides information on sample topography, element distribution by atomic number, actual electron beam spot size and sample area excited by the beam, and a ratio basis for quantitative analysis.

A single set of four electrostatic deflection plates is located at the electron optical objective lens position (near the specimen). A similar set of deflection plates is located in the oscilloscope. Two of the four plates near the specimen are utilized for X-axis and two for Y-axis deflection of the electron beam. The deflection plates are synchronized with their complementary plates in the oscilloscope.

The oscilloscope console provides a variety of functions—display of the signal from any detector on an oscilloscope screen, readout of these signals on the recording console strip chart, pulse height discrimination from proportional counters before the signal is displayed on the oscilloscope or recording console, and photography of the oscilloscope image with a Polaroid Land Camera. A maximum of eight channels can be accommodated in the EBS system: three dispersive X-ray monochromators; three nondispersive X-ray detectors; one for backscattered electrons; one for sample current.

The nominal deflection of the beam at minimum magnification along either axis at the specimen is 360 μ at 30 kv. Magnification of the area viewed is accomplished by reducing both X-axis and Y-axis deflection of the probe by $\frac{1}{2}$, $\frac{1}{4}$, and $\frac{1}{8}$ while the screen area viewed on the oscilloscope (8 cm × 8 cm) remains constant. Magnifications are 222×, 445×, 890×, and 1780×. By changing the standard resistor network, a magnification of 17,800× can be obtained. Figure 3 is a schematic diagram of the EBS system.

BACKSCATTERED ELECTRONS

The quantity of electrons scattered back by the sample is a function of the atomic weight of the elements present. Heavier elements will scatter back more electrons than lighter elements. Thus element identification with regard to location on the periodic table is indicated by the image presented on the oscilloscope screen. Bright areas would be heavy elements and dark or gray areas would be light elements or holes in the specimen

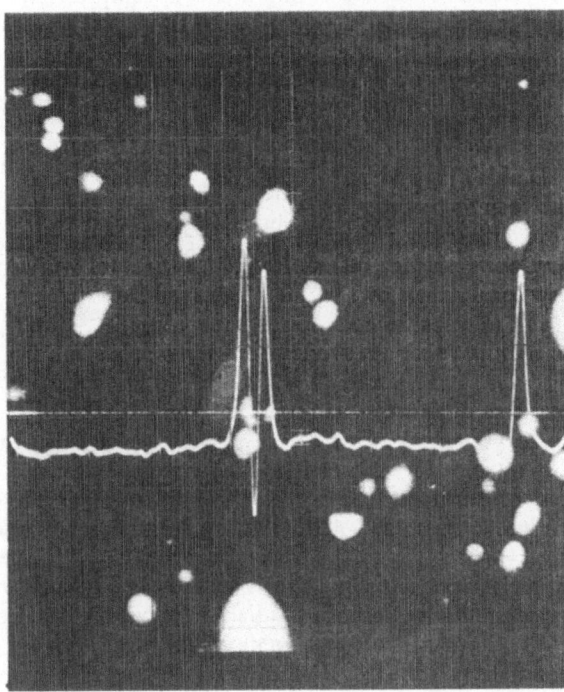

Figure 4. Electron beam scan of
lead inclusions in brass.

surface. Figure 4 is a backscatter picture of lead inclusions in brass. The bright areas are
lead and the dark background is the copper matrix. Another valuable advantage of the
EBS System is the determination of the diameter of the electron beam at the specimen
surface. Again referring to Figure 4, the distance between the graticule marks is $1.1\,\mu$,
and the magnification is $1780\times$. Several lead inclusions can be observed that measure
about $0.4\,\mu$; therefore the diameter of the beam must be smaller than the particle in
order to resolve it on the oscilloscope.

LIGHT OPTICAL SYSTEM

The viewing system has been designed to employ the elements of a Bausch and Lomb
microscope with the following specifications:

Magnification..	$280\times$ standard
Depth of focus	$3\,\mu$
Numerical aperture	0.40
Resolution	$1\,\mu$ or better
Field of view	$500\,\mu$

Magnifications and field of view can be increased or decreased depending upon the choice
of eyepiece. The light optical system permits visual observation of the specimen when
in position for, and during analysis. Precise and positive alignment is achieved, so that
the spot analyzed and the spot at the crosshair of the microscope are identical. In addition,
the depth of focus of the electron beam is $20\,\mu$ (with a $1\text{-}\mu$ diameter beam), and therefore
the visual optics with a depth of focus of $3\,\mu$ assures proper sample position in relation

to electron beam focus. Simultaneous sample viewing offers several important advantages, some of which are:

1. Certain compounds such as oxides, sulfides, and nitrides fluoresce in characteristic colors while under electron bombardment, and can be readily identified by their fluorescence.
2. Some elements of small difference in atomic number, which cannot be distinguished with the EBS system, such as copper and nickel, can be identified through the light optical system.
3. The size and location of the electron beam on a specimen can be determined by observing the fluorescence produced on a material such as alumina. This procedure is also used as a check of beam alignment.
4. Photography of the sample during analysis provides correlation of analytical results with the area on the specimen that has been examined.

X-RAY OPTICAL SYSTEM

A maximum of three scanning monochromators, consisting of crystals, receiver slits, and detectors, are provided. Selected combinations of the type of crystal slit width and type of detector ensure optimum conditions for analysis in the X-ray region of interest. The monochromators are all mounted within the high vacuum (2 to 5 × 10⁻⁵ mm Hg) so that no loss of total X radiation is experienced because of windows between the point X-ray source and the crystals.

In order to obtain maximum intensity from a point source, the crystals are all of the curved and ground focusing type, with the exception of mica, which is curved only. Either 4- or 11-in.-radii focal circles are used. Any wavelength from 0.36 to 18.70 Å can be measured using crystals of LiF, SiO_2, NaCl, EDdT, ADP, and mica. Each monochromator is calibrated directly in Angstrom units and can be read from the operator's position.

Figure 5 illustrates the performance of the 11-in. LiF crystal. Note the excellent resolution of the Ni K_{α_1} and Ni K_{α_2} lines, which are 0.003 Å apart. The intensity ratio

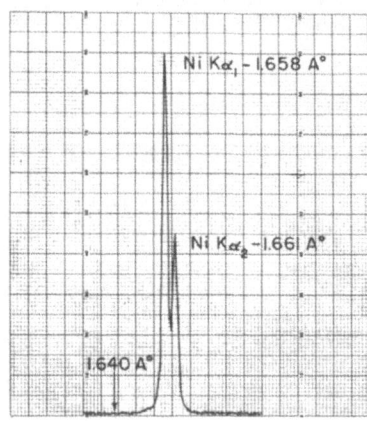

CRYSTAL : 11" R-LiF

DETECTOR : KRYPTON MULTITRON

SLIT : 0.004"

ACCELERATING VOLTAGE : 30 KV

SAMPLE CURRENT : 0.084 μ AMP.

SPOT SIZE : 0.4 μ

SPECIMEN : PURE Ni

HALF WIDTH (Ni K_{α_1}) : 0.0018 A°

PEAK INTENSITY (Ni K_{α_1}) : 7,720 COUNTS / SEC.

BACKGROUND INTENSITY (1.640 A°) : 23 COUNTS / SEC.

LINE / BACKGROUND RATIO : 336/1

Figure 5. 11-in. LiF monochromator trace of nickel.

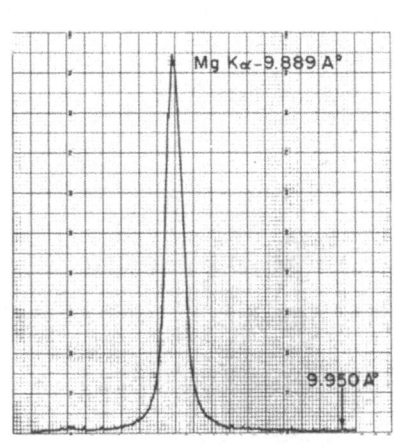

CRYSTAL : 4" R-ADP

DETECTOR : FPC MINITRON

FLOW GAS : P-10

SLIT : 0.010"

ACCELERATING VOLTAGE : 30 KV

SAMPLE CURRENT : 0.03 μ AMP.

SPOT SIZE : 0.3 μ

SPECIMEN : PURE Mg

HALF WIDTH : 0.007 A°

PEAK INTENSITY : 11,500 COUNTS / SEC.

BACKGROUND INTENSITY (9.950 A°) : 31 COUNTS / SEC.

LINE / BACKGROUND RATIO : 371/1

Figure 6. 4-in. ADP monochromator trace of magnesium.

is exactly 2 to 1. It is important to observe the high intensity obtained with the low sample current of 0.084 μa. The ability to work with low sample currents efficiently in conjunction with crystal and detector systems is very important in the microanalysis of organics, films, tissues, and similar materials that would be destroyed by high currents.

Figure 6 shows the efficiency of analysis in the soft X-ray region. The magnesium line to background ratio is 371 to 1 with a sample current of 0.03 μa. The sensitivity of the X-ray optical system is demonstrated in the analysis of sodium, using a mica crystal

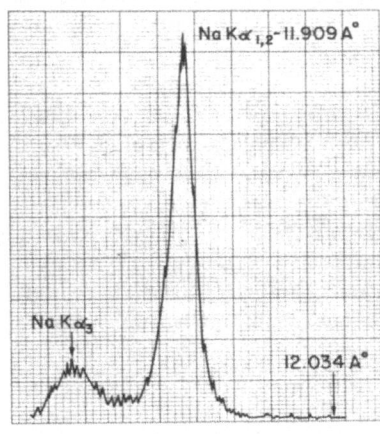

CRYSTAL : 8" R-MICA

DETECTOR : FLOW PROPORTIONAL
                      COUNTER (WITH PHA)

FLOW GAS : P-10

SLIT : 0.010"

ACCELERATING VOLTAGE : 30 KV

SAMPLE CURRENT : 0.40 μ AMP.
                      (MEASURED ON BRASS)

SPOT SIZE : 0.8 μ

SPECIMEN : Na Cl CRYSTAL
                      (TRAVERSED UNDER
                      BEAM AT 96 μ/MIN.)

HALF WIDTH : 0.022 A°

PEAK INTENSITY : 960 COUNTS / SEC.

BACKGROUND INTENSITY (12.034 A°) : 6 COUNTS / SEC.

LINE / BACKGROUND RATIO : 160/1

Figure 7. 8-in. mica monochromator trace of sodium.

Table I. X-ray Efficiency of *K*, *L*, and *M* Series
(Typical EMX Data)

Line	Å	kv	Counts/sec per 0.01 μa sample current from pure specimens		
			11-in. radius LiF crystal 0.020-in. slit Multitron (Kr)	4-in. radius LiF crystal 0.010-in. slit Multitron (Ne)	4-in radius ADP crystal 0.010-in. slit FPC Minitron (P-10)
Sn K_α	0.492	50	4600		
Ag K_α	0.561	50	6800		
Mo K_α	0.710	50	8800		
Bi L_α	1.144	30		2000	
Pb L_α	1.175	30	1700	2200	
Ta L_α	1.522	30	1700	3400	
Cu K_α	1.542	30	3000	5000	
Ni K_α	1.659	30	2600	7200	
Fe K_α	1.937	30		7400	
Mn K_α	2.103	30		7900	
Ti K_α	2.750	30		6000	36800
Sn L_α	3.600	30		860	12700
Ag L_α	4.154	30			5500
Bi M_α	5.118	30			3000
Mo L_α	5.406	30			3800
Au M_α	5.840	30			2400
W M_α	6.983	30			2100
Si K_α	7.126	30			8000
Ta M_α	7.251	30			1700
Al K_α	8.339	30			6700
Mg K_α	9.889	30			4000

which is curved but not ground, so only the radius of curvature (8 in.) is referred to in Figure 7. The NaCl crystal specimen is not conductive, so the sample current was measured on brass.

Table I gives the counting statistics and shows the X-ray efficiency of the three monochromators in various regions of the spectrum. The data were obtained on an EMX during the final production check-out and should be considered typical, although results may vary from instrument to instrument owing to different crystal and detector efficiencies.

The three X-ray monochromators are arranged around the probe column and the emergence angle of the X-rays from the specimen surface is 52.5° for all wavelengths. Monochromator scanning can be either manual or automatic. A 60-cycle, reversible, synchronous motor drive provides automatic scanning speeds of either 0.10 or 0.02 Å/min. Settings are reproducible to within ±0.0005 Å when the setting is approached from one direction.

SAMPLE HANDLING

A selector disk in the airlock chamber accommodates up to eight specimens of 1 in. maximum diameter and $\frac{3}{16}$ in. thick or $\frac{7}{16} \times \frac{7}{16}$ in. It is possible to mount any number of standards or specimens, depending upon their size, in any of the eight specimen holders. As soon as the airlock has been evacuated (normally less than 3 min are required, depending upon the nature and composition of the specimens), any one of the eight specimens can be placed in position for analysis.

Both horizontal and rotational scanning parallel to the specimen surface are provided. A 0.4-in. displacement in two mutually perpendicular directions and a rotation of 360° about the specimen holder axis makes complete coverage of the specimen surface possible. Manual and automatic scanning are provided. Automatic scanning speeds are 8 and 96 μ/min, with other speeds optional. Position reproducibility from the same direction of scan is within ± 1 μ. Automatic step scanning permits a point-by-point integration as the specimen is traversed in adjustable steps between 1 and 10 μ in 1-μ increments.

RECORDING CONSOLE

The recording console provides readout facilities based on the use of a single pen strip chart recorder. Two circuits may be utilized (1) the profile or ratemeter circuit, which gives instantaneous intensity readings for any detector in terms of chart divisions, and (2) the integration circuit, which is used for quantitative analysis utilizing integrated intensity as ratioed to time, sample current, nondispersive radiation, or a monochromator set to a specific wavelength. The recorder characteristics are as follows:

Accuracy	0.25% full scale
Response	1 sec full scale
Chart Speeds	1½ and 6 in./min, other speeds optional
Recorder Range	Sensitivity is 10 to 1
Rate..	Signals from all detectors can be recorded sequentially at the rate of one channel every 2.5 sec.

Pulse height electronics is available and can be used in conjunction with flow proportional counters (ARL Minitrons) and the results can be read out on the recording console or on a scaler or displayed on the oscilloscope. Scaler readout is provided either from any ac pulse amplifier or from pulse height electronics. Crystal-controlled electronic timers and automatic printers can be furnished for use with the high-speed scaler.

Although microprobe analysis finds primary applications in the metallurgical field, many other areas of research can benefit substantially through the use of this unique tool. Some of the fields now utilizing microprobe analysis are atomic energy, oceanography, photographic film industry, chemical industry, geological research, and crystallography. Future applications will probably include studies of catalysts, glasses, and cements, medical and biological research (studies of tissue, bones and teeth, and pathological applications), petroleum research, and metallurgical and ceramic investigations in space age research.

RECENT ADVANCES IN ELECTRON-PROBE ANALYSIS

J. V. P. Long

University of Cambridge
Cambridge, England

ABSTRACT

The electron-probe analyzer has been developed in a period of thirteen years to a point where several different instruments are commercially available and about one hundred are already in use. At the present time it is possible to analyze volumes of about 1 μ^3 at the surface of a specimen for elements above sodium in the periodic table with a concentration sensitivity of the order of 0.1–0.01%. New developments in the field include efforts to extend the range of Z to lower atomic numbers and to increase the spatial resolution of the method so that smaller volumes may be analyzed. Progress in the analysis of the light elements centers at present mainly on the use of nondispersive detectors and computer techniques for the resolution of overlapping pulse-height distributions. With such methods qualitative detection of elements down to beryllium has been achieved in the presence of their near neighbors. Efforts to increase the spatial resolution of the probe have led to the use of thin specimens which are partially transparent to electrons. Partly as a consequence of this approach, apparatus has been developed in which an electron microscope with facilities for selected area diffraction has been combined with an electron-probe analyzer. The ever-widening range of application of the electron-probe technique has inevitably disclosed problems in the interpretation of quantitative data, and it is clear that at the present time more information on the process of X-ray production by electrons is required before semi-empirical methods can be entirely eliminated. Examples of the use of the method in metallurgical and mineralogical studies illustrate some of the current trends in the refinement of measurement technique and interpretation of experimental data.

INTRODUCTION

To gain perspective for a discussion of recent work in the field of electron-probe analysis we should trace briefly the history of the method to the point where the latest developments emerge as new lines of research. Since the original publication by Castaing and Guinier[1] in 1949, many papers describing advances in the technique have appeared in the literature. Outstanding among these are Castaing's thesis[2] (1951), which describes a working instrument and the basic physics of both the instrument and its quantitative application, and the publications of Cosslett and Duncumb[3,4] who were responsible for the development of the scanning technique.

The electron–optical system described by Castaing (Figure 1) has remained essentially unaltered to the present time; two magnetic lenses are used to focus the electron beam from a triode electron gun into an area of about 1 μ diameter at the surface of the specimen. X-rays excited at the point of impact are analyzed by some form of

---

[1] Superscripts pertain to references at the end of the paper.

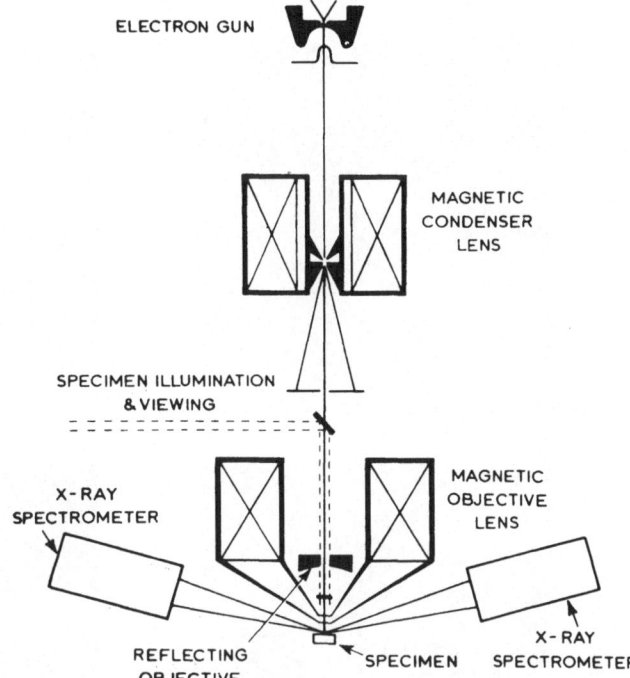

ELECTRON GUN

MAGNETIC
CONDENSER
LENS

SPECIMEN ILLUMINATION
& VIEWING

X-RAY
SPECTROMETER

MAGNETIC
OBJECTIVE
LENS

Figure 1. Diagrammatic arrangement of the static electron-probe analyzer (Castaing[33]).

REFLECTING
OBJECTIVE

SPECIMEN

X-RAY
SPECTROMETER

spectrometer, the intensities of the various spectral lines being compared with those from pure elements or compounds. Since the X-ray intensity is related to composition, quantitative analysis of the irradiated area is possible.

In the scanning instrument (Figure 2) the electron beam may be deflected by electromagnetic coils or electrostatic plates so that it moves in a raster pattern over the surface of the specimen, one frame of the scan occupying about 3 sec and containing about 300 lines. By deflecting a spot on a cathode-ray tube in synchronism with the electron probe and controlling the brightness of the spot either by the output of the X-ray spectrometer or by a signal proportional to the intensity of the backscattered electrons, images may be

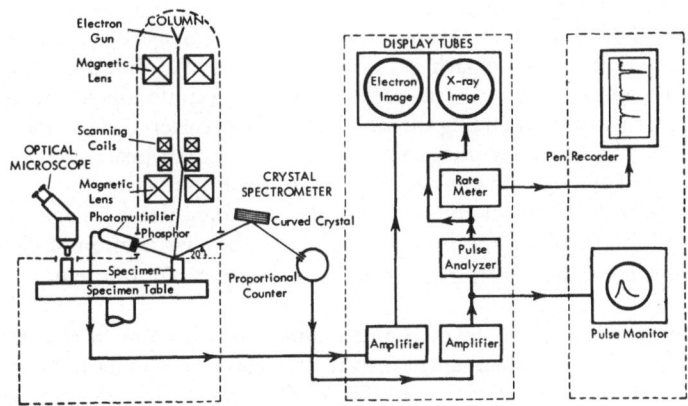

Figure 2. Arrangement of scanning X-ray microanalyzer (Duncumb[4]).

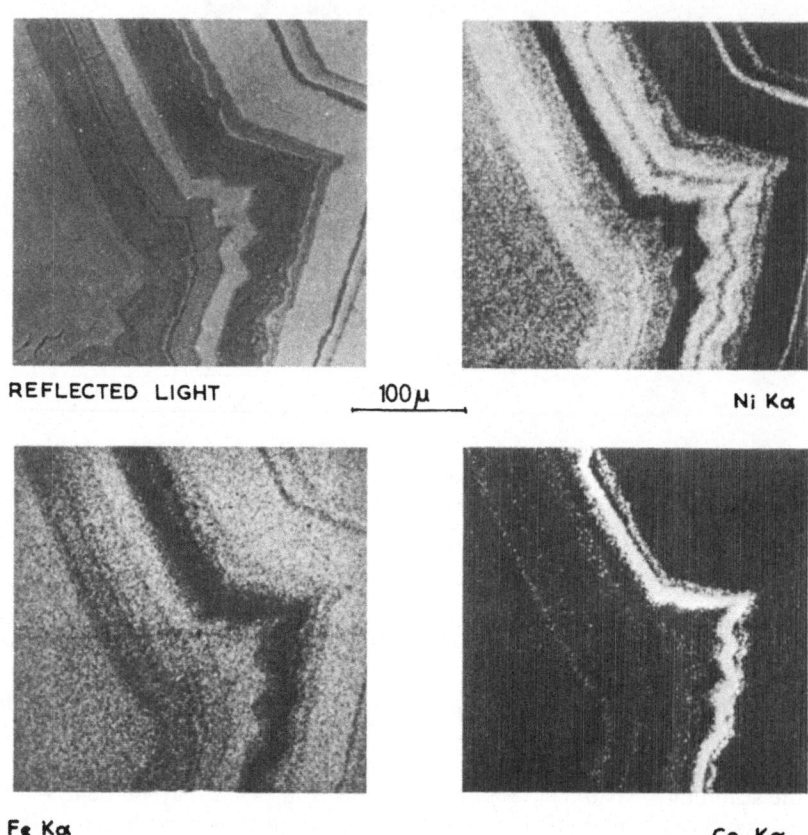

REFLECTED LIGHT   100μ   Ni Kα

Fe Kα   Co Kα

Figure 3. X-ray scanning images of zoned bravoite minerals in the system FeS_2–NiS_2–CoS_2 (Springer and Long[5]).

displayed on the cathode-ray tube screen. When the X-ray signal is used, the image is bright in areas corresponding to regions rich in the element whose radiation the spectrometer is set to detect. With electron modulation of the brightness, the image contrast arises both from variations in the backscattering coefficient over the specimen surface (due to variations in atomic number) and also from topographic detail, which produces shadowing effects by locally reducing the number of backscattered electrons which escape from the surface. The resultant image in this case thus often bears a close relationship to the appearance of the specimen as seen in the optical microscope.

Examples of both X-ray modulated and electron modulated images are shown in Figure 3. In addition to providing information about the distribution of elements at the surface of the specimen, the electron image also provides a means of accurately positioning the probe for point analysis.

The commercial development of the instrument has been such that at the present time at least eight different instruments are available, some of which can be traced directly to research projects in university departments. Generally, the electron beam may be focused to a probe about 0.5–1 μ in diameter and quantitative analyses of elements above sodium or magnesium in the periodic table may be performed. The concentration sensitivity varies with the type of spectrometer used—focusing or nonfocusing,[6] but usually lies in the range 0.1–0.01%.

An obvious deficiency in this performance, particularly from the point of view of the metallurgist, is the restriction to elements above sodium. A considerable part of the research into the instrumentation of the technique in the last few years has been concerned with this problem.

The resolution of the electron probe itself is already in the region of 0.5 μ, but owing to electron scattering within the specimen and to secondary X-ray excitation in the region immediately surrounding the point of impact, the whole of the X-ray production is not confined to a sphere of this diameter. The restriction of electron and X-ray penetration to a volume of the order of 0.5 μ^3 implies the use of low accelerating voltages and hence characteristic emission lines of longer wavelength. The problem of improving the X-ray resolution of the instrument is thus intimately bound up with the extension of the range to low atomic numbers.

The diameter of the electron probe itself is determined by the properties of the electron–optical system and the current i available in a beam of diameter d_0 is given by

$$i = \pi^2 \beta \, (d_0)^{8/3} / (2C_s)^{2/3}$$

where β is the gun brightness (current density/unit solid angle) and C_s is the spherical aberration coefficient of the probe-forming lens. It is clear that the available current in the probe falls very rapidly with d_0 and that large factors of improvement in β or C_s would be necessary, to compensate for a comparatively small-gain electron resolution. At present, practical means of achieving a useful gain in either C_s or β have not appeared in the literature.

THE ANALYSIS OF LIGHT ELEMENTS

A glance at a table of emission wavelengths and absorption coefficients, both of which increase rapidly with decreasing atomic number, indicates immediately the principal difficulties to be surmounted in such an extension of the method. The limited transmission of X-ray windows in counters is a practical but not a fundamental problem, since grids may be used to support thin collodion films against considerable pressure differentials. The major obstacle is the spectrometry of the emitted radiation. To the present time the greatest effort has been directed to the use of proportional counters with various techniques of pulse-height analysis.

Although the resolution of the proportional counter decreases with quantum energy, the spacing of K emission lines increases as the atomic number falls, and it is easily shown[7] that the effective resolution of the proportional counter remains approximately independent of Z. Figure 4, which shows the pulse-height distributions from carbon and aluminum, illustrates the sort of resolution which can be achieved with the proportional counter in the long-wavelength region.

The work of Dolby[8] and Cosslett[9] has shown that the intensity of the characteristic lines of the light elements is considerably greater than had previously been supposed. Although intensity falls quite steeply with decrease of atomic number, the number of quanta, or quantum intensity, varies much less rapidly for a given overvoltage. (Overvoltage = $U = V_0/V_c$, where V_0 is the accelerating voltage of the electrons and V_c is the critical excitation voltage for the characteristic line.) Thus the same quantum output was obtained for carbon at 2.3 kv ($U = 8$) as for aluminum at 5.7 kv ($U = 4$) and for copper at 20 kv ($U = 2.2$). Figure 5 also shows that for any given voltage the quantum intensity decreases with increasing atomic number.

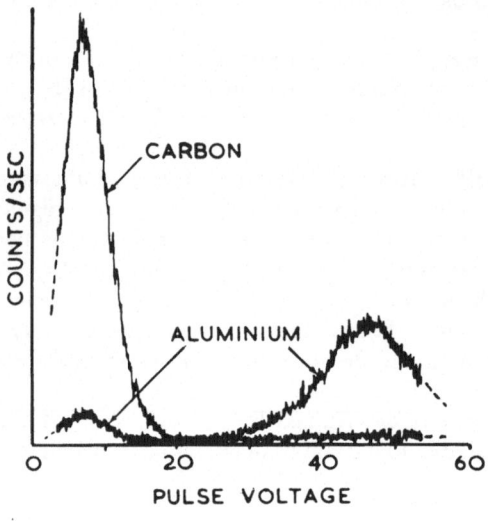

Figure 4. Pulse-height distributions due to carbon K and aluminum K radiations (Dolby[8]).

A simple illustration of the high intensity available from the light elements is given by the scanning picture (Figure 6) obtained by Melford and Duncumb,[10] who used a proportional counter and pulse-height analyzer to detect carbon K radiation from graphite flakes in pig iron. The peak-to-background ratio of the carbon K pulse-height distribution in this case was of the order of 7:1, which should allow the detection of carbon in concentrations down to about 10%, in the absence of interfering lines. Many of the metallurgically interesting problems would, of course, require a considerably higher sensitivity.

Dolby and Cosslett[11,12] have investigated several ways of resolving overlapping pulse-

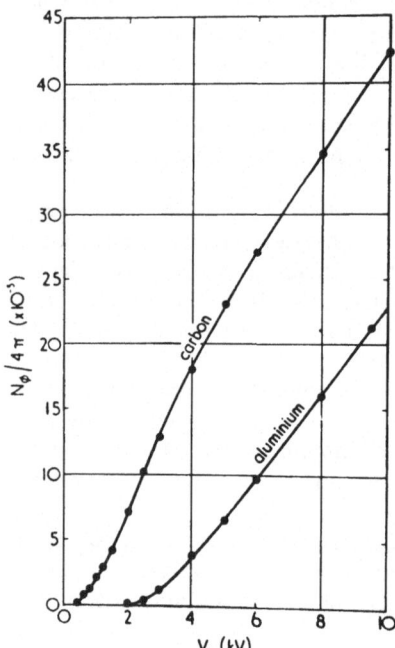

Figure 5. Variation with excitation voltage V_0 of the quantum efficiency for K radiation from carbon and aluminum, in quanta per electron per unit solid angle (Dolby[8]).

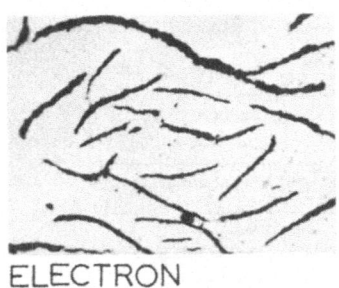
ELECTRON

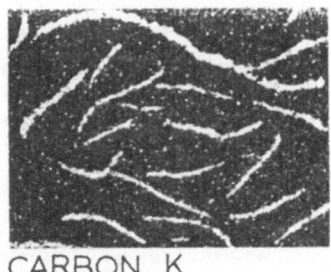
CARBON K

Figure 6. Scanning picture showing graphite flakes in pig iron obtained with a proportional counter and pulse-height analyzer set to select carbon K radiation (Melford and Duncumb[10]).

height distributions from groups of adjacent elements. Of these, the most successful appears to be the simultaneous equations method, in which the intensity of the combined spectrum is measured at a number of points equal to the number of component elements. The intensities at the three points Y_A, Y_B, Y_C (Figure 7), may be represented in terms of the peak heights of the three component distributions, e.g.,

$$Y_A = K_1\alpha + K_2\beta + K_3\gamma$$

Rearrangement of these equations gives expressions for α etc., of the type:

$$\alpha = K_1 Y_A + K_2 Y_B + K_3 Y_C$$

where the constants K_1, K_2, K_3 may be determined by measurements on pure elements A, B, and C.

The solution of these equations is performed automatically by means of a simple analog computer (Figure 8), the output of which may be passed as a modulating signal to a cathode ray tube for the formation of scanning pictures. It can, of course, be read off equally well as a voltage on a suitable meter. In recent work with this technique Dolby[13] has demonstrated the resolution of the light elements down to beryllium, and has obtained

Figure 7. The analysis of overlapping pulse-height spectra by the "simultaneous equation method" (Dolby and Cosslett[12]).

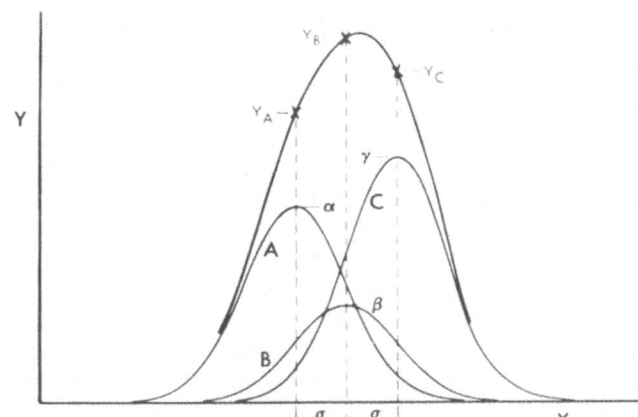

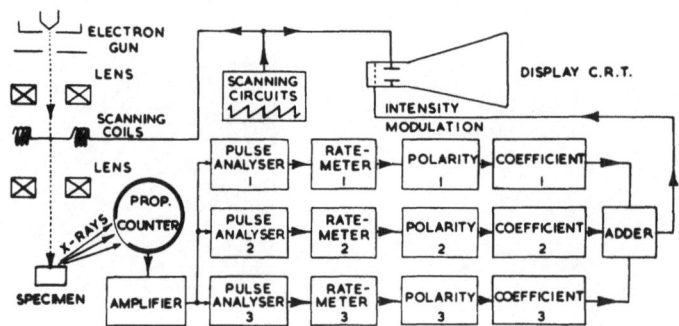

Figure 8. Scanning X-ray microanalyzer incorporating the matrix method of pulse analysis. Intensities of Y_A, Y_B, Y_C (Figure 7) are measured simultaneously by the pulse-height analyzer-ratemeter combinations 1, 2, 3, respectively (Dolby and Cosslett[12]).

scanning pictures and semiquantitative measurements of the distribution of oxygen, carbon, and beryllium in a specimen of oxidized beryllium metal.

Possible alternatives to the nondispersive techniques are monochromators with very large d spacings, or ruled gratings. Although crystals consisting of layers of heavy metal stearates deposited on a curved surface have been employed as monochromators,[14,15] they do not appear to have been used in the microprobe. Similarly, very little work has been reported on the use of gratings for this purpose.

The simplicity of a dispersive technique has much to recommend it, and owing to the inherently higher resolving power, albeit accompanied by a low collection efficiency, it may well provide the only solution to the problem of analyzing the light elements in the presence of overlapping L spectra from heavier elements in the specimen. On the other hand, the very high quantum efficiency of the nondispersive method weighs heavily in its favor in the resolution of simple spectra, particularly when the identity of the component elements is known and only their concentrations have to be determined.

SOME APPLICATIONS OF THE MICROANALYZER WITH "STANDARD PERFORMANCE"

It would be grossly misleading to suggest that the electron probe without facilities for analyzing the light elements and with a resolution no better than $1\,\mu$ is an instrument lacking in versatility. From its multifarious applications, it is possible to select a few which illustrate the general trends in the use of the technique within the confines of the "standard performance" of $1\,\mu$ resolution and an accessible range of elements extending from Mg upward.

Probably the largest single use has been in the analysis of diffusion couples, both for the measurement of diffusion coefficients and for the determination of equilibrium diagrams. The greater part of this work has been carried out in France and the U.S.A., and some of the results of the French school have recently been reviewed by Philibert.[16]

Philibert and his collaborators have carried out an interesting series of experiments on the effect of plastic strain on the diffusion rate in Fe/Ni[17] and Cu/Zn.[18] No differences could be detected between the strained and unstrained couples in either case. On the other hand, the stoichiometry of intermetallic compounds appears in some cases to be dependent on the presence of strain in the phase. For example Adda *et al.*,[19] have demonstrated the presence of a concentration gradient in UCu_5 (Figure 9) in which the

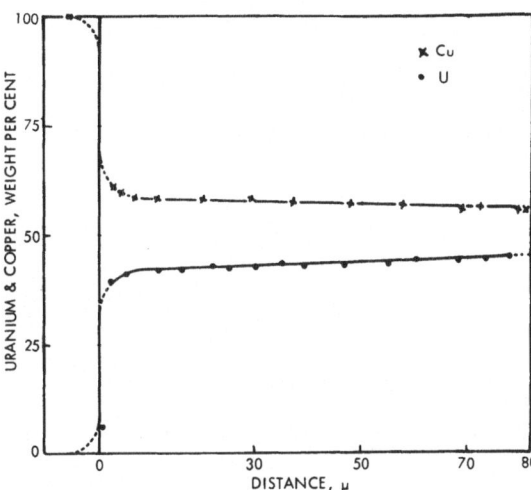

Figure 9. Concentration gradient in intermetallic compound in diffusion zone before application of pressure (Adda *et al.*[19]).

composition varied from $UCu_{4.70}$ to $UCu_{5.25}$ across the diffusion zone. When an external pressure (> 500 kg/mm^2) was applied the gradient was not observed and the exact stoichiometric composition was obtained.

A particularly elegant diffusion study has been carried out by Austin and Richard,[20] who have made detailed measurements on the diffusion of nickel into grain boundaries of various tilt angles in copper bicrystals. Figure 10 shows the nickel isoconcentration contours obtained for a 45° bicrystal. The grain-boundary diffusion coefficient is found to be dependent both on the tilt of the boundary and on concentration. At a temperature of 750°C, the grain-boundary diffusion coefficient is some six orders of magnitude greater that that of the lattice.

Seebold and Birks[21] have recently published an investigation into the precipitates formed in diffusion zones, due to the interaction of impurities with the diffusing metals. Their results indicate that activity of the impurities varies with the nature of the diffusing metal atoms; for example, the same specimen of iron precipitated Nb_2S_5 when diffused against niobium and CrN_2 or Cr_2N_5 when diffused against chromium, although in each case the impurities originated from the iron.

The determination of equilibrium diagrams by the preparation of alloys is greatly facilitated by the electron probe, since it is possible to determine directly the composition of coexisting phases without a knowledge of the bulk composition of the alloy. Melford,[22] for example, has determined part of an isothermal section through the Fe–Sn–Cu ternary diagram by this means.

MEASUREMENTS CLOSE TO PHASE BOUNDARIES

The analysis of small precipitates and fine exsolution lamellae in minerals, even when these have a size greater than the diffusion range of the electrons, is subject to errors due to fluorescence X-rays excited in the matrix or host mineral (Figure 11).

This problem has been investigated in some detail by Reed,[23] and some of the results were presented at the Third International Symposium on X-ray Microanalysis held at Stanford in August 1962. The problem has been approached both theoretically and experimentally, the former work leading to a fluorescence correction equation which is a modification of that given by Castaing.[2] This equation takes account of the fact that the

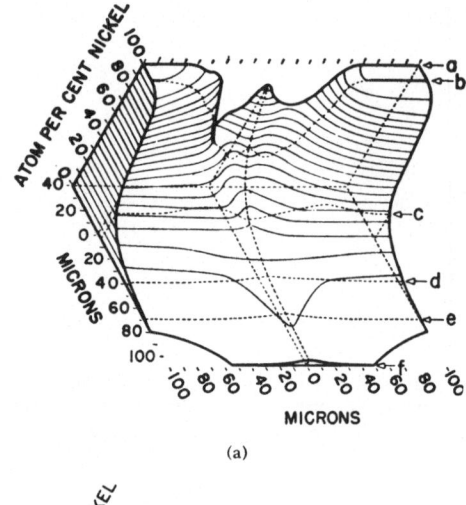

(a)

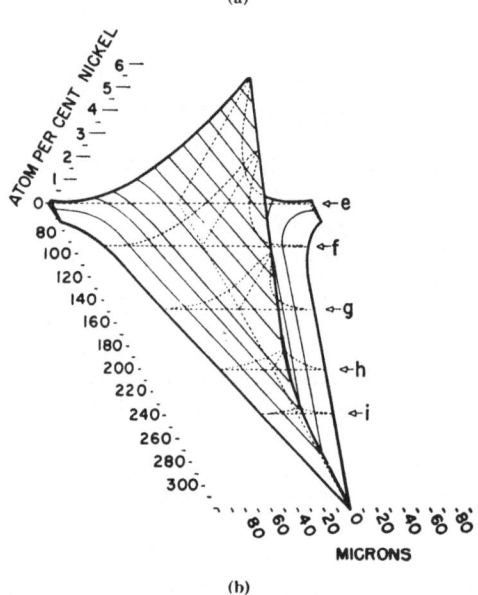

(b)

Figure 10. Nickel isoconcentration contours, resulting from diffusion into 45° copper bicrystal annealed 240 hr at 750°C. (a) Isoconcentration contours at 5% nickel intervals within range of lattice diffusion from original interface. (b) Isoconcentration contours at 0, 0.1, 0.3, 0.5, 1.0, 1.5, 2.0, 2.5, 3.0, 3.5, 4.0, 4.5, and 5.0% nickel from grain-boundary and lateral-lattice diffusion (Austin and Richard[20]).

exciting and excited elements may differ considerably in atomic number and also that the electron beam voltage is not necessarily large compared with the excitation potentials of the characteristic lines. The fluorescence intensities calculated by this equation are always lower than those predicted by the Castaing formula. Although in homogeneous alloys this difference is seldom important; it becomes very significant in the boundary case where the whole of the apparent concentration of a given element observed in one phase may be due to fluorescence excitation of the radiation of that element in the adjacent phase.

The experimental measurements have been carried out by preparing artificial boundaries between pairs of elements and measuring the excitation in one element produced when the probe is placed at different distances from the boundary in the second element. Figure 12 shows a typical result for the boundary between Fe and Ni. In the case of Zn/Fe, for example, where Fe K_α is excited by Zn K_α, both the modified formula and the experimental data indicate a fluorescence intensity which is lower than that given by

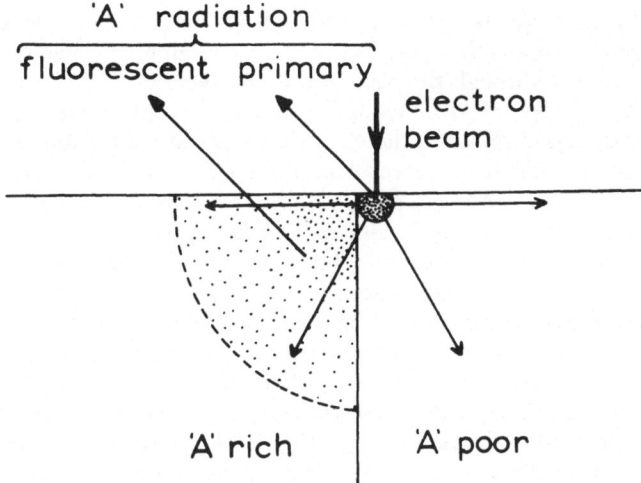

Figure 11. Fluorescence excitation at a boundary between an "*A* rich" phase and an "*A* poor" phase (Reed and Long[23]).

Castaing's formula. These data have been used in the correction of the calcium content of fine exsolution lamellae in pyroxenes.[24]

VISIBLE FLUORESCENCE OF MINERALS

In many transparent minerals a striking visible fluorescence is excited by the electron beam and may be seen with a suitable microscope arranged in the electron probe so that the specimen may be viewed, preferably as a thin section, during electron bombardment.

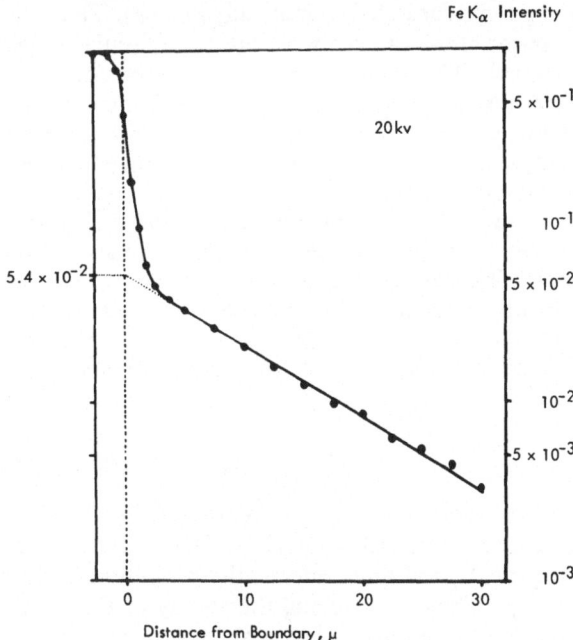

Figure 12. Variation of Fe K_α fluorescence intensity with distance of electron probe from Ni/Fe boundary (Reed and Long[23]).

Since most silicate minerals are electrical insulators and thus have to be coated with a thin metallic conducting layer to prevent charging, the visible fluorescence is most conveniently seen through the back of the thin section.

The electroluminescence of minerals is a well-known phenomenon, but it does not appear to have been used, prior to the electron probe, to aid the study of rocks in thin section. On the microprobe it can serve to show the position of the focused electron beam, which appears in the microscope as a point of light. In addition, when a defocused electron beam is used to illuminate a considerable area of the optical field, variations in fluorescence color and intensity may give useful information on the distribution of trace elements, which are the usual cause of fluorescence.

For example, Figure 13 shows a calcite crystal in which no zoning is visible in a transmitted-light microscope. On electron bombardment the crystal is found to consist of a series of zones which vary in color from brilliant orange-red to black. Electron-probe analyses[25] of these regions show that this results from a rhythmic variation in the Fe/Mn ratio of from about 1:10 in the red zones to 10:1 in the black, the overall concentrations of these elements being about 1%. While it is not generally possible to predict what elements are present by examination of the fluorescence colors, information on their distribution, which may well be difficult to obtain by probe analysis since concentrations are often low, is sometimes clearly demonstrated by this technique.

HEATING BY THE ELECTRON PROBE AND DETERMINATION OF VOLATILES

The determination of water of crystallization and carbonate content of some minerals is an important problem which cannot at present be attacked directly by electron-probe analysis. One possible approach[26] to this problem is to determine the concentration of one element in the hydrated or carbonated mineral and then after dehydration or decarbonation, to remeasure this concentration and also to make a complete analysis (if the general formula is not already known). The volatile content may then be determined in terms of the two analyses and the molecular weight of the dehydrated or decarbonated material. The whole procedure may be accomplished in the microanalyzer by first determining the concentration of the element with a probe current sufficiently low so as to cause no significant temperature rise, then heating the analyzed area with a higher current, and finally returning to the low current to measure the new concentration in the decomposed region.

The accuracy of such measurements will clearly depend on the accuracy with which the concentrations may be determined and also on the proportion by weight of water present. That the technique shows at least some promise is indicated by experiments on gypsum, in which the calcium content was found to rise from 23.5% [characteristic of $CaSO_4 \cdot 2H_2O$ to $29 \pm 1\%$ ($CaSO_4 = 28\%$)] and on calcite where heating in the beam produced an increase of calcium content from 40% ($CaCO_3$) to $73 \pm 2\%$ ($CaO = 71.4\%$ Ca).

ANALYSIS OF THIN SURFACE LAYERS

It has been pointed out by Melford[27] that while the resolution of the electron probe in a direction parallel to the surface of the specimen may be limited by astigmatism, stray fields, and other effects, the resolution in depth is determined by the accelerating voltage. Thus by reducing the energy of the electron beam, primary excitation can be confined to the surface layers and the resolution in depth made substantially better than

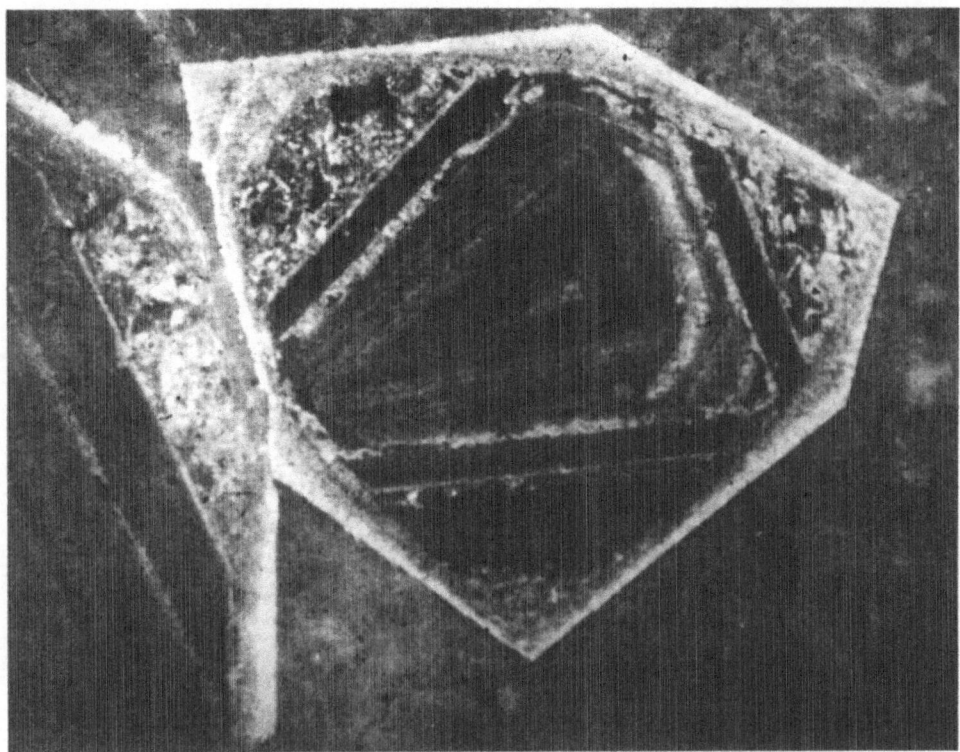

Figure 13. Thin section of calcite crystal in calcite-strontianite rock from Strontian, Scotland, photographed in the light of its own electroluminescence (Long and Iredale[25]).

that which is usually obtained parallel to the surface. Such a technique has been used by Wood and Melford[28] in the examination of oxide films. It is suggested that, provided the requisite ionization function/depth curves for a range of voltages could be determined, the method could, in principle, be used to obtain information about surface films below 1000 Å in thickness.

ELECTRON MICROSCOPY AND X-RAY MICROANALYSIS

Some of the limitations on the X-ray resolution of the microanalyzer have already been noted. While at present the current which can be delivered into a probe of given diameter is limited by the spherical aberration of the electron lenses, the loss of resolution by electron scattering in the target can be eliminated by using very thin specimens in which the electron beam remains essentially undeviated. Under these conditions it is no longer necessary to reduce the beam accelerating voltage since the specimen thickness can be made considerably less than the penetration depth of the electrons.

The examination of thin films in the electron probe suggests the possibility of combining facilities for electron microscopy and microanalysis in one instrument. Recent work in this field includes that of Nixon in the Department of Engineering at Cambridge and the rather different approach by Duncumb at the Tube Investments Laboratories at Hinxton Hall.

Nixon's arrangement[29] (Figure 14) uses the electron–optical system of an electron

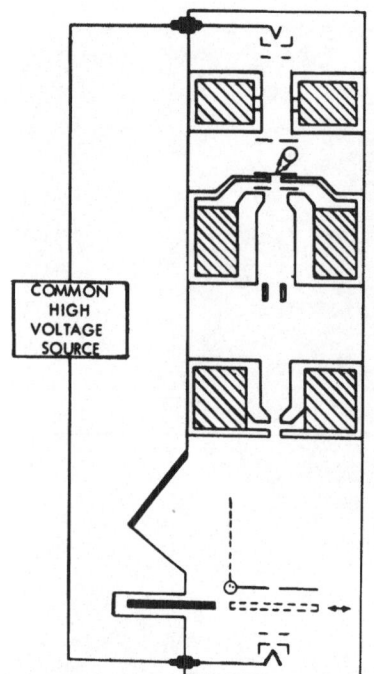

IMAGING ELECTRON GUN

CONDENSER APERTURE AND
BEAM MONITOR
X-RAY COUNTER
SPECIMEN STAGE
ADJUSTABLE APERTURES

OBJECTIVE LENS

COMMON
HIGH
VOLTAGE
SOURCE

STIGMATOR

PROJECTOR LENS
(CONDENSER LENS)

VIEWING WINDOW

FLUORESCENT SCREEN

PHOTOGRAPHIC PLATE

ANALYZING
ELECTRON GUN

Figure 14. Electron microscope with second gun below fluorescent screen for microanalysis (Nixon[29]).

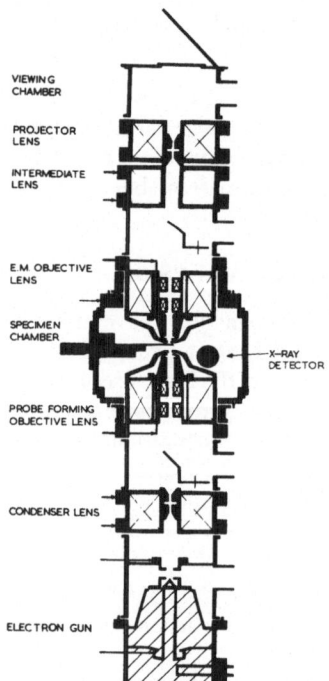

VIEWING CHAMBER

PROJECTOR LENS

INTERMEDIATE LENS

E.M OBJECTIVE LENS

SPECIMEN CHAMBER

X-RAY DETECTOR

PROBE FORMING OBJECTIVE LENS

CONDENSER LENS

ELECTRON GUN

Figure 15. Combined electron-probe microanalyzer and electron microscope (Duncumb[30]).

microscope, with the addition of a second electron gun below the plane of the fluorescent screen. The instrument is operated first as a normal electron microscope and the specimen is moved so that the image of the part of the specimen to be analyzed falls on the aperture in the center of the fluorescent screen. The imaging electron beam is then turned off and the same accelerating voltage applied to the gun below the fluorescent screen. Thus without readjustment of the lenses, an electron beam is brought to a focus on the selected area of the specimen. X-rays emitted from the point of impact are detected and analyzed by means of a proportional counter mounted in the specimen chamber.

In Duncumb's apparatus[30] (Figure 15) only one gun is used and the electron–optical system comprises a microanalyzer followed by an electron microscope. The two objective lenses are identical and the microanalyzer can operate in the normal way or act as a condenser system for the electron microscope. Electron diffraction, microscopy, and microanalysis can thus be performed within the one instrument, and some of the first results obtained with this equipment were described at the Stanford Conference.

CONCLUSION

This paper has been concerned primarily with advances in the technique and mode of application of microprobe analysis. No such discussion would be complete without some mention of present views on the interpretation of quantitative data. Several different formulas[31,32] and methods of deriving absorption and fluorescence corrections have been published recently, but the most difficult outstanding problem at the present time is the analysis of alloys consisting of elements of widely differing atomic number.

The theory of quantitative analysis set out by Castaing[33] suggests that the effect of atomic number is small, that is to say, the intensity of emitted characteristic radiation, corrected for absorption and fluorescence effects, is nearly linearly proportional to concentration. This conclusion is based on the assumption that the effect of electron back-scattering, which increases with atomic number and reduces the characteristic intensity from a given concentration of an element when it is alloyed with a heavy element, is largely offset by an increased mass penetration of the electrons into the heavy alloy. At the present time it is not clear whether the discrepancies which are observed, particularly with alloys containing light elements, can be attributed to a residual atomic number effect or to the use of absorption corrections which are themselves not applicable. The derivation of absorption correction curves, of the type given by Castaing, depends on experiments in which the distribution of X-ray production with depth in the specimen is determined by means of targets consisting of a thin evaporated layer of a "tracer" element which is embedded at various depths within a block of a second element. The characteristic intensity from this layer is then compared with the intensity from an isolated layer of the same thickness. Since the two elements clearly cannot be the same, as it is necessary to distinguish between their characteristic radiations, it is possible that an atomic number effect is implicit in this type of experiment.

REFERENCES

1. R. Castaing and A. Guinier, *Proceedings of Conference on Electron Microscopy*, Delft, 1949.
2. R. Castaing, Thesis, University of Paris, *O.N.E.R.A. Publ. No. 55*, 1951.
3. V. E. Cosslett and P. Duncumb, *Nature* **177**: 1172, 1956.
4. P. Duncumb, *Brit. J. Appl. Phys.* **10**: 420, 1959.
5. G. Springer, and J. V. P. Long, *Proceedings of the 3rd International Symposium on X-ray Optics and Microanalysis*, 1962, Academic Press, Inc., New York, 1963.
6. J. V. P. Long, *Proceedings of the 1st International Symposium on X-ray Microscopy and Microradiography*, Academic Press, Inc., New York, 1957, p. 435.

7. C. F. Hendee and S. Fine, *Phys. Rev.* **95**: 281, 1954.
8. R. Dolby, *Brit. J. Appl. Phys.* **11**: 64, 1960.
9. V. E. Cosslett, *X-ray Microscopy and X-ray Microanalysis*, Elsevier, Amsterdam, 1960, p. 436.
10. D. A. Melford and P. Duncumb, *Metallurgia* **61**: 205, 1960.
11. R. M. Dolby. *Proc. Phys. Soc. (London)* **73**: 81, 1959.
12. R. M. Dolby and V. E. Cosslett, *X-ray Microscopy and X-ray Microanalysis*, Elsevier, Amsterdam, 1960, p. 436.
13. R. M. Dolby, Thesis, University of Cambridge, 1961 (see also ref. 5) (to be published).
14. C. L. Andrews, *Rev. Sci. Instr.* **11**: 111, 1940.
15. B. Lindstrom, *Acta Radiol. Suppl.* **125**: 150, 1955.
16. J. Philibert, *J. Inst. Metals* **90**: 241, 1962.
17. P. Chollet, L. Grosse, and J. Philibert, *Compt. rend.* **252**: 728, 1961.
18. A. Guy and J. Philibert, *Trans. Met. Soc. AIME* (to be published).
19. Y. Adda, M. Beycler, A. Kirianenko, and B. Pernot, *Mem. Sci. Rev. Meto.* **57**: 423, 1960.
20. A. E. Austin and N. A. Richard, *J. Appl. Phys.* **32**: 1462, 1961.
21. R. E. Seebold and L. S. Birks, *Anal. Chem.* **31**: 112, 1962.
22. D. A. Melford, *J. Iron and Steel Inst.* **200**: 290, 1962.
23. S. J. B. Reed and J. V. P. Long, *Proceedings of the 3rd International Symposium on X-ray Optics and Microanalysis*, 1962, Academic Press, Inc., New York, 1963.
24. R. A. Binns, J. V. P. Long, and S. J. B. Reed (Unpublished work).
25. J. V. P. Long and A. Iredale (Unpublished work).
26. J. V. P. Long (Unpublished work).
27. D. A. Melford, *J. Inst. Metals* **90**: 217, 1962.
28. G. C. Wood and D. A. Melford, *J. Iron Steel Inst.* **198**: 142, 1961.
29. W. C. Nixon, *Eur. Reg.*, De Nederlaudse Vereeniging voor Electronen Microscopie, Delft, 1961.
30. P. Duncumb, *Proceedings of the 3rd International Symposium on X-ray Optics and Microanalysis*, 1962, Academic Press, Inc., New York, 1963.
31. L. S. Birks, *J. Appl. Phys.* **32**: 387, 1961.
32. L. S. Birks, *J. Appl. Phys.* 1962 (in press).
33. R. Castaing, in *Advances in Electronics and Electron Physics, Vol. 13*, Academic Press, New York, 1960, p. 317.

DISCUSSION

K. F. J. Heinrich (E. I. du Pont de Nemours and Co.): What complications could one expect in the analysis of the light elements due to interference by L and M radiation of the heavier elements?

J. V. P. Long: This is clearly going to be the major objection to the use of the proportional counter. I think that it is very likely that in such a case one would have to use a dispersive system, rather a crystal or a grating, unless one has prior knowledge as to which L or M radiations are present. For example, if one is analyzing a fairly simple iron carbide one knows that iron L radiation is present. Then it is almost certain that the proportional counter technique can do something about it. If, for example, one is trying to distinguish between an iron carbide and an iron nitride, this would be all right; but if one had a complex spectrum of L or M radiations about which one knew nothing, this might well be a very difficult problem by the nondispersive method.

OSCILLOSCOPE READOUT OF ELECTRON MICROPROBE DATA

Kurt F. J. Heinrich

E. I. du Pont de Nemours & Company, Inc.
Wilmington, Delaware

ABSTRACT

In view of the amount of data per time unit the scanning electron probe is capable of producing, time economy in data presentation is of great practical importance. In this paper, the characteristics and advantages of various previously known techniques of data presentation are described, and some novel procedures are proposed.

THE PROBLEM OF DATA PRESENTATION

The scanning electron probe microanalyzer is capable of producing information at a prodigious speed. At the same time, several types of information and different ways of presenting it are available (Table I). The proper choice of data selection and presentation is, therefore, important for an efficient operation, particularly if the results must be conveyed to persons not connected, and generally not too familiar with the operation of the instrument.

Apart from the obvious need for a thorough discussion of the nature of the sample and of the problem, the following suggestions are offered to increase the efficiency of operation.

a. One should not strive for higher precision than necessary.
b. Area or line scans are to be preferred to point analysis.
c. Electron data are preferred over X-ray data, where possible.
d. Recorder output should be transferred to oscilloscope photos.
e. Multiple oscilloscope exposures should be used liberally.

The importance of quantitative analysis with the microprobe is obvious. Frequently, however, semiquantitative information over an area or along a line will, with less expense of time and effort, tell more about the sample than quantitative analysis of a few selected spots. Hence, quantitative point analysis should only be undertaken when other techniques fail to yield the necessary information.

For a beam of a given diameter, the area from which signals are obtained with electrons is considerably smaller than that corresponding to X-ray signals. Therefore, electron signals (backscattering, target current), are preferred for topographic images (Figure 1). As the amount of backscattering is a function of the atomic number, in many cases target current measurement can also be used for quantitative analysis,[1] with the advantages that the signal is less subject to statistical fluctuations than X-ray signals and is insensitive to electrostatic scanning beam deflections that tend to defocus the X-ray

---

[1] Superscripts pertain to references at the end of the paper.

Table I. Classification of Signals

(a) Source	X-rays	Dispersive / Nondispersive
	Electrons	Backscattered / Secondary / Target current
	Cathodoluminescence	Direct observation / Photomultiplier pickup
(b) Extension in space	Point	One / Step-scan
	Line	Mechanical / Electronics
	Area	Mechanical / Electronics
(c) Registration	Digital	Scaler / Printer
	Visual	Cathodoluminescence / Meters / Oscilloscope
	Recorder	
	Oscilloscope-Camera	
(d) Precision	Qualitative / Semiquantitative / Quantitative	

optics. Therefore, electronic line scans over several hundred microns or mechanical scans over several millimeters can be quantitatively interpreted as well as stationary measurements with the further advantage of extremely simple interpretation. The target current varies linearly with concentration, and the complicated and controversial corrections that must be applied to X-ray measurements (absorption correction, fluorescence correction, etc.) are conspicuously absent, as well as the limitations in the detection of

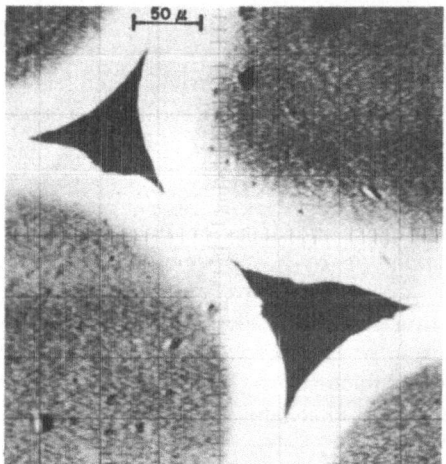

Figure 1. Niobium spheres coated with tantalum. Partial diffusion by heat treatment. Backscattered electron area scan.

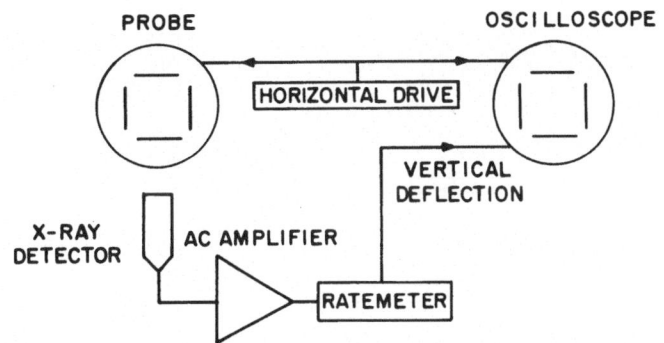

Figure 2. X-ray line scanning.

light elements. The chief requirements for the sample are good surface preparation and sufficient electrical conductivity.

Usually ratemeter readouts of X-ray signals are presented on a recorder. When only one detector function is registered at a time, the ratemeter output can be fed instead into the oscilloscope amplifier (Figure 2). It is common practice to superimpose such line scans, obtained by beam deflection, upon an area scan background (usually target current or backscattered electrons) on which the locus of the scan is marked (Figure 3). When several such line scans are registered on one picture, as in Figure 4, it is advisable to shift and mark the zero line for each signal, avoiding crossing of traces.

The ARL Microprobe also features a readout mode called "multiplexing," consisting of successive ratemeter readout of a programmed sequence of detector signals, and a "step scanning" mode, consisting of a series of readouts of integrated (capacitor) signals collected simultaneously and compared with a standard capacitor charge normally provided by the target current. After readout, the sample can be shifted automatically from 1 to $10\,\mu$, and the cycle is automatically repeated. This mode can easily be converted

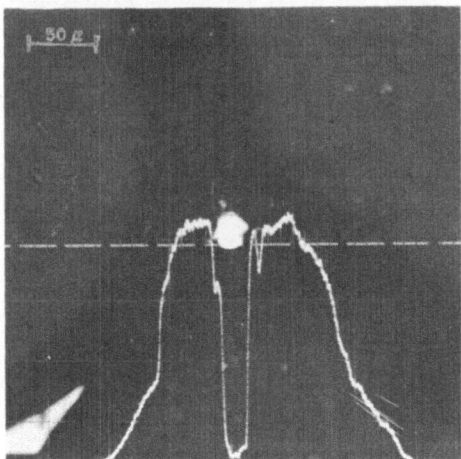

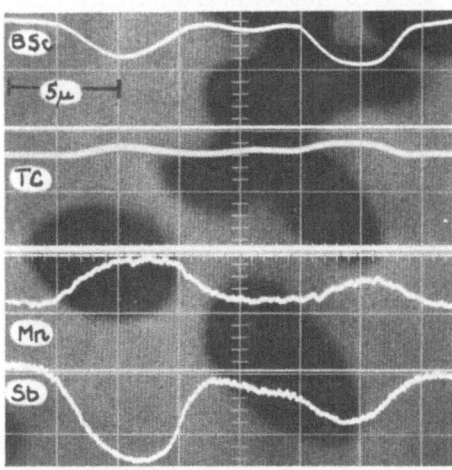

Figure 3. Samples as in Figure 1. Triple exposure. Target current area scan + locus of a horizontal line scan (broken line) + Ta line scan.

Figure 4. Manganese antimonide. Backscattered electron area scan + locus of line scan + four line scans (backscatter, target current, Mn K_α, Sb L_{α_1} with corresponding zero levels).

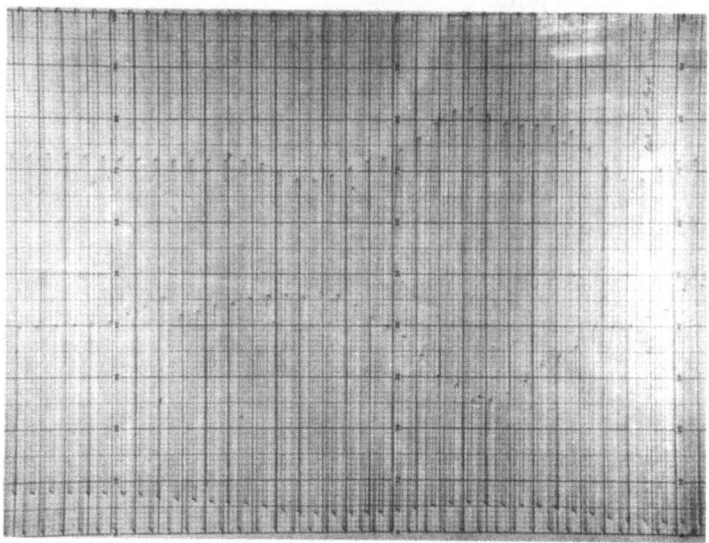

Figure 5. Step scanning trace on recorder chart.

into fixed-time operation, and target current, if so desired, can be registered as well as the X-ray signals. Both systems normally use a chart recorder readout.

The disadvantage of chart recordings is poor reproducibility on photographs and, generally, excessive lengths of chart paper required during an analysis. Besides, unless an XY recorder with penlift is used, the traces connecting successive readings in "multiplexing" and "step scanning" frequently obscure the readout quite considerably (Figure 5). Therefore, for reporting, such charts must be provided with connecting lines or be completely replotted, with considerable loss of time. By use of a second slide-wire system installed on the recorder which, by means of a battery, furnishes a potential proportional

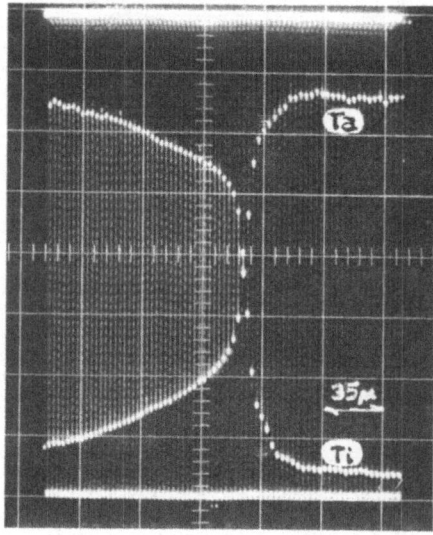

Figure 6. Step scanning trace recorded on oscilloscope.

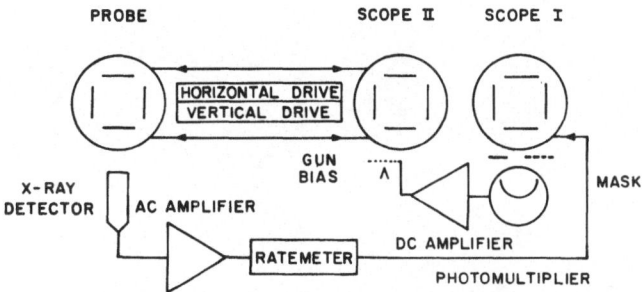

Figure 7. Concentration mapping.

to the pen position, the recorder information can be carried onto the oscilloscope (Figure 6), with considerable gain in clarity. In step scanning, the slow trace corresponding to the capacitor charge period can be blanked out. Photographs can be copied by means of a print copier, attachable to a standard Polaroid camera, obtainable from Polaroid Corp. This procedure expedites considerably the storage and communication of data commonly registered on strip charts.

CONCENTRATION MAPPING

Usually, the analytical task involves determining the concentration of an element which exists at a certain point of the sample. The inverse problem, namely, to find on which areas of the sample the concentration reaches a certain predetermined level, is an interesting alternative, as pointed out recently by Melford, who proposes using a ratemeter and "passing the ratemeter output through a pulse-height analyzer."[2] While a ratemeter output could not be handled by a conventional pulse-height analyzer, the construction of an electronic device suitable to Melford's proposal would be quite feasible. In our work, and independently from Melford's proposal, an optical device is used instead for handling the ratemeter output in order to mark isoconcentration lines or areas[3] and this procedure we call "concentration mapping" (Figure 7).

The amplified X-ray detector output, if desired after passing through a pulse-height selector (single-channel pulse-height analyzer), is fed into a fast ratemeter, and the ratemeter output is fed into the vertical amplifier of an oscilloscope. (The horizontal sweep is normally inactive.) The Z-axis (light intensity) of this scope is kept constant. The front screen is covered by a black mask, except for one or more slits or windows arranged in

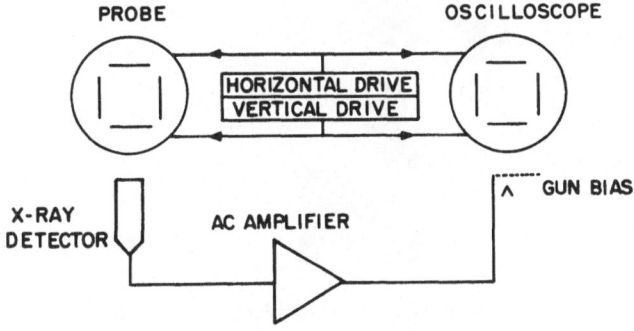

Figure 8. X-ray area scanning.

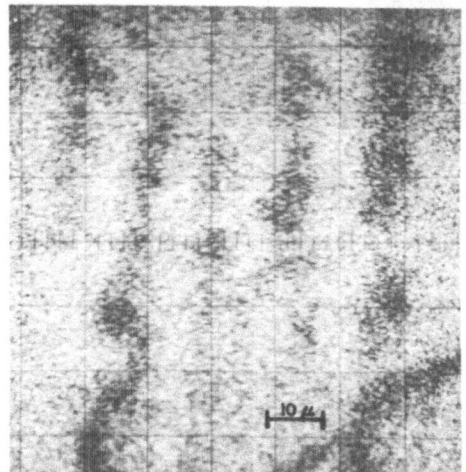 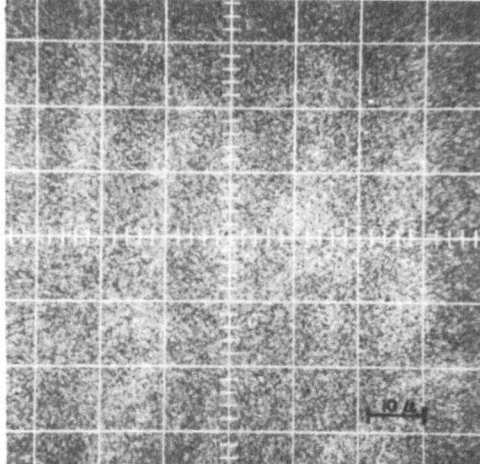

Figure 9. Ilmenite–leucoxene grain. Fe K_α area scan.

Figure 10. Same sample area. Ti K_α area scan.

such a fashion that the light spot is only visible when the X-ray signal intensity indicates a concentration within the range (or ranges) of interest. An encased photomultiplier in front of the mask "reads" the light spot, and its amplified output modulates the light intensity of a second scope sweeping in synchronism with the electron probe beam, as customary in area scans. This optical arrangement is extremely flexible. Several windows or multiple window combinations can be exchanged quickly by horizontal displacement of the beam of the first oscilloscope and gray shades or dotted areas can be added by applying a square wave horizontally.*

The simplest use of this device is the application of bias in order to differentiate

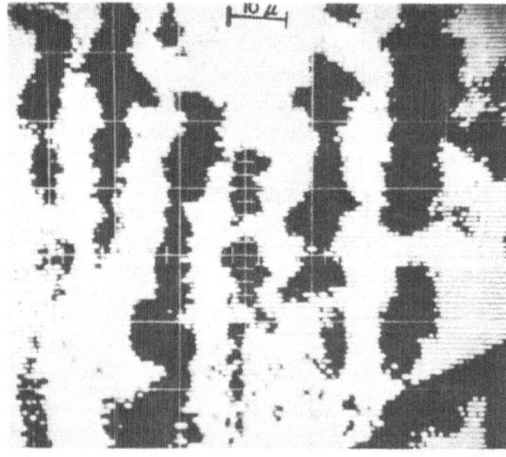

Figure 11. Same sample area. Fe K_α concentration mapping.

* In the case of an instrument having two oscilloscopes, or where an independent second oscilloscope is available, the auxiliary equipment needed represents a very modest incremental investment.

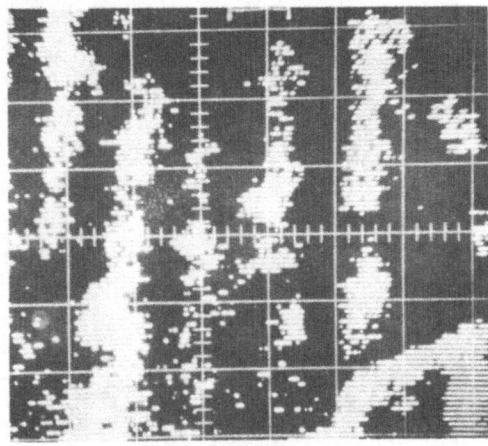 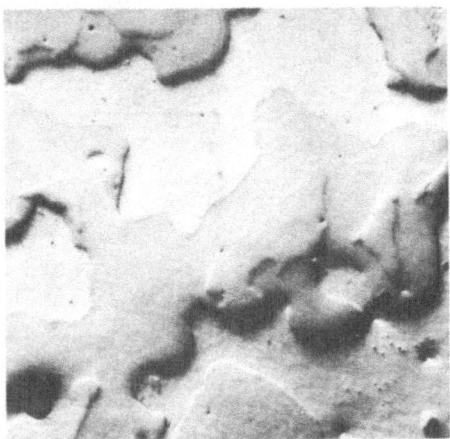

Figure 12. Same sample area. Ti K_α concentration mapping. Bias set to show leucoxene only.

Figure 13. Ni–Cu explosive compaction micrograph.

levels of signals. An example is seen in Figures 8–11. The sample under examination is an altered ilmenite (ferrous titanate) grain with leucoxene (TiO_2) inclusions. A conventional iron X-ray scan (Figures 8 and 9) shows the variation in iron concentration, but the corresponding Ti K_α scan (Figure 10) shows very little difference between the two phases. By using concentration mapping with appropriate biases (Figures 11 and 12), not only can the iron distribution be observed with highly increased contrast, but also the two titanium concentration levels are now clearly differentiated.

Figures 13–15 show a nickel–copper sample obtained by explosive compaction, on which three phases of 0, 57, and 100% of copper are clearly distinguishable. Figure 14

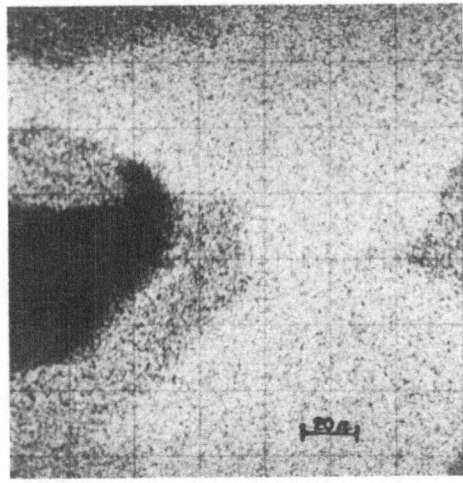 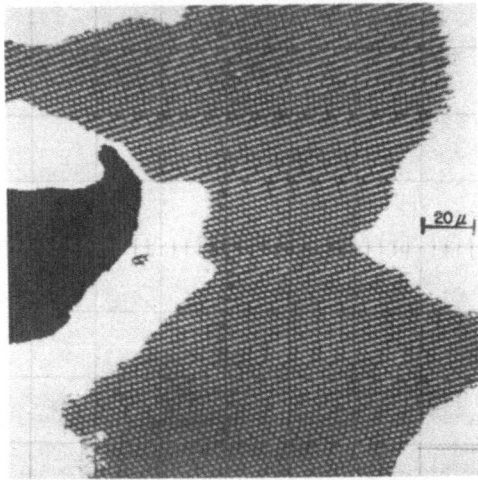

Figure 14. Cu K_α area scan.

Figure 15. Same area. Concentration mapping.

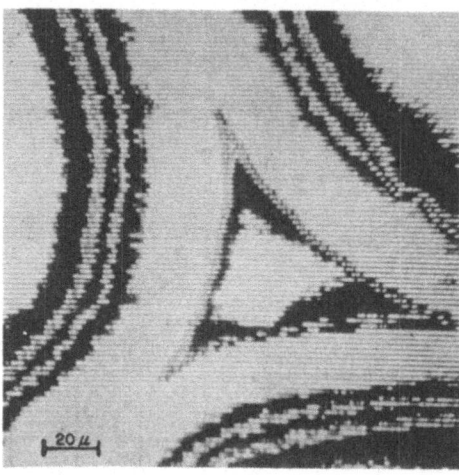

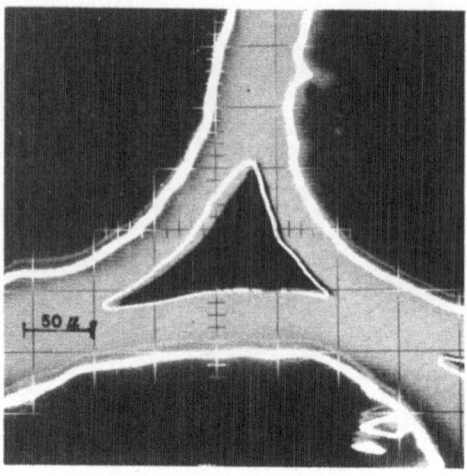

Figure 16. Same sample as Figure 1. Multiple slit Ta L_{α_1} concentration map.

Figure 17. Target current area scan + target current concentration map showing the 50% Ta contour.

shows the conventional representation, and Figure 15 a three-tone concentration map of the same area.

Figure 1 shows an electron backscatter image of three niobium spheres coated with tantalum and then partially diffused by heat treatment. An area showing three such spheres by a multiple slit concentration map (Figure 16) shows both the 0% Ta and 100% Ta zones, white, and five intermediate zones of approximately equal widths marking the diffusion zones.

In many cases, particularly in binary systems, the concentration mapping technique can be applied advantageously to target current. This permits a faster readout and more

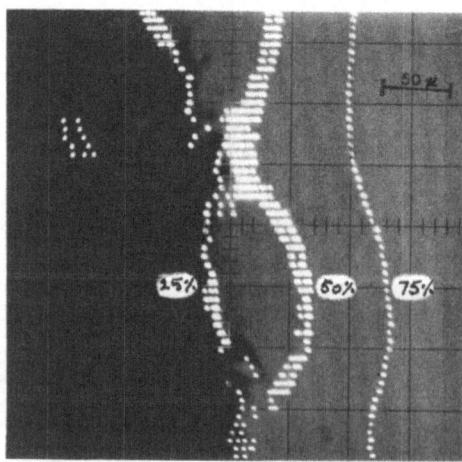

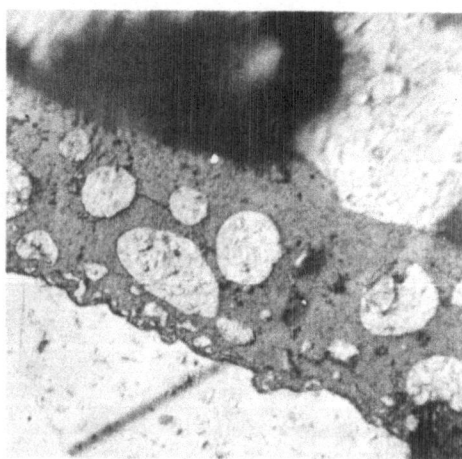

Figure 18. Ta–Ti diffusion. Backscatter area scan + triple slit concentration map with 25% Ta, 50% Ta, and 75% Ta contours.

Figure 19. Area of diffusion. Bi–brass micrograph.

precise delineation. (A complete X-ray concentration map as shown on Figures 15 and 16 requires approximately 30 min, while a similar target current scan is obtained in 5–10 min.) Moreover, the absence of complex corrections, as necessary with X-rays, greatly simplifies the proper setting of the concentration levels to be mapped. Figure 17 shows a conventional target current image of the above-described niobium spheres, with the 50/50% target current concentration map superimposed. Figure 18 shows the area of contact of pieces of tantalum and titanium, respectively, after heat treatment. The background picture is an electron backscatter area scan image, and a target current concentration map showing the 25, 50 and 75% Ta levels superimposed. At two points, the diffusion has been hindered by lack of contact between the two metals. While in the cases described above, the concentration mapping and the background scan have been done successively, the two operations could be done simultaneously, with very minor modifications of the equipment.

CATHODOLUMINESCENCE

The irradiation of many types of samples with electrons produces luminescence of varied colors, and observation of this phenomenon has proven useful, particularly in the study of minerals.[4] Photographic records of photoluminescence can be obtained through the optical microscope of the electron probe, but the quality of these images is generally poorer than desirable. It is, however, possible to get images of good resolution by detecting the luminescence with a photomultiplier tube, and represent the conveniently amplified signal on the oscilloscope in a fashion analogous to the conventional electron backscatter image. In instruments that are provided with electron backscatter scintillation detectors (or any other photomultiplier detector), no additional electronics are necessary.

In the ARL microprobe, the photomultiplier tube can be mounted in the place normally occupied by the illuminator lamp housing, and the connections normally used for the nondispersive X-ray scintillation counter are employed. The signal is amplified with the aid of a dc amplifier, as one does with target current or backscatter signals. Provisions for light filters in the luminator housing can be used for selective reception of color.

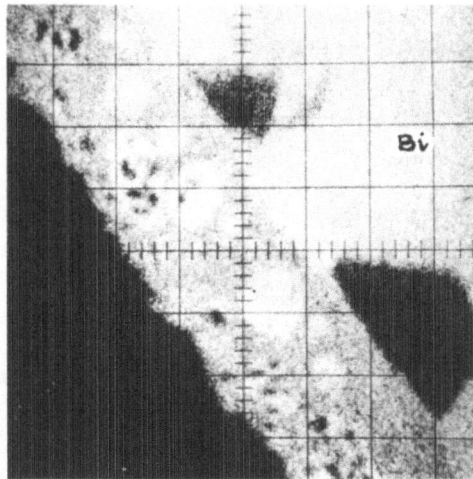 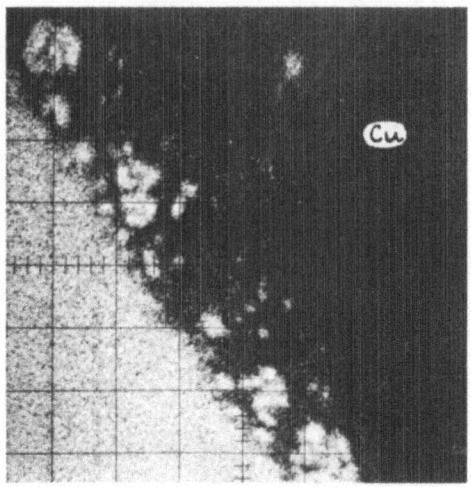

Figure 20. Same sample. Bi M_{α_1}.   Figure 21. Same sample area. Cu K_α.

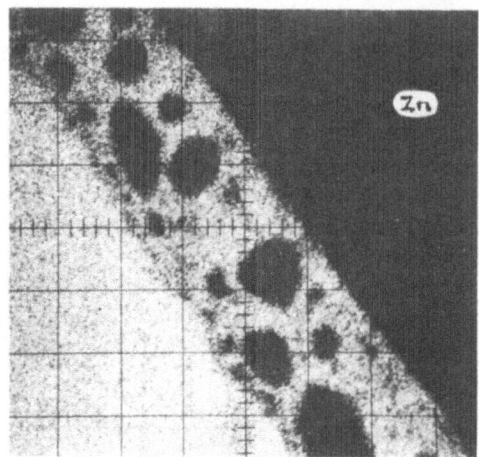

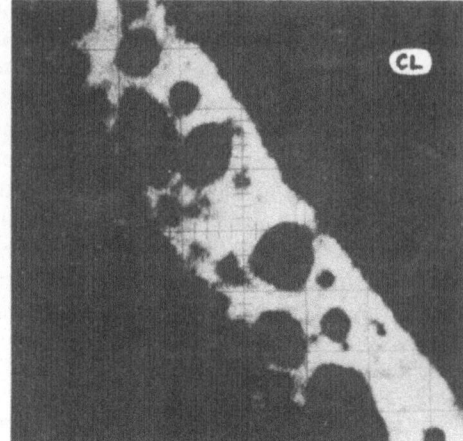

Figure 22. Same area. Zn K_α.   Figure 23. Same sample. Cathodoluminescence.

With such an arrangement, the location of any point of the cathodoluminescence image can easily be found on any other area scan, with the aid of the oscilloscope grating, and a wider range of magnification is available. Multiple exposures can also be used to advantage.

One possible application of this method is the registration of certain oxides in metals and alloys. Zirconium and thorium oxide, for instance, luminesce very intensely and even minute particles can be located easily in this manner.

Figures 19–23 show an alloy consisting of zinc and bismuth formed by diffusion of bismuth into brass.

The area outlined by cathodoluminescence (Figure 13) checks closely with this alloy phase, as visible on the Zn K_α area scan (Figure 22).

It is to be expected that further exploration of this field will result in the discovery of more applications of cathodoluminescence in the microprobe.

REFERENCES

1. D. M. Poole and P. M. Thomas, *J. Inst. Metals* **90**: 228, 1961–62.
2. D. A. Melford, *J. Inst. Metals* **90**: 217, 1962.
3. K. F. J. Heinrich, *Rev. Sci. Instr.* **33**(7): July 1962.
4. J. V. P. Long, M. I. T. Summer Course 1960 (unpublished).

PRACTICAL APPLICATIONS OF FILTERS IN X-RAY SPECTROGRAPHY

Merlyn L. Salmon

Fluo-X-Spec Laboratory
Denver, Colorado

ABSTRACT

Applications of filters are routine and useful techniques in X-ray diffraction and can also be useful in X-ray spectrography to improve analytical results with very simple procedures. Fankuchen demonstrated the use of filters over the window of the X-ray tube to minimize background from the target element and/or elements in other components of the X-ray tube in research programs at the X-ray Laboratory, Metallurgy Division, Denver Research Institute during the summer of 1952. This general procedure has been adapted to routine analyses by modification of the spectrograph to provide for movement of filters in and out of position between the window of the X-ray tube and the sample while the instrument is in operation. Placement of filters in the X-ray beam path of the spectrograph between the sample and the analyzing crystal is also a useful procedure to reduce interferences by elements exhibiting lines at closely adjacent wavelengths.

INTRODUCTION

Filtering the Output of the X-Ray Tube

The spectra obtained from an X-ray spectrographic examination of a specimen may include X-rays emitted by the target element of the X-ray tube and other elements in components of the X-ray tube in addition to X-rays emitted by elements contained in the specimen.

Earlier vintages of X-ray tubes exhibited high levels of emission from several elements in addition to the target element and some of the early research and development work done in the X-ray Laboratory, Metallurgy Division, Denver Research Institute was concerned with interferences of these high levels of unwanted background with indicated intensities of X-rays from elements to be determined in the sample analysis.

Fankuchen[1] devised a scheme of using filters over the window of the X-ray tube to reduce the background from the target element and/or elements in other components of the X-ray tube. The filters were selected on the basis of a particular absorption edge to reduce the intensities of unwanted spectra from the X-ray tube and these filters were selected from a variety of thin metal foils or mixtures of metal oxides prepared in thin layers with Duco cement as a binder.

Improvements in X-ray tube quality have greatly reduced the levels of emitted radiation from elements other than the target element to equivalent concentrations of a few parts per million in most cases; however, there are instances when even the low levels of unwanted background create difficulties in analyses for low concentrations.

A general program for qualitative, semiquantitative analyses in this laboratory is

---

[1] Superscripts pertain to references at the end of the paper.

commenced with preliminary evaluation of the sample by study of an automatic chart recording of the wavelength range to include spectra of all elements with atomic numbers 22 and above in the periodic table. This chart is obtained with a lithium fluoride analyzing crystal and a tungsten target X-ray tube to achieve a good overall balance of intensity and resolution for this wavelength region. The preliminary examination is done with goniometer scanning rates of 8, 16, or 32 deg/min depending on the interest in low concentrations and desired precision of experimental data.

The chart record includes regions of interference of the tungsten lines at wavelengths of analytical lines for selenium, rhenium, mercury, germanium, gold, platinum, arsenic, and tantalum as well as low-level backgrounds for copper, nickel, iron, and chromium resulting from spectra emitted by the tube.

Figure 1 shows a comparison of the intensities of the tungsten lines scattered from a silica sample without a filter (upper trace) and with a filter over the window of the X-ray tube (lower trace). In this case a filter of one layer of 0.001-in. (1-mil) thick shim brass plus a layer of 1 mil nickel foil was used. The copper and zinc content in the brass causes high absorption of the L_β and L_γ tungsten lines while the nickel causes high absorption of the L_α tungsten lines. A background of copper radiation in the lower trace results from excitation of copper in the filter. The use of this filter demonstrates complete absorption of the tungsten radiation from the X-ray tube; however, this effect is accompanied by substantial reduction in excitation capability of the output beam of the X-ray tube, which in turn lowers the intensities of X-rays from elements contained in the sample. It will be shown that proper choice of a filter yields a suitable balance in the degree of absorption of the tube spectra with the degree of excitation by the filtered X-ray beam from the tube.

Figure 2 shows the spectra obtained from a synthetic standard sample containing 1000 ppm each of selenium, arsenic, and germanium compared to the spectra from the

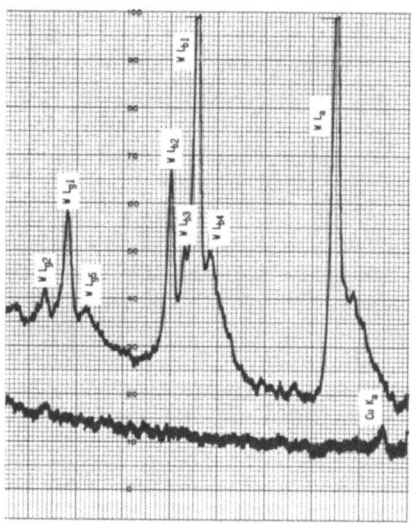 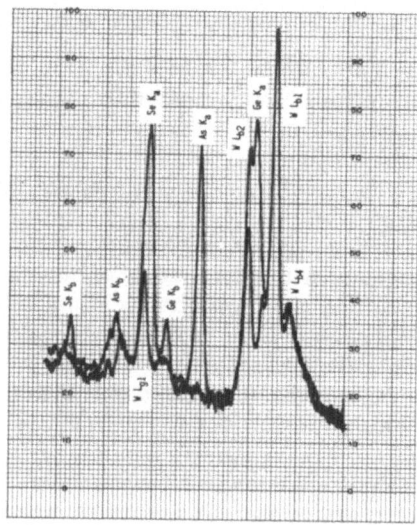

Figure 1. Silica sample. Full-scale intensity = 3200 counts/sec. Upper—no filter. Lower—filter of 1-mil shim brass plus 1-mil nickel foil over X-ray tube window.

Figure 2. No filter. Full-scale intensity = 3200 counts/sec. Upper—synthetic standard sample containing 1000 ppm each of selenium, germanium, and arsenic. Lower—blank matrix sample of synthetic standard.

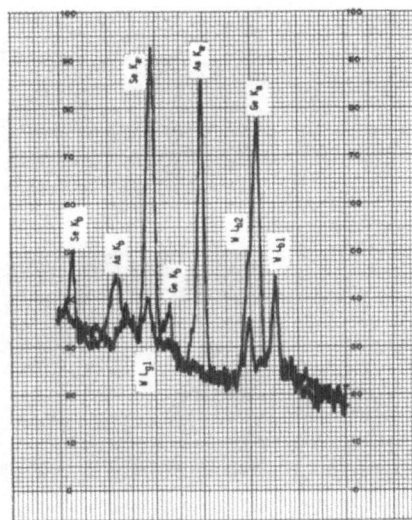

Figure 3. A 1-mil nickel foil filter. Full-scale intensity = 1600 counts/sec. Upper—synthetic standard sample containing 1000 ppm each of selenium, germanium, and arsenic. Lower—blank matrix sample of synthetic standard.

blank matrix of the synthetic standard. The interferences of the tungsten lines indicate unmeasurable backgrounds in routine analysis of unknown samples since precise blank matrix samples are not generally available for practical application.

Figure 3 shows the effects of using a 1-mil nickel foil over the window of the X-ray tube. Spectra from the same synthetic standard sample and blank matrix are compared. The interferences are substantially reduced by attenuation of the tungsten line intensities to about $\frac{1}{7}$, while the intensities of the analytical lines are attenuated to only about $\frac{1}{2}$ the intensities shown in Figure 2 when no filter was used.

A trial program for qualitative, semiquantitative analyses was studied with Fankuchen's scheme, and a filter was continuously used over the window of the X-ray tube for the complete wavelength range of the routine chart-scan technique. This caused reduction in the observed X-ray intensities for all elements in the sample in gaining the advantage of lower background levels of intensity of the tube spectra. A series of methods of studying portions of the chart with and without the filter was tried but additional sample handling was required and it was necessary to disrupt operation of the instrument to add or remove the filter. The methods were impractical for routine use and the need for a mechanism to move the filter in and out of position over the window of the X-ray tube with no loss of operating time was obvious.

The spectrograph was modified to add a movable filter frame that can be rotated in and out of position over the window of the X-ray tube while the spectrograph is in operation. With this modification, removal of some of the material inside the spectrograph tube-sample housing was required, and a rotatable shaft extending through the wall of the housing was installed. The filter frame is attached to the shaft inside the housing, and an index knob on the outside end of the shaft indicates the position of the filter frame with respect to the window of the X-ray tube. The frame is made of 1100 aluminum alloy and is constructed so that the filter material can be easily changed in the frame while the X-ray tube is removed from the housing. When the filter frame is rotated in or out of position over the window of the tube, there is no blockage of the path between the sample face and the aperture of the exit collimator.

The modification is summarized in Figure 4 showing the normal operating condition with the X-ray tube window not covered inside the spectrograph housing (upper

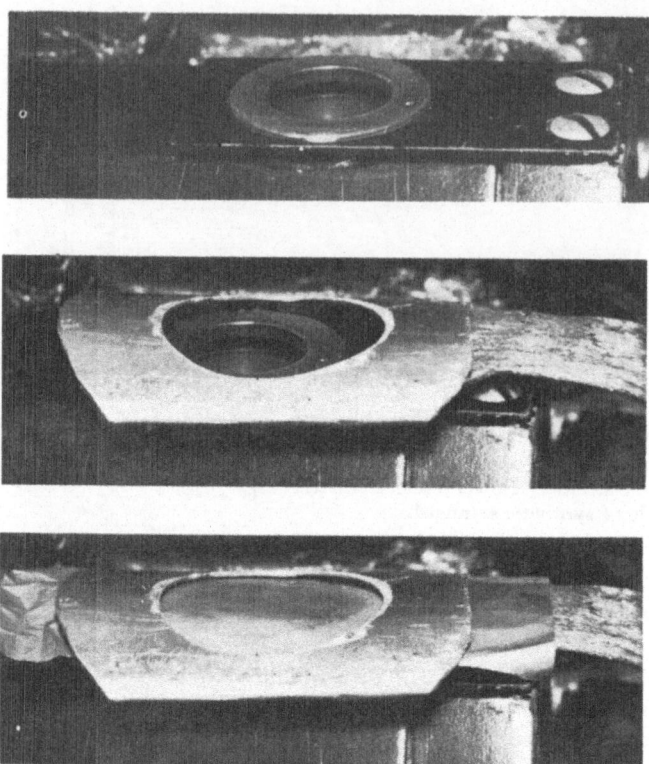

Figure 4. Spectrograph modification. Upper—normal operating condition with no filter over the X-ray tube window. Middle—movable filter frame in position over window of X-ray tube. Lower —nickel foil inserted in filter frame.

view); the movable filter frame rotated into position over the window of the X-ray tube (middle view); and a nickel foil inserted in the filter frame (lower view).

Filtering Adjacent Wavelengths in the Fluorescent X-ray Beam

The spectra obtained from an X-ray spectrographic examination of a specimen may include X-rays of nearly equal wavelengths that are characteristic of two different elements in the sample. In some cases of interference at these adjacent wavelengths, it is possible to improve the apparent resolution by use of an absorption filter to preferentially absorb the radiation with the shorter wavelength. The composition of the filter is critical in this application since the absorption edge of the filter material must be at a wavelength between the wavelengths of the interfering lines. The filter can be placed in the fluorescent X-ray beam between the sample and the analyzing crystal or between the analyzing crystal and the detector.

PROCEDURE

Instrumental Conditions

A modified 100-kv Norelco X-ray spectrograph was used, and the reported results were obtained with a tungsten target FA-100 X-ray tube; a scintillation counter; a lithium fluoride analyzing crystal a 2 × 0.023 in. spacing parallel-blade collimator between the analyzing crystal and the detector; a 4×0.005 in. spacing parallel-blade collimator between the sample and the analyzing crystal; and filters prepared from metal foils and/or mixtures of powder materials and silicone fluid. The X-ray tube was operated at 50 kv and 28 ma and the detector at 900 v. A scanning rate of 8 deg/min was used with a time constant of 0.6 sec and the indicated scale factor adjustments for the chart recordings. The CaPlug sample container system and the sample spinner were both used for all determinations. There was an air atmosphere in the optical path of the spectrograph.

Filter Preparation

Thin metal foils are the most convenient for preparation of filters; however, the ranges available in composition and thickness of desired materials are limited.

A procedure for making thin layers of a mixture of powders and silicone fluid has been described[2] and a variety of filters can be prepared in this manner. The thin layers are mounted on an EC-16 CaPlug with 1-in. curtain rings or on a frame of two matching nylon rings that hold the filter in the style of embroidery hoops. The latter are available from Technical Equipment Corporation, Denver, Colorado.

Preliminary determinations of desirable filter materials can be made by noting absorption edges and absorption coefficients at different wavelengths prior to experimental evaluation of the materials.

Three important considerations in filter preparation from usable materials are: the degree of attenuation of unwanted radiation; the residual intensity of wanted radiation after sufficient reduction of the interference condition has been accomplished; and possible unwanted contributions to background intensity due to excitation of elements in the filter.

RESULTS AND DISCUSSION

Filtering the Output of the X-ray Tube

Results for various metal foils over the X-ray tube window are demonstrated by comparisons of spectra from the same samples with and without the filters. As well as indicating the properties of the different filters, these results also show justification for design of a filter frame that has a variety of filters available for use at times during sustained operation of the spectrograph.

Nickel foil is available in several thicknesses and the results in Figure 5 for a 1-mil foil show attenuation of the tungsten line intensities to $\frac{1}{3}$ to $\frac{1}{12}$ the unfiltered line intensities scattered from a silica sample. The K absorption edge of nickel is evident just slightly to the high-wavelength side of the W L_α peak in the lower trace and some Ni K_β background is also noticeable from excitation of the filter material.

Figure 6 indicates the reduction of interferences from the tungsten lines with more than one-half the original analytical line intensities for arsenic, germanium, and selenium shown with filtering the output of the X-ray tube by a 1-mil nickel foil. The approximate sensitivity ratio is 1 ppm/count per sec for germanium, arsenic, and selenium with the filtered output of the X-ray tube.

The effects of a thicker filter with lower absorption coefficients are shown in Figure 7 for a silica sample and 10 layers of 1-mil aluminum foil as the filter. The tungsten line intensities are reduced slightly less than the reduction with the 1-mil nickel, but a more desirable background condition is noted on the long wavelength side of the W L_α peak than is shown with the nickel filter. This is especially important in analyses for low concentrations of tantalum (10 to 100 ppm). There is also no noticeable background indication due to excitation of the filter material. Unwanted background intensities of copper,

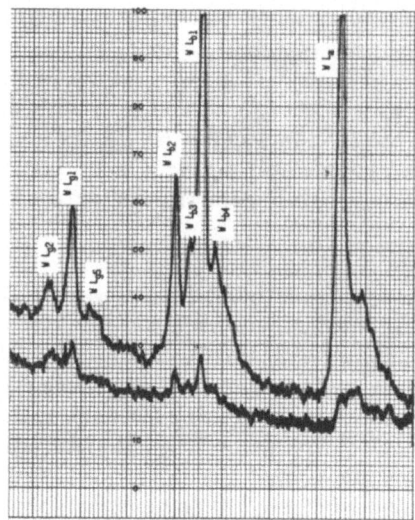 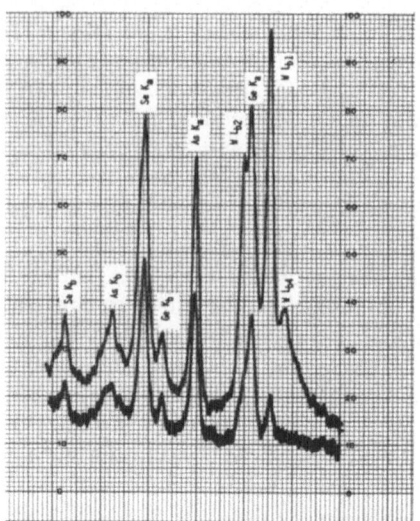

Figure 5. Silica sample. Full-scale intensity = 3200 counts/sec. Upper—no filter. Lower—a 1-mil nickel filter over X-ray tube window.

Figure 6. Synthetic standard sample containing 1000 ppm each of selenium, germanium, and arsenic. Full-scale intensity = 3200 counts/sec. Upper—no filter. Lower—a 1-mil nickel foil filter over X-ray tube window.

nickel, iron, and chromium emitted by the X-ray tube are absorbed by the 10-mil aluminum filter with reduction of the analytical line intensities of these elements to about $\frac{1}{3}$ to $\frac{1}{2}$ the intensities with no filter used. The use of the 10-mil aluminum over the window of the X-ray tube assists in determinations of ppm concentrations of copper, nickel, iron, and chromium in routine unknown samples.

It is evident in Figure 8 that the 10-mil aluminum filter suppresses emission from the tube to reduce sufficiently the interferences from tungsten lines, and that the analytical lines for selenium, arsenic, and germanium exhibit intensities in excess of 50% of the intensities shown when no filter is used.

The effects of a 1-mil shim brass filter are indicated in Figure 9. The higher absorption coefficients of zinc and copper for the L_β and L_γ lines of tungsten are reflected in the complete absorption at these wavelengths in contrast to the lower absorption coefficients and significant intensity at the $W L_\alpha$ wavelength. There are also indications of zinc and copper in the background of the filtered beam as a result of excitation of these elements in the filter.

As shown in Figure 10, the intensity levels of the analytical lines for selenium, arsenic, and germanium are reduced to about $\frac{1}{3}$ the intensity levels with no filter when a 1-mil shim brass filter is placed over the window of the X-ray tube. The sensitivity ratio is about 2 ppm/count per sec for these elements with this procedure.

The composite absorption properties of the brass, nickel, and aluminum cause complete absorption of all tungsten lines as shown in Figure 11. The aluminum foil on the outer faces of the filter minimizes background indications of zinc, copper, and nickel due to excitation of these elements in the filter.

The results in Figure 12 indicate that the analytical line intensities for selenium, arsenic, and germanium are reduced to about $\frac{1}{4}$ the intensity with no filter when the

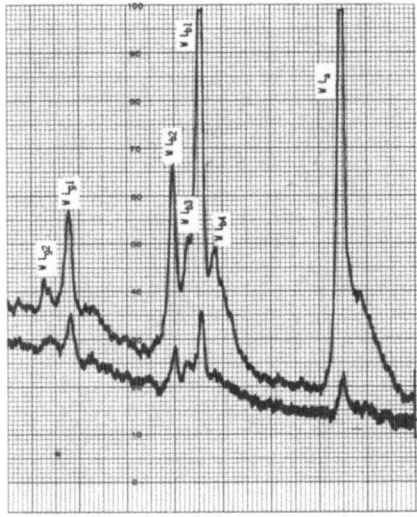

Figure 7. Silica sample. Full-scale intensity = 3200 counts/sec. Upper—no filter. Lower—a 10-mil aluminum foil filter over X-ray tube window.

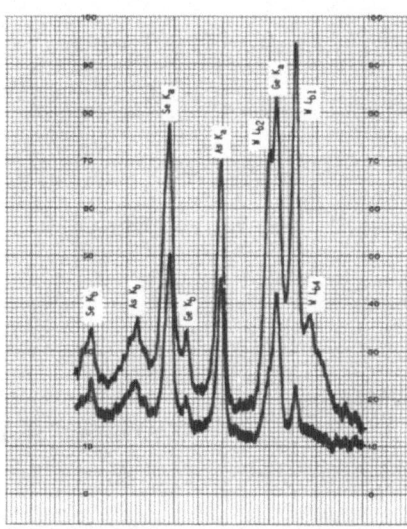

Figure 8. Synthetic standard sample containing 1000 ppm each of selenium, germanium, and arsenic. Full-scale intensity = 3200 counts/sec. Upper—no filter. Lower—a 10-mil aluminum foil filter over X-ray tube window.

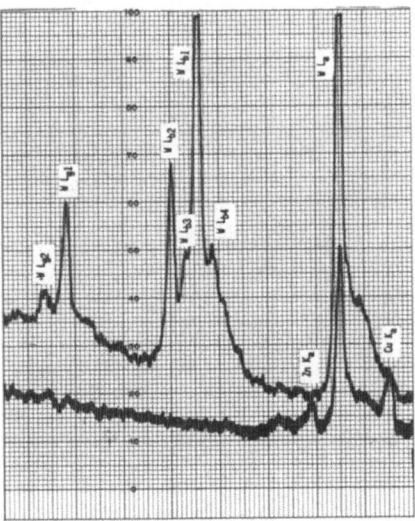

Figure 9. Silica sample. Full-scale intensity = 3200 counts/sec. Upper—no filter. Lower—a 1-mil shim brass filter over X-ray tube window.

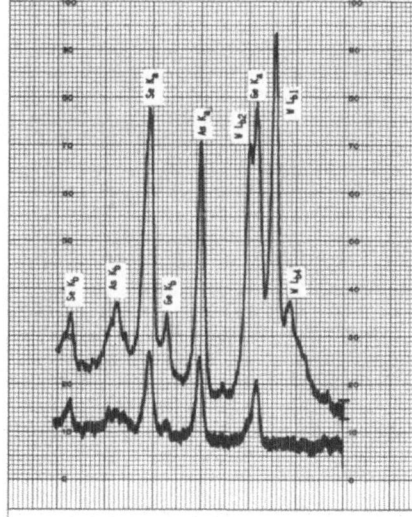

Figure 10. Synthetic standard sample containing 1000 ppm each of selenium, germanium, and arsenic. Full-scale intensity = 3200 counts/sec. Upper—no filter. Lower—a 1-mil shim brass filter over X-ray tube window.

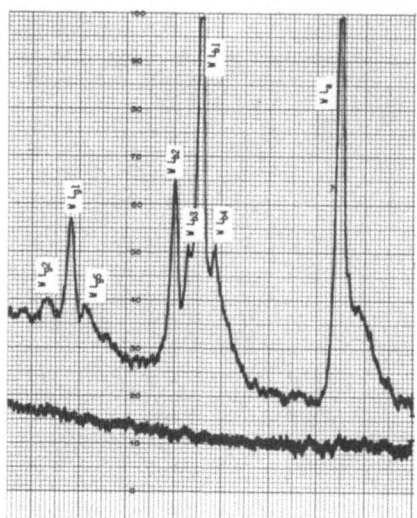

Figure 11. Silica sample. Full-scale intensity = 3200 counts/sec. Upper—no filter. Lower—a filter of 1-mil nickel foil plus 1-mil shim brass sandwiched between 1-mil aluminum foil on each face over X-ray tube.

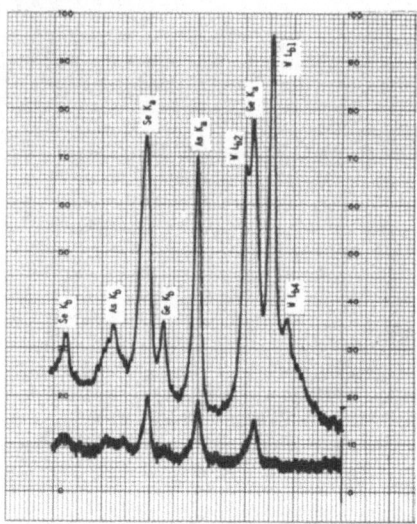

Figure 12. Synthetic standard sample containing 1000 ppm each of selenium, arsenic, and germanium. Full-scale intensity = 3200 counts/sec. Upper—no filter. Lower —filter of 1-mil nickel foil plus 1-mil shim brass sandwiched between 1-mil aluminum foil on each face over the X-ray tube window.

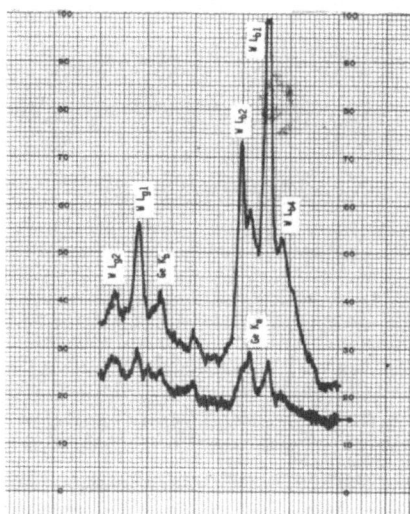

Figure 13. Synthetic standard sample containing 300 ppm germanium. Full-scale intensity = 3200 counts/sec. Upper—no filter. Lower—a 1-mil nickel foil filter over the X-ray tube window.

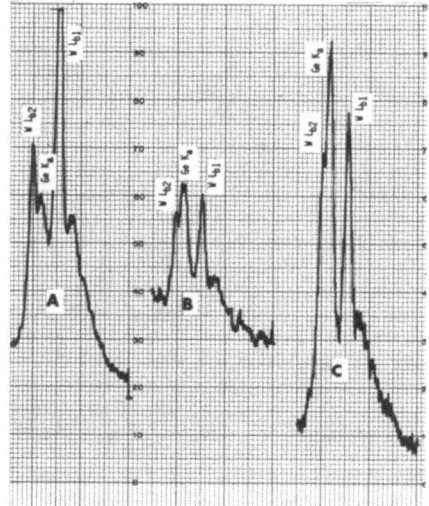

Figure 14. Synthetic standard sample containing 300 ppm germanium. A—full-scale intensity = 3200 counts/sec with no filter. B—full-scale intensity = 1600 counts/ sec with 1-mil nickel foil filter over X-ray tube window. C—full-scale intensity = 400 counts/sec with 1-mil nickel foil filter over X-ray tube window and application of electronic discrimination.

sandwich filter of 1-mil nickel foil, 1-mil shim brass and two layers of 1-mil aluminum foil is used. The sensitivity ratio is approximately 3 ppm/count per sec for the selenium, arsenic, and germanium in the sample.

The upper trace in Figure 13 indicates the difficulty of even the qualitative identification of the 300-ppm concentration of germanium in the sample when no filter is used to reduce the interferences from tungsten lines. The lower trace not only allows for positive identification of the germanium but also provides a measurable intensity of Ge K_α radiation that can be converted to an estimated concentration of 350 ppm when a 1-mil nickel foil is used as a filter.

In Figure 14 the results with no filter, a 1-mil nickel filter, and a 1-mil nickel filter plus electronic discrimination are compared to indicate the information that can be displayed on a chart record for use in the qualitative, semiquantitative analysis of the sample for the 300-ppm concentration of germanium. With the technique using the filter plus electronic discrimination, it is possible to observe a sensitivity ratio of about 4 ppm/chart unit for direct conversion of the Ge K_α peak height to concentration. Analytical determinations for the element used as the target of the X-ray tube in routine examination of unknown samples can also be accomplished with the use of filters over the window of the X-ray tube.

The instance of a blank matrix comparison for a tungsten determination is shown in Figure 15 with the spectra for a sample of silica matrix containing 1% WO_3 compared to the spectra of the blank silica matrix. The intensities of the scattered tungsten X-rays emitted by the tube account for greater than 50% of the total intensities for the scattered X-rays plus those due to the 1% concentration of WO_3 in the sample. With few exceptions, routine unknown samples are difficult to evaluate by blank matrix comparisons with practical procedures.

In Figure 16 the results for use of a 10-mil aluminum filter show substantial reduction

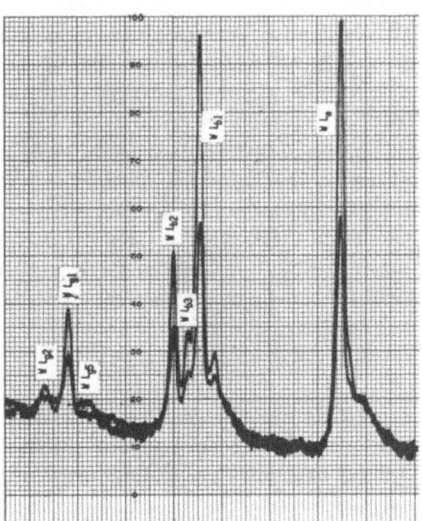 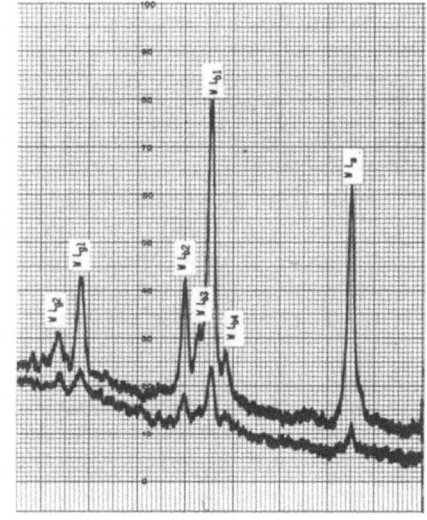

Figure 15. No filter. Full-scale intensity = 6400 counts/sec. Upper—silica matrix containing 1% WO_3. Lower—blank silica matrix.

Figure 16. A 10-mil aluminum foil filter. Full-scale intensity = 3200 counts/sec. Upper—silica matrix containing 1% WO_3. Lower—blank silica matrix and suppression of zero by five negative chart units.

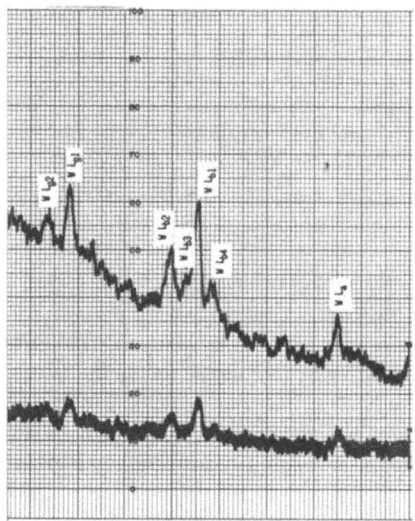

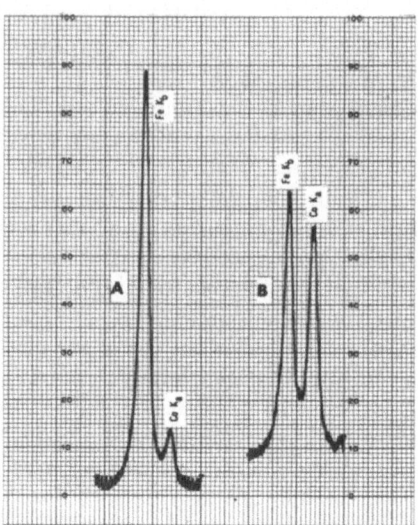

Figure 17. A 10-mil aluminum foil filter. Full-scale intensity = 1600 counts/sec. Upper—silica blank sample. Lower—ferric oxide blank sample.

Figure 18. Sample containing approximately 8% iron and 0.3% cobalt. A—full-scale intensity = 25600 counts/sec, no filter. B—filter of europium oxide in silicone fluid between sample and analyzing crystal in fluorescent X-ray beam, full-scale counting rate = 3200 counts/sec.

of the scattered tungsten lines from the tube with only a 50% reduction of the tungsten X-ray intensities due to the 1% WO_3 in the sample.

To determine blank tungsten intensities without complete absorption of the tungsten radiation emitted by the tube, it is possible to correlate the tungsten background levels with the blank scattered background levels at the wavelengths of the tungsten lines when blank samples with different absorption properties are studied with use of the filter. Two extremes of this general correlation are shown in Figure 17 by comparisons of peak heights of the tungsten lines *vs.* scattered background intensity levels for the silica and iron oxide blank samples. Curves correlating blank tungsten peak intensities *vs.* scattered background intensity levels are prepared for use in the analysis of unknown samples. Measurement of the background intensity levels can be converted to peak intensities for zero tungsten concentrations to be deducted from measured tungsten peak intensities used in semi-quantitative determinations of the presence of tungsten in the sample.

Filtering the Fluorescent X-ray Beam

Proper selection of the filter material can provide for apparent resolution of peaks at closely adjacent wavelengths. The filter is chosen with an absorption edge between the two peaks to cause preferential absorption of the peak with the shorter wavelength.

In practical applications, this procedure is useful when the intensity of the peak at the shorter wavelength is so great that the peak with lower intensity at the longer wavelength is partially or completely obscured on the chart record.

A common example is shown in Figure 18 with a sample containing approximately 8% iron and 0.3% cobalt. The interference of the Fe K_β peak with the Co K_α peak is evident when no filter is used. A filter of europium oxide in silicone fluid is chosen because the L-III absorption edge of europium is at a wavelength between the wavelengths of

Fe K_β and Co K_α radiation. The change in ratio of the two peaks is shown due to preferential absorption of the Fe K_β radiation by the europium oxide filter to demonstrate an improvement in apparent resolution of the wavelengths different by 0.03 Å. The sensitivity ratio for cobalt without use of the filter is about 1 ppm/count per sec and with the use of the filter it is approximately 2 ppm/count per sec. The one-order change of magnitude in sensitivity ratio due to use of the filter is inconsequential in comparison to obscurity of the Co K_α peak for low concentrations in samples containing high concentrations of iron. The use of the filter in the fluorescent X-ray beam makes it possible to complete many other similar determinations that would be much more difficult if not impossible to complete by other practical means.

CONCLUSIONS

The use of filters in X-ray spectrography is a simple way in which to improve the versatility of the instrument. A filter fixed over the window of the X-ray tube during the complete period of sustained operation of the instrument causes suppression of the indicated intensity of the tube spectrum to minimize interferences with X-rays emitted by elements in the sample. The fixed filter, however, does also cause a reduction in observed analytical line intensities for wavelength regions where the tube spectra cause no interferences.

Modification of the instrument is justified to provide a movable filter frame over the window of the X-ray tube to allow for rotation of the filter in and out of position at times during sustained operation of the spectrograph with no loss of operating time. On the basis of results with the single-frame filter, it is evident that it would be desirable to develop a multiple-frame filter to allow a choice from several different filters for intermittent use during operation of the instrument.

The choice of filter material is based on consideration of the following: the degree of attenuation of unwanted radiation; the residual intensity of wanted radiation after sufficient reduction of the interference condition has been accomplished; and possible unwanted contributions to background intensity due to excitation of elements in the filter.

A filter in the fluorescent X-ray beam is used to improve the apparent resolution of peaks at adjacent wavelengths by preferential absorption of the shorter-wavelength region.

ACKNOWLEDGMENTS

Appreciation is expressed to Herb Ochs, Frank de Rose, and Roger Harper for their assistance in supplying thin metal foil samples for evaluation as filters.

REFERENCES

1. Unpublished results, Basic Research Program, X-ray Laboratory, Metallurgy Division, Denver Research Institute, University of Denver, June–September 1952.
2. M. L. Salmon, "A Simple Multielement-Calibration System for Analysis of Minor and Major Elements in Minerals by Fluorescent X-ray Spectrography," *Advances in X-ray Analysis, Vol. 5*, University of Denver, Plenum Press, New York, 1962, p. 389.

DISCUSSION

P. Wittig (Haynes-Stellite Co.): How did you fasten the mylar to the ring?

M. L. Salmon: Actually, I didn't show sufficient detail. There is another ring hidden underneath, and these rings slip over one another like embroidery hoops. We just make a sandwich and slip the ring over it and bolt it in place. It is difficult to get an adhesive to work well. The main thing is a tight fit of the mylar over the foil.

C. Lloyd (IBM Research): An interesting application of this is to the Siemens apparatus. The tungsten tube in this setup contains copper contamination, which makes it a little bit difficult if you are interested in trace copper. There is a primary beam-limiting drawer which just fits between the tube and the sample chamber. If one merely tapes in a piece of appropriate filter material (I think cobalt is the one that works best or maybe nickel; I'm not quite sure) one gets a really tremendous boost. The intensity of copper is hardly cut at all, so it makes it useful for trace analysis, whereas before you simply could not do it. If one wants to make something up from scratch, a slurry of Dow Ethocel in alcohol sloshed around on a microscope slide and then peeled from the slide will work. It's sloppy and crude, but for a first attempt it works rather nicely.

M. L. Salmon: Does it stand up in the X-ray beam?

C. Lloyd: I only use it on a one- or two-shot basis, so whether it does or not doesn't make much difference.

C. N. Schieltz (Colorado School of Mines): How do you select the material for certain filter applications?

M. L. Salmon: We use nickel foil for the tungsten tube because (1) it is easily available in 1-mil thickness; (2) it can be stacked if you want to use thicker filters; and (3) it has a suitable absorption edge; besides it doesn't get all of the Compton scatter. As I have just pointed out, if there is copper contamination in the tube, it is in this Compton scatter region and it is difficult to resolve it. Nickel works very nicely for tungsten, but with a little aluminum you can do the same thing without involving an edge. I think it takes 5 mils of aluminum to get a decent reduction in the tungsten level. Actually, the choice is based first, on what is easily available and then on what will provide the proper absorption. If silver is used, for example, there is the added advantage that the absorption edge is way out at the shorter wavelength end and it doesn't allow you too much in the longer-wavelength elements. Even a thin layer of nickel will reduce the excitation level of the lighter elements. If you don't want to use pulse-height analysis, but instead go to the second order of tungsten lines that interfere with, for example, cerium and barium, nickel will still serve, but actually, it's just hit and miss, based on the anticipated absorption properties of the material.

C. N. Schieltz: I think that carefully studying wavelength absorption tables, like those in the back of Pollack's book, can prove very useful in examining the absorption figures.

M. L. Salmon: Oh, you certainly should refer to tables. However, Frank de Rose happened to have a 1-mil tantalum foil that he had gotten somewhere, and we tried it, thinking that tantalum would be a very excellent absorber for tungsten. It was too good. It took all the tungsten out, even though it was only 1-mil thick, and it reduced the excitation level of everything by a factor of about 20 or 30.

C. N. Schieltz: I would like to make another comment here. If you take the absorption formulas and calculate, say, how much copper characteristic radiation you are going to cut out by $\frac{1}{4}$ in. of aluminum plate, you come up with something like 10^{-39} of the original—some such figure—but actually if you put the copper on copper tube with a $\frac{1}{4}$-in. plate, there are enough X-rays coming through there so that you can, at the standard operation, see the direct line in 1 sec. You wouldn't believe it, but it is there. That is what I use when I make up my step filter standard for getting the intensities of lines in the diffraction pattern.

M. L. Salmon: I want to mention the one thing I did miss on these powder filters when we prepared them. We have a series of microscope slides with holes drilled in them. We fill the cavity —it's a precalculated thickness—and then weigh out the desired amount so we have a mask for a square centimeter type of figure for the filament.

THE DAILY USE OF A BASIC NORELCO X-RAY SPECTROGRAPH IN AN ALUMINUM REDUCTION LABORATORY

W. B. Eastman

Kaiser Aluminum and Chemical Corporation
Ravenswood, West Virginia

ABSTRACT

Three basic methods for the daily use of the X-ray spectrograph are presented with the object of demonstrating simplicity of operation, reliability of results, and ease of use of the basic Norelco X-ray spectrograph by personnel not highly trained in the operation of such equipment. The methods to be outlined briefly are: (1) Determination of calcium fluoride in reduction cell electrolyte; (2) Analysis of cast iron for Si, P, Mn, S; and (3) Analysis of carbon materials for S, K, Ca, V, Mn, Fe, and Ni. The mechanics of the methods will be outlined with the reasons for their selection, and the operating and maintenance problems which have been encountered will be discussed.

INTRODUCTION

In July 1959, a basic Norelco X-ray spectrograph was installed in the Kaiser Aluminum and Chemical Corporation Ravenswood Reduction Laboratory in Ravenswood, West Virginia. Since that time this equipment has been operated by a routine analyst on an almost daily basis. We have had two analysts in this period and neither had formal analytical training other than that received on the job. With these thoughts in mind, the types of analytical work being done, the analytical methods used, and the problems encountered may be of interest to persons contemplating the use of this type of equipment in similar circumstances. Briefly, then, I will discuss three routine analytical methods being used, personnel operating the equipment, and a maintenance history of this particular instrument.

ANALYTICAL METHODS

The analytical methods being carried out daily with a minimal amount of supervision are:

1. The routine determination of calcium in bath (reduction cell electrolyte).
2. The analysis of cast iron for Si, P, S, and Mn.
3. The analysis of carbon materials for S, K, Ca, V, Mn, Fe, and Ni.

Fixed instrument conditions for the above determinations are as follows:

1. Norelco basic X-ray unit with X-ray spectrograph attachment, Type No. 52260, Pulse-height analyzer-timer-amplifier, Type No. 52332, and Timer-amplifier unit, Type No. 52333.
2. FA-60 Tungsten target tube.
3. 0.020 × 1 in. entrance collimator and 0.020 × 4 in. exit collimator.

Table I. Routine X-Ray *vs.* Average of Duplicate EDTA CaF$_2$ Analyses

				% CaF$_2$					
EDTA	Routine X-ray	EDTA	Routine X-ray	EDTA	Routine X-ray	EDTA	Routine X-ray
8.94	8.8	8.74	8.7	8.09	8.1	8.60	8.6
9.12	9.0	7.85	7.8	9.10	9.1	8.58	8.6
8.30	8.3	8.65	8.7	10.30	10.2	8.90	8.8
8.99	9.0	7.54	7.5	7.58	7.6	9.30	9.5
8.42	8.3	8.34	8.4	8.90	8.9	8.98	9.0
7.44	7.4	6.98	6.9	8.28	8.2	8.74	8.7
8.04	8.1	8.24	8.1	8.05	8.1	8.28	8.4
8.50	8.4	7.42	7.5	7.84	7.9	7.36	7.3
8.17	8.2	7.95	8.0	6.61	6.7	8.09	8.1
9.36	9.4	8.66	8.6	8.40	8.2	8.34	8.3
7.46	7.5	9.38	9.4	7.72	7.8	7.96	7.9
8.14	8.1	8.08	8.0	8.32	8.2	10.14	9.8
8.13	8.0	7.21	7.2	7.86	7.9	9.14	8.9
9.14	9.0	7.42	7.3	8.25	8.2	8.54	8.6
8.66	8.6	8.49	8.5	8.34	8.4	9.20	9.2
7.90	7.7	7.71	7.6	9.22	9.2	8.78	8.7
7.10	7.0	7.92	7.8	8.66	8.5	8.88	8.9
8.48	8.4	8.20	8.2	8.63	8.5	9.10	9.1
7.74	7.7	8.72	8.8	8.63	8.9	8.56	8.6
8.36	8.3	8.30	8.4	7.15	7.3	8.98	8.7
8.55	8.4	8.34	8.5	7.17	7.1	8.56	8.6
8.48	8.4	7.75	7.8	8.14	8.2	8.99	9.0
						8.39	8.4
						8.65	8.7

4. Flow proportional counter operated with 2.0 ft^3/hr of P-10 gas flow for a $\frac{1}{2}$-hr warm-up and 0.5 ft^3/hr of gas flow following.[1]

5. Electronics always started up $\frac{1}{2}$ hr prior to use and counting circuits are operated for 10–15 min prior to actual measurements.

Calcium in Bath

This analysis is concerned with levels of CaF$_2$ in the range 6 to 10%. The objective of the method is rapid routine analyses on a large number of samples with reasonable accuracy. The steps of the method are summarized below:

1. Samples are ground on a Bico Braun pulverizer using porcelain plates. These samples are not screened, but checks have shown them to be 95% or more minus 100 mesh.

2. Samples are packed in cold-rolled steel holders* by pouring in an excess, taking off the excess with a coarse screen, and smoothing the surface by a rotary motion with a flat nickel crucible lid.

3. The instrument is peaked on Ca K_α using pulse-amplitude discrimination, LiF crystal, power settings of 21 kv and 25 ma, a fixed time of about 19 sec, and scale factors of 16 × 8, with helium flow set at 10 ft^3/hr. The amperage is then adjusted when

---

[1] Superscripts pertain to references at the end of the paper.

* A large number of steel holders were machined from $1\frac{3}{8}$-in. cold-rolled steel bar and statistically checked for uniformity.

necessary to give the correct reading on a 10.0% CaF_2 standard on the E1T (count register) tubes.

4. The helium path is checked for leakage by taking a series of ten duplicate readings, removing the sample from the path, and reinserting between each set of duplicate readings. The average of first readings will agree statistically with the average of the second readings if conditions are satisfactory and the average of the ten readings will agree with the standard value if the power is set properly.

5. The samples are then run with E1T tube readings being recorded as % CaF_2. The first sample run is used as a reference sample after every fifteenth sample to check instrument stability.

The adequacy of this method of analysis is shown in Table I, which compares routine X-ray results run over a period of several weeks with the average of duplicate EDTA determinations. The X-ray results are typical of what can be expected from routine analysis over a long period of time and include all the normal variations in results from grinding through reporting.

Analysis of Cast Iron for Si, P, S, and Mn

This analytical method is used as an in-plant check on uniformity of cast iron used to hold anodes on steel stubs which, in turn, are bolted to copper rods carrying current to the individual anodes. It was felt that we would be analyzing individual samples daily or a series of samples weekly and that graphical calculations requiring exact conditions being reproduced every day would be too tedious and time-consuming. For these reasons cast iron is analyzed using USBS white cast iron standards as references and results are based on the day-to-day count on these standards using empirical calculations. The steps of this method are listed below:

1. The samples are taken in a copper-core, water-cooled mold[2] or a Spex 3904 mold* modified by drilling to furnish a $\frac{3}{16}$-in. pin for carbon analyses.

2. The samples are polished to a mirror finish, using an 80 grit belt, 240 grit belt and 1/0, 2/0, 3/0, 4/0 emery paper.

3. Individual pulse amplitude discrimination settings are used for Si, P, and S. These settings are not changed unless a marked change in background or net counting rate occurs. Instrument conditions and calculations are listed in Table II. The counting times for the various elements are of the following order: Si—100 sec; P—120 sec; S—130 sec; Mn—42 sec.

4. The samples are analyzed for each element by taking the specified count on the standard background (if required by the calculation) and the standard line, taking the specified count on the samples and then, at the end of a series of samples, running the standard again. A known sample is occasionally included with the daily samples.

5. The percent of each element is then calculated using the equations shown in Table II.

Table III shows results, standard deviations, and confidence limits for cast iron analyses using the X-ray method presented vs. routine chemical analyses of sample drillings, where the samples were analyzed routinely over a four week period. These results contain all the errors present in this procedure from sampling through reporting

---

* The use of two different types of mold resulted from damage to the copper-core mold which was used without water-cooling and was handled roughly. The Spex 3904 mold was much more rugged and easy to handle. However, samples taken with the Spex mold had internal stresses which caused some samples to crack.

Table II. Instrument Conditions and Calculations for the Analysis of Cast Iron

A. Silicon

1180 USBS white cast iron standard (3.04% Si)
EDDT crystal 107.98° 2θ line
              109.00° 2θ background
50 kv 45 ma
1525 counter volts (may be varied to suit PAD requirements)
Count on standard . . .25,600
Count on samples . . . 6,400
Count on background. . . 200 to estimate
                   3,200 if used to calculate net count
Helium—10 ft³/hr

$$\% \ Si = 3.04 \times \frac{\text{Gross cps sample}}{\text{Avg. gross cps std.}} - \frac{(\text{gross cps sample})^2}{100} \times 1.17$$

B. Phosphorus

1182 USBS white cast iron standard (0.85% P)
EDDT crystal 88.66° 2θ line
             90.00° 2θ background
50 kv 45 ma
1525 counter volts (may be varied to suit PAD requirements)
Counts on standard . . .6,400
Counts on samples . . .6,400
Counts on background. . . 200 to estimate
                  3,200 if used to calculate net count
Helium—10 ft³/hr

$$\% \ P = 0.85 \times \frac{\text{gross cps sample}}{\text{Avg. gross cps std.}} + \frac{(\text{gross cps sample})^2}{100} \times 0.528$$

C. Sulphur

1179 USBS white cast iron standard (0.16% S)
EDDT crystal 75.16° 2θ line
             76.50° 2θ background
50 kv 45 ma
1525 counter volts (may be varied to suit PAD requirements)
Counts on standard . . .6,400
Counts on samples . . .3,200
Counts on background . . 1,600
Helium—10 ft³/hr

$$\% \ S = 0.16 \times \frac{\text{Net counts on sample}}{\text{Avg. net counts on std.}} \times 1.18$$

D. Manganese

1179 USBS white cast iron standard (0.64% Mn)
LiF crystal 62.91° 2θ line
30 kv 30 ma (ma adjusted to give (640 ± 4) × 16 × 8 on the 1179 std.)
1525 counter volts integrate
Fixed time—Approximately 42 sec   Count, E1T = 640 ± 4, scale factor 16 × 8
Air path
Instrument is direct reading, i.e., % Mn = E1T/1000

the analysis, including two sampling techniques, i.e., drilling *vs.* use of sample surface. As an in-plant control method, such results are quite acceptable.

Analysis of Carbon Materials for S, K, Ca, V, Mn, Fe, and Ni

In the production of aluminum approximately 0.5 lb of carbon materials are consumed per lb of aluminum produced and the purity of the carbon materials is reflected in operations and the purity of the aluminum produced. Analyses of carbon materials are, consequently, of considerable importance in plant control. In these analyses by X-ray,

Table III. Routine X-ray Analyses *vs.* Routine Chemical Analyses of Sample Drillings Over Four-Week Period

Lab. No.	% Si X-ray	% Si Chem.	% P X-ray	% P Chem.	% S X-ray	% S Chem.	% Mn X-ray	% Mn Chem.
4284	3.03	3.10	1.11	1.16	0.24	0.22	0.67	0.65
4285	2.93	2.94	1.15	1.15	0.17	0.17	0.63	0.62
4286	3.48	3.40	1.18	1.27	0.19	0.17	0.75	0.72
4288	3.46	3.42	1.20	1.22	0.15	0.15	0.78	0.75
4289	3.28	3.24	1.14	1.15	0.19	0.18	0.75	0.67
4290	3.31	3.27	1.15	1.12	0.12	0.13	0.66	0.63
4291	3.31	3.33	1.15	1.12	0.19	0.17	0.75	0.72
4292	3.24	3.24	1.22	1.24	0.13	0.14	0.69	0.65
4293	2.77	2.70	0.78	0.78	0.13	0.12	0.66	0.62
4295	3.13	3.11	1.14	1.11	0.14	0.16	0.70	0.67
4296	3.25	3.15	1.16	1.14	0.16	0.18	0.75	0.75
4297	3.41	3.32	1.19	1.23	0.16	0.19	0.76	0.78
4302	3.21	3.16	1.11	1.20	0.14	0.15	0.68	0.65
4303	3.31	3.31	1.21	1.20	0.12	0.13	0.74	0.67
4301 Cu*	3.28	3.24	1.14	1.20	0.13	0.16	0.63	0.67
4301 Fe*	3.30	3.26	1.18	1.20	0.12	0.15	0.62	0.63
4304 Cu*	3.43	3.33	1.07	1.13	0.14	0.19	0.77	0.78
4304 Fe*	3.40	3.32	1.07	1.12	0.15	0.19	0.78	0.79
4305 Cu*	3.20	3.22	1.15	1.15	0.17	0.18	0.66	0.61
4305 Fe*	3.24	3.22	1.14	1.15	0.17	0.19	0.65	0.60
4310	3.37	3.25	1.17	1.22	0.15	0.18	0.70	0.64
4311	2.57	2.47	0.71	0.76	0.09	0.12	0.60	0.56
4312	3.55	3.35	1.06	1.17	0.15	0.18	0.75	0.71
4313	3.66	3.58	1.08	1.19	0.23	0.25	0.92	0.83
4334	3.34	3.31	1.27	1.28	0.19	0.16	0.67	0.64
4335	3.27	3.34	1.15	1.16	0.21	0.15	0.71	0.66
4337	3.46	3.43	1.10	1.16	0.17	0.17	0.74	0.72
4338	3.33	3.42	1.13	1.15	0.18	0.17	0.74	0.70
4339	2.76	2.62	0.87	0.81	0.16	0.14	0.66	0.62
4340	3.57	3.64	1.14	1.10	0.19	0.17	0.79	0.77
4341	3.47	3.48	1.07	1.06	0.25	0.21	0.96	0.93
4342	3.19	3.18	1.25	1.22	0.20	0.17	0.63	0.60

SD = 0.074 SD = 0.049 SD = 0.026 SD = 0.040

95%C.L. = %Si ± 0.15   95%C.L. = %P ± 0.10   95% C.L. = %S ± 0.05   95%C.L. = %Mn ± 0.08

* At this point sampling was switched from a Cu Core Mold to a Spex 3904 Cast Iron Book Mold. Samples before 4303 were taken in a Cu Core Mold. Samples after 4305 were taken in the Spex Mold.

Table IV. Data for Least-Squares Line, S in Carbon Materials

Lab No.	Description	% Sulfur	Reading on E1T Tubes/100 (Avg/3 individual packings)
5419	Raw Pet Coke, Republic, 11–20–59	1.60	1.72
5420	Raw Pet Coke, Republic, 11–20–59	4.12	4.33
5113	American Gilsonite, 10–21–59	0.22	0.28
4608	Pitch, HSP, Barretts, 9–4–59	0.46	0.55
4573	Pitch, HSP, Koppers, 9–4–59	0.63	0.70
CC-1	Raw Pet Coke Std., Prepared Dec., 1958	1.93	2.00
4829	Fluid Coke, KACC, Purvis, Miss., 9–25–59	5.77	5.77
5036	Raw Pet Coke Std., 3–4–59	1.13	1.26
5507	Raw Pet Coke, Nat'l Carbon, 11–26–59	5.05	5.20
4949	Spect. Pure Graphite, National Carbon, Lot No. 28Z	0.00	0.03

the most direct approach would be to analyze the materials as pulverized powders or pelletized samples, and in the case of sulfur a powder method was used. Because we determine ash on these materials and because silicon is at such a level that we felt photometric methods would be required, a solution method is presently being used, primarily for the determination of V and Fe. The method was established and standards made so that K, Ca, Mn, and Ni can also be determined using the same basic procedure. The steps of these methods are listed below:

A. Sulfur

1. Sample holders machined from $1\frac{3}{8}$-in. nylon rod* are used to pack the samples which have been ground to -100 mesh. The packing is done in the same manner as for calcium in bath.

2. Instrument Settings

        a. Power to X-ray tube—50 kv, 45 ma

        b. EDDT Crystal—75.16° 2θ S K_α

        c. Pulse-amplitude discrimination used

        d. Helium path at 10 ft³/hr

* Nonuniformity in these holders required selection of matched sets.

e. Fixed time—approximately 42 sec

f. Scale factor set so that E1T tubes read 582 average on 5.77% standard

g. Counting rate—1762 cps on 5.77% S standard

3. Small adjustments are made in milliamperes to the tube or fixed time so that the standard reads properly and the gas flow and standardization are checked by taking a series of duplicate readings the same as for Ca in bath.

4. The samples are then inserted and the count from the E1T tubes used to calculate the percent sulfur where % S = 0.996 (E1T reading) − 0.0604.

Table IV shows the data used to calculate the formula used for determining individual sulfur percentages while Tables V and VI show results on individual samples and some comparisons of results by different analytical methods. Grinding does not completely eliminate sphericity of fluid coke, and Table VII demonstrates the reproducibility

**Table V. Sample Results Calculated from Least-Squares Line
(Each E1T Reading Is an Individual Packing)**

Lab No.	Count (E1T)–Bk	% S	
		Calculated	Chemical
5419	171	1.64	1.60
	173	1.66	
	170	1.63	
5420	433	4.25	4.12
	427	4.19	
	431	4.23	
5113	25	0.19	0.22
	25	0.19	
	26	0.20	
4608	54	0.48	0.46
	53	0.47	
	54	0.48	
4573	68	0.62	0.63
	67	0.61	
	68	0.62	
CC-1	197	1.90	1.93
	197	1.90	
	197	1.90	
4829	580	5.72	5.77
	577	5.69	
	583	5.75	
5036	121	1.14	1.13
	123	1.16	
	121	1.14	
5507	515	5.07	5.05
	514	5.06	
	511	5.03	

Table VI. Samples of Calcined Petroleum Coke from Chalmette Analyses by Four Laboratories
(Each E1T Reading Is an Individual Packing)

Date	Count (E1T)–Bk	% S				
		Calculated	Supplier	Ravenswood (Leco)	Chalmette (Eschka's)	Mead (Eschka's)
3–29–59 to	163	1.56	1.62	1.56	1.53	1.63
4–4–59	165	1.58				
	164	1.57				
5–17–59 to	190	1.83	1.82	1.85–1.86	1.79	1.85
5–23–59	186	1.79				
	186	1.79				
5–24–59 to	174	1.67	1.67	1.67	1.63	1.66
5–30–59	175	1.68				
	176	1.69				
5–31–59 to	173	1.66	1.64	1.63	1.64	1.65
6–6–59	175	1.68				
	173	1.66				
6–7–59 to	179	1.72	1.60	1.73	1.70	1.72
6–13–59	180	1.73				
	180	1.73				
6–14–59 to	192	1.85	1.80	1.85	1.78	1.83
6–20–59	191	1.84				1.83
	190	1.83				

of individual packings of this material. The fact that a variety of carbon materials from hard pitch to fluid coke was used for this method demonstrates applicability with the reservation that these materials have low ash contents and high ash materials such as anthracite coal would require compensating calculations for matrix effects. A discussion of the determination of S in carbon materials and also V has appeared in the *Norelco Reporter*.[3]

B. K, Ca, V, Mn, Fe, and Ni

1. A 25.0-g sample is ashed at 750°C, fused at 950°C with 2.0 g of 2:1 Na_2CO_3–H_3BO_3 fusion mix, dissolved in 20 ml of 1:4 HNO_3, transferred to a 50-ml volumetric flask, made to volume, and mixed.* Ten-ml aliquots are taken for X-ray analysis and the same 10-ml zytel holder is used for any series of samples.

2. A calibration standard prepared from the pure metals and the pure carbonates dissolved in an equivalent amount of 1:4 HNO_3 and with equivalent amounts of fusion mix is run before and after a series of sample analyses. A background count is made on a solution containing the acid and fusion mix only.

---

* This is the most general case. If ash contents are too high to fuse adequately, sample size can be reduced, or fusion mix, acid, and volume increased proportionately. In case special sensitivities are required, samples could in many cases be increased in size.

3. K, Ca, and V are run with pulse-amplitude discrimination set so that all three elements are included in the same setting while Mn, Fe, and Ni are run without pulse-amplitude discrimination. The K_α lines are used with LiF crystal.

4. The elements are determined in sequence, switching the instrument from integrate to differentiate at the appropriate element and making changes in factors as specified in the routine method.

5. E1T readings are used to calculate results as follows:

$$\% \text{ of element} = \frac{\text{Net E1T sample}}{\text{Net E1T standard}} \times \frac{\text{Volume}}{10} \times \frac{100 \times 2.5}{(\text{sample weight in mg})}$$

For routine work this is further simplified by making a table of factors for volume/weight.

The following detectability and counting statistics, Table VIII, establish the limitations of the method:

Limit of Detectability[4] $= 3 \sqrt{N_B}$

Limit of Detectability in mg per 10 ml $= \dfrac{3\sqrt{N_B}}{N_L - N_B} \times 2.5$

where N_B is equal to counts/sec for background, and N_L is equal to counts/sec for line and background with 2.5 mg of the element for 10 ml of solution; and

$$\sigma\% = \frac{100 \sqrt{N_L - N_B}}{N_L - N_B}$$

where N_B is equal to counts for background in fixed time t, and N_L is equal to counts for line plus background in fixed time t.

The linearity of instrument response to the elements under consideration was checked, using a standard containing 5 mg of each element per milliliter and calculating a ratio of counting rates of various dilutions. In this case, the limiting concentration is dependent upon counting rates not matrix effects, and the maximum counting rate used

Table VII. SD Due to Packing—Individual Packings, One Holder, Fluid Coke

Count (E1T)	Count (E1T)	Count (E1T)	Count (E1T)
1. 565	6. 569	11. 569	16. 573
2. 567	7. 570	12. 568	17. 569
3. 572	8. 573	13. 566	18. 572
4. 567	9. 566	14. 579	19. 569
5. 568	10. 570	15. 578	20. 571
			21. 571

570.1 Avg.

SD due to packing $= 2.70$ $F_\infty^{20} = (7.29/4.41) = 1.65$
$SD_{Inst} = 0.088\sqrt{E1T} = 2.10$ 95% Conf. Limits, $F_\infty^{20} = 1.71$

These results indicate that the variance due to packing plus the variance due to instrument counting is equal to or not much greater than the variance due to instrument counting alone.

is Ni at 5035 net cps with the lowest counting rate being V in air path at 207 cps. Concentrations of the elements are well below a level at which matrix effects should be
introduced, so the levels of element concentration in the standards are arbitrary and
merely reflect a convenient working arrangement.

Tables IX through XII demonstrate the reproducibility of this method in day-
to-day use while Table XIII shows these results compared to routine analyses by spectrophotometric methods. Table XIV shows results of routine X-ray *vs.* routine photometric
results on a variety of materials. Results on V in Table XIV indicate that one method
or the other is in serious error and confirmation of the X-ray method is given in Table
XV. The causes for the error in the photometric method have not been determined;
however, a set of standards carried through the determination with the titration and
used to plot a standard photometric curve did give the correct results, which are also
listed in Table XV.

Table VIII. Limits of Detectability for K, Ca, V, Mn, Fe, and Ni

Limit of detectability, mg of element/10 ml		σ %	Fixed count	Seconds	% of element, 25 g, 50-ml volume
K	0.078	3.2	6400	305	0.0016
Helium path		2.3	12800	610	
With pad		1.6	25600	1219	
		1.2	51200	2438	
Ca	0.094	4.0	6400	168	0.0019
Helium path		2.8	12800	337	
With pad		2.0	25600	674	
		1.4	51200	1347	
V	0.072	6.2	6400	65	0.0014
Helium path		4.4	12800	129	
With pad		3.1	25600	258	
		2.2	51200	517	
V	0.17	4.8	6400	126	0.0034
Air path		3.4	12800	251	
With pad		2.4	25600	502	
		1.7	51200	1004	
Mn	0.13	11.1	6400	19	0.0026
Air path		7.9	12800	39	
Without pad		5.6	25600	77	
		3.9	51200	154	
Fe	0.100	14.5	6400	11	0.002
Air path		10.3	12800	23	
Without pad		7.2	25600	46	
		5.1	51200	81	
Ni	0.065	17.2	6400	8.2	0.0013
Air path		12.1	12800	16	
Without pad		8.6	25600	33	
		6.1	51200	66	

Table IX. X-Ray Results on Lab. No. 4416, Calcined Coke, Great Lakes, 8–10–59, MON 4262 (25-g Samples, 100-ml Volume)

Element	1–25–61	1–26–61	1–30–61	2–1–61	2–3–61	2–6–61	Average
K	0.003	0.003	0.004	0.004	0.003	0.003	0.003
Ca	0.019	0.019	0.019	0.019	0.019	0.019	0.019
V*	0.019	0.020					
V†	0.018	0.018	0.020	0.019	0.019	0.020	0.019
Mn	0.000	0.000	0.000	0.000	0.001	0.000	0.000
Fe	0.054	0.053	0.056	0.055	0.054	0.055	0.054
Ni	0.007	0.008	0.008	0.007	0.008	0.015‡	0.0076

* Helium Path
† Air Path
‡ 0.015% Ni in this sample is believed to be due to contamination by flakes visible in the sample from the chromel thermocouple in the furnace. This result was omitted from the average.

Table X. X-Ray Results on Lab. No. 4434, Calcined Coke, Gilsonite, 8–20–59, D & RGW 18471 (25-g Samples, 200-ml Volume)

Element	1–25–61	1–26–61	1–30–61	2–1–61	2–3–61	2–6–61	Average
K	0.005	0.004	0.006	0.005	0.005	0.005	0.005
Ca	0.037	0.025	0.025	0.025	0.025	0.024	0.027
V*	0.001	0.001					
V†	0.000	0.000	0.000	0.000	0.001	0.000	0.000
Mn	0.000	0.000	0.000	0.000	0.001	0.000	0.000
Fe	0.031	0.031	0.032	0.030	0.032	0.032	0.031
Ni	0.071	0.070	0.071	0.072	0.071	0.072	0.071

* Helium Path
† Air Path

Table XI. X-Ray Results on Lab. No. 4346, Calcined Coke, General Carbon, 7–31–59, SSW 76070 (25-g Samples, 50-ml Volume)

Element	1–25–61	1–26–61	1–30–61	2–1–61	2–3–61	2–6–61	Average
K	0.001	0.001	0.001	0.001	0.001	0.001	0.001
Ca	0.020	0.019	0.020	0.019	0.019	0.019	0.019
V*	0.002	0.002					
V†	0.001	0.001	0.002	0.001	0.002	0.001	0.001
Mn	0.000	0.000	0.000	0.000	0.000	0.000	0.000
Fe	0.030	0.031	0.032	0.032	0.032	0.031	0.031
Ni	0.006	0.006	0.007	0.007	0.007	0.007	0.007

* Helium Path
† Air Path

Table XII. X-Ray Results on Lab. No. 10,362, Calcined Anthracite, Mountaineer, 12–27–60 B & O 629175 (5-g Samples, 100-ml Volume)

Element	Sample 1	Sample 2	Sample 3	Average
% K	0.092	0.084	0.089	0.088
% Ca	0.042	0.039	0.037	0.039
% V	0.011	0.009	0.009	0.010
% Mn	0.0017	0.0017	0.0026	0.0020
% Fe	0.31	0.31	0.31	0.31
% Ni	0.0016	0.0016	0.0012	0.0015

Table XIII. Comparison of X-Ray Results to Routine Analyses by Spectrophotometric Methods

Element	Lab. No. 4416		Lab. No. 4434		Lab. No. 4346		Lab. No. 10,362	
	X-ray avg, %	Photometric duplicates	X-ray avg, %	Photometric duplicates	X-ray avg, %	Photometric duplicates	X-ray avg, %	Photometric duplicates
K	0.003		0.005		0.001		0.086* 0.088	
Ca	0.019		0.027		0.019		0.041* 0.039	
V	0.019	0.020 0.020	0.000	0.004 0.003	0.001	0.004 0.003	0.010	
Mn	0.000		0.000		0.000		0.002	
Fe	0.054	0.052 0.052	0.031	0.030 0.030	0.031	0.030 0.029	0.31* 0.31	0.31 0.30
Ni	0.008	0.008 0.008	0.071	0.065 0.065	0.007	0.006 0.006	0.0015	

* These results were on 1.0-g samples in 100-ml volumes and a table of these analyses was not included.

Table XIV. Comparison of Some X-Ray Analyses with Routine Photometric Analyses

Lab. No.—Material	Method	% V	% Fe	% Ni
CC-4	X-ray	0.014	0.020	0.010
Std. Calc. Coke	Photometric	0.013	0.017	0.011
CC-4	X-ray	0.014	0.019	0.010
Std. Calc. Coke	Photometric	0.013	0.017	0.011
10,334	X-ray	0.001	0.042	0.066
Cal. Gilsonite	Photometric	0.005	0.043	0.062
10,340	X-ray	0.032	0.003	0.014
Fluid Coke	Photometric	0.026	0.003	0.014
10,347	X-ray	0.020	0.095	
Baked Anodes	Photometric	0.024	0.098	0.025
10,352	X-ray	0.026	0.053	0.008
Low S C. Pet. Coke	Photometric	0.025	0.043	0.008
10,362	X-ray		0.32	
Cal. Anthracite	Photometric		0.30	
10,390	X-ray	0.029	0.15	0.017
K-30	Photometric	0.035	0.13	0.020
10,391	X-ray	0.034	0.032	0.015
All Coke Fines	Photometric	0.031	0.031	0.016
10,394	X-ray	0.033	0.005	0.014
W-1 Disc Feeder	Photometric	0.029	0.005	0.015
48	X-ray	0.027	0.042	0.012
Green Anode	Photometric	0.022	0.039	0.012
49	X-ray	0.000	0.015	0.000
Pitch from Scales	Photometric	0.001	0.015	0.001
50	X-ray	0.031	0.036	0.020
K-30, All Coke	Photometric	0.025	0.037	0.020
51	X-ray	0.034	0.025	0.015
Fines, All Coke	Photometric	0.028	0.020	0.015
52	X-ray	0.028	0.017	0.009
Coarse, All Coke	Photometric	0.022	0.016	0.009
54	X-ray	0.036	0.005	0.015
W-1 Disc Feeder	Photometric	0.027	0.004	0.014
53	X-ray	0.028	0.22	0.024
Oversize, All Coke	Photometric	0.035	0.20	0.22
CC-4	X-ray	0.015	0.020	0.011
Std. Values	Photometric	0.013	0.017	0.011
10,424	X-ray	0.036	0.004	0.014
Fluid Coke	Photometric	0.028	0.003	0.014
3	X-ray	0.028	0.041	0.008
Raw Pet. Coke	Photometric	0.028	0.044	0.010
5	X-ray	0.000	0.027	0.071
Cal. Gilsonite	Photometric	0.003	0.029	0.068
9	X-ray	0.020	0.13	0.024
Baked Anodes	Photometric	0.027	0.12	0.023
10	X-ray	0.022	0.11	0.018
Green Anodes	Photometric	0.026	0.10	0.018
15	X-ray	0.033	0.004	0.014
Fluid Coke	Photometric	0.024	0.004	0.014
20	X-ray	0.036	0.003	0.015
Fluid Coke	Photometric	0.027	0.002	0.014
65	X-ray	0.034	0.003	0.015
Fluid Coke	Photometric	0.027	0.002	0.013
55	X-ray	N.A.*	0.16	0.050
W-2 Disc Feeder	Photometric	N.A.	0.15	0.050

* Not analyzed.

Table XV. Confirmation of X-Ray V Determinations by "Umpire" Volumetric Method

Lab. No.	% V			
	X-ray	Volumetric	Routine photometric	Photometric (titrated standards)
CC-8	0.034	0.034–0.035	0.028	0.035
1841	0.055	0.055	0.044	0.056
2032	0.059	0.060	0.047	0.059
CC-8	0.034	0.034	0.028	N.A.*
2062	0.060	0.061	N.A.	N.A.
1830	0.053	0.056	N.A.	N.A.

* Not analyzed.

Table XVI. Instrument Maintenance and Problem History

Date	Problem	Symptom
July 1959	Leak in water hose connection above milliampere control	Shorted out battery and threw overload
July 1959	Poor connection to P-10 gas	Excessive use of gas
Aug. 1959	Slipping gear in goniometer	Angle calibration lost
Sept. 1959	Fumes from Goodyear Pliobond in room	Reduced and erratic count
Oct. 1959	Dented exit collimator, installation of FPC	
Jan. 1960	High-voltage cable shorted out at entrance into transformer cabinet	Smoke at point of short
June 1960	Dust from bath samples in collimators	Reduced sulfur count
Mar. 1961	Milliampere and line controls oscillating	Harmonics between sola regulator on line and instrument regulator
June 1961	Hose to X-ray tube leaking	Water on samples
July 1961	Broken ADP crystal, wrong position in holder	
Aug. 1961	Flow proportional counter sent in for repair	Excessive electronic background count
Aug. 1961	Main power supply sent in for repair	No power from supply
Sept. 1961	Two bad cylinders of helium	Reduced calcium count
June 1962	High-voltage cable shorted out at entrance into transformer cabinet	Smoke at point of short
July 1962	Slipping gear in goniometer	Angle calibration lost
July 1962	Dust in system	Reduced sulfur count
July 1962	Entrance collimator washed with acetone; had to be rebuilt	Fell apart
July 1959 July 1962	Helium path bellows have been changed four times	Gas flow not adequate at 10 ft³/hr

OPERATIONAL PROBLEMS AND MAINTENANCE

In any routine use of an analytical instrument which is expected to carry a heavy work load trouble-free operation is desired. For routine maintenance we follow a set schedule on changing batteries, water filter, alternating X-ray tubes, checking tubes, and cleaning the instrument. This maintenance is recorded in chart form to keep a continuous record of what has been done and when the next routine maintenance is scheduled. In addition to our routine maintenance, we have made up a table of "Basic Trouble Shooting" for the X-ray spectrograph, which lists the common symptoms and remedies. As a result of these preventive measures, we have not had much down time on the instrument; however, there have been some significant breakdowns and some problems due to our operation. Table XVI is a tabulation of these major and minor difficulties with any tube changes omitted since tube failures are in the category of expected problems.

CONCLUSIONS

Over the period of use of this instrument, our intralaboratory checks and interplant checks have consistently demonstrated that routine results using the procedures outlined are more reliable than results of chemical analyses by the same personnel. This is due mainly to the fact that chemical procedures inherently contain more potential sources of error. With experience, developed skills, and increased understanding of basic principles of X-ray spectroscopy, the value of the work which can be routinely and conveniently handled will be a significant addition to simplified laboratory operation.

REFERENCES

1. W. J. Campbell and J. W. Thatcher, "Determination of Calcium in Wolframite Concentrates by Fluorescent X-ray Spectrography," Bureau of Mines Report of Investigation 5416, USDI, 1958.
2. E. L. Roth and G. Antonic, Motor Castings Company, Milwaukee, Wis., "The Use of the Quantovac for Foundry Control of All Elements in Cast Iron," Gray Iron Research Institute Meeting, Hartford, Conn., October 14, 1960.
3. R. H. Black and W. J. Forsyth, "The Determination of Sulfur and Vanadium in Carbon Materials by X-ray Fluorescence Analysis," *Norelco Reptr* 6(2): 53–54, March–April 1959.
4. L. S. Birks, *X-ray Spectrochemical Analysis*, Interscience Publishers, Inc., New York, 1959, pp. 51–55.

EXPERIENCES OF X-RAY ANALYSES
IN STEEL AND FERRO-ALLOY PRODUCTION

J. Baecklund

Avesta Jernverks Aktiebolag, Sweden

ABSTRACT

Avesta Steel Works manufactures stainless steel, low-alloy steel, alloys for precision casting, welding electrodes, and ferro alloys. When it was decided in 1958 that a new laboratory for production control was to be built, there was an investigation into different spectrographic possibilities. A combination of X-ray and optical methods was selected, that is, determination of C, P, and S was to be made optically while Si, Ti, and heavier elements were to be determined on X-ray equipment. Specimen taking and specimen preparation techniques were worked out. Certain interelement effects were recorded and methods to correct for them found. Methods to correct for surface properties in determination of high concentrations have been taken into use. This laboratory has now been in continuous use for more than two years. Statistical data on accuracy, reliability, and continuity of operation are given.

At Avesta, X-ray analysis has been used in three different forms:

1. Ordinary compositional determination to guide steelmaking and other production processes as well as analysis of materials for delivery. This is the most important application and will be dealt with in more detail.
2. A fixed channel portable instrument for molybdenum determination is used in the field and at the manufacturing shop in Avesta for checks on 18-8-Mo-steels. This instrument, utilizing a dentist's X-ray unit as primary source, has been designed and built at Avesta.
3. A fixed four-channel instrument designed for operation on the line is being used in Avesta today. This instrument will check the composition of sheet material at the packing stage. A code for Mo, Ti, Ni, and Cr contents is printed out from the instrument onto the material itself as well as on the packing slip. The instrument is fully automatic and will be actuated by the arrival of material to be controlled. It was designed and built at Avesta.

In the spring of 1960 a laboratory for rapid analysis was installed and put in service at Avesta. The range of production of the Avesta Steel Works covers metallic materials from the Ferro-Alloy Shop, Steel-Melting Shop, Rolling Mills, Fabricating Shop, Precision Foundry, and Welding Electrode Shop.

At present the routine program of the laboratory comprises most raw materials and end products within these fields of activity, as well as certain intermediate products. During 1958 an investigation was made aimed at learning how to analyze these materials in the most economical way, and in this connection Table I was compiled.

We considered it possible by using the X-ray method to cover a great part of the demand for analyses economically. Ninety-eight percent of the demand could be covered within the range of elements suitable for this method. As a comparison, I can mention

Table I. Arithmetic Means of the Number of Determinations per Month. Taken from Records from the Months of April, June, and December, 1957

Element	Specimens suitable for simple application of X-ray analysis		Number of specimens with too low concentration of element	Total number of specimens	Suitable specimens, %
	Specimens from heats	Specimens from different sources: tubes, sheet, wire, weld seams			
Si	790	90		897	98.1
Mn	425	90		524	98.3
Cr	1268	180		1468	98.6
Ni	797	180		994	98.3
Mo	490	637		1137	99.1
Ti	20	18		38	100.0
Cu	224	21		250	98.0
Pb	246	1		247	100.0
Sn	10			10	100.0
Al	2	5	7	7	0.0
Cb	1	17		18	100.0
Co	149	3		152	100.0
Mg	3		3	3	0.0
Ca	4			4	100.0
Fe	4			4	100.0
Ta	1	12		13	100.0
W	3	11		14	100.0
V	1	3		4	100.0
				6000	

Percent of suitable specimens for the determination of these elements: 98.3.

that on the same occasion, we estimated that an optical vacuum spectrograph for C, P, and S would be able to cover approximately 75% of the demand for analyses for these elements. On the basis of this investigation, we purchased the equipment for a new laboratory for rapid analysis, which was placed close to the most important users of its service, the steel-melting shop and the ferro-alloy shop.

No equipment for temperature control or air-humidity control is required in the laboratory. The equipment has been subjected to a temperature range of 65–100°F without deviation in results. The X-ray instrument—an Autrometer—was chosen, among other things, in consideration of the fact that extreme rapidity was not required, and for this reason a serially working instrument is quite sufficient. Such an instrument is considerably cheaper than a parallel instrument for many channels.

From the character of the works and its products, it is possible to draw certain conclusions concerning types of samples used. The majority of samples consists as can be expected, of ladle samples from melting shops and the foundries. (See Figures 1 and 2.) Other types which occur frequently are sheet samples and certain chip samples from the rolling mills and the manufacturing shop, cast samples from the ferro-alloy shop, as well as crushed brittle ferro alloys (silicon–chromium, ferromolybdenum, raw-ferrochromium), and crushed slags. The grade of crushing (200 mesh) is not fine enough to eliminate grain size effects, but the composition of the individual grains corresponds closely to the average composition of these materials, and, therefore, this preparation is quite sufficient if it can be made in a reproducible way.

Owing to the design of the sample holder, the sample size of solid materials is maximized at a cylinder of approximately 2 in. high and 2 in. in diameter. For very thin or small samples, the normal specimen holder allows determinations to be made on samples larger than a cylinder having $1\frac{1}{4}$ in. diameter and a height that is determined by the penetrating ability of the radiation used, i.e., in practice approximately 0.004 in.

A special sample holder designed by us permits the use of sample diameters down to $\frac{1}{4}$ in., but in this case the instrument cannot give its full sensitivity for trace elements, etc. The chip samples from the rolling mills and manufacturing shop are specimens taken with a file along the edge of a sheet. This procedure is used to check the type of alloy in case a mix-up could have occurred and the separation is carried out through a rough determination of Ti, Ni, Mo, or Cr content. The specimen is taken by hand and transferred to a small gelatin container, which is used both for transport and as a specimen holder. The specimen is taken out of its holder only for the determination of Ti. Accuracy is rather low, but determination and handling are fast. The handling of such samples from the melting and ferro-alloy shops, which have been taken out by means of a ladle and cast in a mold, as well as samples from sheets and bars, starts with grinding on a semi-automatic grinding machine. Depending on how good the surface of the sample is, this procedure takes 1 to 3 min. The surface obtained by this grinding need not be perfect; small holes and cracks can be allowed if the total surface area covered by them is less than 0.5% of the irradiated area, and even more if the accuracy of the determination can be reduced.

When a certain element of an alloy is to be determined it is necessary to set three variables—this is the so-called programming. The first setting means that intensity

Figure 1. Specimen taken in the ferro-alloy shop. Ferrochromium cooling in the ladle.

determination is to take place at a certain wavelength; the second setting covers the statistical accuracy; and the third setting gives the current required to the X-ray tube for this determination. In addition, a number of other factors have been permanently preset for each wavelength by means of soldered connections which are automatically switched in. Among these are choice of crystal and detector and connection of pulse-height analyzer.

Because different materials are to be analyzed with regard to different elements, and because one and the same element can appear at greatly different contents in different materials, it is necessary to use different programs for different types of samples. The Autrometer is based on a comparison betweeen a sample and an external standard. It is also necessary, in order to utilize the method fully, to have different standards—as a rule one for each program. We have named the programs in accordance with the color code by which the standards have been marked, as this system makes it easy to learn the handling of the instrument, and reduces the risk of confusion. When a specimen is placed in the Autrometer a corresponding program and standard are inserted at the same time. The program is selected from a set of program cards which is fixed to a stand on the programming unit of the Autrometer. (Most of the operators know the programs completely by heart.)

The results are reported by the machine in the form of intensity data—the ratio between X-ray intensity from the sample and a standard—for a series of different wave lengths corresponding to the elements in question. With the aid of tables which have been made by means of a series of standards of well-known compositions, it is possible for the operators to transform the intensity data into contents of these elements. By means of a Teleautograph (a type of telescriber) the contents are reported by the operators to the melting shop personnel, and at the same time a copy is automatically obtained at the main laboratory as well as at the transmitter. When all preset elements have been

Figure 2. Specimen being taken in the melting shop. Double specimens are always taken and the best is used for analysis.

reported by the machine, the sample is transferred to the optical spectrograph. Analyzing and reporting from this apparatus takes place in a similar way. The extent to which the calibrations have been carried out is shown in Table II.

Only one operator is required to perform this work at such an analytical load that the different machines can be used one after the other. In addition to the shift-working personnel, we also have a skilled and experienced group supervisor, who has to make sure that gas, grinding wheels, and spare parts are available, to check the instruments in accordance with a detailed program, and to take the responsibility for new calibrations. He is also prepared to be of service when instrument faults occur or when personnel is

Table II. Range of Calibration

Calibration Range for Samples of Minimum Width $1\frac{1}{4}$ in.

Grades	Sn	Ag	Mo	Cb	Pb	Se	Ta	Cu	W	Ni	Fe	Mn	Cr	V	Ti	Si	Co
18-8 and 18-8 Mo	x	x	x	x	x	x	x	x		x		x	x		x	x	x
13-20%Cr	x		x		x			x		x		x	x			x	x
25-5			x							x		x	x			x	
25-20			x		x			x		x		x	x		x	x	
High-Mn steel			x	x		x				x		x	x			x	
C 242			x								x	x	x		x	x	x
Ni alloys			x						x		x	x	x			x	
FeCr											x	x		x		x	
Mild steel and low-alloy steel			x	x			x	x		x		x	x	x		x	x

Calibration Range for Samples of Minimum Width $\frac{1}{4}$ in.

Grades	Mo	Cb	Ni	Mn	Cr	Ti	Cu	W
18-8 and 18-8-Mo	x	x	x	x	x	x		
13–20%Cr	x		x	x	x			
25-5	x		x	x				
25-20	x		x	x	x	x	x	
Ni alloys				x	x			x

Calibration Range for Powder Samples

Type of sample	Mo	Cu	Fe	Cr	Si	Ca
FeMo	x	x	x	x	x	
Quartz		x				
SiCr				x	x	
Raw chrome				x	x	
Slag (FeCr)				x		
Slag (melting shop)			x	x	x	x

reduced by illness. As an extra safeguard, two men working at the main laboratory have been trained as instrument operators. Instrument faults can occur at any time and the following measures have been taken to reduce the effects of an instrument failure and the risk of an interruption:

1. *Spare equipment*, consisting of a manual X-ray spectrograph, is located at the main laboratory and calibrated in the same way as the Autrometer. Operators have been trained to use this instrument and have been instructed to change over immediately to use the manual equipment as soon as Autrometer operation is disturbed.
2. *Skilled service personnel.* Instrument engineer and group supervisor are responsible for instrument repair and maintenance.
3. *Spare parts*
4. *Service equipment* such as oscilloscope and pulse generator.
5. *Comprehensive control and maintenance.* Accuracy is checked every shift by means of a couple of standard specimens, which are selected to cover the most important concentration ranges in our most common alloys. Results are plotted on a graph so that changes can be followed from day to day. (See Figure 3.) To check the statistical accuracy a twenty-times-repeated analysis procedure is carried out every shift for the most important elements in another standard specimen. The number of results obtained that deviate more than would be expected from a Gaussian distribution are plotted from day to day on another diagram. This plotting gives the lower curve in Figure 4; the upper curve in the same figure represents the daily variation of the mean of these twenty determinations. Maintenance schedules have been set up and are carried out every fortnight.

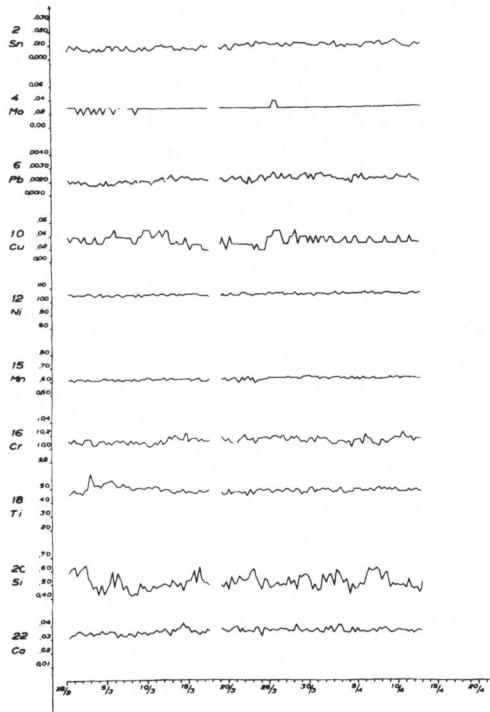

Figure 3. Day-to-day variation in readings. Values with decimal points are concentrations in percent.

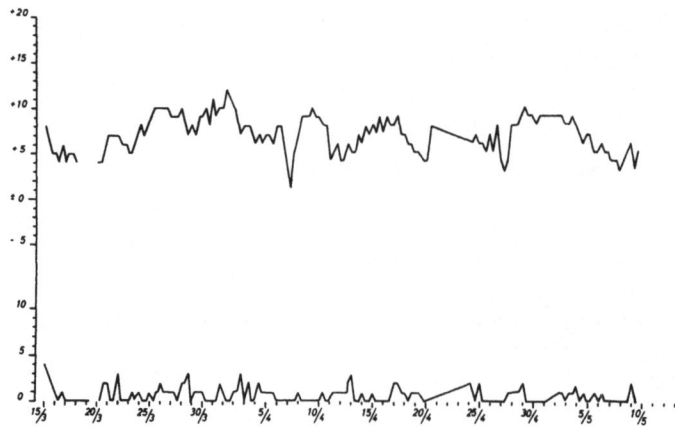

Figure 4. Lower curve: Number of values deviating more than ±0.5%
from the mean. Upper curve: Mean of the 25 readings.

RESULTS

It has been possible to transfer the main part of the analytical load, 93%, to the
X-ray method. As a comparison, it can be noted that 78% has been transferred to the
optical vacuum spectrograph. This difference is mainly due to difficulties involved in
preparation of samples of varying origin and condition.

The improvement in efficiency by the change in analytical methods has resulted
in a considerable saving of personnel. During the first of the two years the equipment
was in full operation, 14 people were withdrawn from routine analysis work. At the same
time the analytical load increased by 50%. This is because our costs for analysis have
decreased, and it is now possible to carry out comprehensive work within reasonable
time. The speed of analysis has been found sufficient. A furnace test specimen analyzed
for Cr, Ni, Mo, Mn, Si, C, P, and S is made in 12 min, including sample preparation
and report writing. It should be noted, however, that reporting of the first element (Mo)
takes place within 4 min after reception of the specimen.

Calibration of different elements in different materials has generally been quite
straightforward. Standard specimens have been made in sets, one set for each element
and each material. Within the set, the element in question has been exchanged for the
base element (generally iron), and the concentration of other elements kept constant.
Chromium in stainless steels has shown dependence on the other elements such as titanium,
molybdenum, columbium, and tungsten. Very large differences in nickel content also
cause changes in chromium intensity. That chromium intensity will be affected by
changes in the stainless steel matrix is evident from absorption data for Cr K_α. Iron,
the base element, has a low mass absorption coefficient, 115; but Ti has 565; Mo, 441,
and Cb, 414; and W, 455. Besides most elements that replace iron will decrease Cr K_α
intensity as Fe K_α increases it by secondary fluorescence.

Chromium is also being determined in high concentration in the ferrochromium
alloy. Accuracy is important here and from Figure 5, the calibration graph for chromium
content in cast and ground ferrochromium, the accuracy can be judged to be ±0.3%
in the 65% range. Variations in surface conditions, however, usually lead to lower intensities than those obtained from the standard specimens. To correct for this, iron,
silicon and nickel contents are determined as well, and an adjustment is made so that a
sum of 99.5% is obtained.

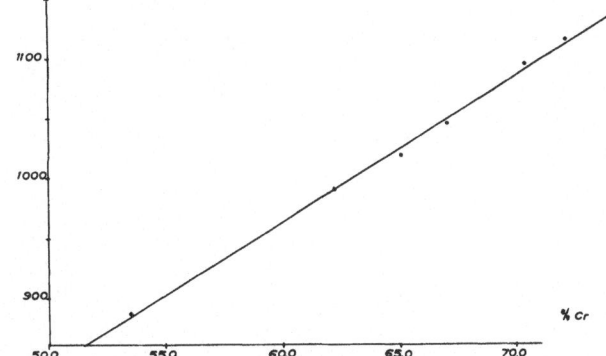

Figure 5. Chromium determination in ferrochromium alloy. Cast samples. Statistical accuracy ±0.5%.

Powdered samples are generally more intricate than cast samples. Powdered ferro alloys, however, are generally quite straightforward. Figure 6 is an example of this.

According to our experience, the sensitivity of the X-ray method has often been underestimated. For most elements, 100 ppm in a steel matrix is easily determinable and 10 ppm is often possible to detect. Lead is an important trace element with regard to steelmaking and is very often analyzed by us in the concentration range 10–50 ppm. Our calibration has been simple, using the L_β line. More difficult has been finding a good wet method for lead in steel, and we have developed a way of using X-ray fluorescence after coprecipitating lead sulfate on barium sulfate with strontium as the internal standard.

The laboratory gives accurate results. A very important point in judging laboratory performance is the accuracy that can be obtained in the whole steelmaking procedure. Some of the modern alloys like 17-7 PH and PH 15-7 Mo require close analytical control. In Table III, a comparison is made between the composition aimed at and the actually obtained result for 12 melts. The extent to which committed errors depend on laboratory equipment and personnel can be estimated from the following data: The melting shop demands analytical checks to be made on approximately 1% of our final determinations on heats. In 0.06% of these cases deviations larger than those statistically expected have been found. During the two years our X-ray equipment has been in operation only one big error has been made, and most probably this arose through a mix-up of materials. In Figure 7 the distribution of 25 measurements on chromium, nickel, and molybdenum in stainless steel can be seen. Deviations larger than 0.01% Mo, 0.05% Ni, and 0.1% Cr are uncommon.

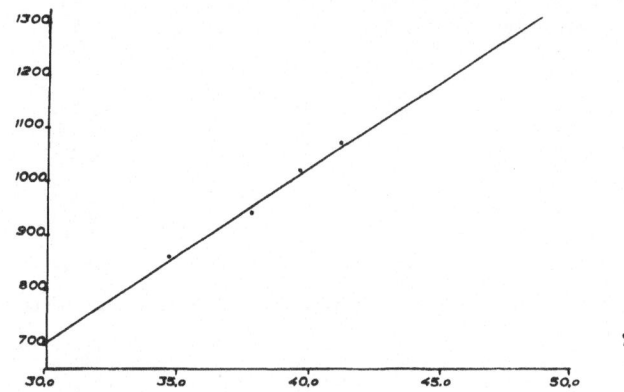

Figure 6. Iron in ferromolybdenum alloy powder. Statistical accuracy ±0.5%.

Table III

	Si %	Cr %	Ni %	Mo %
Heat No. 1477				
Heat adjustment sample composition	0.58	17.6	9.9	—
Final composition, aimed at	0.50	18.4	10.4	—
Final composition, obtained	0.51	18.2	10.6	—
Heat No. 1476				
Heat adjustment sample composition	0.71	17.9	9.0	—
Final composition, aimed at	0.50	18.1	9.4	—
Final composition, obtained	0.51	18.2	9.5	—
Heat No. 1475				
Heat adjustment sample composition	0.59	18.6	10.6	—
Final composition, aimed at	0.50	18.4	10.4	—
Final composition, obtained	0.53	18.5	10.5	—
Heat No. 1474				
Heat adjustment sample composition	0.50	17.2	11.0	2.49
Final composition, aimed at	0.50	17.0	11.5	2.65
Final composition, obtained	0.56	17.0	11.5	2.64
Heat No. 1473				
Heat adjustment sample composition	0.64	17.2	8.5	1.45
Final composition, aimed at	0.50	17.3	9.0	1.50
Final composition, obtained	0.51	17.4	9.1	1.50
Heat No. 1471				
Heat adjustment sample composition	0.52	17.6	9.2	—
Final composition, aimed at	0.50	17.5	9.0	—
Final composition, obtained	0.52	17.6	9.2	—
Heat No. 1470				
Heat adjustment sample composition	0.43	18.1	10.5	—
Final composition, aimed at	0.50	18.4	10.4	—
Final composition, obtained	0.52	18.1	10.5	—
Heat No. 1469				
Heat adjustment sample composition	0.70	16.8	11.8	2.89
Final composition, aimed at	0.40	17.5	12.0	2.80
Final composition, obtained	0.48	17.5	12.2	2.83
Heat No. 1468				
Heat adjustment sample composition	0.49	17.4	12.3	2.23
Final composition, aimed at	0.50	17.0	12.0	2.30
Final composition, obtained	0.62	17.1	12.1	2.30
Heat No. 1467				
Heat adjustment sample composition	0.44	17.5	10.4	—
Final composition, aimed at	0.50	18.4	10.4	—
Final composition, obtained	0.46	18.3	10.5	—
Heat No. 1466				
Heat adjustment sample composition	0.59	16.6	11.3	2.61
Final composition, aimed at	0.50	17.0	11.5	2.65
Final composition, obtained	0.40	17.4	11.7	2.67
Heat No. 1465				
Heat adjustment sample composition	0.50	18.3	11.2	—
Final composition, aimed at	0.50	18.0	11.3	—
Final composition, obtained	0.57	18.0	11.3	—

The following standard deviations of obtained compositions from aimed compositions can be calculated:

Si %	Cr %	Ni %	Mo %
0.06	0.17	0.14	0.02

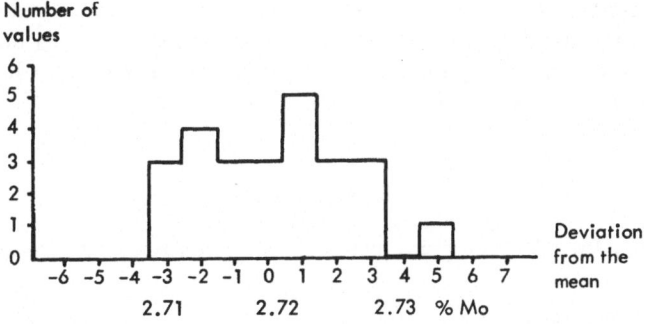

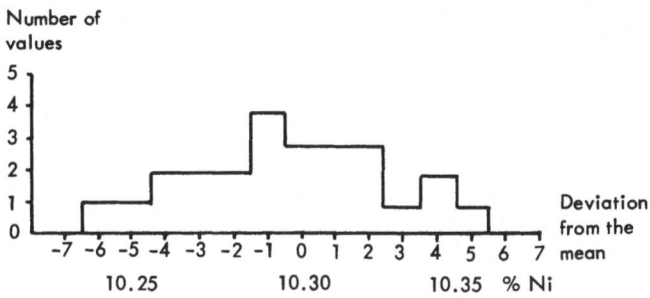

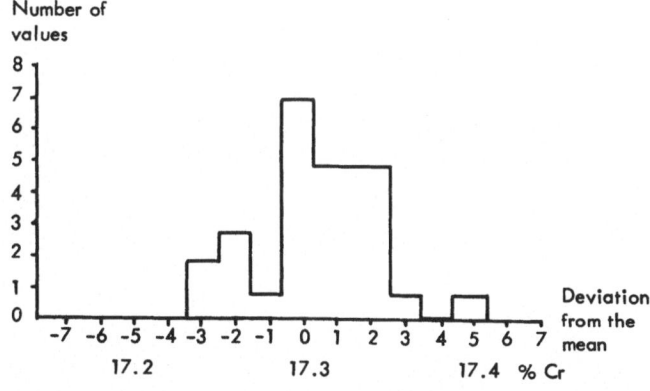

Figure 7. Statistical spread of 25 analyses for Mo, Ni, and Cr determination.

The dependability of the new laboratory has been high. The only specimens which require immediate analyses are samples from heats taken for adjustment of composition. About one such sample in a thousand has been analyzed by other means. All other samples which usually require attention within the same day have been analyzed in the new laboratory. This means that instrument faults have been repaired within a few hours. More important parts which have been replaced on these occasions are crystal changer, X-ray tube, stepping switches, and gear wheel in the gear box for the goniometer movement.

Our experience indicates that the so-called ratio system, i.e., recording the intensities as the ratio between a sample and a standard, is very valuable. It has been possible to replace detectors and crystals, as well as the X-ray tube, and to adjust goniometer settings without recalibration of any element. When the instrument is put to work after a repair the calibrations are quite or almost quite unchanged. This has substantially contributed to the short off-times.

DISCUSSION

R. Larson (Sundstrand): How much metal do you take off from the test samples taken from the heat?

J. Baecklund: That depends on what shape the mold is in, and also how careful the personnel in the shop have been in cleaning the bottom of the mold. Sometimes we may have to take off as much as a $\frac{1}{2}$ mm. But that is a problem we had only at the beginning. Now the shop personnel are more careful, and we have had to grind off only $\frac{1}{10}$ mm.

DETERMINATION OF ALUMINUM IN IRON-ALUMINUM ALLOYS BY VACUUM X-RAY FLUORESCENCE

J. C. Wagner and F. R. Bryan

Ford Motor Company
Dearborn, Michigan

ABSTRACT

Aluminum is determined within the range 2–16% in iron. Repeatability of X-ray determinations is within a standard deviation of 0.07% aluminum for a sample containing 10.2% aluminum. The standard deviation of the chemical values from the straight line calibration curve is 0.3% aluminum. X-ray analytical time is approximately 5% of the time required for chemical analysis. Peak-to-background ratio is improved by using a chromium-target thin-window X-ray tube instead of a tungsten tube.

INTRODUCTION

The development of commercial iron–aluminum alloys includes the problem of providing fast and accurate production control analyses. Chemical methods require several hours. Optical emission techniques do not normally yield results of high absolute precision at high concentrations. The undesirable length of time required by chemical methods and the poor precision of optical emission prompted an investigation of the analysis of these alloys by means of vacuum X-ray fluorescence.

EXPERIMENTAL

Equipment

A Norelco vacuum X-ray spectrograph equipped with an EDDT (ethylenediamine *d*-tartrate) crystal, a flow-proportional counter, and a pulse height analyzer, were used. A tungsten-target tube was used in the initial stages of the investigation. However, most of the data reported herein were obtained using a Machlett thin-window chromium-target X-ray tube rated at 50 kv and 35 ma. A mechanical pump was used to obtain the required vacuum.

Sample Preparation

Samples were polished with 400 grit paper and placed in the sample holder with the polishing scratches always in the same direction. Constant sample area was achieved by using a platinum mask containing a $\frac{5}{8}$-in.-diameter hole in the center.

Operating Details

The chromium tube was operated at 50 kv and 35 ma, and the tungsten tube was used at 50 kv and 50 ma. Pressure was maintained at 25 μ Hg. Counting rate was computed by determining time necessary for 32,000 pulses. Counting times ranged from 2 to 9 min. Aluminum K_α radiation was measured.

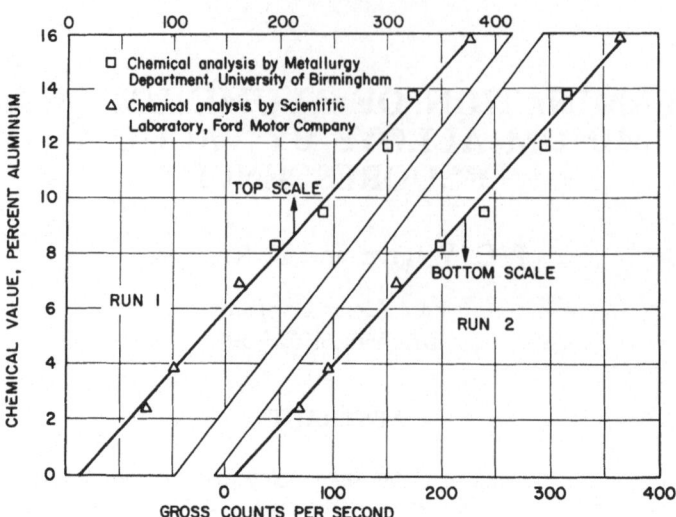

Figure 1. Calibration curves plotted from X-ray data taken several hours apart.

Chemical Analysis of Calibration Standards

At the Ford Scientific Laboratory the chemical analysis is performed in two steps: (1) separation of iron from solution by means of a mercury cathode; (2) precipitation of aluminum by ammonium hydroxide and subsequent gravimetric determination.

Some of the standards had been analyzed at the University of Birmingham. In that laboratory iron was determined and aluminum content was calculated by difference. The method of analyzing for iron consisted of using a chelation procedure and titrating the iron–aluminum solution for ferric ions. Ethylenediamine tetraacetic acid was used as the titrating agent. All standards are binary iron–aluminum alloys, with the exception of the 15.8% aluminum specimen. The latter contains 0.5% manganese.

RESULTS

Calibration curves of aluminum content determined chemically *versus* counting rate determined by X-ray fluorescence are shown in Figure 1, plotted from the data of Table I. Each counting rate value is based on an individual determination, not an average. Most of the points are located close to the same straight line. The position of the latter in the case of Run No. 1 was determined by the method of least squares.[1] The straight line representing Run No. 2 was constructed parallel to the other line, with an attempt to have points fall into the same relationship with the line of Run No. 1.

The proximity of most of the points to a straight line is strong evidence of a linear relationship between counting rate and aluminum content. Differences between chemical values and those taken from one of the straight lines are shown in Table II. The standard deviation of the points from the straight line is 0.3% aluminum. The deviation of the points from the straight line is probably due more to chemical errors than X-ray errors. The basis for this statement is twofold: (1) excellent precision of the X-ray method, and (2) agreement of calibration curves from data collected at different times. The calibration curves of Figure 1 were plotted from data obtained several hours apart. The positions of

---

[1] Superscripts pertain to references at the end of the paper.

Table I. Counting Rate *vs.* Chemical Value

| Sample | Chemical value, % Al | Gross counting rate, counts per second | |
		Run No. 1	Run No. 2
1	2.4	75	69
2	3.8	101	94
3	6.9	163	158
4	8.3	195	198
5	9.5	239	240
6	11.9	299	295
7	13.8	324	316
8	15.8	376	365

individual points relative to the curve are very consistent. These relationships were not altered in spite of a small amount of drift.

Repeatability

Data from sixteen consecutive tests of a 10.2% aluminum alloy are given in Table III. Standard deviation is 1.7 counts per second, equivalent to 0.07% aluminum. Repositioning of the specimen between tests was found to have little influence upon repeatability.

Improved Peak-to-Background Ratio Resulting from Use of a Chromium-Target X-ray Tube

Peak-to-background ratio is improved by using a thin-window chromium-target tube instead of a tungsten-target tube. This is shown in Table IV. Background counting rates for chromium and tungsten targets are 7 and 45 counts per second, respectively.

Length of Time Required for Test

Not considering sample preparation, an alloy containing 2.4% aluminum can be analyzed in about 9 min. If the aluminum content is 15%, analysis time is 2 min. Pump-down time of 30 sec per sample is used in obtaining the above figures. Pump-down time is actually about 2 min, but the average time per sample is reduced to 30 sec if four samples are placed in the revolving turret.

Comparison of Vacuum and Helium Paths

The transmittance of aluminum radiation at a vacuum of 25 μ Hg was found to be about 10% greater than in a helium path. Pump-down time for the vacuum system is approximately 2 min. Purging time for the helium path is about 10 min.

Transmittance of Aluminum K_α Radiation as a Function of Pressure

The effect of pressure upon the transmittance of aluminum K_α radiation was found to agree substantially with published information.[2] At 500 μ Hg, approximately 98% transmittance is obtained, and the corresponding value for 100 μ Hg is greater than 99.5%.

Table II. Comparison between Chemical Values and X-ray Values*

Sample	X-ray value Run No. 1, % Al	Chemical value, % Al	d	d^2
1	2.6	2.4	− 0.2	0.04
2	3.8	3.8	0.0	0.00
3	6.6	6.9	+ 0.3	0.09
4	8.0	8.3	+ 0.3	0.09
5	9.8	9.5	− 0.3	0.09
6	12.4	11.9	− 0.5	0.25
7	13.6	13.8	+ 0.2	0.04
8	15.8	15.8	0.0	0.00
				$\Sigma = 0.60$

$$\sigma = \sqrt{\frac{\Sigma d^2}{n-1}} = \sqrt{\frac{0.60}{7}} = 0.3\% \text{ Al}$$

where σ is the standard deviation from the calibration curve in percent aluminum, d is the deviation of each individual chemical value from the calibration curve for a given counting rate, and n is the number of determinations.

* X-ray value is the percent aluminum as read from the calibration curve.

Table III. Repeatability of Counting Rate, 10.2% Aluminum Sample

Test	Counts per second	d	d^2
1	288	0.6	0.36
2	289	0.4	0.16
3	289	0.4	0.16
4	288	0.6	0.36
5	287	1.6	2.56
6	293	4.4	19.36
7	289	0.4	0.16
8	286	2.6	6.76
9	288	0.6	0.36
10	289	0.4	0.16
11	291	2.4	5.76
12	290	0.4	0.16
13	286	2.6	6.76
14	288	0.6	0.36
15	288	0.6	0.36
16	289	0.4	0.16
Mean	288.6		$\Sigma = \overline{44.16}$

$$\sigma = \sqrt{\frac{\Sigma d^2}{n-1}} = \sqrt{\frac{44.16}{15}} = 1.7 \text{ counts per second (representing } 0.07\% \text{ Al)}$$

where σ is the standard deviation, n is the number of determinations, and d is the deviation from mean.

Table IV. Comparative Peak-to-Background Ratios Resulting from Use of Chromium- and Tungsten-Target Tubes

Target	Peak-to-background ratio	
	2.4% Al sample	12% Al sample
Chromium	$\dfrac{4}{1}$	$\dfrac{13}{1}$
Tungsten	$\dfrac{2}{1}$	$\dfrac{5}{1}$

CONCLUSIONS

Vacuum X-ray fluorescence of iron–aluminum alloys offers the following combination of advantages:

1. Analysis time is only a small fraction of that required by chemical methods.
2. Repeatability of results, especially at high aluminum concentrations, is superior to typical optical spectrographic results.
3. Accuracy is probably at least as good as achieved by the best chemical methods.

ACKNOWLEDGMENT

The X-ray calibration is based upon two different chemical methods. One of these was developed at the Scientific Laboratory, Ford Motor Company, by Mr. G. E. Fisher. The other chemical technique is the result of work performed at the Metallurgy Department, University of Birmingham, England, by Mr. C. Sargent.

REFERENCES

1. B. D. Cullity, *Elements of X-ray Diffraction*, Addison-Wesley Publishing Co., Inc., Reading, Mass., 1956, p. 335.
2. D. C. Miller and P. W. Zingaro, "The Universal Vacuum Spectrograph, and Comparative Data on the Intensities Observed in an Air, Helium, and Vacuum Path," *Advances in X-ray Analysis Vol. 3*, University of Denver, Plenum Press, New York, 1959, pp. 49–56.

DISCUSSION

J. Gilfrich (U.S. Naval Ordnance Laboratory): What were the parameters used on the pulse height analyzer. The reason for the question is that at the U.S. Naval Laboratory the peak-to-background ratio was higher than at your laboratory.

J. Wagner: The baseline voltage was $1\frac{1}{2}$ v and the window width was 15 v, approximately.

J. Gilfrich: This is a little surprising, because these are approximately the same parameters we use and our signal-to-noise ratio was considerably larger. Do you have any thoughts on why this might be true?

J. Wagner: What type of X-ray tube were you using?

J. Gilfrich: The Philips FA-60 tungsten-target tube, 30-mil beryllium window. Our counting rates were of the same order of magnitude as yours.

J. Wagner: One possibility is a malfunction of the pulse height analyzer.

E. L. Gunn (Humble Oil and Refining Co.): I believe neither gentleman has mentioned the voltage on the counter tube.

J. Gilfrich: I assumed that since we were using roughly the same pulse height analyzer parameters we probably had the same order of magnitude of voltage on our counter tubes.

E. L. Gunn: Of course it occurs to me that you might be getting noise in there that you are not excluding, and that he is not getting the same noise level.

J. Wagner: Dr. Gunn has brought up a good subject. I didn't mention the voltage on the counter tube. The voltage that I used was regulated from day to day so that I would obtain the best peak-to-background ratio, but it was always very close to 1625 v.

C. Manning Davis (International Nickel Co.): We have had experience in transmitting the parameters from one laboratory to another within the same company and we run into this same problem. Why is the peak-to-background ratio different? Our general conclusion is that there are a whole host of things. One reason we found was the distance between the X-ray tube window and the sample, because the cone of X-rays hitting not only the sample but the mask around the sample had a great effect on the background. The mask material has an effect on this because in trying to do aluminum we have been using a phenol fabric which is a plastic type of material. We tried methyl methacrylate and we tried copper metal—each mask had a different amount of scatter and the amount of scatter, of course, affected the peak-to-background ratio.

G. Walden (Union Carbide Nuclear): I have experienced the same thing as Mr. Wagner has with this type of tube. The chromium-target tube gave a signal-to-noise ratio considerably larger than the tungsten tube, due to the great decrease in background.

CONTINUOUS DETERMINATION OF ZINC COATING WEIGHTS ON STEEL BY X-RAY FLUORESCENCE

James A. Dunne

Philips Electronic Instruments
Mt. Vernon, New York

ABSTRACT

The measurement of the area density of zinc coatings on steel by X-ray fluorescence is considered from an instrument design point of view. Each of the two general approaches, i.e., measurement of the attenuation of the iron emission line by the zinc coating and measurement of the zinc emission line has limitations. Calculations indicate that contrast requirements are best satisfied by the iron attenuation technique in the coating weight interval 0.2–1.2 oz/ft² per side. Experimental data collected on galvanized and zinc foil samples are presented in support of this contention. Conversely, advantages in the application of the zinc emission method to very thin zinc coatings are pointed out. Spectral resolution is discussed in terms of the ultimate precision and range of coating weight measurements. For the iron attenuation method, it is concluded that a black plate to infinite zinc intensity ratio of approximately 500 permits reasonably precise measurement of zinc foil coating weights in the vicinity of 1.2 oz/ft² per side. This requirement can be met through the use of a LiF monochromator crystal and 5 in. of effective 0.020-in. flat plate collimation.

INTRODUCTION

In response to the growing demand by hot-dip galvanize manufacturers for accurate continuous coating weight data to aid in the further improvement of their product, Philips Electronic Instruments has developed an X-ray gauge for the continuous measurement of zinc and aluminum coatings on a moving steel strip. The instrument, in principle, operates upon the strip sample in the same manner as does the Norelco tin coating weight gauge. In both cases, energy in the form of X-radiation is imparted to the base metal, and a measurement is made in terms of the attenuation in the coating of the resultant iron fluorescent radiation. In the case of the tin gauge, however, the base metal can be selectively excited, whereas the unavoidable simultaneous excitation of both zinc and iron requires a more elegant instrumental approach for the successful determination of zinc area density on steel.

THEORETICAL CONSIDERATIONS

The principle of coating weight measurements by X-ray absorption is well known. The basic relation between coating weight and observed X-ray intensity can be written as

$$I_t/I_o = \exp\left[-\left(\mu_1 \csc \phi + \mu_2 \csc \theta\right) t\right] \tag{1}$$

where I_t is the intensity observed for thickness t, I_o is the intensity observed for zero coating weight, μ_1 is the linear absorption coefficient of the coating for the incident

345

radiation, μ_2 is the linear absorption coefficient of the coating for the fluorescent radiation, ϕ is the incident angle, and θ is the emergent angle.

The relation between coating thickness and coating weight is

$$t = w/\rho \tag{2}$$

where w is the coating weight (area density in g/cm^2) and ρ is the density of the coating in g/cm^3.

Hence, if one substitutes mass absorption coefficients for linear absorption coefficients and area density for thickness in equation (1), coating density drops out, and X-ray intensity can be seen to be related to coating weight directly, without regard to coating density. In a given analytical system, all of the terms in the exponential can be held constant, and a simple relationship

$$I_w/I_o = \exp\left(-Qw\right) \tag{3}$$

can be derived, where Q is a constant containing mass absorption coefficients, incident and emergent angles, and unit conversion factors for w. For any given experimental arrangement, the value of the constant Q can be readily determined experimentally. This approach avoids the ambiguity associated with the exact calculation of Q, namely, the determination of the effect of incident radiation absorption upon the observed X-ray intensity for the divergent polychromatic source.

An alternate method of determining coating weight using X-rays is that which consists of observation of the radiation emitted by the constituents of the coating itself. In this case, the simplified expression can be written

$$I_w/I_t = 1 - \exp\left(-Rw\right) \tag{4}$$

where R is a constant similar to Q in equation (3), containing the absorption, geometric and unit conversion factors peculiar to a given coating and analytical system, and I_t is the intensity observed from a coating of infinite thickness.

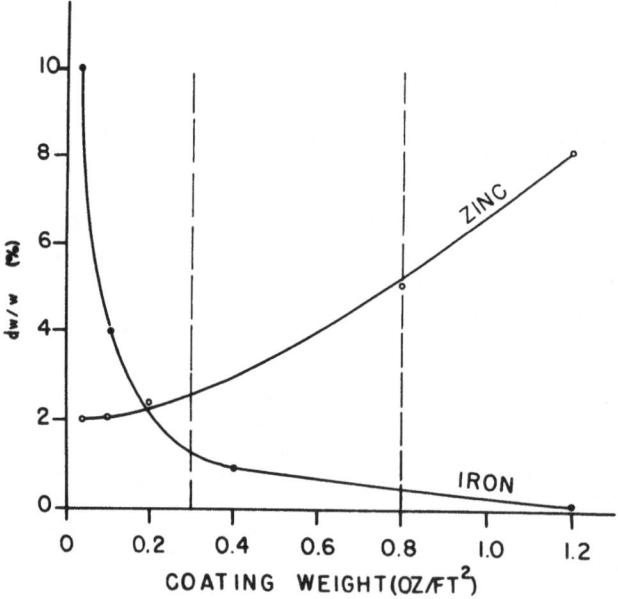

Figure 1. Plot of contrast vs. coating weight for zinc on steel.

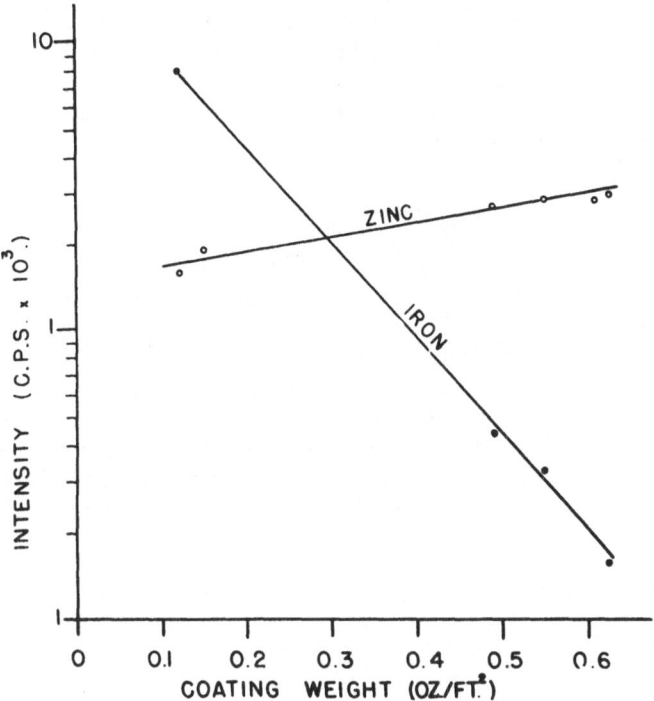

Figure 2. Plot of intensity *vs.* coating weight for hot-dip galvanized steel.

The choice between these two general approaches to the determination of area density is conditioned by the nature of the coating, the base metal, the coating weight interval over which measurements are to be made, and the required precision of the measurement. Since the latter consideration will always be of prime importance in any measurement, a convenient yardstick against which to compare alternate approaches is that of contrast: the percent change in coating weight for a given percent change in signal intensity. From equations (3) and (4), one can derive, by differentiation and substitution, the following relationships:

$$dw/w = dI_w/wQI_w \qquad (5)$$

and

$$dw/w = (dI_w/wRI_w) \left[\exp \left(R_w \right) - 1 \right] \qquad (6)$$

For the case of zinc coatings on steel, calculated contrast in percent is plotted against coating weight for a 2% change in signal intensity ($dI_w/I_w = 0.02$) in Figure 1. Incident and emergent angles of 90° are assumed. The exciting radiation is assumed to be parallel and monochromatic. On the basis of these calculations, it would appear that coating weight measurements in the interval 0.2–1.2 oz/ft² can be made with best precision by observing the attenuation of Fe K_α in the zinc coating. However, it is also evident that for very thin zinc coatings, the zinc emission method is far superior in terms of ultimate precision. The dashed lines in Figure 1 enclose the coating weight interval which was considered to be of most interest in the present hot-dip galvanizing process.

Figure 2 is a plot of experimental data taken on hot-dip galvanized samples using a standard Norelco bulk spectrograph. The coating weight interval covered is approximately equivalent to a range of 0.2–1.2 oz/ft² at an emergence angle of 90°. The observed full-range contrast for the iron absorption technique is about 25 times the contrast shown in the zinc emission data. The experimental data, then, were found to be in good agreement with previous calculations, and the decision was made to utilize the iron absorption method in the proposed Norelco zinc gauge.

X-RAY OPTICS DESIGN

One of the principle design objectives was that the instrument be capable of measuring a uniform zinc coating of 1.2 oz/ft² with a 3 σ precision of $\pm 2\%$ or 0.024 oz/ft². If we assume $I_0 = 80,000$ cps, $Q = 5.5$ (experimentally determined), and a time constant of 5 sec, the minimum acceptable black plate to infinite zinc intensity ratio turns out to be approximately 200 for the case where a background correction is made. Accordingly, a moderate resolution nonfocusing dispersive X-ray optical geometry with about $5\frac{1}{2}$ in. of effective 0.020-in. parallel plate collimation was selected. This arrangement has been found to provide a black plate to infinite zinc intensity ratio of 500. Background subtraction is achieved by means of a zero offset circuit. Figure 3 shows intensity *vs.* coating

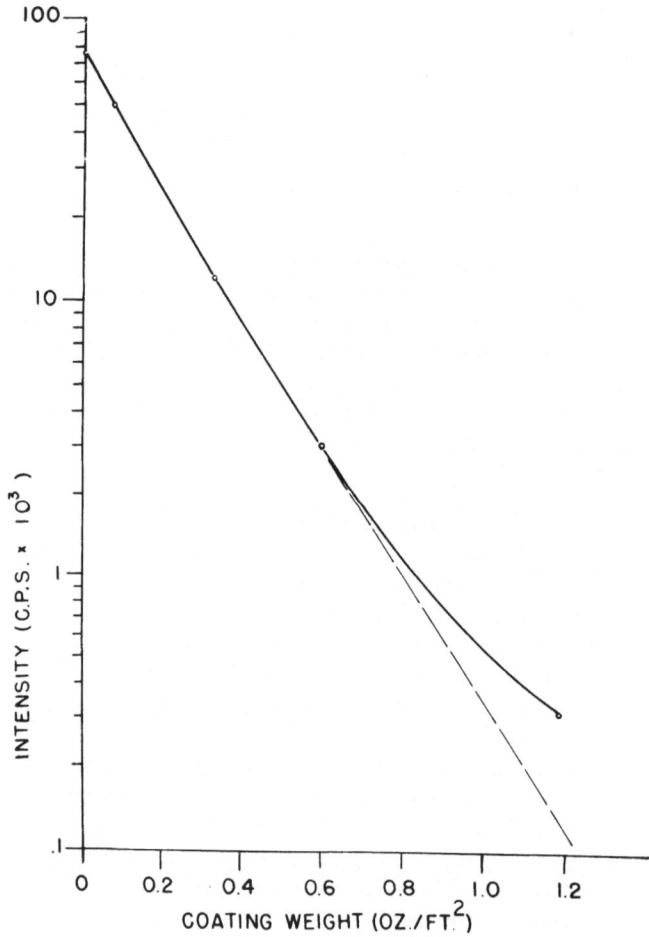

Figure 3. Plot of intensity *vs.* coating weight for zinc coating using iron absorption method.

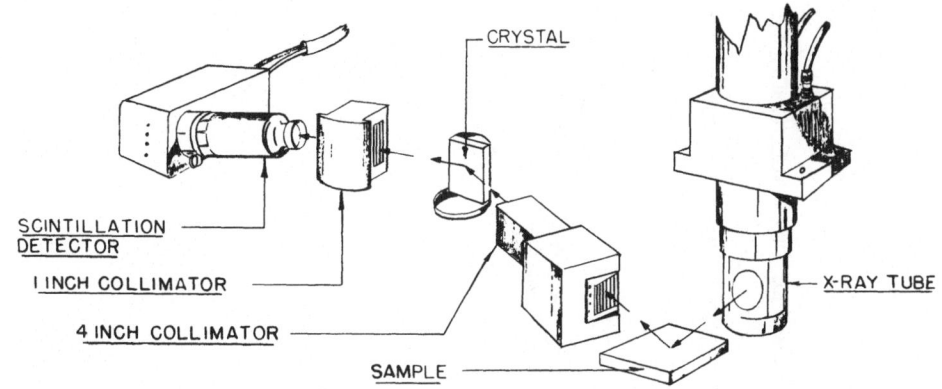

Figure 4. Sensing head X-ray optics.

weight data obtained on an early production model of the gauge sensing head. The dashed extension line is shown to demonstrate curve deflection due to background. Total observed contrast in the interval 0–1.2 oz/ft² is about 700 after background subtraction.

SENSING HEAD

A diagrammatic representation of the sensing head X-ray optics is given in Figure 4. The sample plane is shown rotated 90° from its proper position. X-rays generated in the sample are directed in a relatively parallel bundle onto a $1\frac{1}{2} \times 1$ in. LiF analyzing crystal by the incident 0.020×4 in. collimator. Those rays scattered and diffracted by the crystal are further collimated before entering the scintillation detector by a 0.014×1 in. collimator. The W-target FA-60 X-ray tube, crystal, 1-in. collimator and detector can be independently adjusted and locked. The adjustment and locking mechanisms are located external to the X-ray path.

FLUTTER COMPENSATION

A necessary function which the X-ray optical system must perform is the compensation for small displacements of the sample sheet along a vector normal to the sample plane. A small specified amount of such movement (flutter) must have virtually no effect upon the indicated coating weight if the gauge is to be practical for on-line measurements. The Norelco gauge has been designed to be capable of tolerating a flutter of $\pm\frac{1}{8}$ in. around a nominal pass line located $\frac{1}{2}$ in. from the base of the sensing head. Within these limits, the indicated coating weight will change by an amount that is less than the specified precision of static measurement. This compensation is accomplished by a rotational adjustment of the X-ray tube with respect to the incident collimator in such a manner that less of the most intensely excited area of the sample is seen by the collimator as the sheet moves toward the sensing head from the nominal sheet position. With reference to Figure 4, this means that point of intersection of the X-ray tube maximum intensity vector with the sample plane will not coincide with that of the incident collimator axial vector when the sample plane is in its nominal position with respect to the sensing head. Figure 5 shows data collected on a production model of the sensing head. Head to sample distance is plotted against indicated coating weight for three different rotational adjustments of the X-ray tube. The dashed lines represent 2% of the sample coating weight.

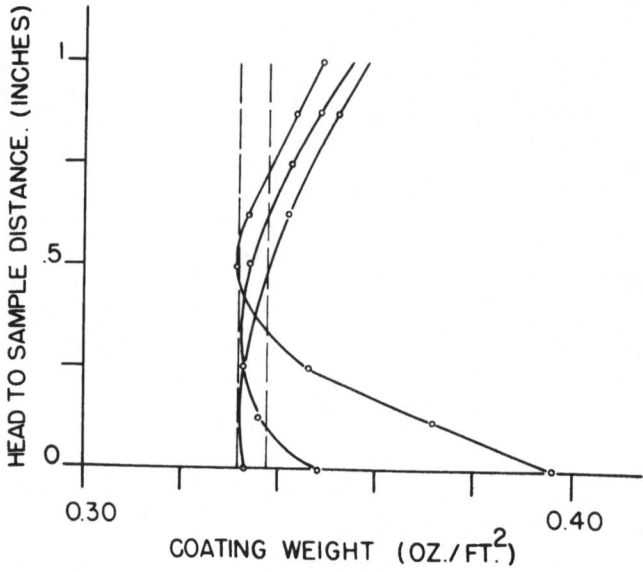

Figure 5. Effect of distance from sensing head to sample.

As can be seen from these curves, a total displacement of $\frac{1}{2}$ in. can actually be tolerated if care is taken in the selection and adjustment of the nominal pass-line position. In addition, it will be noted that, with proper alignment (center curve), no sharp boundary effects are encountered throughout a total flutter range of 1 in., the maximum reading error being approximately $6\frac{1}{2}\%$ of the indicated coating weight.

ELECTRONICS AND READOUT

Pulses generated in the scintillation detector are passed through an impedance-matching cathode follower–preamplifier appended to the tube and on through cabling to the console. There they are amplified and fed through a Schmidt trigger and a pre-selected number of binary stages to a rate meter, the voltage output of which is readout on a potentiometric strip chart recorder calibrated in oz/ft^2 per side. Three ranges are provided for the measurement of zinc coatings and one for hot-dipped aluminum. These in oz/ft^2 per side are as follows:

Zinc: 0.05–0.40
0.30–0.80
0.70–1.50
Aluminum: 0.25–0.50

The number of binaries between the Schmidt trigger and the rate meter is a function of the range selector switch position, as is the millivolt range of the recorder. Two scale bars are supplied, one for the three zinc ranges, and one for the single aluminum range. All scales are color-coded to coincide with indicator lights located on the console just below the recorder cabinet. In order to avoid the necessity of changing chart paper when changing ranges, recorder and desk overlays are provided for each range. The recorder overlays can be changed in a matter of seconds, and are positioned close to the chart paper to minimize chart reading error. The overlays are also color-coded to correspond to the indicator lights energized by the range selector switch.

It will be noted that no monitor detector is present in the sensing head (Figure 4). A standard Philips water-cooled constant potential X-ray generator supplies the required 300–400 w of X-ray power to each of the sensing heads. The amperage stabilizers contained in those units, in combination with the 5 kv-amp regulator on the line voltage provide a degree of X-ray stability which makes monitoring of the X-ray output unnecessary. Drift over a period of 12 hr does not exceed specified coating weight precision.

TRAVERSING MECHANISM

The sensing heads are mounted on the same traversing mechanism as supplied with the Norelco tin coating weight gauge. Strips up to 42 in. in width can be traversed. The small area of view defined by the sensing collimator (0.39 in.^{2}) is of advantage in terms of the spatial resolution it allows on the sample sheet. In order to utilize this feature properly, however, a suitable traversing speed and pattern must be selected. To this end, a speed of 2.8 ft/min (0.52 in./sec) has been provided. A pen marker can be supplied to indicate one strip edge. Photoelectric edge sensors automatically reverse the traverse motors when strip edge is approached. Four calibration standards are mounted on the traversing mechanism. Push buttons on the motor control box allow the operator to position the sensing heads over any of the standards, which consist of zinc or aluminum foil samples prepared and provided by Philips.

PERFORMANCE

Precision figures can be stated for zinc and aluminum foil samples. Zinc precision is $\pm 2\%$ of coating weight or 0.005 oz/ft^2 per side, whichever is the greater. On 99% aluminum foil samples, precision is $\pm 2\%$ of the indicated coating weight. With regard to product performance, no on-line accuracy data have been taken to date. Such data will be reported as they become available. Static tests on hot-dipped galvanized products show a mean deviation of 0.02 oz/ft^2 per side from adjacent weigh-strip-weigh determinations. These data were taken on 29 samples in the coating weight interval 0.34 to 0.73 oz/ft^2 per side. Precision was found to be within the limits stated for foil samples.

DISCUSSION

A. Goldblatt (Angstrom, Inc.): What kind of helium flow do you use when you measure aluminum?

J. A. Dunne: We do not measure aluminum K_α for hot-dipped aluminum. Therefore, the X-ray optical path system is identical to that of the zinc case.

SOME ASPECTS OF NONDESTRUCTIVE X-RAY SPECTROCHEMICAL ANALYSIS OF ALLOYS

J. Robert Rickenbach, Jr.

The Carpenter Steel Company
Reading, Pennsylvania

ABSTRACT

The paper is divided into two parts. The first section presents a review of three basic methods of X-ray spectrochemical analysis, including the internal standard method, the absolute intensity method, and the ratio method. The applications and limitations of each are discussed briefly. The second section presents the ratio method at length. Included are the methods of acquiring standards, sample preparation, curve plotting, and a discussion of the factors contributing to the precision of the analysis. Various analytical curves are presented and discussed and the sensitivity for minor constituents is shown.

INTRODUCTION

The use of X-ray spectrography as a tool for nondestructive and rapid chemical analysis of all types of material has increased manifold in recent years. We at Carpenter Steel, by means of various calibration curves, have been able to use this X-ray technique for chemical analyses of various alloys over a wide range of chemical compositions, with very good precision.

EQUIPMENT

The theory of the generation and properties of X-rays has been elaborately discussed in textbooks such as *Elements of X-ray Diffraction*,[1] and *X-Ray Absorption and Emission in Analytical Chemistry*,[2] so only a brief discussion of the instrumentation is given here.

The simplest arrangement of components of an X-ray spectrograph is shown in Figure 1. Primary X-rays from the X-ray tube strike the specimen and generate the characteristic X-rays of the specimen elements. These characteristic X-rays are emitted in all directions and the first step in analysis is to collimate them in a bundle of parallel rays. These secondary X-rays then strike the analyzing crystal cleaved on a given plane, which is of known chemical composition. The crystal acts very similarly to a prism or grating in emission spectroscopy. For each setting of the crystal, only one wavelength will be diffracted according to the Bragg law, $n\lambda = 2d \sin \theta$.

The diffracted radiation is measured by the detector. The wavelengths of the measured X-ray lines determine the elements present in the specimen, and the intensity of each line is related to the percentage composition. By proper selection of analyzing crystals and detectors, it is theoretically possible to perform X-ray spectrochemical analyses of elements of atomic number 8 and upward. However, our present setup limits us to the range of elements from titanium ($Z = 22$) to tin ($Z = 50$).

---

[1] Superscripts pertain to references at the end of the paper.

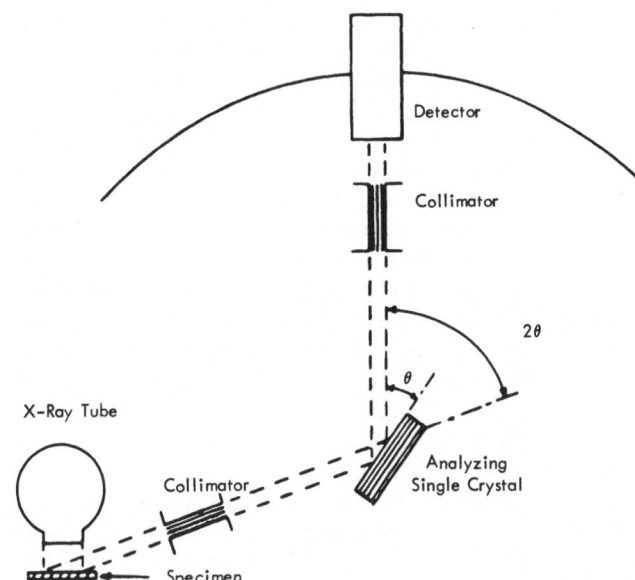

Figure 1. Schematic diagram
of an X-ray spectrograph.

The equipment used in this work was a G.E. XRD-5 X-ray spectrograph with standard components. A lithium fluoride analyzing crystal was used for the analysis of all elements from titanium to tin. A krypton-filled detector tube was used for the detection of the elements from selenium ($Z = 34$) to tin, and a flow detector tube using PR gas was used for the detection of the elements from titanium to copper ($Z = 29$). The K_α first-order reflection was used for all the elements except for cobalt in the presence of high iron content, in which case, the K_α second-order reflection was used.

All specimens were prepared by grinding on a 120-grit belt.

ANALYTICAL PROCEDURES

Three different methods can be used for X-ray spectrochemical analysis: (1) internal standard, (2) absolute intensity and, (3) ratio method.

1. Using the internal standard procedure, we add some element to both standards and unknown material. Then the intensities of the internal standard element and the element(s) of interest are obtained, and a ratio calculated using the equation

$$R = I_e/I_{is} \tag{1}$$

where I_e is the intensity of the element of interest and I_{is} is the intensity of the internal standard element.

The results are read from a pertinent calibration curve of ratio R vs. % concentration. This is not a suitable procedure for solid samples since the addition of the internal standard element cannot be properly controlled, as in the case with liquids and powders.

2. In the absolute intensity procedure, measurements, which should be corrected for background, are obtained for each element of interest, and the percent concentration is obtained from an intensity vs. % concentration curve. This is a suitable procedure for all types of analyses, but requires that standards be run to check each curve to compensate for shifting due to temperature, voltage drift, or electronic instability from one analysis to the next.

3. These variations can be eliminated by using a ratio of intensities instead of an absolute intensity. In the ratio method of quantitative spectrochemical analysis, the following equation can be used to compute the percentage of an unknown:

$$(I_u/I_s) \times \%s = \%u \tag{2}$$

where I_u is the intensity of the unknown, I_s is the intensity of standard, $\%s$ is the concentration of the standard (the same element in a similar matrix but of known composition) and $\%u$ is the concentration of the unknown.

When the ratio I_u/I_s is between 0.990 and 1.010, good agreement of X-ray $vs.$ wet-chemical results can be expected. As the ratio departs from these limits, there is a greater difference between the wet-chemical and X-ray spectrochemical results.

To overcome this problem and simplify the procedure, curves are used in conjunction with equation (2). To draw these working curves, a series of standards for each element in each matrix is required. These are generally referred to as secondary standards. One specimen, with an approximately known chemical composition in a similar matrix is selected as the comparing standard, sometimes referred to as the primary standard. The ratio of intensity of a secondary standard intensity to that of the primary standard is then calculated and plotted against the percent concentration. The percent concentration of an unknown can then be obtained from this curve by comparing the intensity ratio of the secondary to that of the primary standard.

STANDARDS

Care must be exercised when making up standards to avoid absorption and enhancement of X-ray intensity. W. J. Wittig[3] in his excellent paper on automatic X-ray spectrography reported that only four standard samples were required to obtain four working curves for the chemical analysis of "Hastelloy"—a nickel-base alloy—by varying the concentrations of four elements simultaneously at four heats of the alloy.

Vast differences between X-ray and wet-chemical analyses in some of the standards for our alloys were observed when we tried to vary more than one element at a time, possibly due to the fact that absorption characteristics of the matrix were changed. For example, the working curves drawn for three of the four elements of interest in Type

Table I. Counting Statistics for a Gaussian Distribution

No. of Counts	Standard deviation at 99% confidence limits, 3σ
1,000	9.5%
2,000	6.7%
4,000	4.9%
10,000	3.0%
20,000	2.1%
40,000	1.5%
100,000	0.95%
200,000	0.67%
400,000	0.47%
1,000,000	0.30%

Table II. Range of Elements Determined by X-ray Emission Spectroscopy

Element	Atomic number	Range in %
Ti	22	0.01–4.00
V	23	0.01–1.00
Cr	24	4.0–32.0
Mn	25	0.01–6.0
Fe	26	0.01–6.0
Co	27	0.01–20.0
Ni	28	5.0–85.0
Cu	29	0.01–0.60
Se	34	0.01–0.50
Zr	40	0.01–0.50
Nb	41	0.01–2.0
Mo	42	0.01–29.0
Sn	50	0.01–3.0

A-286 alloy (Cr:13.5–16.00, Ni:24.0–27.0, Mo:1.0–1.50, Ti:1.75–2.00), namely, molybdenum, titanium, and nickel, were in good agreement with the individual analytical curves for these three elements, but the same curve for chromium was in disagreement. This discrepancy was found to be due to titanium. Varying the titanium content from its nominal value seemed to have a pronounced effect on chromium excitation. On the other hand, chromium did not seem to have any significant effect on titanium.

We, therefore, adhère to the practice of making a series of heats for standards by varying only one element at a time.

PRECISION

The precision of X-ray spectrochemical analysis depends upon the total number of counts accumulated if all other variables are kept constant. Table I lists the counting statistics for various numbers of counts.[4] From this table it is seen that the greater the total number of counts obtained, the better the precision. Table II shows the number of elements and the ranges most often encountered in this work. In practice, we desire a precision of $\pm 0.5\%$ of the amount present when the concentration of an element is above 5%, and $\pm 2\%$ of the amount present below 5% concentration. Operating conditions of the X-ray unit were adjusted to accumulate 400,000 counts in about 15–25 sec for elements of high concentration. Equation (2) can be rewritten in terms of time unit as

$$R = T_s/T_u \qquad (3)$$

where R is the ratio, T_s is the time required to accumulate X number of counts from the standard, and T_u is the time required to accumulate the same number of counts from the unknown.

A study was made to determine statistical precision while counting 100,000 and 400,000 counts when the ratio R is 1.000. The results of this study are shown in Table III. Similar studies were made to determine the precision of ratios below, above, and at 1.000. The results are shown in Table IV.

From these studies it is quite evident that better precision can be obtained when the ratio is below 1.000 than when it is above; it is also clear that counting for 400,000

Table III. Standard Deviation at 99% Confidence Limits, 3σ, for Ratio $T_s/T_u = 1.000$ When Counting for 100,000 and 400,000 Counts

100,000 Counts			
Individual	*Avg. of 2*	*Avg. of 3*	*Avg. of 5*
±0.015	±0.011	±0.009	±0.007
400,000 Counts			
Individual	*Avg. of 2*	*Avg. of 3*	*Avg. of 5*
±0.009	±0.006	±0.005	±0.004

instead of 100,000 counts affords better precision. It is therefore desirable to have a comparing standard with a higher concentration than that anticipated in the analysis.

Ratios are obtained by measuring the time to accumulate X number of counts for the comparing standard and each of the secondary standards alternately. The average of ten ratios is used to plot each concentration point on the curve, using the above formula. The number of ratios required to produce the precision demanded is determined by studying Table III and the effect that this precision of ratios will have upon a determination for each element and each analytical curve. For a precision of ±0.5% of the amount present of a given element, one generally needs the average of three ratios. Figure 2 shows such a curve for nickel in a Type 302 stainless steel.

CORRECTIONS

In many cases there will be no need for correction factors for a given matrix. As stated before, however, there are occasions when the least-suspected element causes an error in the X-ray spectrochemical analysis. Several things can be done. One is to apply

Table IV. Standard Deviation at 99% Confidence Limits, 3σ, for Low, Medium and High Ratios when Counting for 100,000 Counts

Low Ratio (0.301)			
Individual	*Avg. of 2*	*Avg. of 3*	*Avg. of 5*
±0.006	±0.004	±0.003	±0.003
Medium Ratio (1.009)			
Individual	*Avg. of 2*	*Avg. of 3*	*Avg. of 5*
±0.015	±0.011	±0.009	±0.007
High Ratio (3.280)			
Individual	*Avg. of 2*	*Avg. of 3*	*Avg. of 5*
±0.060	±0.042	±0.035	±0.027

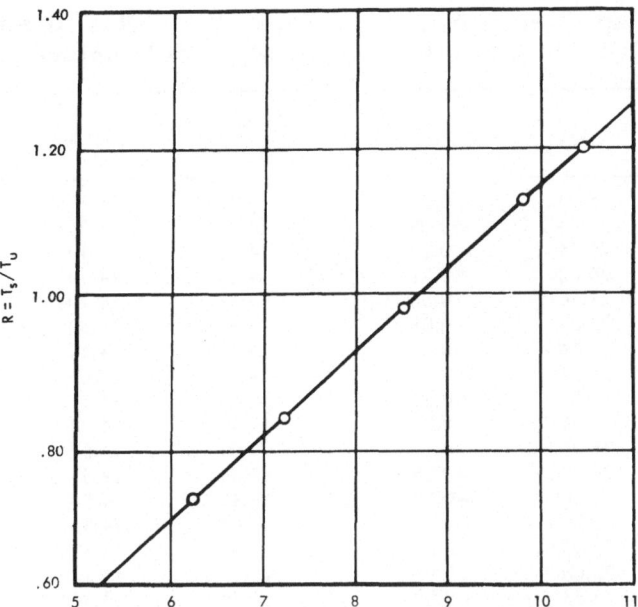

Figure 2. Nickel concentration wt. % Type 302 stainless steel.

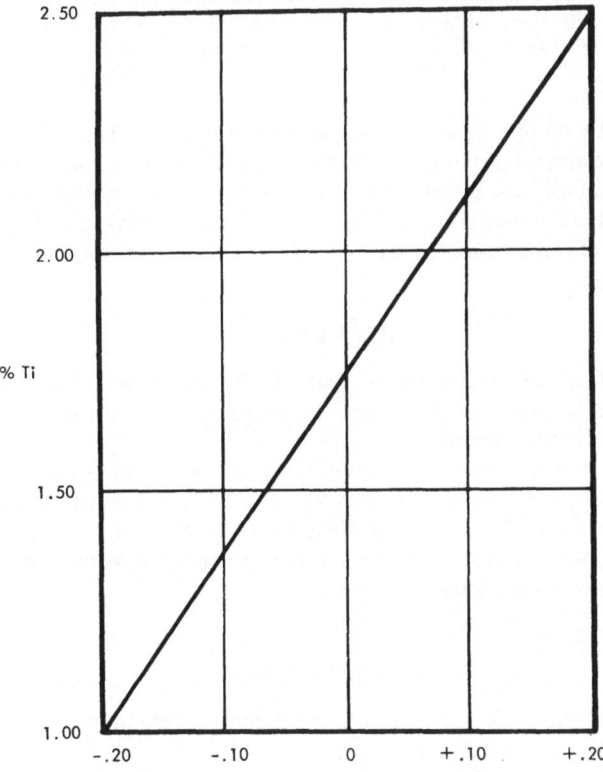

Figure 3. Percent chromium correction in Type A 286 alloy.

Table V. X-Ray Spectrochemical Analysis of Nickel in A-286 Alloy

Wet chemical	X-ray	Difference
25.90	25.57	−0.33
25.86	25.94	+0.08
26.35	26.25	−0.10
26.11	26.62	+0.51
25.72	25.66	−0.06
25.86	25.85	−0.01
25.52	25.44	−0.08
25.51	25.53	+0.02
25.53	25.73	+0.20
25.99	25.94	−0.05
26.02	25.91	−0.11
26.26	26.00	−0.26
26.29	26.10	−0.19
25.91	25.85	−0.06
25.96	25.91	−0.05
25.64	25.68	+0.04
25.61	25.53	−0.08
25.59	25.91	+0.32
25.60	25.80	+0.20

95% limits of the difference between wet and X-ray analysis is equal to ±0.37%.

Table VI. X-Ray Spectrochemical Analysis of Chromium in A-286 Alloy

Wet chemical	X-ray	Difference
14.37	14.37	0.0
14.47	14.28	−0.19
14.51	14.40	−0.11
14.68	14.52	−0.16
14.64	14.79	+0.15
14.51	14.59	+0.08
14.49	14.49	0.0
14.41	14.49	+0.08
14.44	14.51	+0.07
14.40	14.28	−0.12
14.64	14.57	−0.07
14.64	14.57	−0.07
14.47	14.54	+0.07
14.47	14.54	+0.07
14.40	14.32	+0.08
14.43	14.22	−0.21
14.44	14.34	−0.10
14.44	14.47	+0.03
14.40	14.47	+0.07

95% limits of the difference between wet and X-ray analysis is equal to ±0.22%.

a direct correction by adding or subtracting an experimentally determined figure to the ratio or the percent concentration figure. Another idea is to draw a correction curve where percent concentration of the interfering element is plotted on the ordinate and the correction factor on the abscissa. Figure 3 shows such a curve drawn for Type A-286 alloy, where the effect on chromium of variation in titanium from its nominal value is shown.

RESULTS

X-ray spectrochemical results are comparable to wet-chemical analyses, as shown in Tables V and VI. The results shown in these two tables are routine, and no attempts were made to resolve the differences.

Precision of X-ray spectrochemical analyses of nickel and chromium in three specimens of A-286 alloy is shown in Tables VII and VIII, respectively. This study was conducted over a period of 30 days, and during this time the comparing standard was resurfaced several times, but the test specimens were resurfaced only once. All results are within the precision desired by our laboratories.

CONCLUSIONS

1. Chemical analysis by X-ray spectroscopy is comparable to wet-chemical methods, but it is very rapid and less cumbersome.
2. Calibration curves can be plotted in terms of the ratio of the time required for accumulating a certain number of counts from the primary standard to the time

Table VII. Percent Nickel in Specimen			
Analysis	1A	2A	3A
1	26.24	26.27	26.18
2	26.22	26.27	26.22
3	26.22	26.24	26.24
4	26.18	26.22	26.18
5	26.24	26.22	26.18
6	26.24	26.27	26.15
7	26.17	26.22	26.17
8	26.18	26.18	26.21
9	26.29	26.24	26.18
10	26.22	26.24	26.15
Average	26.22	26.24	26.19
Standard deviation at 99% confidence limits, 3σ	±0.12	±0.09	±0.09

Table VIII. Percent Chromium in Specimen			
Analysis	1A	2A	3A
1	14.53	14.52	14.58
2	14.50	14.53	14.58
3	14.50	14.55	14.57
4	14.53	14.58	14.64
5	14.50	14.52	14.60
6	14.55	14.57	14.60
7	14.52	14.58	14.63
8	14.55	14.57	14.62
9	14.53	14.55	14.62
10	14.47	14.55	14.62
Average	14.52	14.55	14.61
Standard deviation at 99% confidence limits, 3σ	±0.06	±0.06	±0.06

for the secondary standard *vs.* % concentration, and these curves are used to determine the composition of unknown.

3. Preparation of standards for working-curve purposes is very critical. Altering more than one element at a time can affect the working curves.

ACKNOWLEDGMENT

Acknowledgments are due to The Carpenter Steel Company for permission to publish this work.

REFERENCES

1. B. D. Cullity, *Elements of X-ray Diffraction*, Addison-Wesley Publishing Co., Inc., Reading, Mass., 1956.
2. H. A. Liebhafsky, *et al.*, *X-ray Absorption and Emission in Analytical Chemistry*, John Wiley & Sons, Inc., New York, 1960.
3. W. J. Wittig, Production Control Analyses Using the Automatic X-ray Spectrograph, *Norelco Reptr.* **6**(1): 17, 1959.
4. P. W. Zingaro, Statistics in X-ray Intensity Measurements, *Norelco Reptr.* **5**(5–6) 99, 1958.

DISCUSSION

F. Bernstein (General Electric Co.): In the curve you showed, there was a correction of titanium on chromium. It was a straight line. You showed no points. I was wondering if this was actually measured, or if it is an approximation.

J. R. Rickenbach: The curve shown was drawn for illustration only. We have one that was drawn from experimentation. Yes, we had titanium results as low as 1% and as high as 2.75, I think, and they did form a rather straight line.

F. Bernstein: Isn't that a chemical correction?

J. R. Rickenbach: Yes.

F. Bernstein: In terms of doing the ultimate chemical analysis originally, is there so much titanium that you could make an error in the chromium?

J. R. Rickenbach: Yes.

F. Bernstein: That is, it is not an X-ray curve.

J. R. Rickenbach: This is right. The titanium figure is read from that curve and the correction at the bottom of the curve is applied to the chromium analysis. This would correct for the difference between X-ray and wet-chemical.

F. Bernstein: Which analysis do you add it to?

J. R. Rickenbach: The X-ray analysis.* It has been suggested by the chairman that I elaborate on this. The difficulty lies in the fact that sometimes there is a difference in results between X-ray and wet-chemical analyses. In most cases, there is very little difference between X-ray and wet-chemical analyses of a given alloy, but it so happens that with this type A-286 alloy we were in hot water for quite a while because we had been receiving specimens from all over the plant—from arc-melting, from vacuum-melting, from induction-melting—and we noticed differences between X-ray and wet-chemical analyses. As we would eventually use X-ray analysis exclusively, a study of X-ray *vs.* wet-chemical analysis of chromium showed that at times we were high and other times low. For the moment we didn't know what was causing this discrepancy. We conducted a study to determine the cause of this disagreement. We then made up the special series of standards of Type A-286 alloy and the plot of the chromium curve was in disagreement with the standard chromium working curve. We found that titanium was the cause of this. We did not suspect titanium to be the cause since the analyses were within 1%, i.e., the range from the highest to the lowest points were about 2%. So we took all the chromium results that had accumulated over several months and plotted the difference between X-ray and wet-chemical analysis of chromium *vs.* % titanium, and from this we obtained a curve which is now used to correct all of our chromium results. Mr. Gunn suggests that we investigate the absorption effect causing the discrepancy in the chemical analysis and we agree that this is causing the difficulty, but in order to reduce it to a minimum amount of work and to make it easy for operation, we did not bother with correction factors. We did it the easy way to get the answer.

* At this point it was apparent that there was a misunderstanding between Bernstein and Rickenbach.

SODIUM AND MAGNESIUM FLUORESCENCE ANALYSIS—PART I: METHOD

Burton L. Henke

Pomona College
Claremont, California

ABSTRACT

As is well known, the fluorescent yield decreases very rapidly with the atomic number with the result, for example, that sensitive sodium and magnesium analysis is extremely difficult if not impossible with conventional X-ray spectrographs. It is demonstrated, however, that analysis for sodium and magnesium can be accomplished with sensitivity comparable to that conventionally obtained for elements such as aluminum, silicon, and phosphorous, providing that the conditions for excitation and measurement of the associated soft X-radiations are optimized. A high-intensity demountable tube using an aluminum anode has been developed which can be used interchangeably with the conventional spectrographic X-ray source. This provides a large amount of incident radiation, aluminum foil filtered, optimally close in wavelength to that of the line radiation being excited. A gypsum analyzing crystal is used along with greatly reduced beam collimation. The standard flow proportional counter and pulse height discrimination is employed. An appropriate filter, such as aluminum foil, is used as a window for the counter to provide further discrimination and enhanced signal-to-background ratio.

INTRODUCTION

By an optimization of the fluorescence analysis method as extended into the long-wavelength X-ray region (5 to 20 Å region), it is possible to gain a considerable improvement in sensitivity for the routine analysis of elements Cl, S, P, Si, Al, Mg, Na, and F. This is accomplished by the combination of recently developed vacuum spectrographs, long-spaced crystals, flow proportional counters, and demountable high-intensity sources of soft X-radiations.

The basic objective in the design of a fluorescence analysis measurement is to gain a maximum in the counting rate for a given line radiation from the sample along with the required signal-to-background ratio. The minimum allowable signal-to-background ratio is set by well-established statistical analysis considerations.[1] This minimum ratio and the minimum peak counting rate which is permissible is dependent upon the time which can be allowed for each measurement and, therefore, ultimately upon the stability of the total analysis system. A good test, of course, is to apply the method to calibrated samples of interest. Results of such measurements will be listed at the end of this paper. A comprehensive evaluation of the method as applied to Na and Mg analysis of silicates is presented in this volume in the paper by A. K. Baird, *et al.*

Some insight regarding the factors which determine the amount of fluorescence signal in a particular measurement can be gained from the intensity equation. An

---

[1] Superscripts pertain to references at the end of the paper.

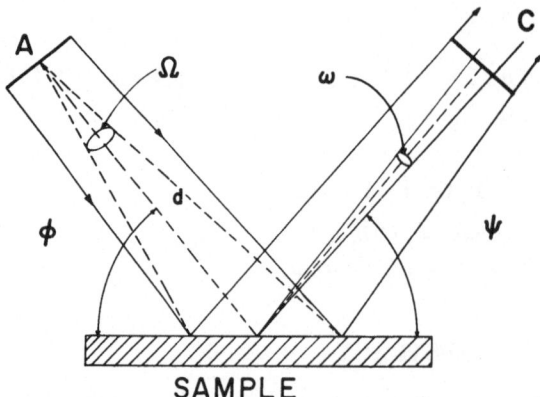

Figure 1. Optics of the fluorescence excitation.

approximate equation (which assumes that all of the exciting radiation strikes a homogeneous sample at grazing incidence ϕ and that all of the radiation accepted by the crystal analyzer leaves the sample at grazing angle ψ) can readily be obtained by extending a derivation similar to that given by Compton and Allison.[2] The result may be written as follows:*

$$Q_x = \left[\frac{q_0 A}{d^2}\right]\left[z_i W_q\left(\frac{r-1}{r}\right)\right]\left[\frac{\rho_x \tau_x}{\mu_1}\right]\left[\frac{\omega RTC}{4\pi}\right]\left[\frac{\sin\phi/\sin\psi}{1+(\mu_i/\mu_1)(\sin\phi/\sin\psi)}\right] \quad (1)$$

In order to obtain a maximum of intensity, it is important to have the geometric ratio, $\sin\phi/\sin\psi$ large compared with unity. It is desirable to have ψ equal to about $90° - \phi$ in order to minimize the contribution to background due to sample scattering—primary scattering is a minimum in the $90°$ direction. Equation (1) may be further simplified by writing it specially for the case of interest here, *viz.* for the analysis of light elements. In the F to Cl range, W_K varies between 0.01 and 0.1, which small values account for the very low intensities that are characteristic of light-element analysis. If W_K is taken as approximately 0.05, and the corresponding values of r and z_i for these light elements are taken as about 10 and 0.6, respectively, (and using the optimally large angles of incidence for ϕ) equation (1) becomes

$$Q_x = 0.002\left[\frac{q_0 A}{d^2}\right]\left[\omega RTC\right]\left[\frac{\rho_x \tau_x}{\mu_i}\right] \quad (2)$$
$$\underset{\text{Source}}{}\underset{\text{Spectrograph}}{}\underset{\text{Matrix}}{\phantom{\left[\frac{\rho_x \tau_x}{\mu_i}\right]}}$$

(For light-element K radiations and for (μ_i/μ_1) $(\sin\phi/\sin\psi) \gg 1$.)

SPECTROGRAPH

The foregoing relation, equation (2), demonstrates that the fluorescent intensity is proportional to the value of the photoelectric absorption cross section τ_x and therefore it increases with approximately the cube of the exciting wavelength. Consequently, for maximum sensitivity the exciting radiation as well as the fluorescent radiation will be of long wavelength, and the use of a vacuum spectrograph is nearly always necessary.

* A complete list of notation is given at the end of the paper.

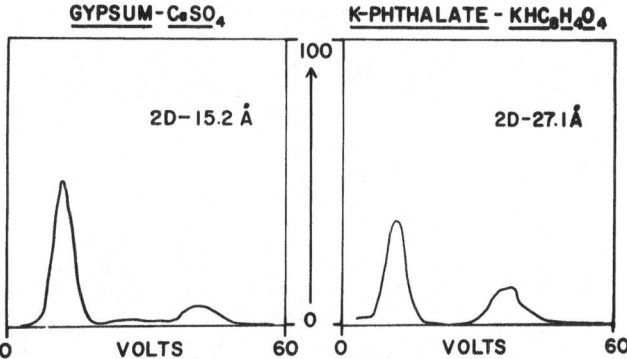

Figure 2. Pulse height distribution from gas flow proportional counter and with spectrometer set on Na K_α (11.9 Å) line. It should be noted that the two crystals used here may or may not be representative of their type. Sample is Standard Rock G-1 (3.26 wt. % Na_2O). Excitation—10 kv Al K_α.

The X-ray optics as conventionally involved in the vacuum spectrograph which was used in this study (Philips) seemed nearly ideal for the light-element work. In this instrument the angle ϕ is made large, and the angle between the exciting radiation and that in the measured fluorescent beam is approximately 90°. The optimum condition that $(\mu_i/\mu_1)(\sin\phi/\sin\psi)$ be large compared with unity is sufficiently well satisfied for most long-wavelength excitation measurements.

The cross-sectional area C of the measured fluorescent beam is limited by the size of the analyzing crystal and by the allowable size for the sample. If larger crystals, samples, and effective counter windows become feasible, a gain in intensity in proportion to the square of the scaling ratio is to be expected.

Three commercially available analyzing crystals were chosen for the wavelength analysis region of interest here (5 to 20 Å). The organic crystal EDT yields very satisfactory signal and signal-to-background ratios and was found to be superior to gypsum for all wavelengths up to its $2d$ limit of 8.81 Å. Fortunately, gypsum becomes satisfactorily efficient for the longer wavelengths as excited, for example, by an Al K source because the background due to its calcium and sulfur fluorescing is greatly reduced. As illustrated in Figure 2, the gypsum crystal was superior in the 12-Å region to the potassium acid phthalate crystal. It should be noted that this comparison was made between only two commercial crystals, which may or may not be representative of their class. After the $2d$ limit on gypsum of 15.2 Å, the KAP crystal was used—$2d$ equal to 27.1 Å. Tables I and II include the principal emission lines of the K, L, and M series (kX units have been converted to Angstrom units), the 2θ values for the several crystals, the photon energies, and the excitation energy in kev. The relatively large dispersion among the fluorescent lines of interest here, as shown in Tables I and II, allows the gain of an appreciable improvement in intensity by reducing the amount of collimation and thereby increasing the solid angle factor ω. Long-wavelength analysis demands a proper matching of the collimation to the obtainable dispersion for optimum efficiency. This subject will be considered in detail in Part II of this work.

EXCITATION RADIATION SOURCE

As stated above, a considerable improvement can be gained in both the signal and in the signal-to-background ratio for light-element analysis by using long-wavelength

Table I. Spectroscopy of X-Rays in the 4 to 70 Å Wavelength

EDT: ethylenediamine d-tartrate, $H_{22}C_{10}N_2O_{10}$. $2d = 8.81$ Å—Gypsum: calcium sulfate,

K_{α_1}	K_{β_1}	L_{α_1}	L_{β_1}	M_α	M_{β_1}	λ, Å	2θ EDT	2θ Gypsum	2θ KAP	Photon kev	Excitation kev
			46Pd			4.15	56.2	31.6	18.0	2.99	3.33
		47Ag				4.15	56.3	31.7	18.0	2.98	3.35
18Ar						4.19	56.8	32.0	18.1	2.96	3.20
		46Pd				4.37	59.4	33.4	18.9	2.84	3.17
			45Rh			4.37	59.5	33.4	18.9	2.83	3.15
	17Cl					4.40	60.0	33.7	19.0	2.82	2.82
		45Rh				4.60	62.9	35.2	19.9	2.70	3.00
			44Ru			4.62	63.3	35.4	20.0	2.68	2.97
17Cl						4.73	64.9	36.2	20.5	2.62	2.82
		44Ru				4.85	66.7	37.2	21.0	2.56	2.84
			43Tc			4.88*	67.3	37.4	21.1	2.54	2.80
					83Bi	4.91	67.7	37.7	21.3	2.53	2.70
	16S					5.03	69.7	38.6	21.8	2.46	2.47
					82Pb	5.08	70.3	39.0	22.0	2.44	2.60
		43Tc				5.11*	70.9	39.3	22.2	2.42	2.68
				83Bi		5.12	71.0	39.3	22.2	2.42	2.59
			42Mo			5.18	72.0	39.8	22.5	2.39	2.63
					81Tl	5.25	73.1	40.4	22.8	2.36	2.49
				82Pb		5.28	73.7	40.7	22.9	2.35	2.50
16S						5.37	75.1	41.4	23.3	2.31	2.47
		42Mo				5.41	75.7	41.6	23.5	2.29	2.52
					80Hg	5.43*	76.1	41.8	23.6	2.28	2.39
				81Tl		5.46	76.6	42.1	23.7	2.27	2.39
			41Nb			5.49	77.1	42.3	23.8	2.26	2.47
					79Au	5.62	79.3	43.4	24.4	2.20	2.29
				80Hg		5.65*	79.7	43.6	24.5	2.20	2.30
		41Nb				5.73	81.0	44.2	24.8	2.17	2.37
	15P					5.80	82.4	44.9	25.1	2.14	2.14
					78Pt	5.83	82.8	45.1	25.3	2.13	2.20
			40Zr			5.84	83.0	45.1	25.4	2.12	2.31
				79Au		5.84	83.0	45.2	25.4	2.12	2.21
					77Ir	6.04	86.5	46.8	26.2	2.05	2.11
				78Pt		6.05	86.7	46.8	26.3	2.05	2.12
		40Zr				6.07	87.1	47.0	26.4	2.04	2.22
15P						6.15	88.6	47.7	26.7	2.01	2.14
			39Y			6.21	89.7	48.2	27.0	2.00	2.15
				77Ir		6.26	90.6	48.6	27.2	1.98	2.04
					76Os	6.27	90.7	48.7	27.3	1.98	2.03
		39Y				6.45	94.1	50.2	28.1	1.92	2.08
				76Os		6.49	94.9	50.5	28.2	1.91	1.96
					75Re	6.50	95.2	50.6	28.3	1.91	1.95
			38Sr			6.62	97.5	51.6	28.8	1.87	2.01
				75Re		6.73	99.6	52.5	29.3	1.84	1.88
					74W	6.76	100	52.7	29.4	1.83	1.87
	14Si					6.76	100	52.8	29.4	1.83	1.84
		38Sr				6.86	102	53.6	29.9	1.81	1.94
				74W		6.98	105	54.7	30.4	1.78	1.80
					73Ta	7.02	106	55.0	30.6	1.77	1.79
			37Rb			7.08	107	55.4	30.9	1.75	1.87
14Si						7.13	108	55.9	31.1	1.74	1.84
				73Ta		7.25	111	56.9	31.6	1.71	1.73
					72Hf	7.30	112	57.4	31.9	1.70	1.71
		37Rb				7.32	112	57.5	31.9	1.69	1.81
				72Hf		7.54	118	59.4	32.9	1.64	1.66
			36Kr			7.57*	118	59.7	33.0	1.64	1.73
					71Lu	7.60	119	60.0	33.2	1.63	1.64
		36Kr				7.81*	125	61.8	34.1	1.59	1.68
				71Lu		7.84	126	62.1	34.3	1.58	1.59
					70Yb	7.91	128	62.7	34.6	1.57	1.58
13Al						7.98	130	63.3	34.9	1.55	1.56

* An asterisk indicates that an L-series wavelength was derived from a photon energy given by Fine and

Region with Gypsum and KAP Crystal Analyzers[3-7]

$CaSO_4$. $2d = 15.2$ Å—KAP: potassium acid phthalate, $KHC_8H_4O_4$. $2d = 26.6$ Å.

| | | | | | | | 2θ | | | Photon | Excitation |
K_{α_1}	K_{β_1}	L_{α_1}	L_{β_1}	M_α	M_{β_1}	λ,Å	EDT	Gypsum	KAP	kev	kev
			35Br			8.13	135	64.6	35.6	1.53	1.60
				70Yb		8.14	135	64.7	35.6	1.52	1.53
					69Tm	8.24*	139	65.6	36.1	1.50	1.52
13Al						8.34	142	66.5	36.5	1.49	1.56
		35Br				8.37	144	66.8	36.7	1.48	1.55
				69Tm		8.46*	148	67.6	37.1	1.46	1.47
					68Er	8.59	155	68.8	37.7	1.44	1.45
			34Se			8.74	165	70.1	38.4	1.42	1.47
				68Er		8.80	175	70.7	38.6	1.41	1.41
					67Ho	8.96	—	72.2	39.4	1.38	1.39
		34Se				8.99	—	72.5	39.5	1.38	1.43
				67Ho		9.16	—	74.1	40.3	1.35	1.35
					66Dy	9.36	—	76.0	41.2	1.32	1.33
			33As			9.41	—	76.5	41.4	1.32	1.36
				66Dy		9.54	—	77.7	42.0	1.30	1.30
	12Mg					9.56	—	77.9	42.1	1.30	1.30
		33As				9.67	—	79.0	42.6	1.28	1.32
					65Tb	9.79	—	80.1	43.2	1.27	1.28
12Mg						9.89	—	81.1	43.7	1.25	1.30
				65Tb		9.94	—	81.6	43.9	1.25	1.25
			32Ge			10.2	—	84.0	45.1	1.22	1.25
					64Gd	10.3	—	84.8	45.6	1.21	1.22
				64Gd		10.4	—	86.4	46.0	1.19	1.19
		32Ge				10.4	—	86.6	46.0	1.19	1.22
					63Eu	10.7	—	89.9	47.4	1.15	1.16
				63Eu		11.0	—	92.1	48.9	1.13	1.13
			31Ga			11.0	—	92.9	48.9	1.12	1.13
					62Sm	11.3	—	95.5	50.3	1.10	1.11
		31Ga				11.3	—	95.8	50.3	1.10	1.12
				62Sm		11.4	—	97.4	50.8	1.08	1.08
	11Na					11.6	—	99.6	51.7	1.07	1.08
					61Pm	11.8*	—	102	52.7	1.05	1.05
11Na						11.9	—	103	53.2	1.04	1.08
			30Zn			12.0	—	104	53.6	1.03	1.05
				61Pm		12.0*	—	105	53.6	1.03	1.03
		30Zn				12.3	—	107	55.0	1.01	1.02
					60Nd	12.4	—	109	55.6	1.00	1.00
				60Nd		12.7	—	113	57.0	0.978	0.978
					59Pr	13.1*	—	118	59.0	0.950	0.955
			29Cu			13.1	—	118	59.0	0.950	0.953
				59Pr		13.3*	—	122	60.0	0.932	0.935
		29Cu				13.3	—	122	60.0	0.930	0.933
					58Ce	13.8	—	130	62.5	0.899	0.906
				58Ce		14.1	—	135	64.0	0.882	0.889
			28Ni			14.3	—	140	65.0	0.868	0.871
		28Ni				14.6	—	147	66.6	0.851	0.853
10Ne						14.6	—	148	66.6	0.847	0.874
			27Co			15.7	—	—	72.3	0.792	0.794
		27Co				16.0	—	—	74.0	0.776	0.779
			26Fe			17.3	—	—	81.1	0.718	0.721
		26Fe				17.6	—	—	82.9	0.706	0.708
9F						18.3	—	—	86.9	0.676	0.687
			25Mn			19.1	—	—	91.8	0.648	0.650
		25Mn				19.4	—	—	93.7	0.637	0.639
			24Cr			21.3	—	—	106	0.582	0.583
		24Cr				21.7	—	—	109	0.572	0.574
8O						23.6	—	—	125	0.524	0.531
			23V			23.8	—	—	127	0.520	0.519
		23V				24.3	—	—	132	0.511	0.512

Hendee; M-series wavelengths so designated, have been interpolated from data given by J. M. Cork.

Table II. 2θ Values for a Barium Stearate Soap Film Analyzer
$2d = 100.38$ Å[8]

K_{α_1}	L_{α_1}	L_{β_1}	λ, Å	2θ	Photon, kev	Excitation, kev
8O			23.6	27.3	0.524	0.531
		23V	23.8	27.5	0.520	0.520
	23V		24.3	28.0	0.511	0.512
		22Ti	27.0	31.2	0.459	0.460
	22Ti		27.4	31.7	0.453	0.454
		21Sc	31.0	36.0	0.400	0.411
	21Sc		31.3	36.4	0.396	0.406
7N			31.6	36.7	0.392	0.399
		20Ca	36.0	42.0	0.345	0.352
	20Ca		36.3	42.4	0.341	0.349
6C			44.6	52.8	0.278	0.283
5B			67.8	85.1	0.183	0.192

excitation providing that a good source for such wavelengths is available. A very efficient source for the soft and the ultrasoft X-ray region has been developed and has been applied successfully to this problem.

Although designed primarily as a low-voltage X-ray tube, the instrument was applied here first as a source for Ti K radiation (2.7 Å), for Ag L radiation (4.2 Å) and finally for Al K radiation (8.3 Å). This was done to establish the principle that a considerable improvement in sensitivity for such light elements as sodium and magnesium can be gained with a practical source of wavelength near the critical absorption edge wavelengths for these elements, *viz.* Al K (8.34 Å). Pulse height distributions, line profiles with optimum pulse height integration, and excitation curves with the anode power held at one kilowatt, were taken with each source. A 6-μ aluminum window was used on the source, a gypsum crystal and a standard flow proportional counter with a 6-μ mylar window and P-10 gas were also used in this test. Results of such measurements for sodium analysis on a powdered sample of granite (USGS Standard-G1 of 3.26 wt.% Na_2O) are shown in Figure 3 and for the oxalates of both sodium and magnesium in Figures 4 and 5. It should be noted that had a 6-μ mylar window been used on the X-ray source rather than the 6-μ aluminum foil, the gain in relative intensity for both the Ti K and the Ag L excitation, as compared to that for Al K radiation, would have been no more than a factor of three so that even in this instance the Al K excitation would be markedly superior. Also, apart from the intensity gain, it is important to note the very great improvement obtained in the reduction of the background from crystal fluorescence due to sample-scattered primary radiation. As shown in Figure 4, with the oxalates for which sample matrix effects cannot contribute significantly to background, the sulfur and calcium fluorescence background from the crystal is essentially absent for the excitation by the aluminum anode source. (So as to eliminate any effects of X-ray tube differences in these measurements, after the titanium and silver anode sources were tested, the anodes were coated with a vacuum-evaporated layer of aluminum and the measurements were retaken in exactly the same manner in order to evaluate the enhancement of signal and signal-to-background resulting from the Al K excitation.)

The design of the excitation source used in this light-element work is illustrated in Figure 6. In this source two relatively large focal spots, $\frac{1}{4} \times 1$ in., are placed along the

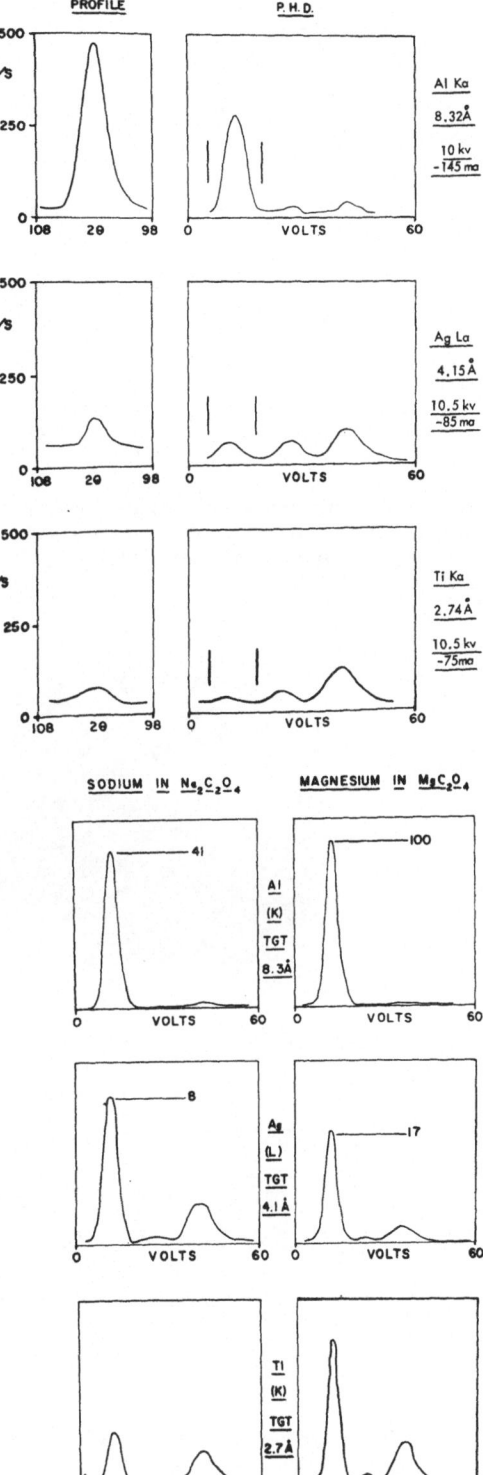

Figure 3. Line profiles and pulse height distributions for sodium in granite (2.4% Na) as excited by sources with a titanium, a silver, and an aluminum anode. A 6-μ window was used on the source and a 6-μ mylar window on the counter. The background, which is displayed in the pulse height distributions and particularly evidenced with the shorter wavelength excitation, is due to the fluorescence of sulfur and calcium in the gypsum crystal used as analyzer.

Figure 4. Measurement conditions were the same as stated in Figure 3 except for the samples being the oxalates of sodium and magnesium. These pulse height distributions illustrate the increased contribution to background due to crystal fluorescence by the shorter wavelengths as separated from any possible sample matrix interference. Peak signals are noted relative to that for magnesium with aluminum source excitation.

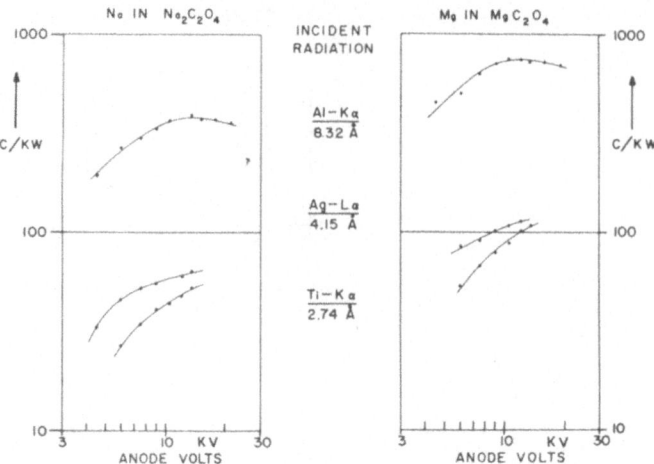

Figure 5. Measurement conditions are the same as were stated for Figures 3 and 4. The counting rate per kilowatt of source anode power was taken as a function of anode voltage for the line radiation peaks (taken with optimum pulse height integration in each case).

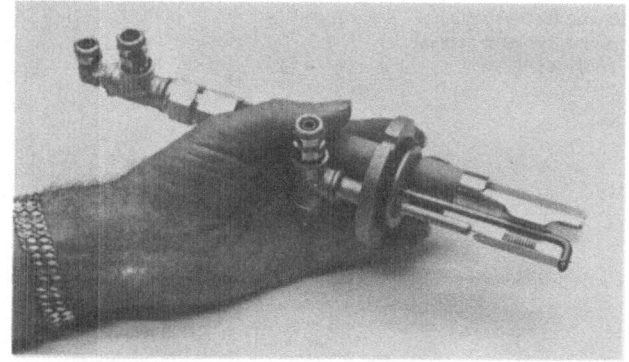

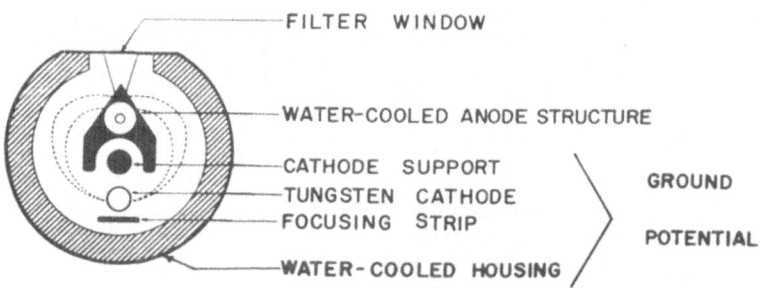

Figure 6. The interchangeable anode–cathode assembly which inserts, with an O-ring seal, into the cylindrical tube housing. This source may be operated in the 500 to 15,000 v range for the generation of high-intensity characteristic radiations in the 5 to 200 Å region.

sides of an anode of triangular cross section. This is accomplished by electrostatically deflecting the beam from a helical tungsten cathode below the anode by the effect of the cylindrical housing which is also at cathode potential. The focusing is sharpened by the effect of the cathode support and a strip placed immediately below the cathode also at cathode potential. Anode "skirts" are brought down on each side to eliminate space charge effects that might limit the desired amount of anode current.

This design permits effective maximization of the first factor in the intensity relation, equation (2), $(q_0 A/d^2)$. In order to have the anode source area subtend as large a solid angle as possible (given by A/d^2) from a point on the sample, it is required that the focal spot area on the anode be large and close to a relatively large window in the X-ray tube. In order to maximize the specific intensity of the source q_0, the anode must be run at nearly its loading limit and therefore very large anode currents are required (from 100 to 1000 ma depending upon the excitation voltage employed, and upon the anode material). High-current operation, particularly in a demountable tube, results in a relatively large amount of tungsten evaporation from the cathode. Tungsten contamination of the target cannot be tolerated in soft X-ray sources because of the very high absorption of the exciting electron beam within the contaminating layers. Therefore, to eliminate the tungsten contamination and to permit placing the anode very close to the window, the cathode is placed below the anode structure. Also, in this design, the window is at cathode potential and thus electron beam bombardment of the thin foil window is prevented.

Contamination of the anode from condensables present in the vacuum system is nearly completely eliminated by operating at maximum anode power so as to maintain the anode at an elevated temperature. A relatively clean vacuum, at about 10^{-6} torr, is gained through the use of a 15-liter/sec VacIon pump.

A 6-μ aluminum foil window seems to be an ideal filter for the aluminum anode source. In order to permit its supporting atmospheric pressures, the window area is separated into five sections by four webs. A 60% transmission perforated screen backing can also be used for supporting this and thinner window materials. (Available from Buckbee Mears Co., St. Paul, Minn.)

The most serious source of background signal with soft X-ray excitation seems to originate from the fluorescing of nearby elements either in the sample or in the crystal which cannot be eliminated completely through pulse height discrimination with the proportional counter. In principle, such background could be eliminated by having a monochromatic source of wavelength very close to the line being excited. However, in practice, it is not practical to have the characteristic radiation from the X-ray source completely separated from the continuous radiation. Fortunately, with soft X-ray excitation, the proportion of the continuous radiation component is greatly reduced, and properly chosen window materials as filters for the X-ray source can provide very effective improvement of signal-to-background ratios. Raising the tube voltage and thereby increasing the hard continuous radiation output is ineffective in increasing signal intensity because the resulting X-ray generation lies too deeply within the sample. The background is increased because of increased scattering of the primary radiation and fluorescing of the higher atomic number elements. This result is illustrated in Figures 7 and 8, in which the signal per kilowatt of anode power and the signal-to-background ratio are both given as a function of the anode voltage for magnesium and sodium analysis in a rock sample (Standard granite-G1) using an aluminum anode tube with a 6-μ aluminum foil window. For Al K excitation of sodium and magnesium, the optimum voltage is in the 10 to 15 kv range. The maximum anode power as set by the loading limit for the particular anode described here is 2 kw. A silver-plated copper anode is used for the Ag L excitation of the higher atomic number elements. Its loading limit is probably between 3 and 4 kw.

FLOW PROPORTIONAL COUNTER

A standard, side window flow proportional counter of 2 cm ID is used with P-10 gas filling (90% A–10% methane). Sufficient gas amplification can be obtained with this flow counter to bring the pulse height distribution of $0 K$ (23.6 Å) radiation and shorter-wavelength emissions out of the amplifier noise region. It is therefore possible to improve appreciably the signal-to-background ratio by pulse height integration techniques in a manner similar to that employed for the more conventional X-ray analysis. As can be noted from Figure 9, the absorption of the radiation of wavelengths longer than about 8 Å, which is essentially due to argon absorption, is practically 100%. The total counter efficiency, taking into account the transmission characteristics of the 6-μ aluminum window (as used for sodium and magnesium analysis) is shown in Figure 10. A 6-μ mylar

Table III

λ-Angstrom		Beryllium 25 μ — 4.6 mg/cm²	Mylar— $C_{10}H_8O_4$ 6 μ — 0.87 mg/cm²	Aluminum 6 μ — 1.62 mg/cm²	Teflon— $(CF_2)_x$ 3 μ — 0.65 mg/cm²	2 cm — Argon 1 atm — 25°C
Ag L_{α_1}	4.15	0.898	0.897	0.298	0.860	0.362
A K_{α_1}	4.19	0.896	0.893	0.292	0.856	0.554
Pd L_{α_1}	4.37	0.878	0.882	0.252	0.839	0.512
Rh L_{α_1}	4.60	0.853	0.863	0.200	0.812	0.467
Cl K_{α_1}	4.73	0.841	0.853	0.176	0.804	0.440
Ru L_{α_1}	4.85	0.828	0.851	0.157	0.796	0.415
Tc L_{α_1}	5.11	0.803	0.822	0.120	0.763	0.361
S K_{α_1}	5.37	0.777	0.797	0.091	0.732	0.314
Mo L_{α_1}	5.41	0.771	0.793	0.087	0.727	0.304
Cb L_{α_1}	5.72	0.738	0.764	0.061	0.692	0.257
Zr L_{α_1}	6.07	0.699	0.725	0.039	0.646	0.198
P K_{α_1}	6.15	0.687	0.717	0.035	0.636	0.186
Y L_{α_1}	6.45	0.651	0.683	0.024	0.596	0.150
Sr L_{α_1}	6.86	0.597	0.634	0.013	0.542	0.105
Si K_{α_1}	7.13	0.562	0.599	0.009	0.509	0.083
Rb L_{α_1}	7.32	0.535	0.583	0.007	0.483	0.071
Kr L_{α_1}	7.81	0.472	0.521	0.003	0.417	0.041
Al K_{α_1}	8.34	0.408	0.459	0.593	0.353	0.022
Br L_{α_1}	8.37	0.406	0.453	0.590	0.350	0.022
Se L_{α_1}	8.99	0.326	0.384	0.530	0.228	0.010
As L_{α_1}	9.67	0.249	0.310	0.461	0.217	0.004
Mg K_{α_1}	9.89	0.225	0.278	0.417	0.196	0.003
Ge L_{α_1}	10.4	0.176	0.237	0.368	0.156	0.001
Ga L_{α_1}	11.3	0.110	0.165	0.284	0.098	0.000
Na K_{α_1}	11.9	0.077	0.125	0.257	0.071	
Zn L_{α_1}	12.3	0.058	0.102	0.206	0.057	
Cu L_{α_1}	13.3	0.030	0.057	0.137	0.029	
Ni L_{α_1}	14.6	0.010	0.028	0.079	0.013	
Ne K_{α_1}	14.6	0.010	0.028	0.079	0.013	
Co L_{α_1}	16.0	0.002	0.011	0.062	0.005	
Fe L_{α_1}	17.6		0.003	0.015	0.002	
F K_{α_1}	18.3		0.002	0.009	0.312	
Mn L_{α_1}	19.5		0.001	0.004	0.247	
Cr L_{α_1}	21.6		0.000	0.001	0.157	
O K_{α_1}	23.6		0.002		0.096	
V L_{α_1}	24.3		0.001		0.082	
Ti L_{α_1}	27·4				0.033	

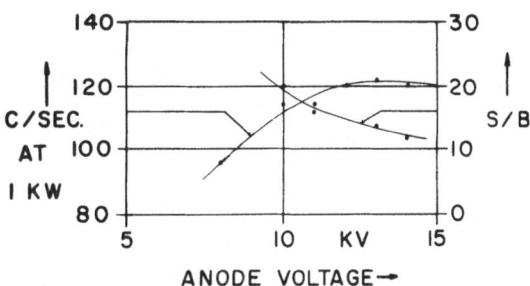

Figure 7. Effect of excitation voltage on signal and on signal-to-background for magnesium analysis in granite. An aluminum anode source with a 6-μ aluminum window was used. Optimum pulse height integration with a 6-μ mylar window on the flow proportional counter was employed. Sample is Standard Rock G-1, (0.39 wt.% MgO).

window is used for the heavier element analysis, and a 3-μ teflon window is used for the fluorine analysis. The transmission characteristics of these windows are presented in Table III. It is usually not advisable to attempt to support thinner windows of these or other materials as counter windows in order to gain intensity because of the consequent loss in the filtering property which is so effective in the improvement of the equally important signal-to-background ratio.

The absorption band for argon, shown in Figure 9, at the shorter wavelengths is far enough removed so as not to affect sodium and magnesium analysis very seriously. Unfortunately this band embraces such emissions as Ca K (3.4 Å) and K K (3.7 Å) which often constitute troublesome background radiations. Nevertheless, these radiations are not excited by Ag L (4.2 Å) line radiation so that the argon counter remains practical and reasonably efficient for the analysis of the elements Cl through Al when this excitation is employed. A neon filled counter (temporarily sealed off, but refillable) would not have this enhanced sensitivity in the 2 to 3.9 Å region as is shown in Figure 11.

MEASUREMENT

An accurate analysis for the light elements within a given sample demands that two overall conditions be satisfied: (1) that a sufficiently high signal and signal-to-background be obtained and (2) that this information be correctly related to the mass fraction ρ_x which is sought for the particular element. The last factor in the intensity relation, equation (2), predicts that the success of the complete measurement is dependent ultimately upon the sample matrix even for the idealized case of a homogeneous, geometrically plane sample.

For example, the factor, τ_x/μ_i, (the ratio of the photoelectric absorption cross section for the given element and the exciting radiation to the total absorption cross section of the sample for the fluorescent line radiation being measured) will be favorably very large for

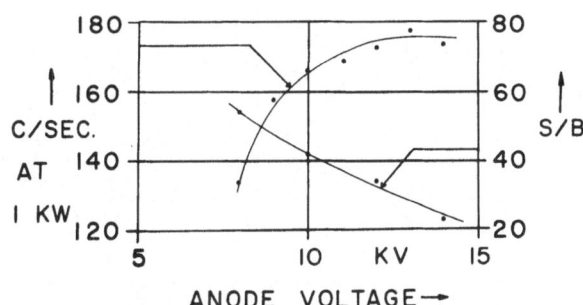

Figure 8. Effect of excitation voltage upon signal and signal-to-background for sodium analysis in granite. Measurement conditions were the same as for those described in Figure 7.

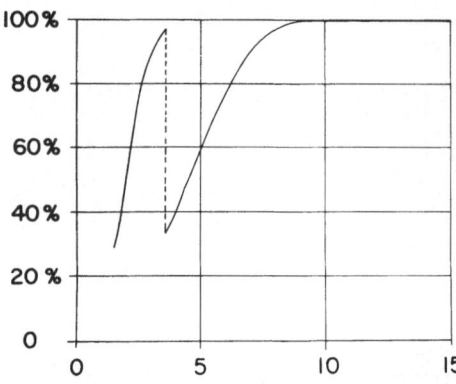

Figure 9. Absorption as a function of wavelength for argon within a counter as used in this study (absorption essentially as obtained for P-10 gas). Two-cm path of argon at 1 atm and 25°C.

the case of an element such as sodium or magnesium within an organic matrix and will be very small for the case of an element such as fluorine in a matrix of heavier elements. Measurements on fluorine were made from a smooth block of teflon (CF_2) and from a polished sample of fluorite (CaF_2). A potassium phthalate crystal was used with an aluminum filtered aluminum anode source and a flow counter with a 3-μ window of teflon. The signal and signal-to-background were so low for the fluorite sample as to make this measurement unfeasible. However, as shown in Figure 12, the results on the teflon were very encouraging, yielding a signal and background of 46 and 4 counts/sec, respectively. It should be noted that the teflon window was supported upon a grid of only about 16% optical transmission. By using the perforated grids which have now been obtained and mentioned above, this signal on fluorine can now be increased by a factor of four.

A small drop of blood serum, 50 μl in volume, can be dispersed into a smooth, homogeneous layer of 1 or 2 cm^2 on a substrate such as the "dull" side of a thin aluminum foil, or the surface of tracing paper. The analysis for phosphorous, magnesium, and sodium in this sample is of very great importance in routine medical analysis* and, as is shown in Figure 13, a preliminary measurement of these elements in 5-mg samples of dried human serum demonstrated that the method presented here can be very successfully applied to this problem. These represent what are probably the first such measurements on magnesium and sodium and a large improvement in signal and signal-to-background

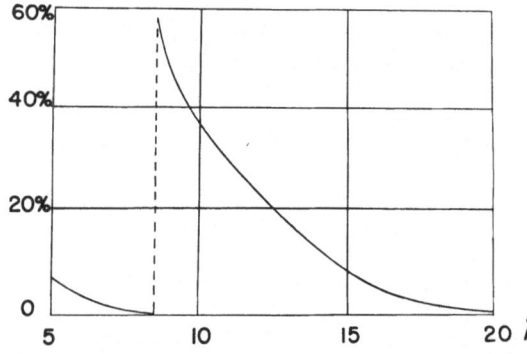

Figure 10. The combined counter efficiency for Al, Mg, and Na analysis for the P-10 gas-filled flow proportional counter (absorption assumed as for pure argon) —argon absorption × transmission of the 6-μ aluminum foil window.

* This problem was suggested by the medical research group of the Swedish Hospital, Seattle, Washington—Dr. Paul K. Lund, Dr. James Mathies, and Dr. Douglas Morningstar.

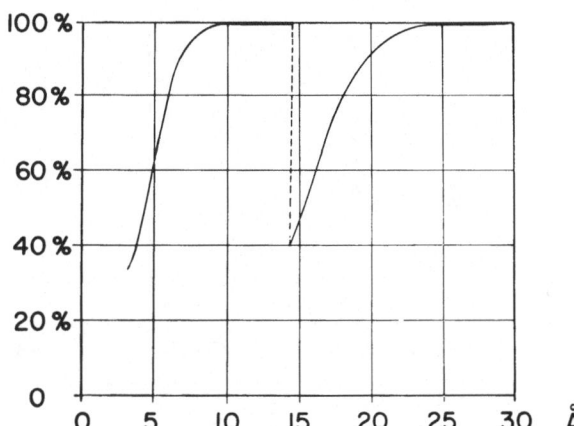

Figure 11. The absorption characteristics for a neon-filled counter which can provide strong discrimination against wavelength shorter than 5 Å. Two-cm path of neon at 1 atm and 25°C.

for the measurement of phosphorous in the sample over that by conventional excitation methods. A 25-μ beryllium filter was left on the Ag L anode source for the phosphorous measurement as reported here. An aluminum foil support was used for the sodium and magnesium measurement and tracing paper was used as support in the phosphorous measurement. Since the protein matrix absorption in this particular sample maintains the value of μ_i essentially constant, there exists a direct proportionality between signal-over-background and the concentration.

In Table IV are listed some typical values achieved to date by the methods described above for both signal and background (counts/sec) from the measurements described above and from measurements on phosphorous, silicon, aluminum, magnesium, and sodium in three rock standards. As will be shown in Part II of this work, these results

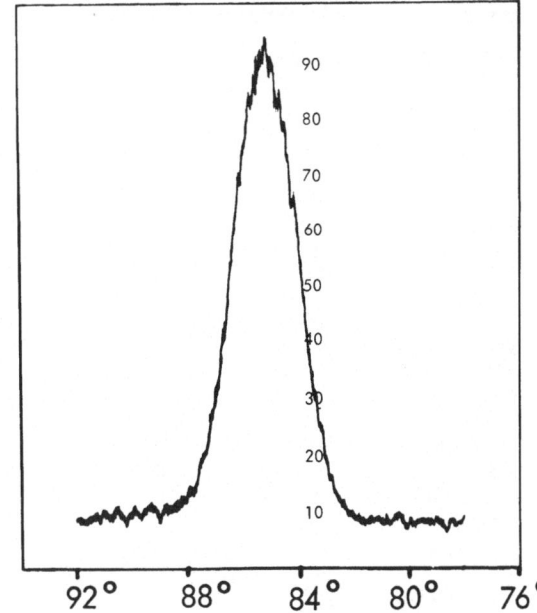

Figure 12. Fluorescence line profile for F K radiation from a teflon block as sample. Excitation was by an aluminum anode source (Al K_α-8.34 Å) with a 6-μ aluminum foil window. Optimum pulse height integration was used with a flow proportional counter P-10 gas-filled and with a 3-μ teflon window. The window was supported upon a 16% transmission grid. This signal may be increased by a factor of four by the use of a more transparent, perforated screen grid. Crystal-K-phthalate.

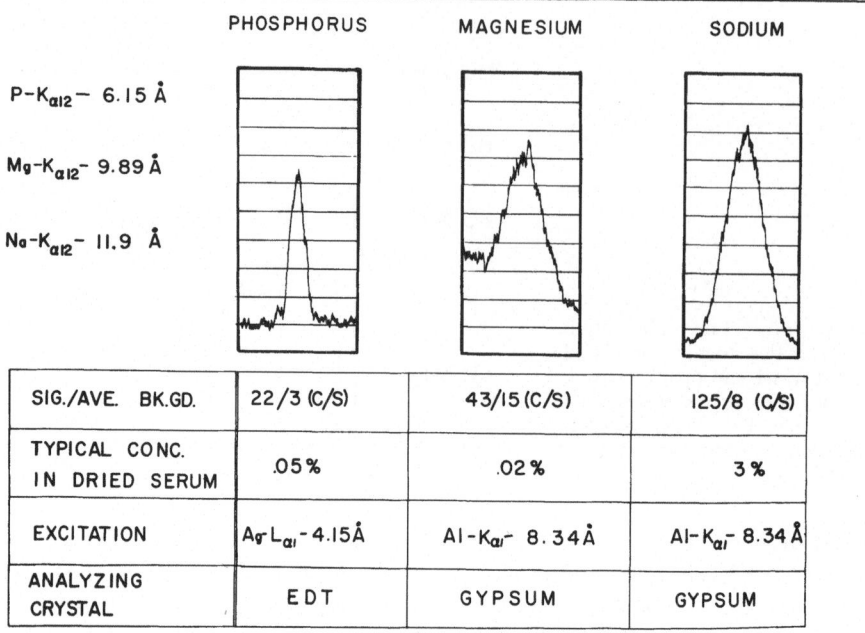

Figure 13. X-ray fluorescence analysis of human blood serum. Optimum pulse height integration was employed in each measurement with the P-10 gas-filled proportional counter. A 6-μ aluminum foil window was used on both the counter and on the X-ray source window for the Na and Mg measurements. For the P measurement, a 6-μ mylar window was used on the counter and a 25-μ Be window, self-supporting, was used on the X-ray source.

Table IV. Typical Values of Counting Rates from Several Standards for Light Element Fluorescence Analysis as of August, 1962

Standard	Element	Wt. %	Signal-background counts/sec	Excitation	Crystal
Granite G-1 (USGS)	Mg	0.235	197—28	Al K_α 8.34 Å	Gypsum
	Na	2.42	347—23	,,	,,
Syenite-1 (CAAS)	Al	4.9	3636—212	Ag L_α 4.15 Å	EDT
	Si	27.8	16732—91	,,	,,
	P	0.083	94—13	,,	,,
	S	0.07	37.4—23.7	,,	,,
Sulfide-1 (CAAS)	Al	4.8	3368—202	,,	,,
	Si	16.4	8678—50	,,	,,
	P	0.044	51—16	,,	,,
	S	12.1	77—51.6	,,	,,
Dried Blood Serum (Human) 5-mg Sample	P	0.05	156—21	,,	,,
	Mg	0.02	43—15	Al K_α 8.34 Å	Gypsum
	Na	3	125—8	,,	,,
Teflon (CF$_2$)	F	76	46—4	,,	KAP

represent a great improvement (20 to 30 times more signal and greatly improved signal-to-background) over what has been obtained to date for rock analysis with more conventional excitation means.

Relating the signal-over-background to actual concentrations in, e.g., these rock samples can be a very difficult problem if the value of μ_i is not a constant, and if grain size effects are present. The effect of a variable μ_i might be accounted for by semiempirical corrections based upon an approximate knowledge of the sample matrix. Or μ_i can be made nearly a constant value by introducing heavy absorbers into the matrix. Many new techniques may also need to be tried in order to achieve the homogeneity and uniformity of sample which is of particular importance in light element analysis.

A new dimension in the light element analysis is the possibility of using milligram amounts of material with little or no loss in intensity so that methods of thin sample preparation as commonly employed in other kinds of analysis might well be employed here.

NOTATION

A = the effective projected source area
C = the effective cross-sectional area of the beam as accepted by the collimation–crystal system
d = the source-to-sample distance
μ_i = the total mass absorption coefficient for the sample and for the fluorescent radiation being measured
μ_1 = the total mass absorption coefficient for the sample and for the exciting radiation
ω = the solid angle of radiation measured from a point on the sample which is accepted by the collimation–crystal system
ϕ, ψ = as defined in Figure 1
Q_x = quanta/sec of fluorescent line radiation entering counter
q_0 = the quanta/sec emission rate per unit source area and per unit solid angle from the exciting radiation source
R = the fraction of incident line intensity reflected by the crystal
r = the appropriate absorption jump ratio for the given element
ρ_x = the mass fraction of the element being analyzed
T = the combined transmission of the source and detector windows for the line radiation being measured
τ_x = the mass photoelectric absorption coefficient for the element being analyzed
W_q = the corresponding fluorescent yield
z_i = the relative strength of the series line being measured

ACKNOWLEDGMENT

This work has been in collaboration with Dr. A. K. Baird, Dr. Donald B. McIntyre, and Mr. Edward E. Welday of the Pomona College Department of Geology, and the technical assistance of Mr. William K. Hart and Mr. James P. Campbell. The project has been supported by a grant from the Office of Scientific Research, U.S.A.F.

REFERENCES

1. See for example: H. A. Liebhafsky, *et al.*, *X-ray Absorption and Emission in Analytical Chemistry*, John Wiley & Sons, Inc., New York, 1960.
2. A. H. Compton and S. K. Allison, *X-rays in Theory and Experiment*, second edition, D. Van Nostrand Co., Inc., Princeton, N. J., 1935, p. 482.
3. A. E. Sandstrom, "Experimental Methods of X-ray Spectroscopy," *Handbuch Der Physik*, Springer-Verlag, 1957.
4. A. H. Compton and S. K. Allison, *X-rays in Theory and Experiment*, second edition, D. Van Nostrand Co., Inc., Princeton, N.J., 1935, pp. 783–790.

5. S. Fine and C. F. Hendee, A Table of X-ray K and L Emission and Critical Absorption Energies for All the Elements, Philips Laboratories, Irvington-on-Hudson, New York.

6. M. A. Blokhin, *The Physics of X-rays*, second revised edition, Translation Series, United States Atomic Energy Commission, Office of Technical Information, AEC-TR-4502.

7. J. M. Cork, *Handbook of Chemistry and Physics*, Chemical Rubber Publishing Co., 1945.

8. T. C. Furnas, Jr. and E. W. White, WADD Technical Report 61–168, 1961.

9. A very useful work on the problem of excitation of the elements from atomic number 19 (K) to atomic number 13 (Al) has been recently accepted for publication by the Journal of Applied Physics—"The Target and Inherent Filtration as Factors in the Fluorescence Excitation of X-rays," by Nathan Spielberg, Philips Laboratories, Irvington-on-Hudson, New York.

SODIUM AND MAGNESIUM FLUORESCENCE ANALYSIS—PART II: APPLICATION TO SILICATES

A. K. Baird, D. B. McIntyre, and E. E. Welday

Pomona College
Claremont, California

ABSTRACT

Moderate counting rates, in excess of 400 cps at a signal/noise ratio of 18 on 3.5% Na_2O in rocks, permit high-precision quantitative analysis for light elements in silicates. Special advantages of fluorescence analysis in the 10 Å region, with adequate excitation provided by an aluminum target tube, will be described. These advantages include an ease of discrimination by excitation potential and by high dispersion using large $2d$ space crystals. Minor and simple modifications of the optic path combined with electronic discrimination make the method ideal for sodium and magnesium. Routine runs over periods of days have been made possible by a new technique of calibration involving computer corrections for systematic drift from any source. Repeated tests show that the total analytical error, including specimen preparation, closely approaches that of the counting statistics used, and that the drift corrections are complete. In granitic rock analyses resulting precisions (standard deviation/mean) are 1% for 2–5% Na_2O and 2% for 0.25–1.0% MgO.

INTRODUCTION

In recent years it has been demonstrated that X-ray fluorescence spectrography is a powerful tool in many analytical problems, often yielding quantitative results with higher precision at greater speed than other methods. Such applications in the field of silicate analysis (ceramics, glass, clay, cement, etc.) have proven to be very rewarding as most "wet methods" require long and laborious techniques which must be performed by skilled analysts if high precision and accuracy is expected. A serious limitation to the application of X-ray fluorescence to silicate analysis, however, has long been the special problems associated with the lightest elements, often in low concentration, but of critical importance (e.g., magnesium in cement materials, alkalies in clay, etc.). This paper describes the application of new techniques, which have overcome many of the light element restrictions, and presents analytical procedures for silicates which should cover all common elemental constituents above atomic number 10 in quantities down to at least 0.1 wt.%.

Because the development of these techniques has involved the adaptation of a new X-ray tube to commercially available equipment, both general considerations of the problems of light-element analysis in silicates and details of the procedures and equipment will be given.

GENERAL CONSIDERATIONS

As a rule, silicates presented to the spectrographer for analysis are heterogeneous materials composed of 6–12 major "oxides" usually ranging from atomic number 11 (sodium) through atomic number 26 (iron). Even though a drastic reduction in fluorescent

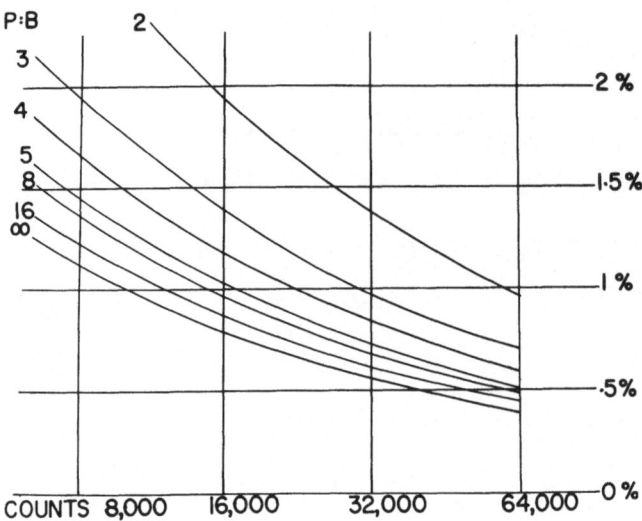

Figure 1. Counting statistic errors in terms of percent relative deviations (standard deviation/mean × 100) of cps for various peak/background ratios.

yield occurs with decreasing atomic number, satisfactory quantitative results have been obtained previously down to and including atomic number 13 (Al)[1] and even with number 12 (Mg) in some cases of moderate to high concentration.[2,6] Only recently has it been possible to obtain a usable signal from Na.[3] For Na, Mg, and Al, and for very small quantities of somewhat heavier light elements, the basic problem is to increase the signal with respect to the noise. Even if an analytical line is detectable, a low signal/noise ratio in practice limits the usefulness of the X-ray method because the basic error, which cannot be overcome, is that of the counting statistics. For example, Figure 1 shows that for the spectrographer to achieve a precision in Si analyses comparable to wet chemistry (0.5%), requires at least 64,000 fixed count at a signal to noise of at least 16. Thus, the time required to collect counts may become excessive and may exceed equipment stability unless operation is under optimum conditions. In an attempt to achieve optimum signal and signal/noise the following have been considered:

1. X-ray tube target materials and corresponding excitation efficiencies in counts/kw *vs.* operating voltage for various elements sought.
2. X-ray tube window materials for both transmission and filtration.
3. Collimation between specimen and analyzing crystal and between crystal and detector.
4. Reflectivity and dispersion and spectral interference of analyzing crystals.
5. Detector operating conditions combined with pulse height selection and detector window materials for transmission and filtration.

Most of these points have been discussed in Part I of this study, and the details of application to commercially available spectrographic equipment are presented here with emphasis upon the equipment modifications, operating conditions, and results obtainable. Some data and conclusions concerning overall quantitative analysis in the light-element range are also considered.

---

[1] Superscripts pertain to references at the end of the paper.

Figure 2. X-ray laboratory. Equipment (left to right): electronic circuit panel; X-ray power supply for diffraction and conventional W-target FA-60 spectrograph tube; power supply for VacIon pump used with soft X-ray source and gauges for measurement of X-ray tube and spectrograph vacuum; Universal Vacuum Spectrograph with soft X-ray tube mounted; power supply for soft X-ray tube. Mechanical pumps for starting the VacIon pump and for the spectrograph are housed in the cabinet beneath the spectrograph.

INSTRUMENTATION

Basic equipment used is the Philips Universal Vacuum Spectrograph and associated electronics panel. Figure 2 shows the laboratory layout and indicates the major pieces of additional equipment required for light-element analysis. A close-up of the soft X-ray source mounted in the spectrograph is given in Figure 3. Figure 4, a diagrammatic cross section of the spectrograph in the plane of the optic path, indicates the geometric arrangement.

Soft X-Ray Source

The demountable, X-ray tube described in Part I is placed in the conventional position in the spectrograph housing. The longer length of the tube housing necessitated removing the specimen spinner shaft. Water cooling to the housing only is provided by the regular Philips system using flexible Tygon tubing. Water cooling to the target is applied by a separate external system (see Figure 3). The tube housing is pumped to 2 or 3×10^{-6} mm Hg by a Varian Associates 15LS VacIon pump. Pump-down time following change of target or window is usually accomplished in 3 to 5 hr. Atmospheric pressure on the tube during specimen changing, while the spectrograph is not under vacuum, is supported by a window of 6-μ-thick aluminum foil epoxy cemented to a frame of $\frac{1}{8}$-in. aluminum plate having an opening $1 \times \frac{3}{8}$ in. with four cross-rib supports. Foil and frame are clamped to the opening in the tube with an indium seal between. Filtration properties of the window are described in Part I. The entire tube, housing, and pump

Figure 3. Soft X-ray tube mounted in vacuum spectrograph. View of right-hand side of spectrograph with goniometer to the left and VacIon pump system to the right. Water cooling lines to anode at left front.

system can be removed as a unit, while still under high vacuum, when changing to the conventional FA-60 or other source for harder excitation.

High-Voltage Power Supply

Power is supplied to the X-ray tube by a specially built constant–potential unit of 2-kw rating at 10 kv, adequate for use with the aluminum target. Other anode structures (e.g., silver-plated Cu for Si and Al analysis) require 3 kw at 15–20 kv for optimum

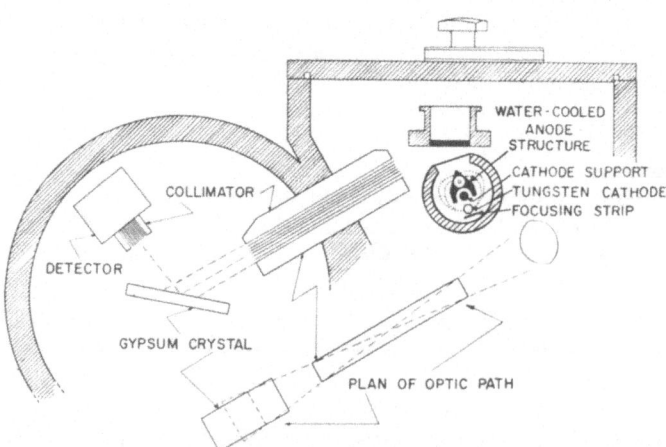

Figure 4. Diagrammatic cross section of spectrograph in the plane of the optic path. Soft X-ray source shown in place of conventional spectrograph X-ray tube.

excitation. Voltage regulation is maintained by the conventional stabilizer incorporated in the Philips panel. No amperage stabilization is used nor does it seem necessary under the operating conditions.

Primary Collimation and Analyzing Crystal

Gypsum ($2d = 15.2$ Å) disperses Mg K_α and Na K_α22.9° 2θ. This large dispersion of neighboring element K lines, combined with discrimination against lines of energy higher than that of Al K (because of the excitation source), and with the use of pulse

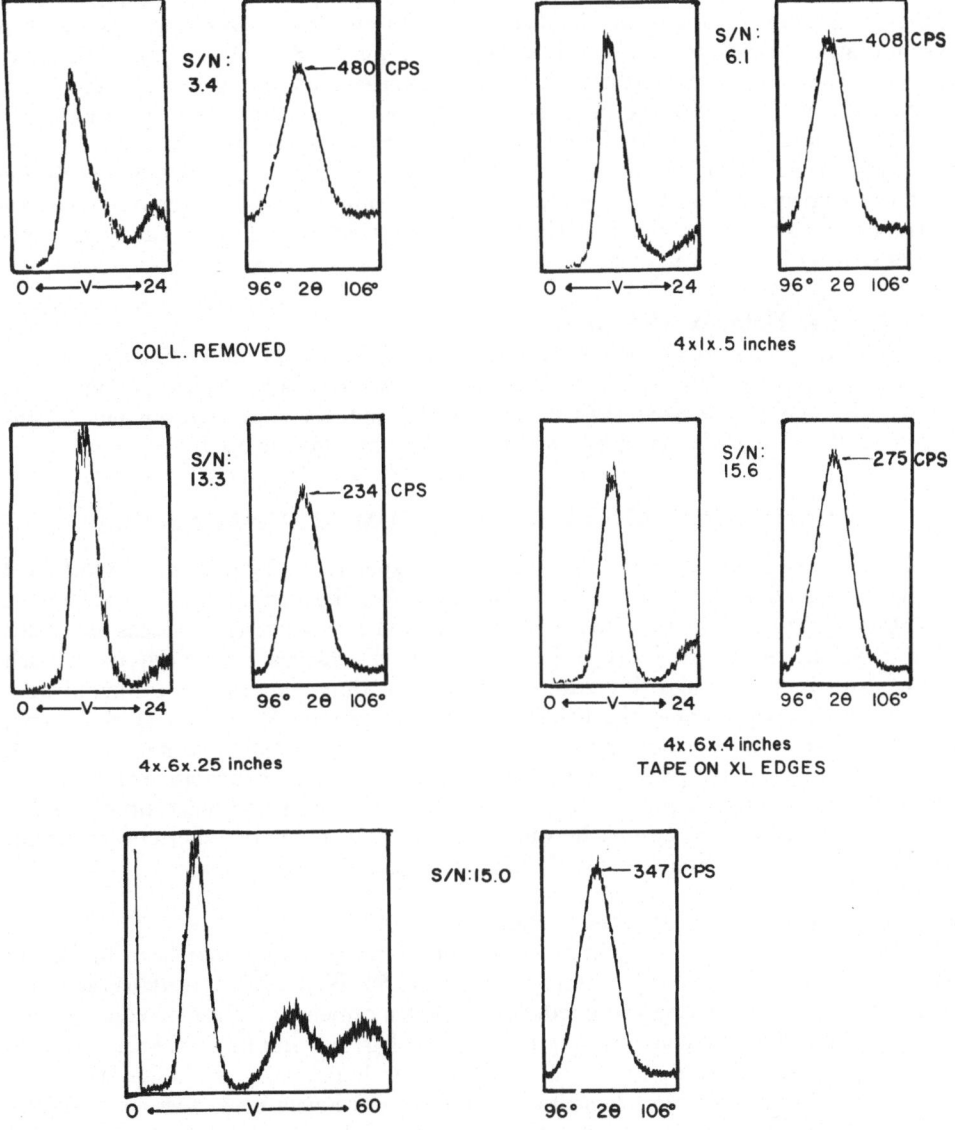

Figure 5. Effect of collimation on pulse height distribution and line profile of 3.26% Na$_2$O in granitic rock. Excitation Al K at 10 kv—150 ma. Lower distribution and profile selected as optimum.

height selection, enables very coarse collimation to be employed between specimen and crystal. Figure 5 shows the effect of various collimator arrangements on total signal and signal/noise. The conventional collimator supplied with the Philips instrument can easily be altered to suit the particular analytical problem. For Na and Mg in silicates, optimum collimation is $4 \times 1 \times 0.5$ in., that is, removing all blades and spacers from the collimator housing. Additional improvement is obtained by covering the metal edges of the crystal holder with teflon or electrician's tape, apparently reducing scatter to the detector.

The interdependence of excitation source, collimation, and crystal should be noted. Although gypsum proves to be satisfactory for Na and Mg analysis, and better than K-phthalate (27.1 Å) (see Part I), it is not optimum for Al, Si, or higher atomic number elements. Harder excitation radiation used for these elements causes excessive secondary fluorescence of Ca and S from the crystal. The 500-mil collimation used for Na and Mg must be reduced to 35 mils for Al and Si fluorescence dispersed from an EDDT crystal and further reduced to 15 mils to avoid K K_β and Ca K_α interference when analyzing for these elements. When blades are placed in the collimator it is, of course necessary to use spacers. Thus, the collimator width is reduced to 0.6 in. and there is also the reduction of spacing in the optic plane.

Detector and Detector Collimation

The flow proportional counter and associated 0.5 in. $\times$ 11 mil collimator supplied with the spectrograph are used. The flow counter window is 6-μ-thick aluminum foil and is adequately supported against the spectrograph vacuum environment by the detector collimator. P-10 gas, at a flow rate of 0.5 ft^3/min (air) is used.

QUANTITATIVE ANALYSIS OF SODIUM AND MAGNESIUM

Quantitative analyses by the method of X-ray spectrography present very different problems depending upon the elements analyzed for, the materials submitted to the laboratory, and the results expected. For those light elements presently accessible by the X-ray method, several radically different, analytical schemes have been used; each scheme with its merits and limitations of precision, time, and expense. The advance reported in this paper opens the field of silicate analysis by X-rays to the remaining, heretofore inaccessible, major constituents. However, the quantitative methods which have been used so far with this new application must be considered tentative only. Extension of the X-ray method into the 12 Å region not only raises additional problems but also indicates some new approaches to quantitative work in the shorter-wavelength regions. These will be discussed briefly in a later section.

Specimens and Specimen Preparation

The present quantitative application of soft X-ray excitation has been limited to granitic rocks with a content of up to 1.5% MgO and 5% Na$_2$O. Initial work has assumed an essentially constant matrix and a reduction in enhancement complications due to variable quantities of heavy constituents, and has emphasized simplicity of specimen preparation. An attempt has been made to avoid fusion techniques as used in our laboratory for analysis of Al and heavier elements in silicates.[1] Obviously borax flux could not be used for Na analysis. Rocks to be analyzed are crushed, pulverized, and ground to a mean particle size of 2μ. A 1-g split is pressed without internal binder into a pellet at 28,000 psi with a bakelite jacket which fits the standard Philips circular specimen holder.[4]

Standards, for comparison by the calibration curve technique, are prepared in a

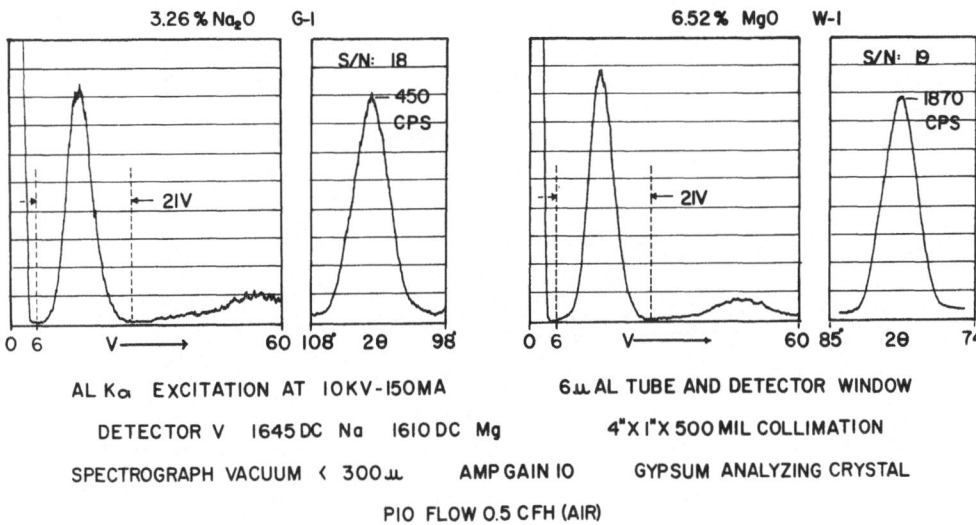

Figure 6. Pulse height distributions and line profiles for Na₂O and MgO in typical standards.

similar fashion. These consist of the U.S. Geological Survey standard granite G-1, standard diabase W-1[5] and a secondary standard quartz monzonite of our laboratory (PC 84) calibrated from G-1. Specimens of each standard have been diluted with SiO_2 (optically clear quartz) in known proportion to encompass the range of unknowns now being considered. Diluted oxide values are indicated on the examples of calibration curves used in this paper.

Analytical Technique

Our procedures are designed for analysis of unknowns in batches of 30 to 150 or more by comparison with three or four standards assumed to define simple polynomial calibration curves. Preliminary work with the soft X-ray source (as well as with the conventional tungsten target FA-60 tube) has shown detectable drift over periods of hours. In an attempt to avoid either elaborate stabilization or too frequently repeated readings on standards, drift corrections are made. The analytical procedure followed is:

1. Continuous operation of electronic circuitry, including flow counter preamplification. One-half-hour warm-up of X-ray source at operating power.
2. Establishment of detector voltage, amplifier gain, and pulse height selector and goniometer settings using pure MgO and NaCl. Resulting operating conditions, along with line profiles and pulse height distributions of typical standards, are given in Figure 6.
3. A total fixed count is selected from Figure 1 based on the measured signal/noise and a compromise between the time available for an analytical run and a counting error as low as possible. In practice this usually means 16,000 fixed count yielding 1% relative deviation for Mg and 0.8% relative deviation for Na in terms of cps.
4. An elapsed-time meter, recording to the nearest minute, is started and five replicate readings of seconds to accumulate the fixed count are made for each of the three or four standards to be used. Elapsed time is noted for each reading.
5. All unknowns are run, recording seconds of fixed count and elapsed running time for each.

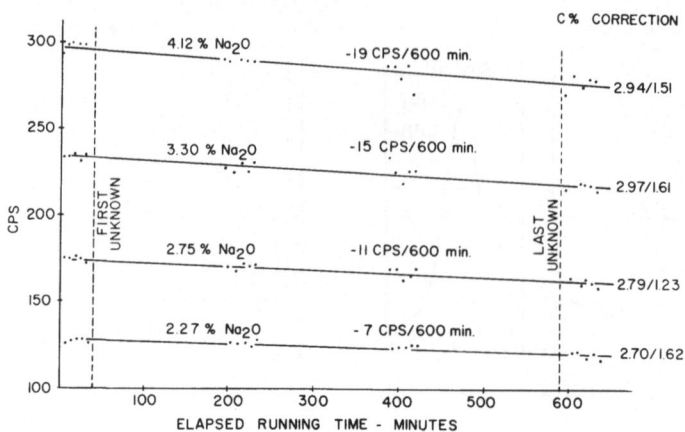

Figure 7. Drift analysis for a typical quantitative run on 150 specimens of granitic rock for Na2O. Counting error 0.9%. Points represent readings on standards. Uncorrected and drift corrected values of relative deviation (C%) are indicated.

6. Replicate readings on each standard are repeated (as in 4) after the last unknown is run.

Resulting data are handled on a small digital computer (Clary DE-60), which performs the following operations:

1. Determines a drift curve for each standard and tests for significance and magnitude of slope.
2. Dedrifts the standards to the elapsed running time of a given unknown.
3. Computes a quadratic or cubic calibration curve, by the method of divided differences, for the particular unknown and prints oxide percentage. Thus a new calibration curve is determined for each unknown, based upon the time the unknown was run and the magnitude and character of the drift.

A recent run of 150 specimens for Na2O provides an example of this method of computation. Identical procedures are used for MgO (and other oxides). In practice the drift and calibration curves are not graphed (except as checks), but are presented here as illustrations (Figures 7 and 8). Figure 7, drift analysis over 600 min elapsed running time, shows the regression results for each of four standards with the basic cps counting error of 0.9% relative deviation. For the purpose of this illustration standards were run between the first and last unknown to check the systematic linearity of the drift curves. Relative deviation (C%) improvement introduced by correcting for the drift is indicated at the right of the graph with the first figure shown the total deviation of C% for the standard and the second figure, C% with drift removed. It should be noted that the drift-corrected values bring the relative deviations much closer to the basic counting error of 0.9%. Total magnitude of drift in cps/600 min is indicated. Conditions causing a decrease in cps with time undoubtedly include systematic shift in the pulse height distribution with respect to the window and a gradual coating of the X-ray tube anode by impurities within the system.

Two limiting calibration curves, of the 150 computed in this example, are shown in Figure 8. These correspond to the elapsed times when the first and last unknowns were read. For each elapsed time in minutes, cps values for each standard are computed from the drift regression and a cubic curve passing through the four points is generated. In practice, one such curve (for a midelapsed time) is always plotted to make certain

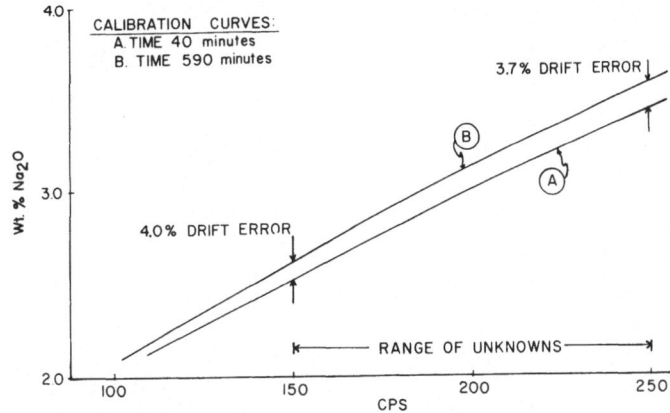

Figure 8. Calibration curves at beginning and end of quantitative run of Figure 7. A 4% error is indicated if drift corrections had not been made.

that ill-behaved curve conditions do not result from the process of forcing the polynomial through the points. These two limiting curves show that an error close to 4% of the amount of Na_2O present can result from drift during the 550 min required to run 150 specimens; this has been detected and eliminated.

Precision of the Analyses

In the run of 150 specimens used as an example above, 30 rocks (diamond-drilled cores) were split and each half submitted to separate preparation and analysis. The Na_2O and MgO values determined were studied for total analytical error by analysis of variance. Results and comparison with other methods are indicated in Table I. Expressed in terms of 95% confidence, in our routine analyses of granitic rocks we can quote single Na_2O and MgO results as 3.00 ± 0.12 wt.% and 0.50 ± 0.06 wt.% respectively; in practice the precision can be improved by replication.

QUANTITATIVE ANALYSIS OF ELEMENTS HEAVIER THAN MAGNESIUM

As shown in Part I of this study, selection of optimum conditions of excitation, collimation, dispersing crystal, and window material applies equally well to elements heavier than Na and Mg in silicates. In particular, the use of Ag L excitation, wider collimation, and appropriate window filters has greatly improved results obtainable on Al and Si. These improvements are summarized in Table II along with the operating

Table I. Precision—Relative Deviation in Percent (Standard Deviation/Mean × 100)

| Compound | Wt.% | X-ray | Emission* | Wet Chemistry | | |
				24 Labs†	Intralab*	Flame*
Na_2O	2–5	1–2	2.5	10	4–7	1.5–3
MgO	0.2–7	2–6	6.5	30	6–10	—

* Fairbairn, *et al.*,[5] Table 14.
† Fairbairn, *et al.*,[5] Table 18.

Table II. Conditions of Quantitative Analysis and Results for Ten Oxides in Silicate Rocks: August 1962

Element	Na	Mg	Al	Si	P	S	K	Ca	Ti	Fe
Approximate element wt.% in specimen	3.0	0.4	3.0	13.6	0.02	0.01	1.3	0.4	0.06	0.6
Approximate oxide wt.% in specimen	4.0	0.7	5.7	29.0	0.04	0.01	1.6	0.5	0.1	0.6
Specimen preparation	Ground to 2 μ and pressed		40% rock in borax fusion, ground to 2 μ, pressed							
Excitation	Al K		Ag L				W (Philips FA-60)			
Power	10 kv—150 ma		50 kv—35 ma				35 kv 25 ma			
Tube window	6-μ Al Foil		25 μ Be				1500 μ Be			
Collimation	4 × 1 × 0.500 in.		4 × 0.6 × 0.035 in.				4 × 0.6 × 0.015 in.			
Crystal	Gypsum		EDDT							
2θ	103°	81.1°	142.5°	108.0°	88.7°	75.2°	50.3°	44.9°	36.4°	25.4°
Flow ctr window	6-μ Al foil		6-μ aluminized mylar							
Best signal	450	190	1960	9340	20	45	1390	520	245	1070
Best signal/noise	18	9	17	185	4	3	50	30	10	40
Previous signal W excitation	0	?	70	290	not anal.	not anal.	same	same	same	same
Previous signal/noise W excitation	1	1.5	12	32	not anal.	not anal.	same	same	same	same

conditions. With all other conditions held constant, the soft X-ray source using Ag L excitation has improved the signal by a factor of 30 over W excitation for both Al and Si. These signals, combined with improved signal/noise ratios, now enable quantitative analyses to be performed with much greater precision and speed than is possible by any other analytical methods known to us. Table II also shows operating conditions and results obtained for all elements currently being studied in our laboratory. Obviously there is still room for improvement in elements heavier than sulfur through the use of more efficient excitation than W. Present instrumentation is still a compromise between obtaining high counting rates at low signal/noise and the problem of too frequent changes of crystal, tube, collimator, detector, etc. For example, above K, a LiF crystal would provide much higher intensities, but the realignment necessary to change from the EDDT or gypsum crystal does not justify the improvement.

CONCLUSIONS

From the data above, we conclude that adequate instruments and procedures are now available to obtain usable counting rates on all of the major elements in silicates. Analytical precisions on quantities down to 0.2 wt.% of the oxides of Na and Mg are considerably better than wet chemistry and optical emission spectrography. Analytical speed of the X-ray method, especially when combined with fast-calibration techniques, is unquestionably better than other methods known to us. The improvements noted in analysis for elements heavier than Mg only serve to increase further the already well-known high precision and speed of the X-ray method.[1] Without question, X-ray spectrography can now be considered the universal and most powerful tool available for silicate analysis.

Even though some additional improvements in instrumentation and technique are obvious and could be easily made, our study of light elements has raised some problems to which no solutions are presently available or, at least, have not been thoroughly investigated. These include:

1. The development of a very large spacing organic analyzing crystal of high reflectivity, which would greatly help analysis for the lightest elements.

2. For quantitative analysis, a study of specimen preparation in light of the special problems of essentially "surface" excitation encountered in the 10 Å region. The relatively high precisions found in our work, using ground materials excited by long-wavelength radiation, seem to be inconsistent with the heterogeneity effects noted by Claisse and Samson.[6]

3. For pronounced matrix difficulties, a further study of quantitative calibration techniques to avoid narrow working ranges and multiplicity of standards: a problem which, when ignored, may lead to gross inaccuracy even with high precision.[7] In our experience, dilutions of rocks with only moderately varying compositions do not lie on the same calibration curve even after elaborate fusion techniques have been used. Possibly the addition of heavy absorbers will aid solution of this problem.[8]

ACKNOWLEDGMENTS

This study has been supported by a research grant (G19075) of the National Science Foundation and was done in cooperation with Dr. B. L. Henke of the Pomona College Physics Department. The authors are indebted to Mrs. Kathleen Madlem and Mr. James Williams who have contributed greatly to the work under a National Science Foundation Undergraduate Research Participation Program. Thanks are also due Mrs. Shirley Bolton for preparation of manuscript.

REFERENCES

1. A. K. Baird, R. S. MacColl, D. B. McIntyre, "A Test of the Precision and Sources of Error in Quantitative Analysis of Light, Major Elements in Granitic Rocks by X-ray Spectrography," *Advances in X-ray Analysis, Vol. 5,* University of Denver, Plenum Press, New York, 1962, p. 412.
2. A. A. Chodos and C. G. Engel, "Fluorescent X-ray Spectrographic Analyses of Amphibolite Rocks and Constituent Hornblendes," *Advances in X-ray Analysis, Vol. 4,* University of Denver, Plenum Press, New York, 1961, p. 401.
3. B. L. Henke, "Microanalysis with Ultrasoft X-radiations," *Advances in X-ray Analysis, Vol. 5,* University of Denver, Plenum Press, New York, 1962, p. 285.
4. A. K. Baird, "A Pressed-Specimen Die for the Norelco Vacuum-Path X-ray Spectrograph," *Norelco Reptr.* 8(6): Nov.–Dec. 1961.
5. H. W. Fairbairn, *et al.,* "A Cooperative Investigation of Precision and Accuracy in Chemical, Spectrochemical and Modal Analysis of Silicate Rocks," *U.S. Geol. Survey, Bull. No. 980,* 1951.
6. F. Claisse and C. Samson, "Heterogeneity Effects in X-ray Analysis," *Advances in X-ray Analysis, Vol. 5,* University of Denver, Plenum Press, New York, 1962, p. 335.
7. G. Andermann and J. D. Allen, "The Evaluation and Improvement of X-ray Emission Analysis of Raw-Mix and Finished Cements," *Advances in X-ray Analysis, Vol. 4,* University of Denver, Plenum Press, New York, 1961, p. 414.
8. H. J. Rose, *et al.,* "Use of La_2O_3 as a Heavy Absorber in the X-ray Fluorescence Analysis of Silicate Rocks," *U.S. Geol. Survey, Prof. Papers* 450–B, 1962, p. 80.

DISCUSSION

R. B. Kelsey (Pratt and Whitney Aircraft): I would like to know what the window material is in the demountable tube, what the thickness is, and also if there has been any consideration of possibly going to a windowless X-ray tube.

A. K. Baird: The tube has, as we use it for Na and Mg, 6-μ aluminum foil stretched over an area that is roughly $\frac{3}{8}$ in. wide by almost 1 in. long with four cross-rib supports. This seems to give optimum conditions for $\mathrm{Al}\,K_\alpha$ excitation. With a windowless tube you would have to pump the whole system down, and that would mean you would not be able to change the samples.

A. Lutts (Centre National de Recherches Metallurgiques): Do you use baffles in the diffusion pump?

A. K. Baird: We did use a diffusion pump to start with, but we had trouble with the oil back-flowing and coating the anode. We now use a Varian Associates ion pump which is a very clean system—no oil involved in this.

X-RAY SPECTROGRAPHIC ANALYSIS OF RARE EARTHS IN YTTRIUM–IRON GARNET POWDERS

J. C. Lloyd and J. D. Kuptsis

International Business Machines Corporation
Thomas J. Watson Research Center
Yorktown Heights, New York

ABSTRACT

Analysis schemes for eleven rare earths (Ce, Pr, Nd, Sm, Eu, Tb, Dy, Ho, Er, Tm, Yb) taken singly in Y–Fe garnet powders are presented. Straight-line calibrations result in all cases, usually over a range from below 0.5 wt.% to better than 6 wt.%. Errors were held to ±5–10% even using extraordinarily simple sample preparation techniques. Data are presented relating to the effects of particle size and sample preparation on precision and accuracy of the resulting analysis.

INTRODUCTION

Recently published studies of the magnetic properties of rare earth substituted yttrium–iron garnets[1] (YFeG) led to the need for development of suitable X-ray spectrographic schemes for their analysis. The samples were grown by conventional techniques in a $PbO–PbF_2$ flux and submitted for analysis as small crystals of widely varying sizes.* Rare earth concentrations ranged from 0.1 wt.% to 7.3 wt.%. Accuracy of better than ±10% was required.

SAMPLE PREPARATION

All samples were ground to fine powders (below 325 mesh) using a Spex Model 5000 mixer-mill with an agate mortar and ball pestle. This procedure minimized errors caused by variations of matrix effects with sample state. The possibility of using fusion techniques was considered, but not carried out because of time limitations. As later results demonstrated, it is sufficient to pack the powdered material loosely into a suitable holder and analyze it directly. This led to near maximal intensities and sensitivities, yet required relatively short sample preparation times. During the analysis, the samples were covered with $\frac{1}{4}$-mil mylar to avoid contamination of the sample drawer. Figure 1 shows the disposable powder holders used. These were made by punching $\frac{1}{2}$-in.-diameter holes in $\frac{1}{32}$-in. linen phenolic; the phenolic was then glued to 1 × 1 in. pieces of microscope slide with a dilute ethanol solution of Ethocel and allowed to dry overnight. These holders were simple, reproducible, and disposable. A clean microscope slide was used for each sample to press the powders into their holders.

---

[1] Superscripts pertain to references at the end of the paper.

* Sizes ranged from several hundred microns to over a millimeter across.

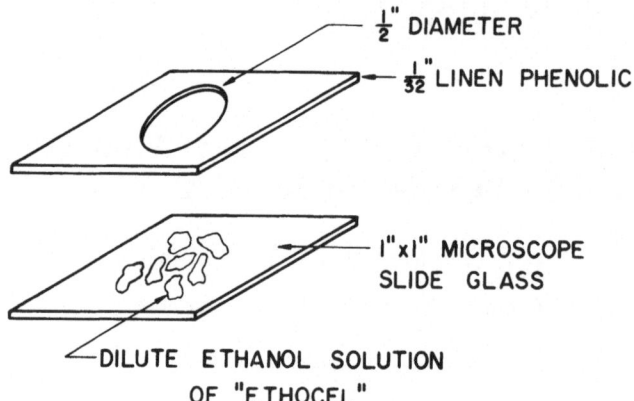

Figure 1. Disposable powder holder.

PREPARATION OF STANDARDS

Finely powdered (below 400 mesh) high-purity oxides of yttrium, iron, and the rare earths were used in preparation of the standards. Careful weighing and blending yielded standards accurate to better than $\pm 0.5\%$ of rare earth present at the 1.0 wt.% level. Concentrations of each rare earth were varied in steps of two from 0.4 to 12.8 wt.% for all elements except Er(0.8–25.6%), Ho(0.2–6.4%), Tm(1.0–16.0%) and Yb(0.6–19.2%).

OPERATING CONDITIONS

All analyses were conducted on a General Electric XRD–5S spectrograph equipped with a Machlett AEG–50S tungsten target X-ray tube operated at 50 kvp and 50-ma tube current. Various combinations of collimators and analyzing crystals were used depending on the problems involved. Table I summarizes the various operating conditions. To cope best with higher-order interferences, a sealed xenon-filled proportional counter (General Electric #6) was used, coupled with a Hamner #N–302 linear amplifier, pulse height analyzer.

EXPERIMENTAL DATA

Representative count rates for the rare earths, iron, and background at the rare earth peak location are presented in Table II. Also included are values for the calibration constants discussed below. A typical calibration curve is shown in Figure 2.

DISCUSSION OF DATA

As shown in Table I, a range of crystals and collimators was required to deal with line interferences and peak/background problems. Wherever possible a LiF analyzing crystal was used in conjunction with a 0.020-in. collimator, since this combination leads to maximal line intensities. Unless line interferences were encountered, the L_{α_1} was used as the rare earth analytical line; second-order Fe K_β was generally used as an internal standard, combining reasonable intensity with excellent peak-to-background ratios.

For each of the rare earths analyzed, a straight-line calibration, up to about 6–8 wt.%, was obtained for a plot of $N_{\text{R.E.}}/N_{\text{Fe}}$ vs. wt.% R. E. on log/log paper. Consideration of

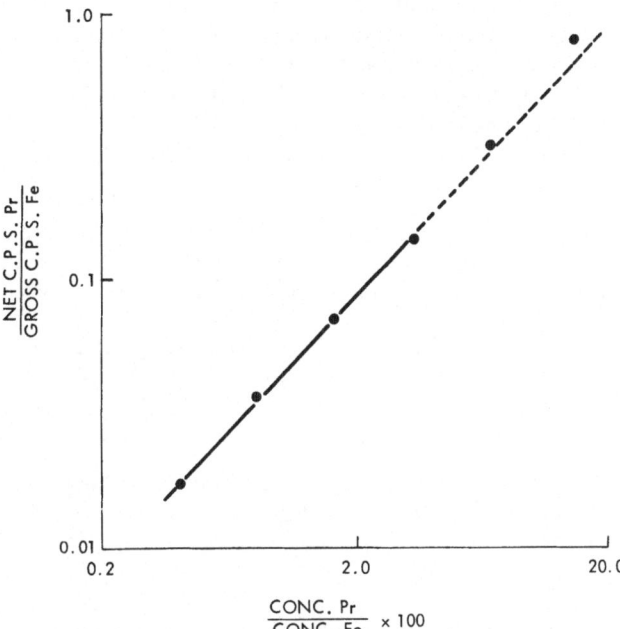

Figure 2. Calibration for 0.4–12.8 wt.% Pr in YFeG.

this led to adaptation of an artifice introduced in an earlier publication.[2] Consider a binary system sufficiently thin as to be free from matrix effects. For such a system it is obvious that the following equation must be obeyed:

$$N_1/N_2 = \alpha_{12}(C_1/C_2) \qquad (1)$$

where $N_i \equiv$ net X-ray peak; $C_i \equiv$ concentration of element i; and $\alpha_{ij} \equiv$ calibration constant.

In the rare earth studies none of these requirements was rigorously met; however, over a limited range of compositions ranging up to 10%, calculation showed $N_{\text{R.E.}}/N_{\text{Fe}}$

Table I. Operating Conditions

Element	Line	Order	Crystal	Collimators, in.	Background	Internal Standard	Order
Ce	L_{β_1}	1	Li F	0.020	Type B*	Fe K_β	2
Pr	L_{α_1}	1	Li F	0.020	83.0°	Fe K_β	2
Nd	L_{α_1}	1	Li F	0.020	90.0°	Fe K_β	2
Sm	L_{α_1}	1	Li F	0.020	70.0°	Fe K_β	2
Eu	L_{α_1}	1	Li F	0.020	66.0°	Fe K_β	2
Tb	L_{α_1}	1	Topaz	0.005	Type B	Fe K_α	1
Dy	L_{α_1}	2	Li F	0.020	Type B	Fe K_β	2
Ho	L_{α_1}	2	Li F	0.020	130.0°	Fe K_β	2
Er	L_{α_1}	2	Quartz	0.005	Type B	Fe K_α	2
Tm	L_{α_1}	2	Li F	0.005	Type B	Fe K_β	2
Yb	L_{β_2}	2	Li F	0.020	86.0°	Fe K_β	2

* Obtain $\dfrac{\text{counts/sec background}}{1000 \text{ Fe counts/sec}}$ for pure YFeG and multiply by N_{Fe} sample.

to be linear, generally to within $\pm 3\%$, with $C_{\text{R.E.}}/C_{\text{Fe}}$ (for convenience we take $C_{\text{R.E.}} + C_{\text{Fe}} = 1$). For the narrow band of compositions covered this is not surprising: The fixed sample geometry and particle sizes coupled with dilution of the elements of interest with oxygen provide a reasonable approximation to a "thin film," at least as far as the above is concerned.

Values of $\alpha_{\text{R.E.,Fe}}$ are tabulated in Table II. Several benefits are derived from the above.

1. Examination of equation (1) shows $\alpha_{\text{R.E.,Fe}}$ to be an index of the relative analytical efficiency of a particular rare earth line compared to that of the internal standard; hence, overall system efficiencies for the various rare earths are characterized by $\alpha_{\text{R.E.,Fe}}$.
2. Once the fact that $N_{\text{R.E.}}/N_{\text{Fe}} \propto C_{\text{R.E.}}/C_{\text{Fe}}$ over a particular range is known, a single standard suffices to establish the calibration over the entire range.
3. Since (2) above is true, rechecks of the calibration may be made at intervals very conveniently, thus ensuring accurate analysis of unknowns.

The above does not take into account the potential difficulties associated with differing particle sizes and degrees of compaction. Such studies were made and are described below.

STUDIES OF THE EFFECTS OF PARTICLE SIZE AND COMPACTION

Two series of studies were made with controlled particle sizes and degrees of compaction for selected limiting cases. One series covered lines on the low-wavelength side of Fe K_β; the other, lines on the high-wavelength side.

Pure YFeG

Large crystals of YFeG were carefully cleaned to remove traces of residual Pb (from the flux in which the crystals were grown) and broken into small pieces. The

Table II. Experimental Results

Sample*	Wt.% Rare earth	Rare earth counts/sec	Fe counts/sec	Background counts/sec	$\alpha_{\text{R.E., Fe}}$
Ce	1.6	198.8	4357.9	15.2	2.47
Pr	1.6	335.8	4558.0	11.8	4.44
Nd	1.6	448.2	4430.0	13.7	6.52
Sm	1.6	915.4	4830.1	32.8	11.4
Eu	3.2	2353.1	4422.1	59.5	15.5
Tb	1.6	71.9	4305.2	$\dfrac{3.64}{1000}$	0.740
Dy	1.6	785.0	4642	$\dfrac{31.4}{1000}$	11.4
Ho	1.6	566.2	4456	64.3	6.99
Er	1.6	47.6	1995.1	8.7	1.17
Tm	1.0	32.2	1182.3	$\dfrac{5.5}{1000}$	2.34
Yb	1.2	38.5	4837.0	15.4	0.439

* All samples in $\frac{3}{8}$-in.-diameter sample holder.

mixer mill and agate mortar and pestle used in sample preparation were used to pulverize the pieces. Next, the residue was passed through a series of sieves of 100, 200, 325, and 400 mesh segregating it into five fractions:

Above 100: Particles above 149 μ
100: Particles of 74–149 μ
200: Particles of 44–74 μ
325: Particles of 37–44 μ
400: Particles below 37 μ

Only the last four portions were used.

In order to establish an upper limit for the effects of particle size and compaction on the short-wavelength side of Fe K_β, a comparison was made of $N_{\text{Fe } K_\alpha \text{ (first order)}}/N_{\text{Y } K_\alpha \text{ (first order)}}$ for hand-packed "loose" samples of 100, 200, 325, and 400 YFeG with pressed pellets made from the same materials. The pellets were made by adding 2 wt.% of Ethocel (as a binder) in an ethanol solution to the YFeG. After it had been dried in air and mixed thoroughly, 400 mg of the material was placed in a $\frac{1}{2}$-in.-diameter die and pressed at 12,000 psi. Two pressed pellets and three loose-packed samples were made for each particle size; three determinations were made for each. The results are summarized in Table III. A deviation of less than 1% was observed between loose packs of 325 and 400. This closely approximates the samples and standards originally tested.

Table III. Experimental Results: Loose vs. Pressed YFeG

Sample	Type	$N_{\text{Fe}}/N_{\text{Y}}$*	Standard deviations, %	Δ	$\lvert L - P\rvert/P$†	$\lvert L_{325} - L_{400}\rvert/L_{400}$
B–1	pressed 100	3.072	0.7	0.6%		
B–2	pressed 100	3.090	0.4			
C–1	pressed 200	3.268	0.6	1.2%		
C–2	pressed 200	3.306	0.6			
D–1	pressed 325	3.461	0.5	0.6%		
D–2	pressed 325	3.481	0.2			
A–1	pressed 400	3.373	0.4	0.4%		
A–2	pressed 400	3.386	0.2			
100P1	loose 100	2.897	0.9			
100P2	loose 100	2.837	0.7	3.2%	8.3%	
100P3	loose 100	2.806	0.1			
200P1	loose 200	3.215	0.3			
200P2	loose 200	3.231	0.2	1.6%	2.4%	
200P3	loose 200	3.179	0.4			
325P1	loose 325	3.372	0.5			
325P2	loose 325	3.330	0.2	1.9%	4.0%	0.7%
325P3	loose 325	3.308	0.6			
400P1	loose 400	3.294	0.2			
400P2	loose 400	3.361	0.7	2.4%	2.0%	
400P3	loose 400	3.282	0.5			

* N_i is X-ray line intensity.

† L refers to average of loose packs; P to average of pressed samples for a fixed particle size range.

Note: All above data taken at 33 kvp and 1 ma.

Pressed pellets led to higher line intensities than did corresponding loose samples. Smaller sample to sample deviations were generally observed for the pressed pellets. Both pressed and loose materials above 325 in particle size led to poor reproducibility.

Praseodymium 3.2 Wt.% in YFeG

Establishment of limits for particle size and compaction effects on the long-wavelength side of Fe K_β was the next requirement. Study of the analytical lines and absorption edges involved led to selection of 3.2 wt.% Pr. This was added (as a powder finer than 400 mesh) to portions of the YFeG fractions. Careful mixing to ensure homogeneity without changing the YFeG particle sizes was essential. This was accomplished by placing the powders in an agate mortar and adding sufficient ethanol to produce a slurry. This was stirred gently and allowed to air dry. The residue was next placed in a polystyrene vial with a polystyrene ball and mixed in the mill for about 30 sec. Previous experience had shown this to produce no grinding action. Two pressed pellets and two "loose" compacts were made for each particle size as was done above. The analytical lines measured were Pr L_{α_1} first order and Fe K_β second order. The results are presented in Table IV. Deviation between loose packs of 325 and 400 was observed to be 2%. Again pressed pellets led to higher line intensities and smaller sample-to-sample deviations. Materials above 325 mesh showed poor reproducibility.

Table IV. Experimental Data: Loose *vs.* Pressed 3.2 Wt.% Pr in YFeG

Sample	Type	Counts N_{Pr}/N_{Fe}	Standard deviation, %	Δ	$\lvert L - P \rvert$*$/P$	$\lvert L_{325} - L_{400} \rvert / L_{400}$
BP–1	pressed 100	0.2942	0.1	42%		
BP–2	pressed 100	0.4181	0.5			
CP–1	pressed 200	0.2245	0.3	4.6%		
CP–2	pressed 200	0.2147	0.2			
DP–1	pressed 325	0.1684	0.2	3.3%		
DP–2	pressed 325	0.1739	0.5			
AP–1	pressed 400	0.1725	0.5	1.3%		
AP–2	pressed 400	0.1748	0.2			
P100P1	loose 100	0.4138	0.4	15%	25%	
P100P2	loose 100	0.4757	0.5			
P200P1	loose 200	0.2520	0.3	2.8%	13.1%	
P200P2	loose 200	0.2451	0.5			
P325P1	loose 325	0.1664	0.3	5.6%	0.6%	2.0%
P325P2	loose 325	0.1747	0.4			
P325P3	loose 325	0.1757	0.8			
P400P1	loose 400	0.1667	0.3	2.7%	2.5%	
P400P2	loose 400	0.1712	0.5			

* L refers to average of loose packs; P to average of pressed samples for a fixed particle size range.

Note: All above data taken at 50 kvp and 25 ma.

CONCLUSIONS

Analysis of Ce, Pr, Nd, Sm, Eu, Tb, Dy, Ho, Er, Tm, and Yb in YFeG was accomplished to an accuracy no worse than ± 5–10% of the rare earth present using simple sample and standard preparation techniques. Calibrations were linear in all cases up to about 6–8 wt.% or better. A calibration constant $\alpha_{R.E.,Fe}$ was defined which serves as an index of overall analytical efficiency for the rare earth line concerned, as well as providing a calibration for the entire linear range.

Studies of loose compacts and pressed pellets of pure YFeG and Pr-doped YFeG for several particle size ranges demonstrated the following:

1. For rare earth analytical lines shorter than that of the internal standard (Fe K_β or K_α) the use of standards finer than 400 mesh and samples finer than 325 mesh (both analyzed as loose compacts) leads to particle size errors less than $\pm 1.0\%$. For rare earth lines longer than those of the internal standard, errors of less than ± 2–3% were found under the same conditions.

2. In each of the above cases, pressed pellets yielded higher X-ray peaks than loose compacts, as well as generally giving better sample-to-sample reproducibility.

3. Both loose and pressed samples of particle size above 37 μ (325 mesh) gave results which were highly erratic. For accurate analytical work in the systems studied they should be avoided.

REFERENCES

1. P. E. Seiden, *J. Appl. Phys. Supp.* **33**: 1234, March 1962.
2. J. C. Lloyd and A. Segmüller, *Z. Naturforsch'g* **16**a: 1097, 1961.

DISCUSSION

C. N. Schieltz (Colorado School of Mines): Were the rare earth standards checked for the possibility of carbonate formation which apparently could occur over a period of time?

J. C. Lloyd: In general, the self-consistency of all results and the linearity and reproducibility of the curves which were obtained would indicate that there was probably no problem in this respect, because a lot of this was done several times so the curve could be started and continued several days later or rechecked several days later. In all cases when I had to run a sample I would recheck at least one standard, and the rechecks on the standards—day to day, week to week, month to month—were well within the accuracy requirement. Of course, $\pm 10\%$ is a rather crude requirement, but I think that most of the results shown in the final two portions here would indicate what you would have to do if you wanted highly accurate results.

A. E. Bernhard (Applied Research Labs): I notice in your experiment with loose *vs.* pressed samples that you ran the yttrium at 33 kv and praseodymium at 50. Now, is there any reason for this?

J. C. Lloyd: Yes. I am an overcautious soul, who gets a little worried when the count rates get out of the ball park, because I have run into some problems there, and I think I have either a counter tube that's getting sick or something along the line that is getting tired, so this was selected solely to get the right count rate. I was generally operating around two or three thousand counts per second—something on this order: High enough to get good statistics, low enough so there was no problem on the effects that can come in as a result of high count rate.

THE USE OF X-RAY EMISSION SPECTROGRAPHY FOR PETROLEUM PRODUCT QUALITY AND PROCESS CONTROL

J. L. Caley

Standard Oil Company of California
Richmond, California

ABSTRACT

A Norelco Autrometer (automatic multielement X-ray spectrograph) is being used by a petroleum refinery control laboratory for process control and product quality control. As many as 100 analyses covering 13 elements are run per day. Concentrations range from a few ppm to 25%.

The instrument has reduced the elapsed time for many analyses from 8 hr to 5 min, made possible in many cases greater accuracy than previously used wet-chemical methods, reduced laboratory manpower and increased refinery production efficiency in many instances.

INTRODUCTION

The use of X-ray fluorescence for the elemental analysis of various contaminates and additives associated with petroleum products has been demonstrated many times. The development of the automatic multielement X-ray spectrograph has made it possible to use this method of analysis for rapid control testing for a variety of elements with a single instrument. The requirements for the instrument in this type of work are rapid analysis, desired precision, and operability by nontechnical personnel.

INSTRUMENTATION

The instrument meeting these requirements and used by the Richmond Refinery Control Laboratory is the Norelco Autrometer[1,2] with the five-channel pulse height analyzer. The 100-kv tungsten X-ray tube, helium optical path, and universal detector[2] are standard with the instrument. The automatic programming includes 24 preset channels for 2θ angles, choices of three crystals and two detector systems, and use of the pulse height analyzer on five channels if desired.

Size EP-16 polyethylene CaPlug inserts are used as disposable sample holders for both liquids and powders. A $\frac{1}{4}$-mil mylar window is held in place with a flat aluminum ring approximately $\frac{1}{4}$ in. thick with an outside diameter such that it will slide easily inside the standard Autrometer solid sample cup. Figure 1 shows the sample holder assembly.

Table I lists the programming of the instrument, concentration ranges utilized, and analytical precision obtained.

---

[1] Superscripts pertain to references at the end of the paper.

Table I. Programming, Concentration Range, and Precision

Element	Line	Order	Crystal	2θ, deg	Detector	PHA	Concentration range	Average precision
Barium	L_{α_1}	1	LiF	87.13	Flow		0-1%	±0.01%
Bromine	K_{α_1}	1	LiF	29.93	Universal[2]		0-2%	±0.01%
Calcium	K_{α_1}	1	LiF	113.0	Universal		1-5%	±0.02%
							0.5-1.0%	±0.01%
							0.05-0.50%	±0.002%
							0-100 ppm	±5 ppm
Chlorine	K_{α_1}	1	EDDT	64.93	Flow	Yes	0-3%	±0.05%
Cobalt	K_{α_1}	2	LiF	125.32	Flow		5-8%	±0.05%
Lead	L_{α_1}	1	LiF	33.92	Universal		0-1%	±0.02%
Lead (as TEL)	L_{α_1}	1	LiF	33.92	Universal		0-5 ml/gal	±0.05 ml/gal
Manganese	K_{β_1}	1	LiF	56.61	Universal		5-8%	±0.1%
Molybdenum (powders)	L_{α_1}	1	EDDT	75.72	Flow		5-25%	±0.1%
(aqueous solutions)	K_{β_1}	2	LiF	36.58	Universal		5-25%	±0.1%
Nickel (powders and aqueous solutions)	K_{α_1}	2	LiF	110.82	Flow		5-15%	±0.05%
(aqueous solutions)	K_{α_1}	1	LiF	48.61	Universal		0-3000 ppm	±10 ppm
Phosphorous	K_{α_1}	1	EDDT	88.66	Flow	Yes	0.5-2.0%	±0.1%
Sulfur	K_{α_1}	1	EDDT	75.16	Flow	Yes	1-5%	±0.05%
			Quartz	106.94	Flow		0.1-1.0%	±0.01%
Vanadium	K_{α_1}	1	LiF	76.84	Flow	Yes	0.01-0.10%	±0.002%
							0-500 ppm	±10 ppm
Zinc	K_{α_1}	1	LiF	41.74	Universal		0-1%	±0.005%
							0-50 ppm	±2 ppm

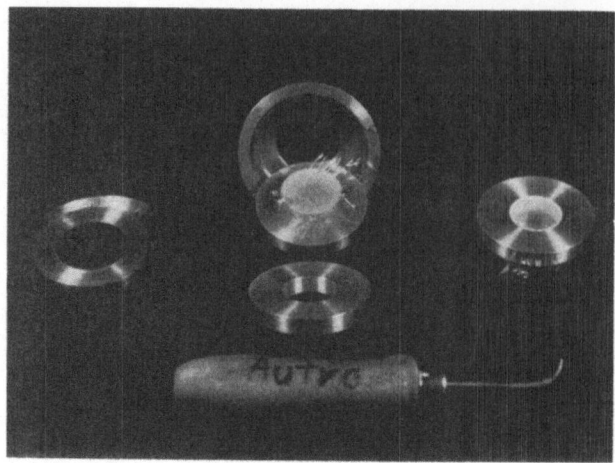

Figure 1. Sample and standard
holder assembly.

CALIBRATION AND STANDARDS

Analysis by the Autrometer is accomplished by the ratio of counts accumulated on a fixed time basis on a standard and a sample. Since the best precision is obtained by keeping the ratio relatively near and preferably less than 1,000, a series of standards for each element is necessary. Table II shows typical standards in use for sulfur, calcium, and lead (TEL).

Standards are either liquid or solid, the solid being preferred. Liquid standards are usually a sample of the material being tested that has been set aside for this purpose. Solid standards are made by mixing appropriate amounts of inorganic salts containing the desired elements with 10 g of transoptic or bakelite powder and fusing in a Buehler Specimen Mount Press with a $1\frac{1}{4}$-in. mold. The formed disc is then mounted in an aluminum ring similar to that used for the samples except that the hole is $1\frac{1}{4}$ in. in diameter, as shown in Figure 1. Duco cement is used to hold the disc in place. We have found that bakelite standards will withstand greater X-ray exposure than the transoptic, which will crack and become erratic in X-ray count after heavy usage.

Calibration curves are then made against the standards, using samples containing known amounts of the desired elements. If matrix effects are evident for certain types of materials, separate curves are made as necessary. This has been especially noticeable in lubricating oil compoundings where concentrations are relatively high. Figure 2 shows typical calibration curves for sulfur and calcium. Matrix effects between the solid standards and petroleum products will often require different amounts of an element in the standard than in the sample to obtain a counting ratio near 1.000. As an example, solid sulfur standards require only about one-fourth the amount of sulfur in them as in the samples.

In some cases the range of concentrations for a particular element is so wide that it is not possible to use the same line and/or crystal for the entire range and obtain satisfactory precision. In these cases we use lines of different intensities or different crystals; e.g., for sulfur below 0.10% we use a quartz crystal with 2θ angle of 106.94° for the K_α line. A slightly higher intensity is obtained with the quartz crystal than with the EDDT crystal, thus giving better precision in the very low range. Table I lists the other combinations used.

A calibration curve is only as good as the analytical results used to establish it, and these are not always easy to obtain, especially in the lower ranges. In cases where a few

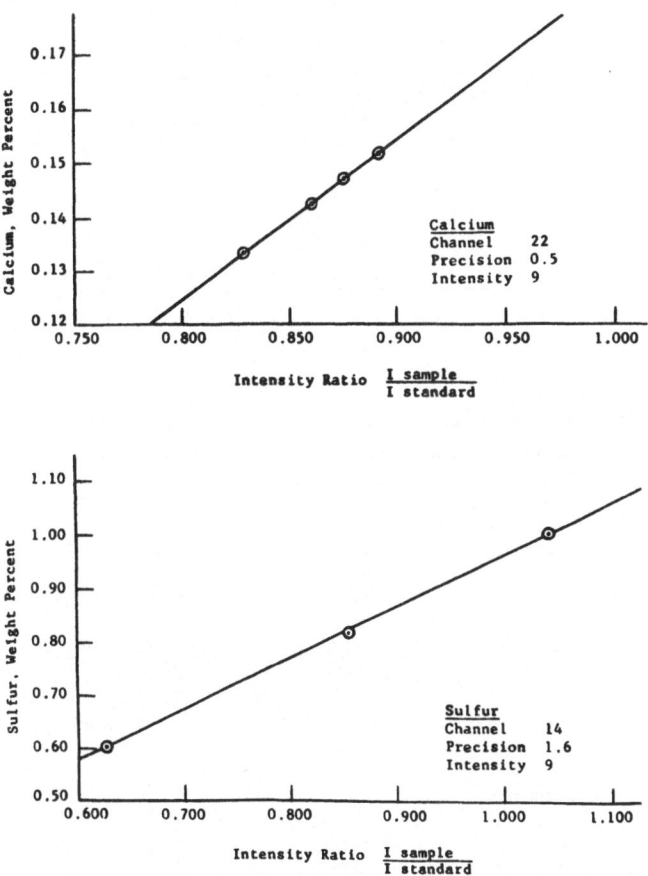

Figure 2. Typical calibration curves.

accurately analyzed standard samples are not available due to inadequacies of the analytical procedures, we will use a large number of less accurate results and draw the best apparent curve, which will be modified as more data are accumulated. Another procedure used is to make up hand mixes by adding known amounts of the desired element to the base materials.

PROCEDURE

Since our purpose is to use the instrument for rapid control testing, we have striven to keep sample preparations to a minimum consistent with the desired precision. Normally, liquid samples are poured directly into the sample cups. Materials such as tar or asphalt are heated until fluid, then poured, immediately after which the mylar film is placed over the cap, the ring slipped on, and the cup inverted so that sample will harden with a flat surface on the window. Granular or powdered samples are ground to finer than 200 mesh and poured into the cup. After the samples are poured and the mylar film affixed, the cup is inverted and the bottom is punctured with a sharp instrument, such as shown in Figure 1, so that the mylar window will not bulge as the sample becomes

Table II. Standards for Sulfur, Calcium, and Lead (TEL)

Sulfur, %	Calcium, %	Lead, ml/gal
	0.004	1.0
0.10	0.100	5.0
0.50	0.50	
1.0	1.0	
4.0	2.0	
	4.0	

Table III. Comparison of X-Ray, Chemical and Emission Spectrograph Analyses

Element	Type of sample	X-ray	Chemical	Emission spectrograph
Calcium	Lubricating oil	0.056%	0.052, 0.052%	
	Lubricating oil	0.062%	0.062%	
	Lubricating oil	0.188%	0.190, 0.193%	
	Lubr. oil concentrate	2.07%	2.04%	
	Lubr. oil concentrate	2.08%	2.08, 2.09%	
	Tar	36 ppm	37 ppm	
	Tar	25, 28, 30 ppm	31, 31 ppm	27, 18 ppm
	Tar	32, 39, 40 ppm	39, 40 ppm	23, 32 ppm
Chlorine	Lubricating oil	1.6%	1.5%	
	Lubricating oil	1.7%	1.8%	
Lead	Lubricating oil	0.28%	0.26%	
	Lubricating oil	0.32%	0.32%	
Lead as TEL	Aviation gasoline	0.39 ml/gal	0.41, 0.41 ml/gal	
	Motor gasoline	0.42 ml/gal	0.40, 0.41 ml/gal	
	Motor gasoline	0.93 ml/gal	0.90, 0.91 ml/gal	
	Motor gasoline	2.90 ml/gal	2.93, 2.94 ml/gal	
	Aviation gasoline	4.20 ml/gal	4.23, 4.24 ml/gal	
Sulfur	Motor gasoline	0.010%	0.011%	
	Motor gasoline	0.039%	0.038%	
	Transformer oil	0.067%	0.064, 0.064%	
	Motor gasoline	0.071%	0.075%	
	Diesel oil	0.22%	0.23%	
	Tar	1.20%	1.22%	
	Fuel oil	1.48%	1.50%	
	Lubricating oil	1.5%	1.6%	
Vanadium	Fuel oil	110 ppm	100 ppm	
	Fuel oil	123 ppm	130 ppm	
Zinc	Lubricating oil	0.292%	0.295%	
	Lubricating oil	1 ppm	1 ppm	1.8 ppm
	Lubricating oil	2 ppm	2 ppm	
	Lubricating oil	42, 42 ppm	42, 41 ppm	

warm while in the instrument. The sample cup is then pushed down in the ring until the window is flush with the bottom of the ring.

The program data for each standard are taken from the calibration curve (see Figure 2) and set up on the programming console. This usually consists of only the channel number, desired precision, and the amount of current to the X-ray tube, which is set to give a counting rate as near 200 counts/sec as possible, or maximum count of less than 200. The tube is set to operate at 66 kv at all times, but the voltage is automatically adjusted as the current is increased so as not to exceed the 2.5 kv-amp power limitation of the X-ray tube. Two or three counting ratios only are obtained on each element unless the counts are not within the statistical accuracy required, in which case more counts would be taken. The average of the ratios obtained is taken to the calibration curve and the answer in percent, ppm, etc. obtained.

RESULTS

The results obtained by X-ray fluorescence have proven to be highly accurate and reliable when proper sample-handling techniques and instrument operating conditions are used. In the majority of cases where differences between chemical and X-ray results have occurred, the chemical results have been proven to be in error. The precision of the X-ray results has proven to be better than that required by many ASTM methods using other procedures.

Comparative chemical and X-ray results are given in Table III.

The average time to obtain a result is approximately 5 min as compared to 6 to 8 hr for a lead (TEL) in gasoline by ASTM D526 and 4 hr for a sulfur by ASTM D1266.

ECONOMIC ADVANTAGES

The use of the automatic-programmed X-ray spectrograph for process control and product quality has proven to have a number of economic advantages for both the laboratory and the manufacturing plants. For the laboratory these are as follows:

1. Two laboratory technicians working together with the instrument make an average of about 80 analyses per 8-hr day. This has resulted in a manpower savings of approximately four technicians over previous requirements.
2. It is possible to provide night-shift coverage on required work with less manpower.
3. Less training and experience is required to operate the instrument than is required for many chemical procedures and a man has to learn only the operation of the instrument rather than a large number of different methods. This makes it possible to use personnel of lower classification.

Manufacturing plants obtain the following advantages from this rapid and, in many cases, more precise method of analysis:

1. For continuous operating plants, faster results allow corrections to be made sooner, and more accurate results allow closer control to the most efficient operating conditions.
2. For batch operations, faster results give faster turn around on equipment, and ultimately more production and more accurate results allow closer control to the most economical specification limits.

CONCLUSIONS

The X-ray fluorescence method of analysis has been a valuable addition to our laboratory for obtaining better elemental analyses. The automatic multielement spectrograph has made possible rapid and accurate process control for refinery operations. The economic advantages have more than justified the cost of the instrument.

We have by no means reached the limits of sensitivity or variety of anlyses that the instrument is capable of and we have many more potential applications that we are working toward adapting to this method of analysis.

ACKNOWLEDGMENT

The author wishes to thank Ralph C. Vollmar for his guidance and help in making this paper possible, and John P. Steiner, Wesley T. Woolley, and George W. Zurilgen who did most of the calibrating and experimental work.

REFERENCES

1. D. C. Miller, "Results Obtained with the Modified Norelco Autrometer," *Advances in X-ray Analysis, Vol. 1*, University of Denver, Plenum Press, New York, 1957, p. 283.
2. W. R. Kiley, "A Universal Detector for the X-ray Spectrograph," *Advances in X-ray Analysis, Vol. 2*, University of Denver, Plenum Press, New York, 1958, p. 293.

DISCUSSION

D. Holden (Universal Oil Products): Did you get different response curves for the different standards?

J. L. Caley: We get different slopes between the transoptic and the bakelite, and we also get different responses from the standard and the sample, in many cases with the same amount of the element in each. In setting up our standards we put in the theoretical amount of the element for the concentration range desired into the standard to compare with the sample that had been analyzed and determine the ratio. Then we adjust the amount of the element in the standard to get us in the right ratio range of around 1.0.

T. P. Schrieber (General Motors Corp.): Have you made corrections for variations in the specific gravity of the gasoline in the tetraethyl lead analysis?

J. L. Caley: We haven't found it necessary. The results through the instrument have been within the precision required without any gravity corrections.

T. P. Schrieber: The specific gravity of gasoline is considercd constant, then?

J. L. Caley: Yes.

ABSORPTION EFFECTS IN X-RAY FLUORESCENCE MEASUREMENT OF ELEMENTS IN OIL

E. L. Gunn

Humble Oil and Refining Company
Baytown, Texas

ABSTRACT

The influence of absorption effects in measuring elements in petroleum oil by X-ray fluorescence can, under certain circumstances, be predicted approximately from theory. Exact information on specific systems and instrumental conditions must practically be provided by experiment, however. The influence of absorption on fluorescent intensity under specific conditions of interest has been studied from three viewpoints: varying the depth through oil of a solid elemental specimen, varying the depth of a solution containing the element dissolved in oil, and varying the amount of the dissolved element in oil over a broad range in concentration. The group of elements selected for study illustrate characteristic fluorescent radiations often measured from oil. The experimental results are expressed graphically and are compared between elements as well as with theory. The results have analytical implications which pertain to the quantitative determination of elements in oil by X-ray fluorescence.

INTRODUCTION

The influence of absorption on the X-ray fluorescent intensity of elements measured in petroleum oil has special significance when compared with X-ray measurements made in most other substrates.[1-5] The low absorption properties of oil for the entering excitation and the emergent fluorescent X-rays cause the penetration depth to be considerably greater than that for most of the inorganic powders or metallic materials usually analyzed by this technique. Hence, a relatively greater volume of the oil sample participates in fluorescence than does that of many other substances. A knowledge of the effect of absorption and sample depth is a practical consideration (1) in the case of a limited sample amount; (2) where a particular design of X-ray optics is employed, (3) where absorption by oils of very different composition and density is involved, (4) in analytical applications where departure from linearity is a consideration, or (5) in cases where possible homogeneity problems may exist. One may easily and readily calculate the penetration depth for X-rays of specified wavelengths in oil from well-known physical relations. However, since the effective wavelength of the excitation beam is unknown and possible change in the optical path of the instrument is involved, it appeared feasible that direct experimentation be employed to obtain the specific information desired on absorption effects. The present study was carried out on a selected oil, and the elements chosen for measurement represented typical characteristic X-ray wavelengths often measured in oils.

---

[1] Superscripts pertain to references at the end of the paper.

EXPERIMENTAL

The influence of change in depth on absorption was investigated from two aspects: first, measurement of the intensity of an element positioned in the form of an adjustable solid specimen through known depths of oil; second, measurement of the intensity of an element in oil solution in which the depth was varied by the addition of known increments of the sample to the sample cell. The instrument was the inverted, geometry type, i.e., the X-ray beam impinges onto the mylar bottom of the sample cell and the fluorescent beam thereby generated in the oil emerges through the bottom, thence into the detector system.

The oil was a high-purity white mineral oil having a carbon content of 85.5%, 14.5% hydrogen, and a specific gravity of 0.8412 at the temperature of the laboratory (77°F).

For measuring the absorption effect through oil, specimens in the form of discs were prepared of molybdenum, nickel, iron, manganese, and titanium from the respective metal elements. A disc containing bromine was prepared as potassium bromide powder with DuPont cement as a binder. The diameter of the mylar window in the bottom of the cell and that of the disc was 3.0 cm. The cell cover was made of lucite plastic arranged as a center-threaded holder so that the disc attached onto the end of a screw post could, by screw action, be raised or lowered by known distances between its plane surface and the surface of the mylar cell window.

For measuring the absorption effect in solution, the element was added to the mineral oil in organic soluble form to prepare a solution of known concentration in that element. The element bromine was added as dibromobenzene. Nickel, iron, manganese, and vanadium were added as the respective naphthenates of each to prepare the oil solution of the element. Incremental portions of the oil containing the trace element were added to the sample cell as tared weights with an analytical balance. The depth could, of course, be calculated from the known density and weight of the oil and the area of the cell bottom.

The influence of changes on concentration of intensity was studied by measuring the element in soluble form in oil over a broad range in concentration. Zinc and vanadium dissolved as naphthenates were measured, and care was taken that the amount of sample used in the cell at each concentration level exceeded the critical depth.

The primary X-rays were generated by a tungsten target X-ray tube and the analyzing crystal was lithium fluoride. A scintillation counter employing pulse height discrimination was used as the detector. The exit and detector collimators had 0.02-in. leaf spacings.

RESULTS

Element as Solid in Varying Depth of Oil

The elements prepared in the form of solid discs were measured in air as well as in oil. Intensity changes produced by cell shielding of the X-rays or by the modification of the optical path should be expected to become significant when the disc is retracted from the cell window a few millimeters distance. The relationship between disc position and intensity can be best shown graphically. This has been done for each element, and the results are plotted on a relative intensity scale so that intercomparisons between different measurements can more easily be made. The experimentally observed intensity in counts per second is indicated on each curve at the relative level of 50 units so that any point on it can readily be converted to the observed counts per second. Intensity has been corrected for background in each set of measurements.

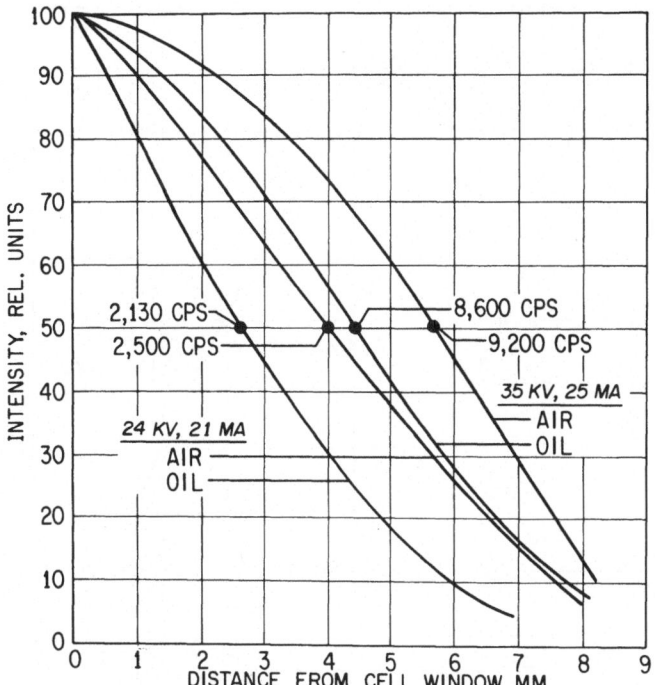

Figure 1. Influence of molybdenum disc position on Mo K_α intensity.

Molybdenum. The results for measuring molybdenum are given in Figure 1. As expected, the intensity decreased as the disc was withdrawn from the window. At a distance of 8 mm, for example, the measured intensity is only about 10% of its initial value, i.e., the measurement at which the disc was in contact with the cell window. Measurements were made at different tube excitation powers. The results show that, for either air or oil, the initial rate of intensity decrease with distance is less at the higher tube power.

As one would expect, it was observed that at either tube power the intensity decrease with distance is less for air than for oil; however, the differences between the two media are not marked. The fact that the intensity decrease is not markedly less for oil is explained by the low absorptive property of the hydrocarbon oil for the relatively hard Mo K_α radiation. The linear absorption coefficient of the oil for this radiation is only about 0.6 cm^{-1}.

Bromine. Figure 2 shows the influence of the disc position on the fluorescent intensity of Br K_α in the potassium bromide—DuPont cement matrix. Bromine fluorescence is absorbed by other elements in the disc as well as by itself, thus differing from the other elemental disc specimens which were examined. The difference between air and oil in rate of intensity decrease is very large, but a significant comparison between them cannot be made because of the difference between the excitation tube powers employed.

Nickel. The influence of disc position on the intensity of Ni K_α is shown in Figure 3. The lower tube power, 15 kv–21 ma, was selected to provide moderate counting rates for the highly emissive nickel disc. The high emissivity of nickel using tungsten target

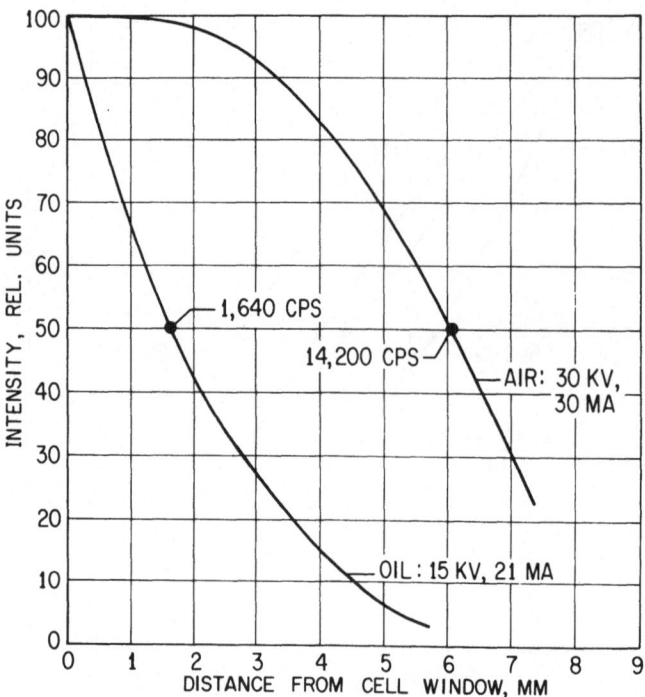

Figure 2. Influence of potassium bromide disc position on Br K_α intensity.

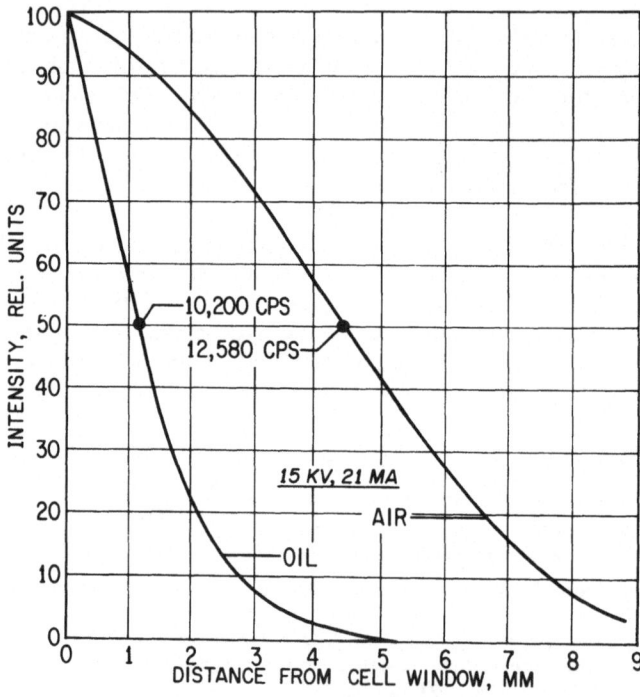

Figure 3. Influence of nickel disc position on Ni K_α intensity.

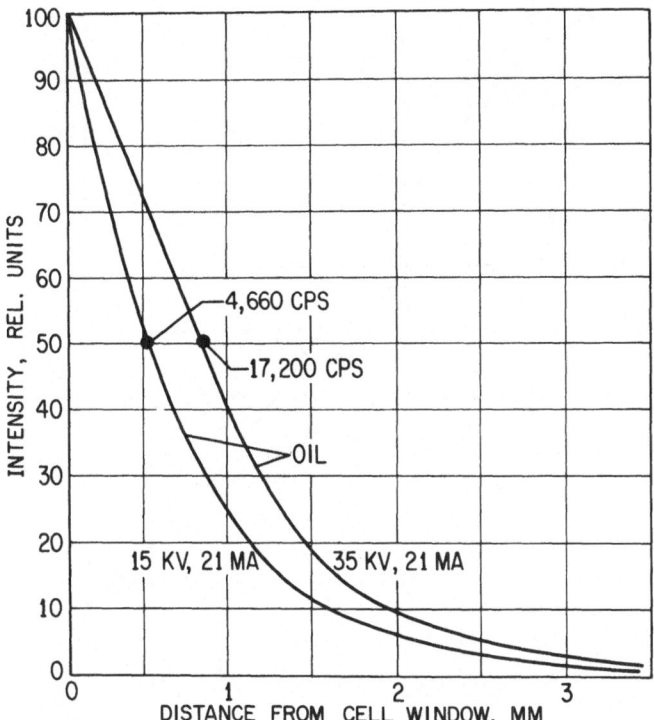

Figure 4. Influence of iron disc position on Fe K_α intensity.

excitation is in part accountable to the occurrence of the Ni K absorption edge at the 1.488-Å position and of the W L_{α_1} peak at 1.476 Å. At a distance in oil of 3 mm, the initial intensity is reduced by more than 90%; for the same reduction in air the distance is 8 mm.

Iron. In Figure 4 the graphical plots again show that intensity decreases more rapidly with distance at the lower tube power. An increase in tube voltage from 15 to 35 kv produces an intensity increase by a factor of approximately 3.7 as shown by the values for the half-intensity levels on the plots.

Manganese. The effects of absorption of Mn K_α by the oil are exhibited in the plots of Figure 5. A marked difference between air and oil absorption is shown, but this difference would be even greater if the same tube excitation power had been used in both cases. In contrast with the results shown for Mo K_α in oil in Figure 1, the difference here between oil and air is highly significant. As an explanation, the absorption coefficient of Mn K_α X-rays in the oil is approximately 20 times as great as it is for Mo K_α.

Titanium. The effect of absorption by the oil is further shown by the plot for titanium in Figure 6. The linear absorption coefficient of the oil for Ti K_α is about 25 cm^{-1}. At a distance of 1 mm, the intensity has been reduced in the oil by more than 90%. The corresponding distance in air is greater than 7 mm.

Comparison of Elements. A comparison of the influence of absorption in terms of cell sample depth of the disc on the intensity for different elements can be made

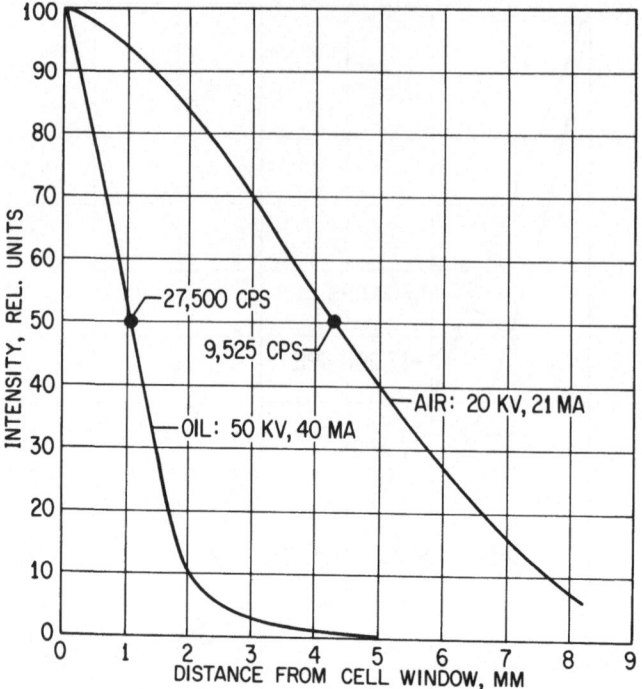

Figure 5. Influence of manganese disc position on Mn K_α intensity.

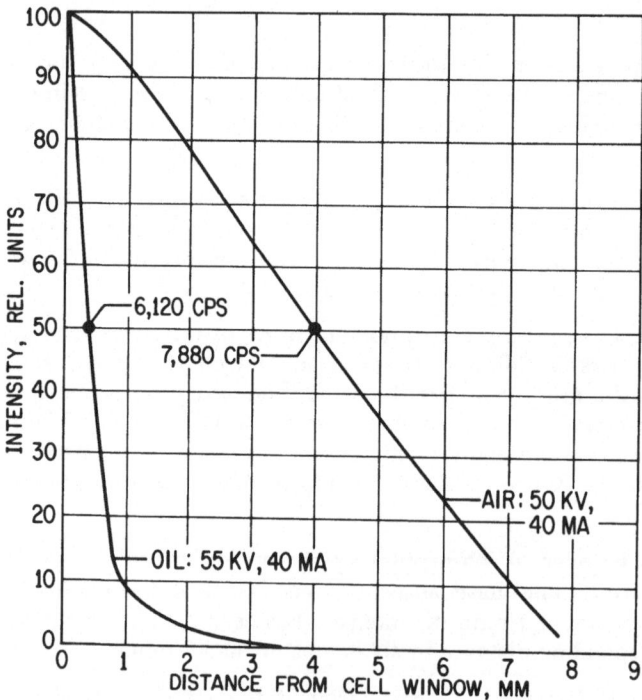

Figure 6. Influence of titanium disc position on Ti K_α intensity.

by determining the depth at which the characteristic intensity is reduced by the same factor for each element, e.g., the depth at which the intensity is reduced by one-half. This has been done for each element and the results are summarized in Table I. Comparisons at the same tube excitation power did not appear to be feasible because of the extreme differences in the efficiency of generating and detecting the characteristic X-rays between elements, e.g., Ni K_α and Ti K_α. It is also possible to compare the observed half-depth with that theoretically calculated in each case. The well-known expression for this calculation is

$$I/I_0 = \exp[-(\mu_1 \csc \theta_1 + \mu_2 \csc \theta_2)\, \rho d]$$

where I_0 is the incident and I the emergent intensity; μ_1 is the mass absorption coefficient for the incident excitation beam and μ_2 that for the emergent fluorescent beam; θ_1 and θ_2 are the angles made by the incident and emergent beams of the instrument (66 and 35°, respectively, for this specific case); ρ is the density of the oil sample; and d is the oil depth. The effective wavelength of the incident beam is not known and therefore μ_1 cannot be assigned a firm value. However, the somewhat higher value of μ_2 for the longer-wavelength emergent beam is the controlling factor in this expression so that the $\mu_1 \csc \theta_1$ term may be dropped with a relatively minor effect on the calculated value. Hence, the half-depth expression becomes

$$d = \frac{-2.303\,(\log 50 - \log 100)}{\mu_2 \csc \theta_2\, \rho}$$

The theoretically calculated half-depth of the oil is also included in the data of Table I.

It is apparent that the observed half-depth becomes less as the characteristic wavelength increases, even though the specimens are not all compared at the same excitation level. This trend is not observed for air, for which the absorption loss is negligible. For oil, the greatest half-depth measured is for molybdenum (4.4 mm) and the least is for titanium (0.40 mm).

Table I. Disc Distance for Reduction of Intensity by One-Half

Element	Wavelength, Å	Tube power		Half depth, mm		
		kv	ma	Air	Oil$_{obs}$	Oil$_{theor}$
Mo	0.71	24	21	4.00	2.60	7.9
		35	25	5.65	4.40	
Br	1.04	15	21	—	1.65	3.1
		30	30	6.05	—	
Ni	1.66	15	21	4.40	1.15	0.82
Fe	1.94	15	21	—	0.47	0.52
		35	21	—	0.80	
Mn	2.10	20	21	4.30	—	
		50	40	—	1.10	0.40
Ti	2.74	50	40	3.95	—	
		55	40	—	0.40	0.19

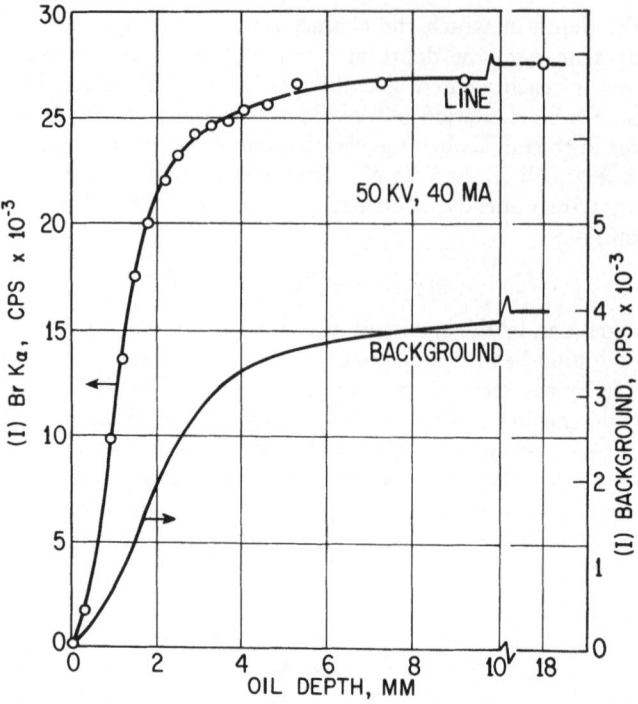

Figure 7. Sample depth *vs.* fluorescent intensity for dibromobenzene in oil (5000 ppm Br).

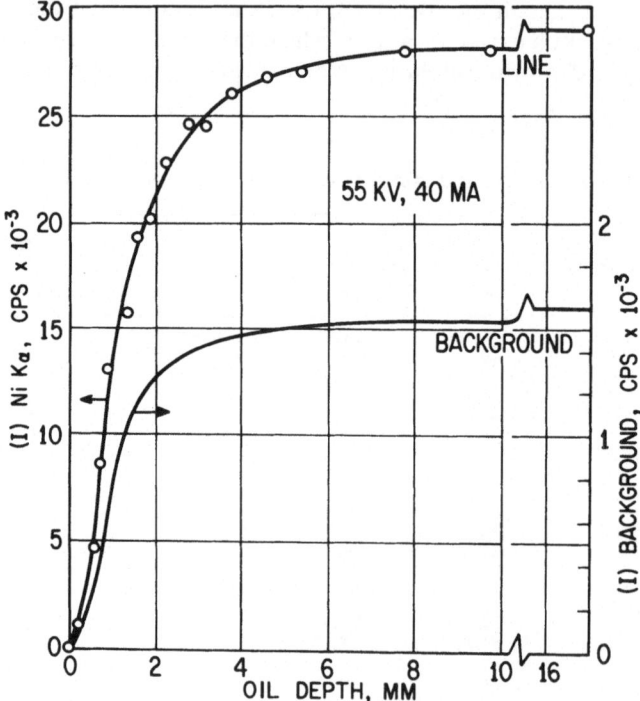

Figure 8. Sample depth *vs.* fluorescent intensity for nickel naphthenate in oil (100 ppm Ni).

As pointed out in the foregoing discussion, the intensity decreases as the disc is retracted from the cell window for two reasons; first, the fluorescent X-rays are absorbed more by the greater depths of oil; second, shielding and modification of the optical path at greater depths cause intensity loss. For long-wavelength X-rays, the absorption effect accounts for almost all of the loss; for short wavelengths the second effect becomes highly significant. Thus, when the observed and theoretical half-depths are compared for molybdenum and bromine, it is observed that the theoretical values are two- to threefold greater than the observed values. It is probable that a correction for optical loss would bring these values into better agreement, but the uncertainty in such corrections did not make this appear to be an attractive effort. For iron and nickel, theory and experiment are closer together, especially at the lower tube powers. Manganese was not measured in oil at low tube power. The observed depth for titanium at high tube power is actually twice the theoretical value. It is quite probable that low tube power measurement would have brought the values into better agreement. The result for titanium, plus the foregoing observations that the rate of intensity decrease is an inverse function of excitation power, indicates that the amount of scattered fluorescent radiation which enters the optical path and is detected becomes relatively much greater at high excitation levels than at low. Thus, at a low tube power, the surface of the disc is the origin of essentially all the fluorescent beam, with the consequence that the theoretical and observed paths in oil are in reasonable agreement for the longer wavelengths. At high excitation levels the considerable increase in scattered radiation of the same wavelength which is measured significantly increases the intensity values at given depths on the plot.

Element in Oil Solution of Varying Depth

In the measurement of varying sample depths of elements synthetically dissolved in oil high excitation tube powers were used, i.e., 50 or 55 kv and 40 ma. Both the K_α line of the element and an adjacent background reference were measured. The graphical plots are for intensity in counts per second, corrected for background, vs. the depth of the oil in millimeters. In each case the critical depth was achieved., i.e., the depth beyond which no further change in intensity was detected. The fixed concentration of the measured element dissolved in mineral oil and the excitation tube power are given on each graph.

Bromine. The results of measuring the Br K_α intensity for varying depths of oil containing dibromobenzene are shown in Figure 7. The critical depth is approached at 10 mm for both the line and the background. The initial rate of intensity change with oil depth is greater for the line than for the background. This also held true for the measurements which follow.

Nickel. The results for nickel are given in Figure 8. The efficiency of generating and detecting Ni K_α radiation is very high with the instrument used in these measurements. The critical depth again is approached at 10 mm.

Iron. The results for iron are presented in Figure 9. The critical depth is approached at 5 mm.

Manganese. Figure 10 presents the plots for manganese measurements. The critical depth is approached at a value of 4 mm.

Vanadium. The efficiency of generating and detecting V K_α radiation is relatively low, as shown in Figure 11. The critical depth is approached at 3 mm.

Comparison of Elements. Comparisons between elements of the influence of oil depth on intensity are given in Table II. These are given for points on the curves which are reasonably easy to define, *viz.*, the depth at a slope of unity and at one-half the critical depth. In each case there is a systematic decrease of depth with increase in wavelength, as should be expected. For the unit slope point the range for the line intensities is from 1.4 to 3.2 mm; for the half-critical depth the range is from 0.8 to 1.2 mm. The concentration of the measured elements in these oil synthetics varied considerably. Although the line-to-background ratio may provide an index of sensitivity for a given element for the specific instrumental conditions used, intercomparisons between these ratios are not valid because both the concentration level and the manner of excitation vary.

Element in Oil Solution in Varying Concentration

The influence of concentration on absorption effects as indicated by fluorescent intensity was determined by preparing a series of synthetic blends to cover several orders of magnitude of concentration of the element, then measuring them. As pointed out above, the cell sample depth exceeded the critical depth in each blend so that depth was

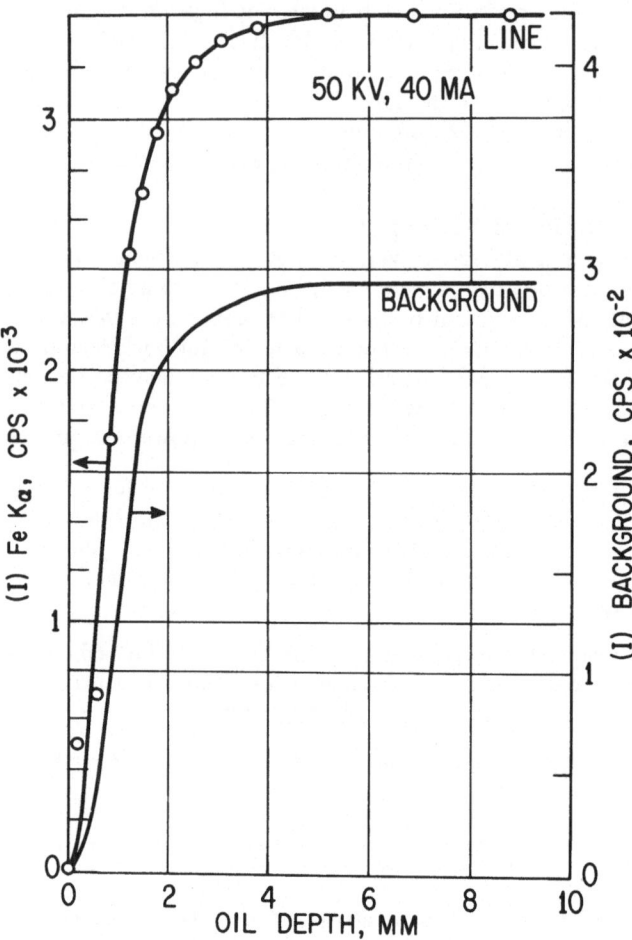

Figure 9. Sample depth *vs.* fluorescent intensity for iron naphthenate in oil (500 ppm Fe).

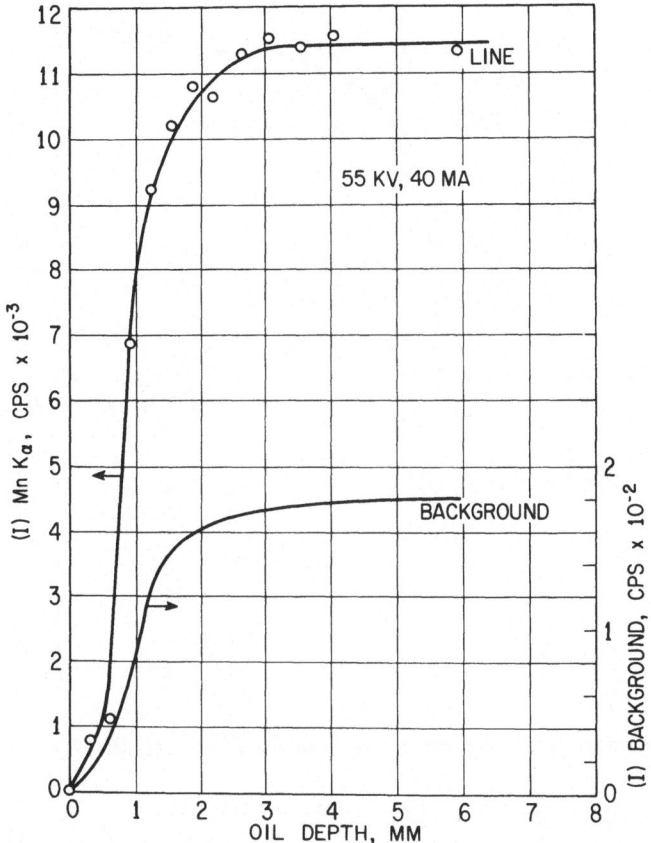

Figure 10. Sample depth *vs.* fluorescent intensity for manganese naphthenate in oil (3000 ppm Mn).

Table II. Values Obtained from Intensity–Depth Curves for Trace Elements in Oil

Element	Figure	Depth of slope of unity, mm		One-half critical depth, mm		Line–background ratio at critical depth
		Line	Background	Line	Background	
Br	7	2.4	2.7	1.2	2.1	7
Ni	8	2.6	2.1	1.15	1.1	18
Fe	9	2.4	2.3	0.85	1.2	12
Mn	10	2.2	1.6	0.80	1.0	64
V	11	1.4	1.3	0.80	1.0	22

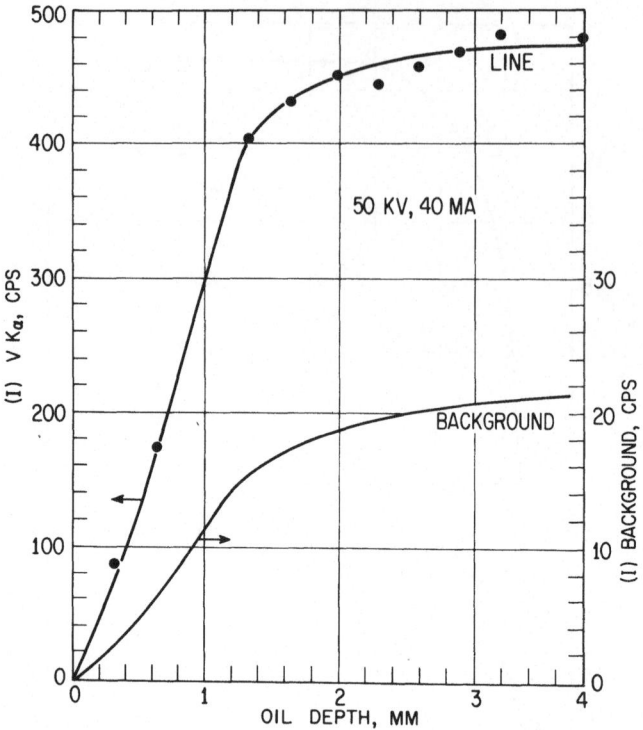

Figure 11. Sample depth *vs.* fluorescent intensity for vanadium naphthenate in oil (1500 ppm V).

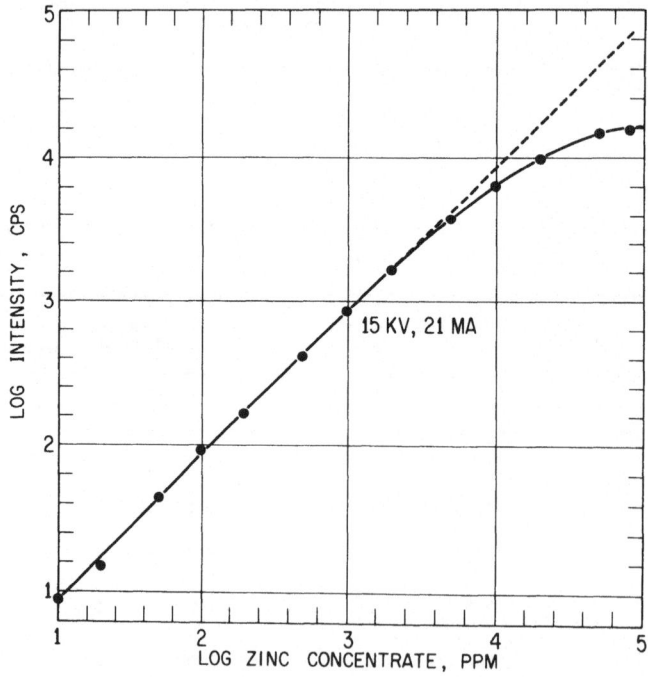

Figure 12. Intensity of Zn K_α fluorescence for zinc naphthenate in oil.

not a measured variable in the series. The intensity was corrected for background in each case.

Zinc. The results for measuring zinc are presented in Figure 12. Linearity between intensity and concentration is observed from 10 to 2000 ppm zinc. Departure from linearity then becomes greater until at 80,000 ppm the observed intensity is only 23% of that predicted from the sensitivity factor for the region of linear behavior. The use of tube powers greater than that shown in no way changed the character of the curve except to shift it to higher values on the intensity ordinate.

Vanadium. The results for measuring vanadium are shown in Figure 13. Linearity is indicated up to a concentration of 5000 ppm vanadium. At 30,000-ppm concentration, the intensity is 56% that predicted from the linear portion of the curve.

Increases in concentration of the element in the two foregoing examples result in two competing effects on the fluorescent emission of the element. An increase in the number of emitting atoms per unit volume of oil tends to increase the emitted intensity, whereas concurrently, the sample absorption coefficient is increased by these same atoms, which tends to reduce the emitted intensity. Presumably, the concentration of the element could be increased to approach a point where an additional increase in concentration would result in no detectable change in intensity. The absorption coefficient of the metal element is somewhat greater for its characteristic K_α emission than is that of the oil in which it is dissolved. But the sample absorption effect does not appear to produce a detectable change until the metal element comprises 0.2 to 0.5% of the total sample composition in these samples. This signifies that calibrations at or above these levels must be established and applied at narrower concentration intervals than they might be in the linear region below these levels to provide precise results.

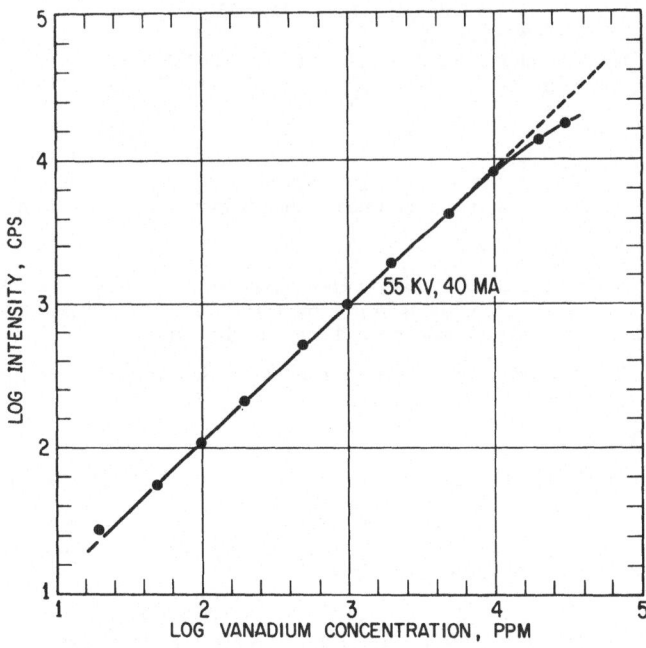

Figure 13. Intensity of V K_α fluorescence for vanadium naphthenate in oil.

CONCLUSIONS

The conclusions which follow apply to the specific instrumental conditions employed and the elements selected for measurement in this study. However, the effects of absorption on the characteristic fluorescence of other elements having atomic numbers falling within the atomic number range herein considered may be estimated with reasonable accuracy from these results. Those outside this range or those measured under somewhat different experimental conditions should be experimentally measured to establish firm values for the absorption effect.

1. Optical path or shielding loss was relatively large and absorption loss small for short penetrating wavelengths such as that of Mo K_α. The reverse was true for long wavelengths, e.g., Ti K_α.

2. The rate of decrease in intensity as the depth is increased is a function of the excitation power; the rate becomes less with increased tube power. The oil depth for a reduction in intensity by one-half ranged from 4.4 mm for Mo K_α to 0.4 mm for Ti K_α.

3. For elements dissolved in oil the half-depth value generally was greater for the background than for the line. The half-depth values for lines varied from 1.2 mm for Br K_α to 0.8 mm for V K_α. Infinite depth desirable in an analysis varied approximately from 3 mm for vanadium to 10 mm for nickel and bromine.

4. A linear relationship between intensity and concentration is exhibited over several orders of magnitude change in concentration—up to a few thousand ppm of the element in the oil. Beyond this concentration, absorption loss produces a reduction in intensity which increases in degree as the concentration is increased. Calibrations should be applied cautiously in concentration regions where absorption loss is significant.

REFERENCES

1. E. N. Davis and R. A. Van Nordstrand, *Anal. Chem.* **26**: 937, 1954.
2. C. W. Dwiggins, Jr. and H. N. Dunning, *Anal. Chem.* **32**: 221, 1960.
3. R. A. Jones, *Anal. Chem.* **31**: 1341, 1959.
4. C. C. Kang, E. W. Keel, and E. Solomon, *Anal. Chem.* **32**: 221, 1960.
5. F. W. Lamb, L. M. Niebylski, and E. W. Kiefer, *Anal. Chem.* **27**: 129, 1955.

DISCUSSION

J. Baecklund (Avesta Jernverks): Have you noticed a dip in the calibration curve as lower and lower concentrations of a given high atomic number element are contained in the oil?

E. L. Gunn: No.

F. Bernstein (General Electric Company): Have you made any attempt to determine how far down you can go into the sample without losing intensity due to the fact that you begin to view a smaller and smaller area if you go far enough down into the sample?

E. L. Gunn: Does the beam become more conelike with increased depth, is that what you mean?

F. Bernstein: Yes.

E. L. Gunn: I think the effect you have in mind is approached. For want of a better term I referred to this in the paper as shielding or optical displacement.

G. Walden (Union Carbide Nuclear Company): Did you determine the point at which your detector or counter ceased to be linear in its response?

E. L. Gunn: Yes. We were very careful in this consideration because some of the counting rates were quite high. As I mentioned in the paper, we used pulse height discrimination; also we verified the linearity of counting. The departure from linearity shown in the experimental curves projected in the slides is real and is not due to nonlinear counting.

DETERMINATION OF CATALYST RESIDUES IN POLYOLEFINS BY X-RAY EMISSION SPECTROSCOPY

G. D. Smith and R. L. Maute

Monsanto Chemical Company
Texas City, Texas

ABSTRACT

X-ray fluorescence methods for the direct determination of aluminum, chlorine, titanium, and iron in polyolefins have been developed. The procedures require no preconcentration of elements, are satisfactory for any polyolefin, and are free from matrix effects. The accuracy and precision are generally comparable with or better than chemical procedures. In addition, the speed of the analysis is six to eight times faster.

INTRODUCTION

The production of Ziegler-type linear polyolefins normally requires metal catalysts containing halogen. However, the presence of these residues has a deleterious effect on the polymer properties and color.[4] In addition, high concentrations of chlorine can cause corrosion of extrusion equipment. Iron contamination from process equipment can produce yellowness in the polymer. Hence, knowledge of catalyst residues and iron is essential.

Sensitive chemical methods are available for ppm analysis of trace residual catalyst as well as iron contamination in polymers. Bolleter[1] determined aluminum and titanium in Ziegler polyethylene using the color complex of titanium-chromotropic acid and aluminum 8-quinolinolate after wet or dry ashing of the sample. Similarly, bathophenanthroline has been used in our laboratory to determine iron. For chlorine, a titrimetric or a colorimetric procedure has been used following decomposition of the sample in an oxygen atmosphere. Although these methods are fairly sensitive, ashing of the sample is time-consuming and careful technique is required.

The speed and simplicity of X-ray emission spectrography or X-ray fluorescence (XRF) should make it an effective tool for these analyses, provided the instrument has sufficient sensitivity for direct analyses of the polymer. Utilization of X-ray fluorescence in trace work is not new or novel; however, there is some indication that insufficient emphasis has been placed on its use in this field. Examples have been given by Thatcher[5] and Gunn[2] regarding trace analysis.

Included in Gunn's discussion was the direct analysis of titanium in polypropylene by XRF. Also, a similar method is reported for the direct analysis of chlorine in polyethylene.[3] However, both methods lack the sensitivity desired for chlorine and titanium at less than the 100 ppm level.

The XRF method described here not only permits the direct determination of titanium and chlorine but also of aluminum and iron in polyolefins at this level. The

---

[1] Superscripts pertain to references at the end of the paper.

procedure is rapid and accurate enough to provide excellent process control. An example of the increased sensitivity is seen in the chloride analysis, where the counting rate (above background) is about two and a half times that previously reported.[3]

The improved sensitivities are due to a combination of improvements in equipment and technique. The equipment includes: (1) high-resolution vacuum spectrograph with a pulse height analyzer (P.H.A.); (2) vacuum rather than helium path; and (3) constant potential X-ray generator with highest tube power.

Our studies on new techniques coupled with improved instrumentation revealed that: (1) the preparation of samples in the form of discs cut from pressed slabs precludes errors due to variation in particle size or packing of powders or pellets, and eliminates absorption effects of mylar film on weak radiation; (2) the constant background for both titanium and chlorine simplifies and shortens the procedure by eliminating background count; (3) normal variation in thickness of the disc sample has negligible effect upon background except for aluminum and chloride, where some error is introduced; (4) the use of compounds of each element which are compatible with the polymer facilitates the preparation of accurate standards; and (5) the use of a mixed standard containing a relatively high concentration of each element permits correction for day-to-day instrument fluctuations.

EXPERIMENTAL

Equipment

The instrument used is a Norelco (Philips Electronics, Inc.) inverted-sample, four-position vacuum spectrograph, with P.H.A., and a constant potential X-ray generator with a FA-60 tungsten tube and voltage and current stabilization. This is coupled with the new Norelco circuit panel with decade scalers. The spectrograph is used with a flow proportional counter, sodium chloride and ethylenediamine d-tartrate (EDDT) crystals, a port collimator having a plate spacing of 0.020 in., and a standard collimator on the flow counter.

Procedure

The polymer discs ($1\frac{3}{16}$ in. diameter $\times$ 0.125 in. thick) are rinsed with acetone and dried. Samples are handled with tweezers to prevent possible chlorine contamination. The standard and sample discs are then placed in the specimen cells without mylar windows and counted at the correct angle for each element after reaching a vacuum of $300\,\mu$ or lower. The P.H.A. is used for all elements. Counts are made at the K_α first-order line for all four elements, using the conditions shown in Table I. The base resin is counted

Table I. Instrument Conditions

Element	Angle-2θ	Excitation	Counter	Crystal	Baseline V	Window V	Fixed time/sec
Ti	58.35°	50 kv–50 ma	Flow	NaCl	10.0	15.0	100
Cl	113.95°	50 kv–50 ma	Flow	NaCl	7.5	15.0	200
Al	142.50°	50 kv–50 ma	Flow	EDDT	7.5	17.5	500
Fe	40.20°	50 kv–50 ma	Flow	NaCl	11.0	11.5	100
			Scint.		6.0	15.0	100

at the aluminum and iron angle for background counts. The concentrations are then found from the predetermined calibration curve as a function of counts per second (cps).

Calibration

Calibration standards are prepared by blending weighed quantities of titanium dioxide, aluminum stearate, iron stearate, and tetrachloro-bis-phenol A with the base resin, high-pressure polyethylene. Homogeneous mixing is accomplished by adding the compounds to molten polymer in a Banbury mixer. This molten mass is mixed, cooled, extruded, and chipped into small pellets, which are molded into slabs 125 mils thick. Discs $1\frac{3}{16}$ in. in diameter are cut for calibration standards, care being taken to avoid surface iron contamination. The standards are prepared to contain 0–100 ppm titanium, 0–100 ppm iron, 0–300 ppm chlorine, and 0–500 ppm aluminum.

Daily fluctuations in the instrument are compensated for by a correction factor (ratio of the counting rate for each element in the mixed standard at the time of calibration to the daily counting rate). Sample cps are corrected by multiplying the daily count by the proper correction factor before reading the concentrations from the calibration curve. One correction factor plus adjustment of detector voltage is made daily for each element. The mixed standard contains approximately 1000 ppm chlorine, 500 ppm titanium, 25 ppm iron, and 2000 ppm aluminum, and gives a reasonably high counting rate for each element.

Background counts (cps) taken on the base resin disc at the proper angle are subtracted from the total cps found in the iron and aluminum standards to give the calibration curves for these two elements. A background count is required for iron because of the high tube contamination compared with the sample concentration and for aluminum because minor fluctuations in the background will have a serious effect on the apparent aluminum concentration. Slight variations in background do not appreciably affect the titanium and chlorine concentrations found, and hence no background counts are required here. This further expedites the analyses. Over a period of several months no change in background was noted for either chlorine or titanium.

The calibration curves for all four elements are linear for the ranges covered. Early work was done with a full-wave rectified X-ray generator and binary scaler circuit panel. Table II gives a comparison of the full-wave rectified and constant potential

Table II. Comparison of Data from X-ray Generators

	Full-wave rectified			Constant potential		
	cps/ppm (above background)	cps background	ppm precision 2σ	cps/ppm (above background)	cps background	ppm precision 2σ
Al	0.026	2.5	12	0.029	2.7	10
Cl	0.29	105	12	0.36	91	8
Ti	5.74	135	1–2	7.40	145	0.4
Fe	20.5	940–1000	0.4–0.6	31.0	1114	0.51
				24.0 (scint.)	799	—

generator. Fixed counts were used with the older circuit panel and fixed time with the new one. The results found with the constant potential generator include the latest pulse height analyzer design and low-background amplifier stage.

These values have varied only slightly during the period the method has been in use. These variations are due primarily to changes in alignment and aging of electronic components. Values quoted represent the optimum found.

Conditions

The K_α first-order line is counted for all elements using P.H.A. The differential peak for each element is set at 15.0 v with a 1-v window by varying daily the dc voltage applied to the detector. Baseline level and window width values are chosen to permit a 3-v drift of the peak without loss of intensity. This permits some minor variation in setting the detector voltage as well as some possible fluctuation of the peak without loss of accuracy during a long run of analyses.

The length of time counted represents a compromise of time and precision. The total counts obtained at the stated counting time at the 100-ppm range are about 90,000 for titanium, 26,000 for chlorine, 400,000 for iron, and 2200 for aluminum. The calculated statistical counting error for all elements is about two-thirds of that found.

The signal/noise ratio is constant for chlorine and aluminum with a tube power of 50 kv and 20–50 ma. However, the cps above background is highest and the counting error lowest at the maximum tube power as expected. For example, an increase of 10 ma raises the cps above background by 11 for chlorine in a particular sample. Hence, the maximum tube power is used for all elements.

A study of variations in background revealed negligible difference in the cps taken at slightly off angle from the four elements in high or low pressure polyethylene or polypropylene. A second study at the same off angles showed that slight changes in thickness of the sample produce some background variations (about 0.2 cps per mil for chlorine, 0.03 cps per mil for aluminum, and a negligible amount for iron and titanium). Since these variations influence the apparent concentration of chlorine and aluminum, efforts are made to control the sample thickness to within ± 5 mils. The critical thickness is exceeded by the sample; it is the scattered radiation only which is increased by greater thickness.

Matrix

Another advantage of trace analyses of elements in polyolefins by XRF is absence of matrix effects. No matrix effects are noticeable at less than 500 ppm of all elements except iron. The minor effects of density of polyolefins on elemental radiation are assumed negligible and within the precision of the method.

RESULTS

This technique has been used routinely for the analyses of catalyst residues in polyolefins. Precision values (95% confidence level) for the XRF procedure with the constant potential generator and a flow proportional counter are: iron—± 0.51 ppm at a 1.2-ppm level, titanium—± 0.4 ppm, aluminum—± 10 ppm, and chlorine—± 8 ppm at 20–100-ppm level. These precision values were obtained by changing the angle detector voltage, tube voltage and current, P.H.A. window and baseline, and resetting before each count. These values should represent day-to-day variation including instrument fluctuation.

The precision values by the XRF method are superior to the chemical methods

Table III. Comparison of Chemical and XRF Data

Ti		Fe		Cl		Al	
Chem.	XRF	Chem.	XRF	Chem.	XRF	Chem.	XRF
< 1	1	< 2	0.8	20	17	< 5	12
7	4	5	2.1	38	32	20	18
6	5	4	2.3	72	34	29	23
9	9	8	5	54	51	5	25
8	11	10	6	94	60	22	27
11	13			72	75	48	29
25	23			100	82	31	34
18	29			146	97	69	43
55	51			109	103	32	47
72	79			157	130	87	75
				132	174	181	110
				243	181	157	130

for all elements except aluminum, where it is about the same; however, careful technique and extreme precaution must be taken with all of the chemical procedures to obtain worthwhile results. The precision (95% confidence level) previously found by our routine chemical analyses is titanium—± 3–4 ppm, aluminum— ± 9 ppm, chlorine— ± 40 ppm, and iron— ± 2 ppm at less than 150 ppm. Frequently, erratic results were obtained by the chemical methods particularly for aluminum and chlorine, due to contamination of apparatus or reagents, losses during decomposition, or analyst's error. Erroneously high chloride or iron values were occasionally encountered by both chemical and XRF techniques due to surface contamination and were eliminated by proper sampling and handling. Since XRF is nondestructive, any apparently erratic values can be rechecked to ensure that the proper instrumental conditions were used.

Table III gives some typical comparison values for each element by the XRF and chemical procedures. Each pair of analyses (chemical *vs.* XRF) is for a separate sample and is not related to the other values given on the same horizontal line.

The speed, sensitivity, and directness of the XRF procedure make it particularly valuable. The methods have been used with equal success to analyze low-pressure polyethylene, polypropylene, polybutenes, and copolymers from various sources. By this technique one analyst can routinely perform 40–50 analyses in the time previously required to make 6–8 analyses by chemical techniques.

REFERENCES

1. W. T. Bolleter, *Anal. Chem.* **31**: 201, 1959.
2. E. L. Gunn, Thirteenth Annual Summer Symposium on Analytical Chemistry-Trace Analysis, Houston, 1960.
3. Norelco Application Data Sheet No. 104.
4. A. Renfrew and P. Morgan, *Polythene*, Interscience Publishers, Inc., New York, 1960, p. 20; N. G. Gaylord and H. F. Mark, *Linear and Stereoregular Addition Polymers*, Interscience Publishers, New York, 1959, pp. 158–159.
5. John Thatcher, Twelfth Annual Mid-American Symposium on Spectroscopy, Chicago, May 1961.

DISCUSSION

G. Brown (Siemens New York, Inc.): How many times do you analyze these samples before they become too embrittled to use?

R. L. Maute: Some of the standards have been used for 6 to 8 months and they are still usable.

IRON OXIDE DETERMINATION BY X-RAY FLUORESCENCE FOR IN-PROCESS CONTROL OF SOLID PROPELLANT AND PREMIXES

Reuel E. Lamborn and Foster J. Sorenson

Thiokol Chemical Corporation
Brigham City, Utah

ABSTRACT

Finely divided iron oxide is used as a burning-rate catalyst in several solid rocket propellants. The concentration is critical and must be accurately determined as a quality control point before the propellant is cast in the motor case and cured. In addition to the iron oxide, the propellant used for ignition of the Air Force Minuteman first stage contains a polymeric binder system, a solid oxidizer, and a metal powder. This composition makes it difficult to determine accurately the iron content by wet methods in the time available during the propellant processing cycle. The use of X-ray fluorescence has been investigated as a means of satisfying the analysis time requirements while meeting the prescribed accuracy of $\pm 1\%$ of the amount of iron oxide present. Procedures for preparing test specimens have been developed and instrument operation conditions chosen which yield satisfactory precision. When ten specimens from each of three premixes were analyzed for iron content, the observed within-mix mean relative standard deviation was 0.28%; for propellant analyzed under the same conditions, the mean relative standard deviation was 0.35%. Factors affecting mix-to-mix accuracy, such as particle size and shape and interelement absorption and enhancement effects, have been investigated. Accuracy is adequate for in-process control of the iron oxide level in the premix, but further work is required before satisfactory control of propellant is achieved.

INTRODUCTION

The igniter for the first stage of the Air Force Minuteman ICBM incorporates a solid composite propellant which contains an organic binder system, aluminum powder (Al), iron oxide (Fe_2O_3), and ammonium perchlorate (AP). The iron oxide is a burning-rate modifier and the concentration must be controlled with $\pm 1\%$ at the 66% confidence level or within $\pm 2\%$ at the 99 % confidence level.

This propellant is manufactured in two major steps. In the first step, a premix is prepared which contains all of the iron oxide, aluminum, and organic polymer required for the propellant mix. In the second step, the propellant mix, the solid AP, and curing agents are added.

Formulation control analyses are effected separately for each step, and consists in part of a total solids determination. This determination has two weaknesses. It requires approximately 40 min to accomplish and does not give a component analysis. To remedy these shortcomings, the X-ray fluorescence analysis technique was investigated as an improved method of solids determination. X-ray fluorescence analysis requires shorter analysis time and determines the iron, aluminum, and chlorine contents separately.

This investigation was undertaken by Thiokol Chemical Corporation's Wasatch

Division primarily to determine whether X-ray fluorescence could meet analysis requirements for Fe_2O_3 in both premix and propellant batches, and secondarily, to estimate X-ray capabilities for determination of Al and AP. This report covers the work on the premix and some preliminary work on the propellant.

EXPERIMENTAL

The instrument used for this work was an Applied Research Laboratories (ARL) Model VXQ Spectrometer, with a tungsten target Machlett OEG–60 tube which was operated at 50 kv and 30 ma. Other conditions were as shown in Table I.

An integration time of approximately 60 sec was required for the integrated fluorescence radiation from the external standard to reach the selected preset level. Simultaneously, the fluorescent radiation from each of the desired elements is integrated in a separate channel. Integration for all channels is terminated when the charge on the external standard capacitor reaches the preset level. The instrument then sequentially reports a value for the integrated radiation for each channel, either on a recorder chart or digital printer. Units for the printer are chart divisions multiplied by ten; these units were used throughout this study. Optimum settings for zero suppression and sensitivity controls were determined and periodically checked against a pellet of composition simulating that of the samples to be analyzed. No background readings or corrections were made as such.

Scattered radiation at various wavelengths has been used as an internal standard[1,2] to compensate for sample variables, such as particle size variations, interelement effects, and density differences. A preliminary evaluation of the use of scatter radiation in this application is included in this study.

Standards

The chemical and physical instability of the premix and propellant involved precluded the use of analyzed samples of these materials as standards. An extensive investigation in various substitutes was conducted, and the best premix standards found were pellets made from powdered Fe_2O_3 and Al, with powdered H_3BO_3 as diluent and binder. To simulate propellant, dichlorobenzoic acid (DBA) was added. The composition of the pellets, in order to give readout values comparable to or slightly higher than those from the premix and propellant is shown below.

Standard	Fe_2O_3	Al	DBA	H_3BO_3
Premix (%)	8	8	—	84
Propellant (%)	4	3	56	37

The ingredients were mixed as a slurry in hexane, vacuum dried, and pressed with a backing of H_3BO_3 to 160,000 psi at room temperature. The pressed surface was sanded, using Tri-M-Ite Wetordry 320A (silicon carbide) paper. The pellets had to be resurfaced periodically to maintain consistent readings. The standard pellets were used to check for instrument drift, and the sensitivity controls were reset as necessary to give the same readout throughout the program. Since the accuracy for analysis is thus limited by the stability and reproducibility of the standard pellets, further improvement in pellet preparation is necessary. The variation in readings associated with resurfacing was more

---

[1] Superscripts pertain to references at the end of the paper.

Table I. Instrumental Conditions

	Al	Fe	Scatter	Cl
Crystals	EDDT	LiF	LiF	NaCl
Detector	Minitron 1-cm	Multitron	Multitron	Minitron 2-cm
Flow gas	Ne–He–C_4H_{10}	Sealed	Sealed	P-10
Atmosphere	He	He/Air	He/Air	He
Analysis peak	K_α (8.34 Å)	K_α (1.94 Å)	2.2 Å	K_α (4.73 Å)

than would be desired. The relative standard deviation for Fe was 1.1%; for Al, 2.5%; for Cl, 0.6%; and for scatter, 0.6%.

Sample Preparation

Premix. The premix, consisting of a polymer mixed with powdered Al and Fe_2O_3 in the 10–35-μ size range, is a thin paste. The solids settle too rapidly to obtain reproducible X-ray fluorescence readings from the surface.

Several methods for dissolving the sample for solution analysis were evaluated and discarded for reasons of excessive time requirements or insufficient precision. The alternative of stabilizing the sample to prevent settling was investigated. It was found that blending one part of premix to 1.8 parts of powdered boric acid produced a stable paste. The standard ARL sample cup was overfilled with the paste, and the excess was carefully scraped off with a straight edge to obtain a smooth and reproducible surface.

Table II. Precision of Premix Analysis

	Fe_2O_3			Al		
	$\bar{X}$	s	Relative Standard Deviation, %	$\bar{X}$	s	Relative Standard Deviation, %
Operator A						
Premix 1	1650	4.59	0.28	940	8.72	0.93
2	2118	4.23	0.20	1034	9.63	0.93
3	2392	5.75	0.24	928	18.4	1.98
Operator B						
Premix 1	1653	5.06	0.31	940	9.55	1.02
2	2114	8.24	0.39	1044	18.8	1.80
3	2386	7.0	0.29	924	20.4	2.21
Mean			0.28			1.48

$\bar{X}$ is the arithmetic mean and s is the standard deviation

$$\sqrt{\frac{\Sigma x^2 - (\Sigma \bar{X})^2/N}{N-1}} \text{ or } \sqrt{\frac{\Sigma(x - \bar{X})^2}{N-1}}$$

a measure of the spread or variation of the individual measurements about the mean. Relative standard deviation is the standard deviation of a series of test results as a percentage of the mean of this series. (This term is preferred over "coefficient of variation.")

For determining the precision of this method of sample preparation, three premixes were prepared with different levels of Fe_2O_3 and Al. Each of two operators prepared a premix–H_3BO_3 paste for each premix, and obtained X-ray readings on ten aliquots of each paste. The observed mean relative standard deviation was 0.28% for Fe_2O_3 and 1.48% for Al (Table II).

Another reason for selecting the boric acid paste method for analysis of the premix was the intriguing possibility that a measure of polymer content could be derived from the readings from the Fe, Al, and SR (scattered radiation) channels. Since the Al radiation at 8.34 Å is absorbed much more by the light elements present in this polymer–H_3BO_3 matrix than is the Fe radiation at 1.94 Å, the interrelation of the radiation intensities should be related to polymer content. If a reliable estimate of polymer content could be thus obtained, this estimate could in turn be used as a correction factor to aid in the Fe_2O_3 and/or the Al determinations.

Propellant. The propellant was sufficiently viscous for analysis in the same manner as the premix–H_3BO_3 paste. Although variations in AP particle size introduced another variable, this direct analysis was chosen over solution techniques for preliminary evaluation because of favorable time factors and good within-mix precision. The mean relative standard deviation for replicate samples taken from the same mix was 0.35% for Fe, 1.4% for Al, and 0.21% for Cl (Table III).

Mixes for Analysis

Premixes. Thirteen 1000-g premixes were prepared with composition ranges comparable to the production premixes. Weights were accurate to ±0.1 g. One lot of polymer was used throughout. Premixes 1, 2, and 3 were replicates at the specification formulation. In premixes 4 and 5, the ratio of Al to polymer was the same as in 1–3, while the Fe_2O_3 was varied 1% higher and lower than in 1–3. In 6 and 7, the ratio of Fe_2O_3 to polymer was held the same as in 1–3, while the Al was varied 1% higher and lower than in 1–3. In 8 through 11, the ratio of Fe_2O_3 to Al was maintained as in 1–3, while the polymer was varied 1 and 3% higher and lower than in 1–3. Premixes 12 and 13 were provided to determine the effect of differences in particle size of Al and Fe_2O_3.

Table III. Precision of Propellant Analysis

	Fe_2O_3	Al	AP(Cl)
Mix A			
$\bar{X}$, Chart Value	456	491	2064
s, Chart Value	1.56	6.41	6.63
Relative Standard Deviation (%)	0.35	1.30	0.32
Mix B			
$\bar{X}$, Chart Value	647	467	2030
s, Chart Value	1.41	5.20	1.89
Relative Standard Deviation (%)	0.22	1.11	0.09
Mix C			
$\bar{X}$, Chart Value	829	461	1979
s, Chart Value	3.90	8.35	4.21
Relative Standard Deviation (%)	0.47	1.81	0.21
Mean Relative Standard Deviation (%)	0.35	1.4	0.21

Table IV. Composition of Propellant Mixes

| | AP | | |
Mix	Unground	Ground	Polymer ratio
1	Medium	Medium	Normal
2	Coarse	Coarse	Normal
3	Fine	Fine	Normal
4	Medium	Medium	High
5	Medium	Medium	Low

Number 12 contained coarser Al, and number 13 contained coarser Fe_2O_3 than the lots used in 1–11.

Propellant. Five mixes of propellant were made, at specification formulation, to obtain a preliminary estimate of the effect of particle size variation of the ammonium perchlorate, and the effect of changing polymer ratio.

The same lots of raw materials were used throughout. The AP was present in two types, unground and ground. By sieving and by suitable selection of ground batches of AP, each type was subdivided to prepare fractions that were near the middle, upper, and lower acceptance limits for each. The amount of Fe_2O_3, Al, total binder, AP, and the ratio of unground to ground were held constant throughout. Formulation was adjusted as shown in Table IV.

RESULTS AND DISCUSSION

Premix

For each of the 13 premixes, one premix–H_3BO_3 paste was prepared, and duplicate X-ray readings were obtained for each of six samples of the paste, for a total of 156 observations.

A multiple regression analysis[3] was designed to summarize the data and to calculate multiple regression equations for Fe_2O_3, Al, and polymer. The form of each equation was

$$\% \text{ Component} = b_0 + b_1 (\text{Fe}) + b_2 (\text{Al}) + b_3 (\text{SR})$$

where b_0 is the Y intercept and b_1, b_2, and b_3 are partial regression coefficients. Symbols in parentheses represent observed chart divisions $\times$ 10.

The resulting equations, with associated correlation coefficients (R^2) and standard errors, are as follows:

$$\% \text{ Fe}_2\text{O}_3 = -0.930 + 0.0128\,(\text{Fe}) + 0.00377\,(\text{Al}) - 0.0115\,(\text{SR}) \qquad (1)$$
$$R^2 = 0.98$$
$$Sy.x = 0.105$$

$$\% \text{ Al} = -4.358 + 0.00338\,(\text{Fe}) + 0.0148\,(\text{Al}) + 0.00502\,(\text{SR}) \qquad (2)$$
$$R^2 = 0.88$$
$$Sy.x = 0.253$$

$$\% \text{ Polymer} = 105.29 - 0.0162\,(\text{Fe}) - 0.0186\,(\text{Al}) + 0.00645\,(\text{SR}) \qquad (3)$$
$$R^2 = 0.96$$
$$Sy.x = 0.279$$

where R^2 is the coefficient of correlation, which is the portion of the dependent variable that is accounted for by the regression equation; $Sy.x$ is the standard error of the estimate, which is the standard deviation along the regression line; the symbol ** indicates that the value is significant at the 0.01 level of probability or 99% confidence level; and N.S. means this is not significant at the 0.05 level of probability or 95% confidence level.

Three facts of primary interest are apparent. First, the coefficients of correlation obtained indicated that the linear equations fit the data well enough to meet our requirements. More complex curvilinear equations did not improve the correlation or fit.

Second, the SR value was significant at the 0.01 level in influencing percent Fe_2O_3. It was not significant, even at the 0.05 level, for the Al and polymer equations. The wavelength used may not be optimum for this particular application; further investigation will be conducted. It was also noted that Fe and Sr values had a correlation of 0.88. The majority of significance was attached to the Fe value.

Third, the percentage of polymer can be calculated from chart values with somewhat more confidence and a smaller standard error than can the percentage of Al. This was surprising, but has been substantiated by a complete rerun on a second full set of premixes made and analyzed in the manner reported here.

The standard errors about the lines of regression $(Sy.x)$, converted to relative standard deviation, are well within the limits currently being followed using wet chemical methods, and are adequate to meet present process control requirements for the premix.

The effect of changes in particle size of the Fe_2O_3 and Al was determined by using a multiple regression analysis similar to the one above, with equations in the form:

$$\text{Fe or Al chart value} = b_0 + b_1 \text{ (WMD Fe)} + b_2 \text{ (WMD Al)}$$

where WMD is weight mean diameter as determined by a Coulter Counter. (Weight mean diameter is the effective screen opening size, in microns, that retains 50 wt.% of the sample analyzed.

The resulting equations were as follows:

$$\overset{**}{} \qquad\qquad \overset{N.S.}{}$$
$$\text{Fe chart value} = 1484.8 + 1.729 \text{ (Fe WMD)} + 1.054 \text{ (Al WMD)} \tag{4}$$
$$R^2 = 0.28$$

$$\overset{**}{} \qquad\qquad \overset{**}{}$$
$$\text{Al chart value} = 1764.3 + 14.04 \text{ (Fe WMD)} - 49.05 \text{ (Al WMD)} \tag{5}$$
$$R^2 = 0.92$$

$$Sy.x = 17$$

That portion of the variation in Fe chart value accounted for by the regression equation is small, so little confidence may be placed in equation (4). Through the range of data studied, Fe chart value apparently is only slightly dependent on change in particle size. The slight effect that is observed is related to changes in WMD of Fe_2O_3.

In equation (5), the R^2 value of 0.92 indicates that most of the changes observed in Al chart values are accounted for by changes in both Fe WMD and Al WMD. The $Sy.x$ value of 17 indicates, however, that precision of measurement in Al is difficult to achieve.

In applying this analysis system to process control, the effect of lot-to-lot variations in particle size is minimized by preparing new calibration curves using premixes made from each new combination of raw materials. Insufficient work has been done at the time of this writing to determine the reliability of equation (5) in correcting the observed Al chart values used in equations (1), (2), and (3) for particle sizes in a particular premix.

Propellant

The primary objective of this portion of the investigation was to determine what effect changes in oxidizer particle size and binder composition would have on Fe chart values. It was also of interest to determine the effect on Al and Cl chart values in preparation for further work on the determination of these two elements in various propellants. Since the concentration of Fe_2O_3, Al, and AP were held constant throughout the set of five mixes, no direct relationship between observed chart values and concentrations was established for propellant.

$$\text{Fe chart value} = 334.9 + 3.002 \overset{**}{(A)} - 9.209 \overset{**}{(B)} + 14.60 \overset{**}{(C)} - 3.540 \overset{**}{(SR)} \qquad (6)$$
$$R^2 = 0.93$$

$$\text{Al chart value} = 349.5 - 1.869 \overset{**}{(A)} + 21.11 \overset{**}{(B)} - 18.71 \overset{**}{(C)} + 5.617 \overset{**}{(SR)} \qquad (7)$$
$$R^2 = 0.94$$

$$\text{Cl chart value} = 3030.9 - 1.960 \overset{**}{(A)} - 1.629 \overset{**}{(B)} - 0.0033 \overset{**}{(C)} - 0.412 \overset{**}{(SR)} \qquad (8)$$
$$R^2 = 0.98$$

where A is the weight mean diameter of unground fraction of AP in microns, B is the weight mean diameter of fine fraction of AP in microns, C is the % polymer present in binder fraction, and SR is the scatter radiation chart value, chart divisions × 10.

Although these equations are considered only preliminary in nature, two items of interest are apparent. First, the Fe and SR chart values are related, with Fe chart values increasing as SR values decrease. Second, the particle sizes of both the ground and the unground oxidizer significantly affect the observed chart values for all three elements sought, but not in the same direction. The magnitude of these effects, when superimposed upon changes in propellant composition, is yet to be determined. The high correlation coefficients (R^2) for these equations, together with the errors associated with determining the oxidizer particle sizes in any particular mix, indicate that difficulty will be experienced in obtaining a sufficiently accurate measure of mix composition by means of this system of analysis.

Work is underway toward the design and conduct of an experiment, similar to the one reported here for premix, to determine the value of X-ray fluorescence analysis of propellant.

ACKNOWLEDGMENT

The authors are indebted to Henry Ashcroft of the Statistical Control Section, Wasatch Division, Thiokol Chemical Corporation, for assistance with the experimental design and statistical computations.

REFERENCES

1. G. Andermann, J. W. Kemp, and M. F. Hasler, "The Application of a Vacuum X-ray Quantometer to Nonmetallic Matrices," presented at 1961 International Colloquium on Spectroscopy, June 1961.
2. T. J. Cullen, *Anal. Chem.* **34**: 812, 1962.
3. O. L. Davis, ed., *The Design and Analysis of Industrial Experiments*, Hafner Publishing Co., New York, 1956.

DESIGN CONSIDERATIONS FOR ON-STREAM X-RAY ANALYSIS

W. R. Kiley and R. W. Deichert

Philips Electronic Instruments
Mt. Vernon, New York

ABSTRACT

Several design concepts were evaluated for on-stream X-ray analysis, chief among which were: (1) sample presentation, (2) X-ray spectrograph and (3) data handling. The sample is presented to the X-ray spectrograph through a vertical cell having a mylar window and a device to measure variations in sample density. The X-ray spectrograph is of fixed-channel, modular design, with a capacity of up to six channels, and is designed for use in industrial environments. The complete spectrograph can be translated in a horizontal plane to position itself and determine the elemental composition in a series of sample cells. The data handling uses digital circuitry to permit normal use of analog readouts or digital data logging systems. Operational data are given for on-stream analysis.

INTRODUCTION

X-ray spectrographic techniques have been established for many years which give rapid, accurate analysis for a static sample. The extension of the outstanding features of the X-ray spectrograph for the analysis of raw materials on a continuous basis requires a thorough evaluation of the needs of the process engineer, and thus the incorporation of design features in an on-stream X-ray analysis unit which will meet the most exacting needs. The major design parameters of such equipment are therefore accuracy, speed and system considerations where:

1. The required accuracy of the output data varies from industry to industry and from plant to plant, ranging from trend information to accuracy on the order of 1% relative or less.
2. The speed of computation is dictated by the response of the process which will vary from 30 min to 30 sec.
3. The systems considerations include operational simplicity, flexibility for future growth, ease of maintenance, and instrument reliability.

All of these features were studied by Philips Electronic Instruments and the instrument to which they were applied is the subject matter of this paper.

Figure 1 shows this on-stream X-ray analyzer which has recently been installed in a zinc ore benefication plant for the determination of zinc content in mill feeds, middlings, concentrates, and tailings. The units, from left to right, are the X-ray generator, console, X-ray sensing heads and sample cells, and data readout, in this case, four strip chart recorders. These components are shown diagrammatically in Figure 2.

The three main functional sections of this system are sample presentation, X-ray spectrograph, and data handling; the important points of these will now be discussed.

Figure 1. On-stream X-ray analyzer.

SAMPLE PRESENTATION

If it can be assumed that, in the case of liquids and slurries, the by-pass sampling stream is representative of the main stream, then the problem resolves to a sampling system which will provide the following:

1. Continuous agitation of the sample, especially slurries.
2. Prevention of interference from entrapped air and bubbles.
3. A self-washing action on the window of the sample cell so that the X-ray spectrograph is always looking at a continuously different sample.
4. Minimized sanding of sample feed lines, especially when slurry samples are intermittently fed through the sample cell.

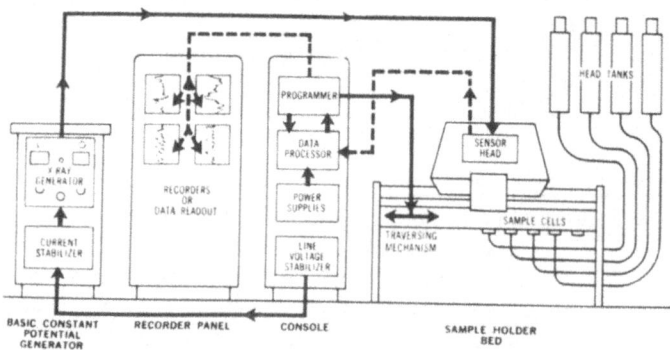

Figure 2. Schematic diagram of the analyzer.

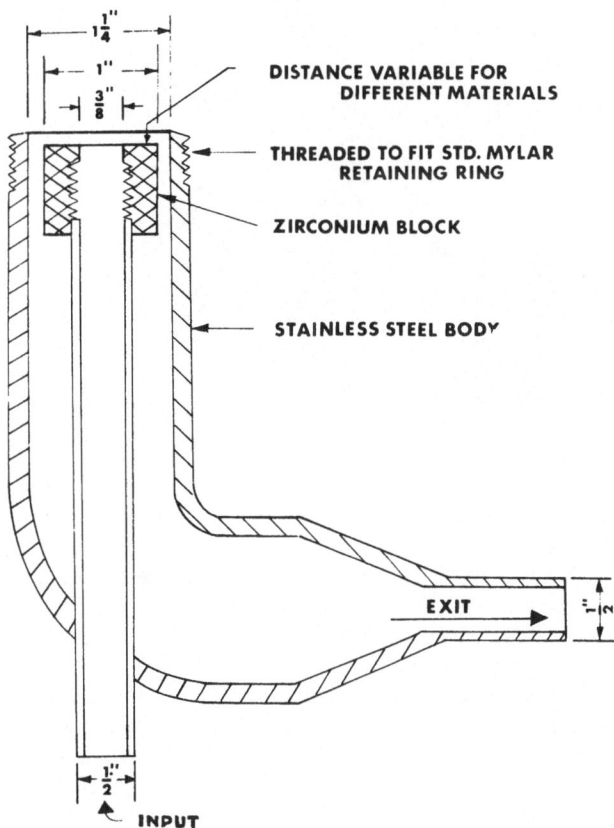

Figure 3. Liquid sample cell with density measuring block.

5. Correction data when samples vary in pulp density.
6. Ease of maintenance.

Figure 3 shows a cross section of a liquid sample with a density measuring device. The input to the cell is supplied from enclosed head tanks which provide a constant head and therefore a constant rate of flow in the sample cell. Continuous agitation gives a representative sample and prevents sanding. These head tanks are sealed at the top and are equipped with a manually operated shut-off valve and an electrically operated solenoid valve.

The output from the head tank is fed by gravity to the input of the sample cell, which is normally fed at the rate of 4 quarts/min. The liquid sample input to the cell impinges against the mylar window over the zirconium block and then out the exit as shown. The X-ray spectrograph looks at the emission from the elements present in the liquid or slurry sample as well as radiation coming from the zirconium block. As the density of the material passing over the zirconium block changes, the zirconium intensity will be absorbed accordingly. This technique of monitoring the zirconium intensity as well as those of the other elements of interest provides a method of correcting for changes in sample density.

Figure 4 shows the actual cell with its replaceable mylar window, which can be replaced in less than a minute's time. The zirconium block is visible close to the mylar window with a center hole which provides passage of the liquid or slurry sample.

Figure 4. Liquid sample cell.

Figure 5 shows the head tanks in position and the lines from the head tank feeding the sample to the liquid cells. In this particular application, four different samples representing four different streams of interest are shown. Actually, the sample holder bed can handle up to 15 separate samples representing 15 different sample locations in the process system. This method of handling the various samples prevents sanding of the sample feed lines, diminishes the requirement for flushing out the cell, and provides ease of maintenance.

If any one of the mylar windows on the sample cups ruptures due to X-ray embrittlement or abrasion, the instrument continues to analyze all of the other sample streams. The sample cells are equipped with a mylar-rupture sensing device which actuates the electrical solenoid valve, shutting off the sample stream from its particular head tank.

Studies have also been made on presentation of powder materials to the X-ray spectrograph. Basically, the general procedure is to grind the powder sample, usually to -100 or -200 mesh particle size. The loose powder can be transported to the X-ray spectrograph on a sample conveyor.

Another procedure which has been evaluated is the pelletizing of certain materials and conveying the compacted pellet to the X-ray spectrograph.

Figure 5. Constant-head tanks.

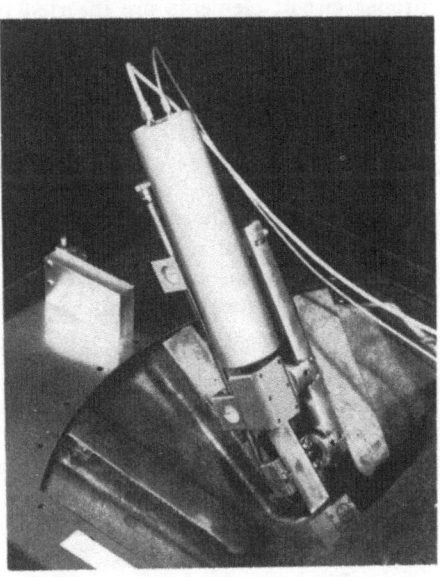

Figure 6. Monochromater channel.

X-RAY SPECTROGRAPH

The X-ray sensing head, as shown in Figure 1, is constructed so that it can be mounted in a fixed position with the sample continuously moving to it or it can be easily attached to a traversing mechanism and positioned sequentially over individual sample cells. The sensing head is composed of fixed channels of modular construction, capable of handling up to six channels. Each channel is composed of flat crystal optics consisting of source collimator, flat crystal, arm collimator, and appropriate detector. The crystal and detector are mounted at the appropriate θ and 2θ angles in a fixed manner, and also have fine adjustments for optimum alignment. A monochromater channel is shown in Figure 6. The linear amplifier and amplitude discriminator are located adjacent to the detector in the sensing head.

The X-ray tube is powered by a Norelco 2.5-kw constant potential source, thus providing necessary power for a wide variety of X-ray tubes and applications.

DATA HANDLING

The X-ray intensity data from the element of interest, in this case zinc, and the density monitoring element, in this case zirconium, are fed to the data processor for analysis. The system under the control of a digital timing clock in the programmer causes the X-ray sensing head to traverse in sequence from a calibrate position with a synthetic pellet through the four sample positions. The programmer successively senses when the sensing head is on a station, permits it to take data, energizes the proper recorder, de-energizes the recorder, and causes the head to move to the next station. In the calibrate position, which is the start of the sampling cycle, a fixed count determination is made for zinc thereby determining a time base which is stored in a digital memory and used in multiples for fixed time determinations of zinc and zirconium in the four following analytical positions. The zinc and zirconium count data are accumulated in linear and logarithmic count registers, respectively, to which are connected digital-to-analog converters. The outputs of the converters are in turn fed to the programmer, where the proper circuit elements are inserted for solution of the equation of the type $mx + b$ which is then presented to the proper recorder.

The number of counts accumulated is chosen to provide the system accuracy required by the problem. The system described here has a 1-min cycle per station; thus the total calibrate plus four sample stream analysis takes approximately 5 min.

After establishment of the feasibility of the on-stream X-ray analyzer to provide a solution to a process measurement problem through proper choice of specimen handling and spectrograph, there yet remain the system considerations to create a fully optimized system. Attainment of operational simplicity, flexibility for future growth, instrument reliability, and ease of maintenance in connection with the instrumentation system under discussion was accomplished as follows:

1. Operational simplicity—the basic philosophy of requiring operator attention only during turn-on, and automatic indication of equipment malfunction is practically mandatory, especially in process plants which are approaching unattended operation. Consequently, a main power switch is provided, setup controls for programming the traverse mechanism and a built-in automatic calibration station are provided. Automatic indication of counting circuit and specimen cell mylar window malfunction are provided. System adjustment controls are located within the cabinets and are used only during routine maintenance.

2. Flexibility for future growth is provided through the use of modular construction centered in the main console unit, a portion of which is shown in Figure 7. The data processor drawer as well as the programmer drawer located above it consist of plug-in component boards, which can be changed as system requirements expand. Thus reprogramming of the system to include more sample streams or the addition of data processing circuitry to accommodate additional analytical elements may easily be accomplished by including additional modular elements.

The second feature which ensures flexibility for future growth is that digital circuitry utilizing fixed-time or fixed-count operation is used throughout the programmer and data processor. Thus digital readout may be made directly or the data converted through digital-to-analog converters for analog presentation on a recorder as in the present system.

3. Instrument reliability is provided through the use of solid state circuitry as shown in Figure 8. These plug-in boards utilize military-type components which have

Figure 7. A portion of the main console unit showing modular construction.

Figure 8. View of circuitry.

been subjected to worse case design and wide limit environmental testing for the utmost in reliability.

The data processor in this system is required to solve an equation of the type $mx + b$ in order to correct the zinc reading for changes in pulp density as monitored by the zirconium radiation absorption. Since the zirconium information is a logarithmic function of pulp density, the conventional approach to the solution of the equation would be to use a logarithmic rate meter or a shaping network on the output of a linear rate meter. Instead, a logarithmic digital counter was utilized for its extreme reliability.

4. Ease of maintenance—while maintenance is expected to be at a minimum through the use of solid state circuitry and adequate dustproofing and ventilation, it was kept in mind in the equipment design. An indicator fuse panel in the bottom of the console reveals the malfunction of all major power circuits. Modular subassemblies together with slide pullout drawers provides for minimum effort maintenance.

This paper has described one typical X-ray on-stream analyzer application. Undoubtedly there are many more. Use of the above enumerated design considerations will result in systems which not only meet system requirements, but which are optimized as well.

DISCUSSION

R. E. Michaelis (U.S. Bureau of Standards): I might ask how long do you feel that the mylar window might last?

R. W. Deichert: It depends on the X-ray embrittlement, which in turn is related to the X-ray power used, as well as the consistency and state of the slurry. Experience has shown that, depending upon the slurry present, it has a life up to 24 to 36 hr.

G. Brown (Siemens New York, Inc.): Have you encountered any problem with liquid precipitation on the mylar window affecting intensity lines?

R. W. Deichert: No. Once the mylar ruptures you normally don't get any precipitation on the outside. On the inside, our solute is basically self-washing. We have not experienced any difficulty with the fogging on the inside.

PARTICLE SIZE AND MINERALOGICAL EFFECTS IN MINING APPLICATIONS

F. Bernstein

General Electric Company
Milwaukee, Wisconsin

ABSTRACT

A relationship is derived between X-ray intensity and particle size for a minor constituent in a powder sample. Theoretical and experimental curves are compared for several elements which are of importance in mining applications. Based upon these curves, some general guides are established relating to a method of predicting the particle size required for a given application. It is also shown that the relative intensity of a minor constituent is independent of the absorption coefficient of the matrix, if the particle size of the matrix is constant. Mineralogical effects, which relate to the occurrence of an element in two or more forms of chemical combination, are discussed and several examples in mining applications are cited.

INTRODUCTION

In the period since World War II, rapid strides have been made in the development of laboratory instruments for X-ray diffraction and emission techniques. These developments are attributable to advances in electronics, X-ray analyzing crystals, special X-ray tubes, goniometers, and detectors. The availability of improved equipment has made it possible to extend the range of applications to an increasing number of industries and processes.

The introduction of instrumental analytical methods to mining and process industries has reduced the time lag between sampling and report of analysis from as much as days to as little as minutes. This has often resulted in reduced costs of operation. For example, exploratory drillings can be analyzed and classified rapidly. When a process goes out of control, the fact is revealed quickly through instrumental analysis. Finally, greater progress toward optimization of process variables has been achieved. Laboratory X-ray instrumentation has proven of considerable value in bringing about the improvements listed above. X-ray emission gauges, which are currently operating continuously on line, have gone a step further than the laboratory units in improving process control. The remaining step toward achievement of automatic, continuous process control is to link the X-ray and other sensors with a computer. Programs of this nature are currently being carried out in several industries.

It is difficult to obtain successful X-ray control of mining materials without some understanding of the effects of the two most important variables influencing X-ray intensities in powder samples, namely, particle size and mineralogical effects. It has long been recognized that particle size is important in X-ray emission measurement, as evidenced by a paper by Fonda[1] in 1933 which deals with the subject.

In a previous paper,[2] it was shown that the fluorescent intensity from a pure material

---

[1] Superscripts refer to references at the end of the paper.

would increase as the particle size of the material was decreased. This increase in intensity was attributed to a decrease in the percent of voids in the surface of the material with the reduction in particle size. In addition, it was shown in two-component systems that the following effects occur:

1. If the particle size of either component is maintained constant and the particle size of the other is decreased, the intensity from the component which is constant will decrease and the intensity from the component whose particle size is decreased will increase.

2. If the particle size of both components is reduced uniformly, the intensities from both components may increase or the intensity from one component may increase while the intensity from the other may decrease. This latter behavior was attributed to the relative absorption coefficients of the two materials. Finally, it appears that the intensities of both components appear to become relatively stable when the particle size is reduced sufficiently.

The subject of particle size effects in mixtures was discussed by Claisse and Samson.[3] They made a fundamental quantitative study of heterogeneity effects in X-ray fluorescence which predicts that the grain size effect appears only in a limited region of grain sizes. This depends on the wavelength of the primary radiation and the nature of the compounds in the mixture. With monochromatic radiation, the fluorescent intensity increases or decreases by a factor of few units as the grain size is decreased. Usually, the grain size effect can be eliminated by intensive grinding. For the light elements, they conclude that fine grinding is disastrous if long wavelengths are used. In addition, it is predicted that by appropriate choice of wavelengths it is possible to eliminate the effect of particle size without grinding. Smithson, Eager, and Van Cleave,[4] in a paper on the analysis of uranium in flotation concentrates, show that the fluorescent intensity of uranium increases with grinding time to some maximum value, beyond which no change is observed with increased grinding time. The present paper deals with a study of the relationship between particle size and intensity for minor constituents. Mineralogical effects which have been encountered in a number of applications are also discussed.

THEORY

If we consider a powder sample containing the element of interest in spherical particles of uniform diameter D, then at any depth x from the surface there are likely to be N particles, where N is proportional to the concentration of the element. If the original particle diameter is reduced to $D/2$, then at the same depth x there will be $8N$ particles. If the particles are cubic in shape then the particle size reduction again results in an eightfold increase in number and can be described as a lateral displacement of the lower half of the cube, as shown in Figure 1. The exciting radiation is absorbed by the matrix until it reaches the particle and the fluorescent radiation is again absorbed by the matrix until it reaches the surface and is measured. Assuming the average depth of particles is the same for the different particle sizes, then the total matrix absorption is virtually independent of particle size. It is evident from Figure 1 that for the cubic particle, reducing the dimension of a side by two is equivalent to exposing twice the original area of the material being measured at half the original thickness. In the spherical particle, the same is true since the surface area of a sphere is proportional to the diameter squared. Thus, each of the smaller particles has one-fourth of the surface of the larger particle.

It is a well-known fact that the intensity of a characteristic line emitted by a thin layer of material increases as the thickness is increased up to a point which we will define

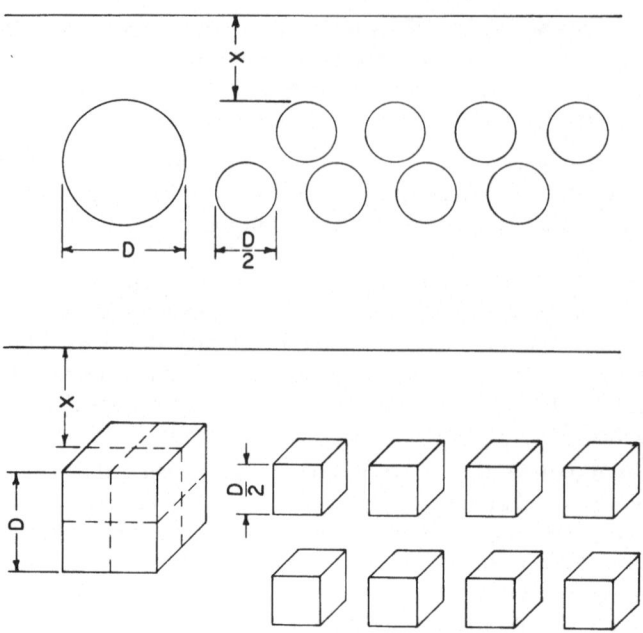

Figure 1. Effect of particle size reduction.

as "infinite thickness." Beyond this infinite thickness value any increase in thickness will result in no observable gain in characteristic signal from the sample. The relationship between the fluorescent intensity of a sample of thickness x to the intensity from the same sample which is of infinite thickness can be shown to be as follows:[5]

$$I_x/I_\infty = 1 - \exp(-a\rho x)$$

In the above equation I_x and I_∞ are the intensities of fluorescent radiation per square centimeter leaving samples of thickness x and ∞, respectively, a is a function involving the absorption coefficients for the incident and fluorescent radiation and the angle of incidence and angle of take-off from the sample, ρ is the density of material, and x is the depth under consideration in centimeters. In order to determine infinite thickness from the above equation, we arbitrarily choose a value of 0.99 for the ratio of I_x/I_∞. The relationship between fluorescent intensity and thickness, based upon the above equation, is shown in Figure 2. We can now define particle size in terms of the infinite thickness value defined above. A particle is said to have infinite thickness when the length of the side of a cubic particle or the diameter of a spherical particle is equal to infinite thickness as calculated above. It is now possible to construct a curve for low concentrations of particles in a matrix based upon the analysis shown in Figure 1 and the intensities which are shown graphically in Figure 2. Thus, for example, if we arbitrarily assign a value of 1 to a particle which has infinite thickness, then reducing the particle diameter to half of infinite thickness and maintaining the same elemental concentration, the fluorescent intensity from the sample which has the smaller-diameter particles will be 1.8 times that of the sample with particles of infinite thickness, since 50% of infinite thickness will produce 90% of the original intensity (Figure 2). In this manner, it is possible to generate the relationship shown in Figure 3, which shows the relative intensity as a function of particle size when the particles are smaller than infinite thickness. It

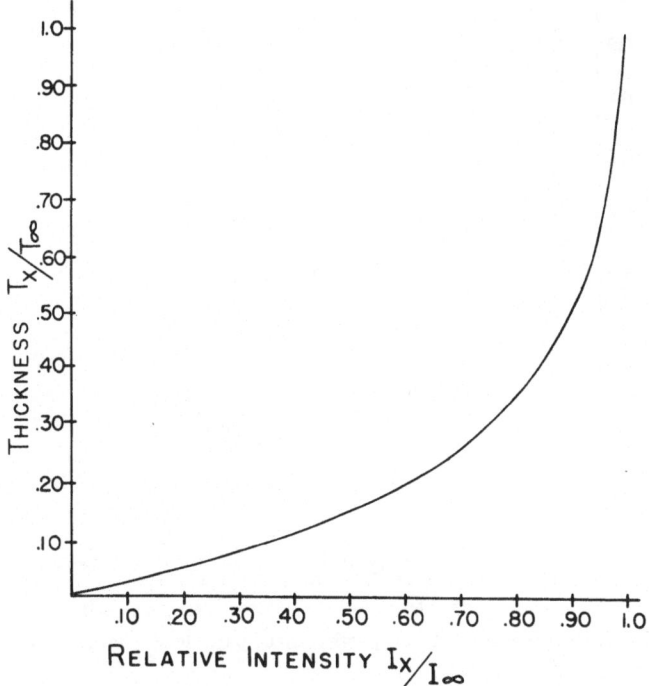

Figure 2. Relationship between fluorescent intensity and thickness.

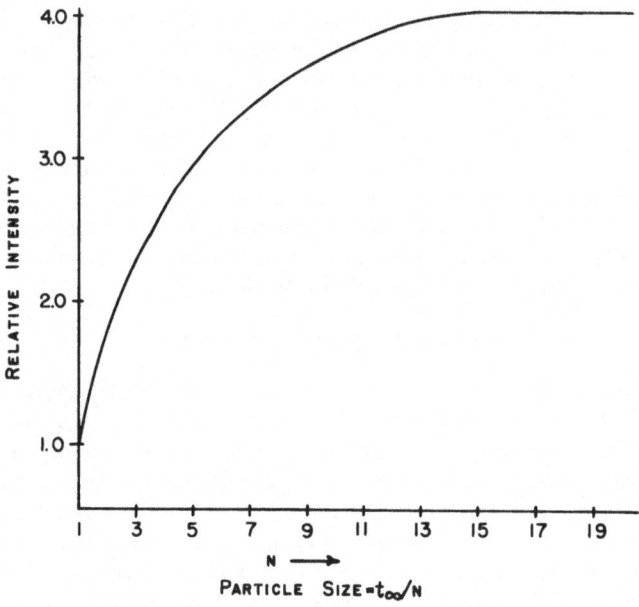

Figure 3. Relative intensity *vs.* particle size; particle size smaller than t_∞.

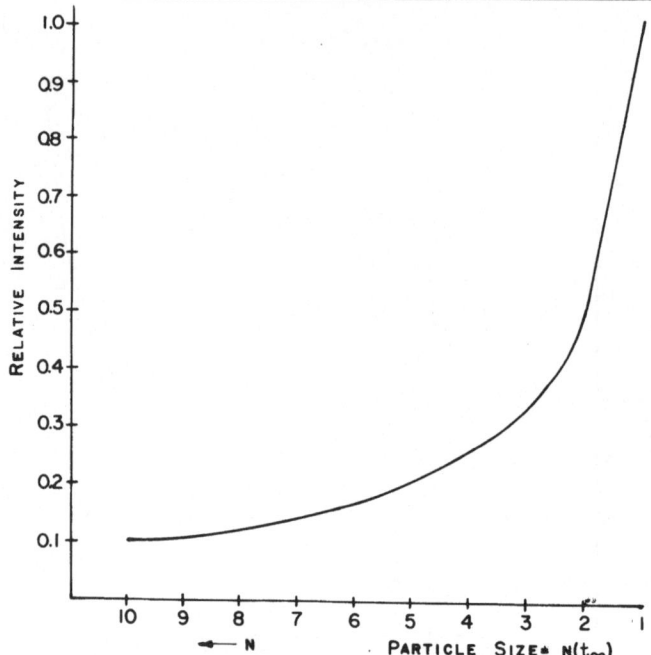

Figure 4. Relative intensity *vs.* particle size; particle size larger than t_∞.

appears from the curve in Figure 3 that the intensity–particle size relationship does not level off until the particle size of the material being studied approaches approximately $\frac{1}{11}$ of infinite thickness particle size.

If we now consider going in the opposite direction insofar as particle size to that which was discussed above, namely, to particles which are larger than infinite thickness, then we must consider a reverse of the analogy which was discussed above. With reference to Figure 1, if the smaller particles are considered to be of infinite thickness, then by doubling the particle size we produce one particle from eight original ones. Now, the lower half of the larger particle will not contribute anything to the characteristic intensity measured from the sample. Thus, the change in particle size from infinite thickness to twice infinite thickness results in a 50% loss in intensity. In this manner, we can generate the points shown in the curve in Figure 4, which shows the relationship between particle size which is larger than infinite thickness and intensity. It can be seen that the intensity falls off very rapidly in the region of infinite thickness and then begins to level off at a point which is approximately ten times infinite thickness.

EXPERIMENTAL RESULTS

It was decided to check the validity of the curves shown in Figures 3 and 4 using two elements which are of importance in the mining industry, namely, molybdenum and copper. For molybdenum, crystals of pure ammonium paramolybdate were used. These were separated into fractions of uniform particle size covering a range from 10 to 1000 μ; the separation was accomplished by a combination of grinding, sieving, and selective flotation. Five percent by weight of each of the particle sizes of paramolybdate was blended with a relatively transparent matrix of sodium carbonate. The particle size of the carbonate was -325 mesh and was maintained constant in all of the samples. The blending was carried out using lucite vials with lucite balls, which combination did not

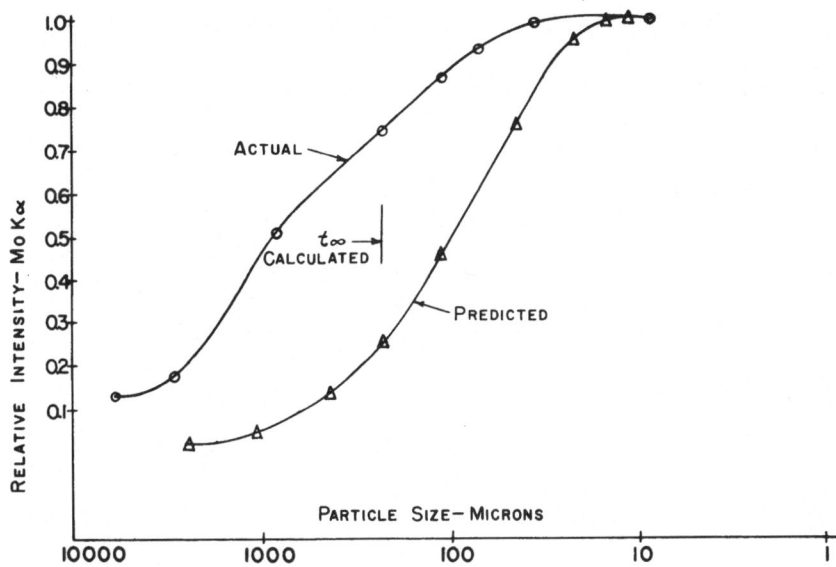

Figure 5. Effect of particle size on intensity; 5% molybdenum in Na₂CO₃.

appear to cause any particle size reduction in the samples during mixing. Molybdenum K_α was measured on compacted samples using tungsten radiation at 50 kv and 50 ma as the excitation. The results are shown in Figure 5, which plots relative intensity of molybdenum as a function of particle size. Also shown in Figure 5 is the theoretical relationship between particle size and intensity which was made by combining the results of Figures 3 and 4 into one curve. Infinite thickness, which is indicated on the curves at a value of 243 μ, was calculated using absorption coefficients for the exciting radiation at a wavelength just below the K edge for molybdenum. While the two curves shown in Figure 5 do not agree as far as intensity levels are concerned, it is important to note that they are of the same general shape. The leveling-off points in the intensity–particle size relation at both the fine-particle and the large-particle diameters, as predicted from Figures 3 and 4, agree fairly well with those of the actual paramolybdate particles.

A similar experiment was conducted using copper oxide powder, which was separated into a number of fractions of various particle sizes and then blended with sugar. Five per cent by weight of copper oxide was used; the sugar was fine powdered sugar and constant in all cases. Copper K_α was measured on compacted samples using tungsten excitation, and the results are shown in Figure 6. Again, both the theoretical curve based upon the results of Figures 3 and 4 and the actual experimental curve are shown. As was the case with the molybdenum curves in Figure 5, the theoretical curve shows somewhat lower predicted relative intensity than the actual curve. However, the general shapes are again quite similar and the leveling-off points of intensity with particle size show good agreement with the theoretical curve. Infinite thickness value for copper oxide, calculated in a manner similar to the calculation for ammonium paramolybdate, was found to be 19.2 μ as shown in Figure 6. A series of samples of manganese dioxide of varied particle sizes in 5% concentration level shows almost the identical relationship of intensity with particle size as was found with the copper. This is of interest because the manganese dioxide has nearly the same infinite particle thickness as copper oxide, manganese dioxide being 20 and the copper oxide 19.2.

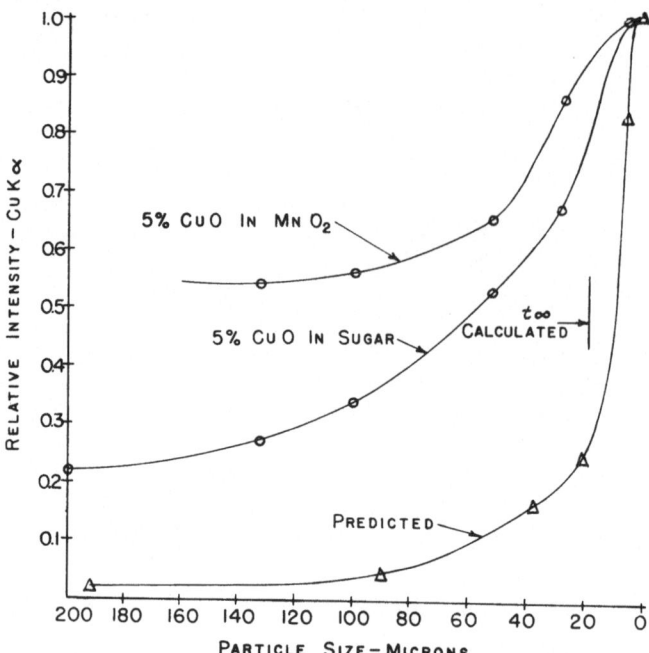

Figure 6. Effect of particle size and matrix on intensity; 5% CuO.

In the experiments which have been described to this point, only the particle size of the constituent which is of interest has been varied. The particle size of the matrix has been maintained constant. As was shown in a previous paper,[2] the intensity from a component in a mix is a function not only of its own particle size but that of the particle size of the material with which it is mixed. In order to evaluate the effect of particle size and matrix upon intensity of the element being measured, the following experiments were conducted. The first involved the above-mentioned copper oxide and manganese dioxide powders. Mixtures of 5% copper oxide in a matrix of manganese dioxide were prepared in such a way as to match the particle sizes of both components as closely as possible. Copper K_α radiation was measured with tungsten excitation and the results are shown in Figure 6. The upper curve represents the copper oxide–manganese dioxide mixtures of equal particle size; the middle curve represents the relationship obtained previously for copper oxide–sugar mixtures. The two curves have the same general shape but, as might be expected, the manganese dioxide has the effect of reducing the spread between the upper and lower levels of curves considerably, as compared to the sugar. This might be expected since the manganese dioxide has a linear absorption coefficient of 900 for copper K_α radiation, while sugar has an absorption coefficient of 13, so that the manganese dioxide represents approximately a 70:1 increase in absorption coefficient. However, it should be noted that the change in intensity level does not appreciably alter the leveling-off points in relationships between particle size and intensity.

In order to establish the mechanism for the intensity shift which was noted in Figure 6 between the copper oxide–manganese dioxide mixtures and the copper oxide–sugar mixtures, the following samples were prepared. A series of 5% copper oxide samples of varying particle size were blended with sugar. Copper K_α measurements were made on these samples in a manner similar to that described above and with similar results,

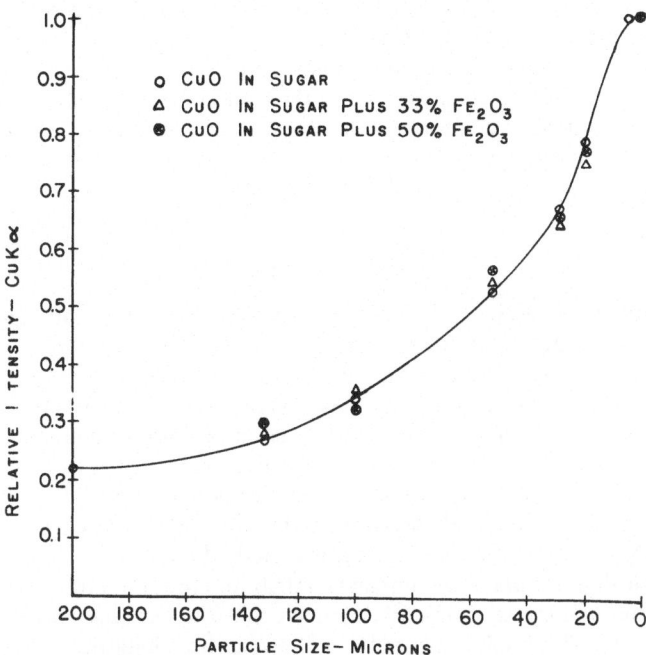

Figure 7. Effect of matrix on intensity; particle size relationship.

as shown in Figure 7. These samples were then blended in the following proportions: first, two parts of sample to one part of iron oxide (Fe_2O_3) and, second, sufficient iron oxide was added to bring the mixture up to one part of sample and one part iron oxide so that the resulting blends represented approximately 33 and 50% iron oxide by weight. Of course, the percent copper oxide would be correspondingly reduced when the iron oxide was added. Intensity measurements were made on all of these samples, and the relative intensities were plotted against particle size, Figure 7, along with the points shown for the copper oxide–sugar mixtures. It is quite interesting to note that the points for the samples with the iron oxide additions are practically coincident with those of the sugar samples. This occurs in spite of the fact that the absorption coefficient of the matrix went from approximately 13 for sugar to 287 for 33% iron oxide and 424 for 50% iron oxide, an increase of approximately 33 times in matrix absorption coefficient. The results shown in Figures 6 and 7 would indicate that the relative intensity of a component in a mixture as a function of its particle size is dependent upon the particle size and absorption coefficient of the matrix rather than the absorption coefficient of the matrix alone. This conclusion would tend to confirm the observation in the previous paper[2] that as the particle size of both components in a mixture were reduced simultaneously, the interactions, namely, absorption and enhancement, between the particles became more pronounced.

The results of the above work can be summarized as follows:

1. The relationship between intensity and particle size of a minor constituent in a mixture shows a leveling off at two different particle size ranges, namely, relatively large particles as compared to the infinite thickness value of a particle, and at relatively small particle sizes as compared to the infinite thickness value. The

approximate numerical relationship is ten times the infinite thickness value and one-tenth of the infinite thickness value.

2. The relative intensity of a minor constituent in a mixture is a function of both the particle size of the constituent and the particle size and absorption coefficient of the matrix material. If the constituent being measured is suspended in a matrix of invariant particle size, then the relative intensity of the minor constituent is independent of the absorption of the matrix.

OBSERVATIONS

The intensity–particle size relationship shows plateaus for both large and small particle sizes, so it might appear feasible to work with relatively large particles as well as with small ones. However, it has been our experience that there are very few instances where samples, particularly mining samples, occur only as relatively large particles. Furthermore, large-particle sizes are difficult to handle; with a laboratory unit, there is the additional problem of preparing a representative sample. In an on-line unit, with continuous presentation, the latter would not be a consideration.

The agreement between the theoretical particle size–intensity curves and the actual ones is reasonably good. In the three examples cited, the actual curve is shifted to the left of the theoretical one. This behavior can be attributed to some extent to an assumption made in calculation of infinite thickness, namely, that the exciting wavelength for each element is just short of the K edge. Actually, at 50 kv, all wavelengths in the tube spectrum from 0.25 Å up to the particular edge are participating in the excitation. This would cause an increase in infinite thickness value, so it is reasonable to expect shifts similar to those observed in Figures 5 and 6. It would be quite difficult to establish the effective exciting wavelength in most cases. Since the assumption used above to calculate infinite thickness gives reasonable results, there does not appear to be any particular need to refine and, at the same time, complicate the calculations.

MINERALOGICAL EFFECTS

Mineralogical effects arise in X-ray emission work due to the occurrence of an element in two or more forms of chemical combination. This has been discussed in other papers by Claisse[3,6] and Campbell and Thatcher.[7] The latter authors present as an example the problem of measuring calcium in wolframite when the calcium may be present as the carbonate, tungstate, or phosphate. Different intensities are observed for equal concentrations of calcium in three forms owing to the fact that when the particle size is large compared to the depth of penetration, a local matrix effect exists. It was shown that extensive grinding will cause the intensities from the different mineral forms to approach a common value, presumably by reducing the absorption within the individual particles to a very small value.

A similar example was reported[2] in the analysis of copper ores, where the copper occurred as chalcopyrite ($CuFeS_2$), infinite thickness 24 μ, and conellite (CuS), infinite thickness 32 μ. As in the above example, grinding the ores to about 5 μ resulted in considerable improvement in X-ray results. The grinding accomplishes two purposes. First, it reduces absorption within the particles and, second, it brings minerals of different infinite thickness values to the upper plateau of the intensity–particle size relationship as described in an earlier section. Obviously, it is desirable to grind both mineral forms to the particle size required by the smaller infinite thickness value.

Another example of a mineralogical effect can be found in the analysis of iron ores where the iron can occur as hematite or magnetite. These two minerals in pure form yield virtually identical iron intensities, although hematite contains 69.9% iron and magnetite 72.4%. Thus it can be expected that equal concentrations of iron as hematite or magnetite would give different iron intensities, the former being slightly higher. In actual samples which have been studied, errors due to the changes in relative concentrations of the minerals have been found to be very small, but it is conceivable that this might not always be true. Grinding would not correct this particular mineralogical problem, but fusion or dilution techniques would probably be applicable.

Regarding the particle sizes of powder samples for X-ray analysis, it should be pointed out that the examination of a large number of samples submitted by mining, cement, and other process industries has shown many instances where normal process control is capable of producing samples which give quite satisfactory results. This occurs in spite of the fact that the particle sizes do not conform to the requirements of the plateaus shown in the earlier sections. This can only be attributed to the samples having a relatively constant particle size distribution, which is not surprising since process efficiencies are often related to particle size control. Cases have occurred where further grinding was necessary, and in these it is extremely helpful to know what degree of particle size reduction is required for successful X-ray analytical control.

REFERENCES

1. G. R. Fonda, *J. Am. Chem. Soc.* **55**: 123, 1933.
2. F. Bernstein, *Advances in X-ray Analysis, Vol. 5*, University of Denver, Plenum Press, New York, 1961, pp. 486–499.
3. F. Claisse and C. Samson, *Advances in X-ray Analysis, Vol. 5*, University of Denver, Plenum Press, New York, 1961, pp. 335–354.
4. G. L. Smithson, R. L. Eager, and A. B. Van Cleave, *Advances in X-ray Analysis, Vol. 2*, University of Denver, Plenum Press, New York, 1958, p. 175.
5. H. A. Liebhafsky, H. G. Pfeiffer, E. H. Winslow, and P. D. Zemany, *X-ray Absorption and Emission in Analytical Chemistry*, John Wiley & Sons, Inc., New York, 1960, p. 154.
6. F. Claisse, *Norelco Reptr.* **3**: 3, 1957.
7. W. J. Campbell and J. W. Thatcher, *Advances in X-ray Analysis, Vol. 2*, University of Denver, Plenum Press, New York, 1958, p. 313.

DISCUSSION

E. W. Franklin (Owens-Illinois Glass Co.): During this morning's papers no one has given consideration to packing density. A weight percent of material is added to the base matrix and then it is loose packed or compacted into a sample holder with no reference to packing density. Couldn't this be troublesome in calculating absorption factors, or is it an unwritten rule that this is to be ignored?

F. Bernstein: The packing density becomes a problem only when you are working with softer X-rays, where absorption and packing densities are critical. Working with harder X-rays the packing density can be pretty much ignored. The problem was met in my lab by the use of standardized procedures in the preparation of samples.

R. McClatchey (Dept. of Defense): I am interested in knowing the reason for the decrease in intensities for the thicknesses beyond infinite thickness.

F. Bernstein: With reference to Figure 1, consider that the element being measured is dispersed in a matrix as cubic particles of infinite thickness. It is assumed that the element is present as a minor constituent and therefore there is little likelihood of superposition of particles at different depths. Under these conditions, every unit volume will contribute toward the total measured intensity as per the relationship shown in Figure 2. If the particle size were to be doubled while maintaining the same weight percent of the element, then the total number of particles would be reduced by a factor of eight. Since the cube depth is now twice infinite thickness, only the upper half contributes to the measured intensity. This means that the measured intensity is now half of the previous intensity since exactly half the volume of sample containing the element of interest is excited as compared to the previous example. By extending this analogy to larger particles, the curve of Figure 4 can be generated.

X-RAY ANALYSIS OF MINING AND MINERAL PROCESSING MATERIAL

H. T. Dryer

Applied Research Laboratories, Inc.
Dearborn, Michigan

ABSTRACT

The X-ray analysis of the materials associated with the mining and mineral processing industries offers a wide variety of potential applications and a challenge to the X-ray analyst. X-ray methods are rapid, reproducible and, when suitably applied, accurate for elements of atomic number 12 or greater. These characteristics make it an ideal process control system. In general, the materials for each type of process require individual analytical techniques to provide reliable analytical data. A wide variety of materials have been studied on the ARL Vacuum X-ray Quantometer and their performances evaluated. An evaluation of the sample preparation and analytical methods required are given.

INTRODUCTION

Last year at this conference, I described the analysis of materials for the iron and steel industry including ores, sinters, slags, iron, and steel by X-ray fluorescence techniques. The excellence of these results plus the need for rapid and complete analyses in the nonmetallic fields have led us to make a number of investigations for a variety of materials.

Experience with the iron and steel materials has shown that the analysis of non-metallic materials, although ideally suited for X-ray techniques, presents difficulties not generally encountered with metallic samples. One of the major problems of these analyses is illustrated in Figure 1, representative of the ore shipment (or body) in terms of hundreds or thousands of tons, the laboratory sample in terms of pounds and the sample for X-ray analysis in terms of grams or milligrams. This represents a sampling factor of about 1:1,000,000,000 and requires that sampling techniques be studied thoroughly to obtain a representative analysis of the material.

To further complicate this analytical problem, these materials may be composed of: (1) massive chunks of separate minerals intermixed and forming a very heterogeneous material, (2) microcrystals of minerals blended to give a uniform material, (3) mixed compounds of the various metals to form a heterogeneous and unpredictable material. Several examples of this are shown in Figure 2; where (a) is a specimen of bauxite ore and (b) a specimen of cryolite with galena and pyrite, both showing massive structure and heterogeneity; (c) a specimen of dolomitic limestone (microstructure), and (d) a specimen of calcite crystals having approximately the same composition but different properties.

In addition to these problems, most nonmetallics are complex mixtures with widely, varying compositions with regard to (1) elements, (2) concentration ranges, (3) minerals, and (4) grain sizes.

The results obtained during our investigations of the nonmetallics used in the iron and steel industry, plus those of numerous other authors, clearly indicated that the correct sample preparation technique is of the utmost importance for the nonmetallics.

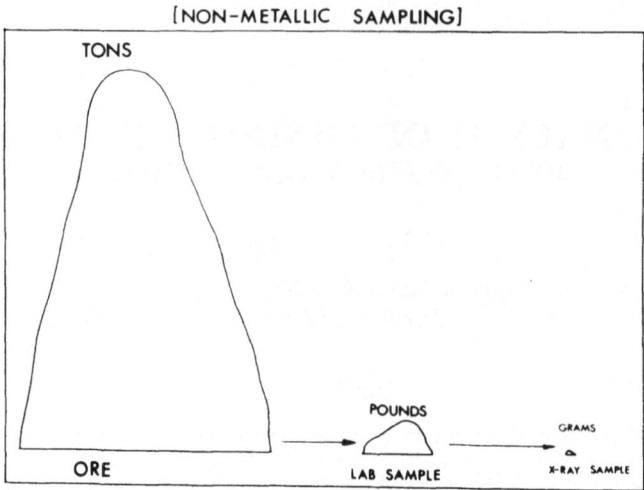

Figure 1. Sampling factor from ore to analytical sample.

Figure 2. Types of nonmetallic samples, (a) bauxite—massive segregation, (b) cryolite—massive segregation, (c) dolomitic limestone—microstructure, and (d) dolomite and calcite crystals.

SAMPLE PREPARATION

Powders

Samples received in the analytical laboratory have been previously reduced to -50 mesh or better. After proper sampling, the material is then prepared by one of the following methods:

I. Samples with Small Mineralogical Differences. A portion of the laboratory sample (10–50 g) is ground in the Bleuler mill shown in Figure 3, for a predetermined time with a suitable binder. The resulting mixture is then briquetted on top of a boric acid backing material—the boric acid backing provides ruggedness and can be used for

Table I. Sample Preparation of Nonmetallics

	I	II	III
Sample weight	10–50 g	1.00 g	0.125 g
Flux	—	2.00 g	1.000 g
Buffer	—	—	0.125 g
Temperature	—	1050°C	950°C
Time	—	5 min	8 min
Grinding	1.5 min	1.5 min	1.5 min

I—General nonmetallics.
II—Borax flux.
III—Lithium tetraborate flux and lanthanum oxide.

sample identification. Two briquettes are shown in Figure 4, (a) cement sample as-received, and (b) cement sample after grinding in Bleuler mill.

II. Samples with Large Mineralogical Differences.[1-4] A portion of the laboratory sample is fused with lithium tetraborate or sodium borate; the temperature and time of fusing depends upon the sample and fluxing material. After solidification, the resulting glass bead is ground and briquetted as in I.

III. Samples with Widely Varying Compositions and Mineralogical Differences.[2,5] A portion of the laboratory sample is fused with lithium tetraborate or sodium borate and a buffer (lanthanum oxide or barium oxide); the temperature and time of fluxing depends upon the material and fluxing agent. After solidification, the resultant bead is ground and briquetted as in I.

The buffer serves to minimize the absorption effects of one element upon another in the sample and eliminates the need for interelement corrections. This method was developed by Adler and Rose of United States Geological Survey and was described at the International Colloquium of 1962.

A summary of these methods and typical conditions are shown in Table I.

Solutions

Two types of holders which we have used are shown in Figure 5; (a) commercially available from ARL and (b) constructed from a polystyrene vial. In both cases, the holders are filled with solution and covered with $\frac{1}{4}$-mil mylar.

Synthetic standards can be prepared for solution techniques and thus offer an advantage from this consideration, especially for new materials where suitable chemical standards are not available. There are, however, several sources of error in solution techniques and must be considered; (1) pH of standards and solutions, (2) reagent mixtures, and (3) density and/or salt content.[6]

ANALYTICAL PROCEDURE

All of these investigations were conducted using an ARL Vacuum Production X-ray Quantometer shown in Figure 6 and described by Davidson et al.[7] in 1959. The

---

[1] Superscripts pertain to references at the end of the paper.

Figure 3. Bleuler rotary mill and components.

polychromator design of this instrument provides maximum analytical data in a minimum of instrument time.

Since X-ray fluorescence analytical methods are comparative in nature, the analytical curves are drawn from reliable standards for the material to be analyzed. After establishment of suitable curves, unknown or production samples are analyzed using sample preparation and analytical methods identical to those used for the standards.

Powders. All solid nonmetallics were prepared by one of the three sample preparation methods described above. The VPXQ was used in the vacuum mode of operation

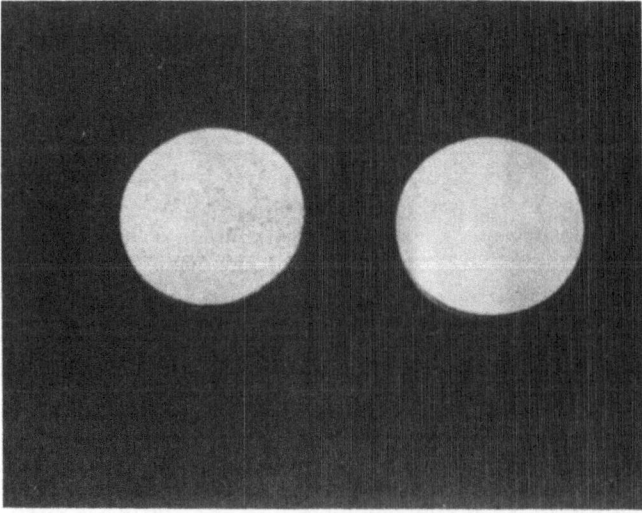

Figure 4. Briquetted finished cements, (a) as-received and (b) after Bleuler grinding.

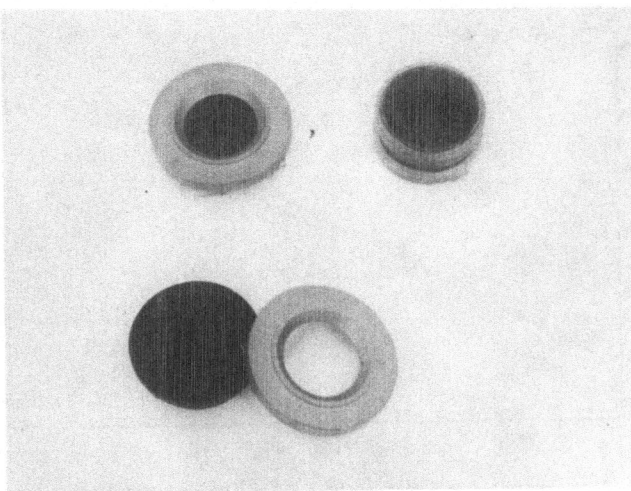

Figure 5. Solution holders for X-ray analysis.

to analyze the briquetted samples with the results provided as a ratio of the element intensity to that of an external standard or scattered radiation.[8]

Solutions. The VPXQ was used in the helium mode of operation for all of the solution samples and results are provided as for powders above.

ANALYTICAL DATA

Bauxite. Two sets (British Guiana and Arkansas ores) of bauxite samples were analyzed for Al_2O_3, Fe_2O_3, SiO_2, and TiO_2. These samples were prepared by Method I using a synthetic soap as the binder. The analytical curve for Al_2O_3 is shown in Figure 7

Figure 6. VPXQ installation.

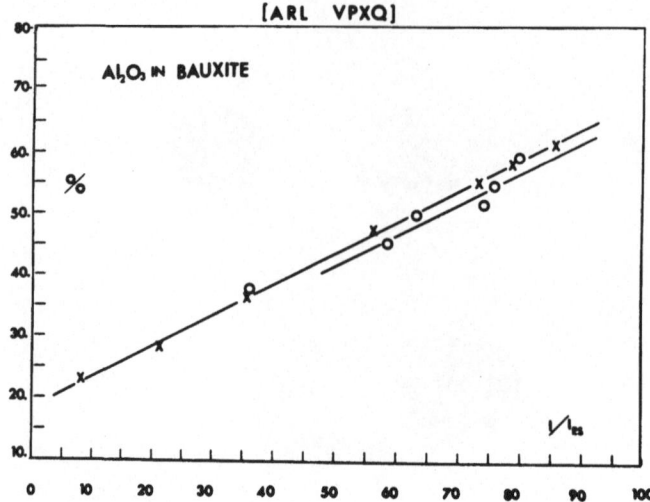

Figure 7. Al₂O₃ in bauxite; × reliable chemical values and ○ routine chemical values.

and that for SiO₂ in Figure 8. Samples designated O had routine chemistries and the errors shown are approximately equal to those expected. A summary of the analytical data for bauxite is shown in Table II.

Limestone. Quarry survey samples of limestone with widely varying concentration ranges were analyzed for CaO, Al₂O₃, SiO₂, Fe₂O₃, and MgO. These samples were prepared by Method I using boric acid as a binder. A typical analytical curve for this material is shown in Figure 9 for CaO. The summary of analytical data is provided in Table III.

Raw Mix (Cement). The raw mix for the production of cement clinker is a blend of various materials such as limestone, sand, shale, iron ore, etc. Raw mix samples, chosen to provide a large variation in blend and blending materials were prepared by Methods I and II and analyzed for CaO, SiO₂, Al₂O₃, and Fe₂O₃. A comparison of the analytical curves for these two methods is illustrated in Figures 10 and 11. A summary comparison of the analytical data is shown in Table IV.

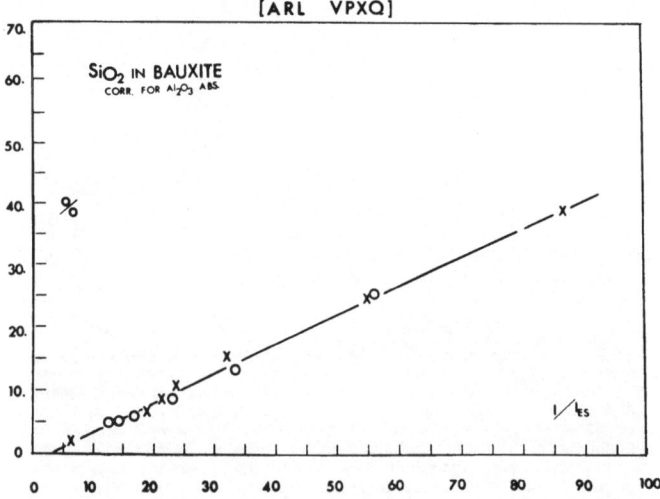

Figure 8. SiO₂ in bauxite, corrected for Al₂O₃ absorption.

Table II. Analytical Data for Bauxite

Element	Concentration range	Average deviation*	Precision	Concentration
Al_2O_3	20–65%	0.95%	±0.39%	51.4%
Fe_2O_3	0.5–50%	0.30%	±0.05%	10.6%
SiO_2	1–40%	0.50%	±0.28%	6.02%
TiO_2	0.8–2.8%	0.10%	±0.04%	2.38%

* Average of duplicate analyses: ARL—VPXQ, 2-min integration, 50 kv and 35 ma.

Table III. Analytical Data for Limestone

Element	Concentration range	Average deviation*	Precision	Concentration
CaO	35–52%	0.36%	±0.055%	41.2%
Al_2O_3	0.5–3%	0.065%	±0.057%	2.0%
SiO_2	0.5–30%	0.26%	±0.15%	10.8%
Fe_2O_3	0.5–1.5%	0.016%	±0.003%	0.95%
MgO	1–6%	0.08%	±0.08%	1.8%

* Average duplicate analyses: ARL—VPXQ, 2-min integration, 50 kv and 35 ma.

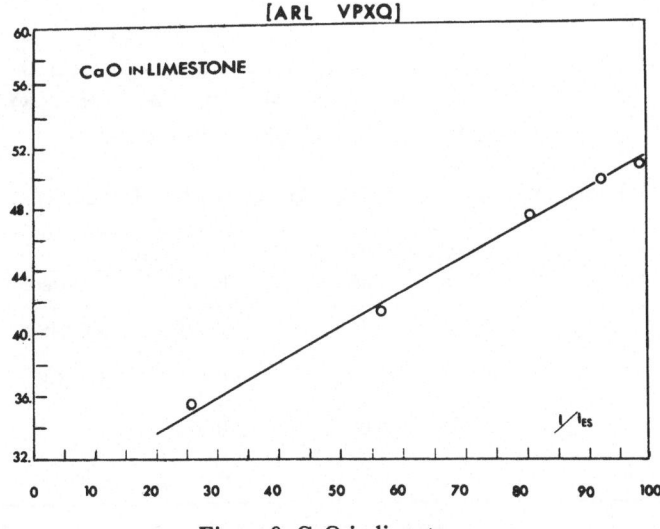

Figure 9. CaO in limestone.

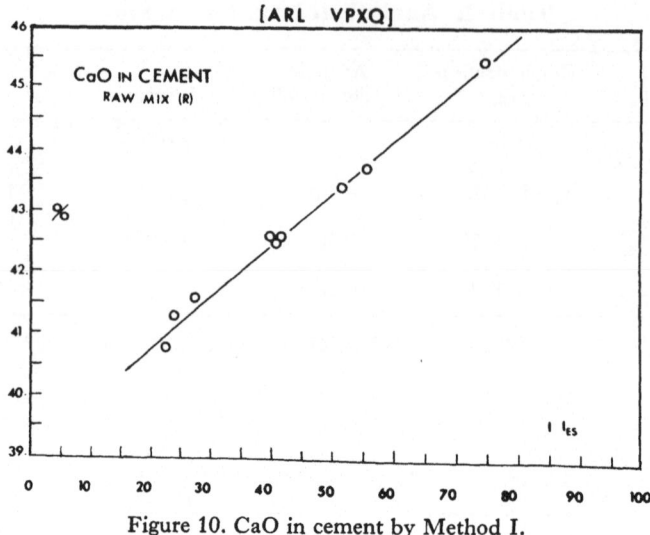

Figure 10. CaO in cement by Method I.

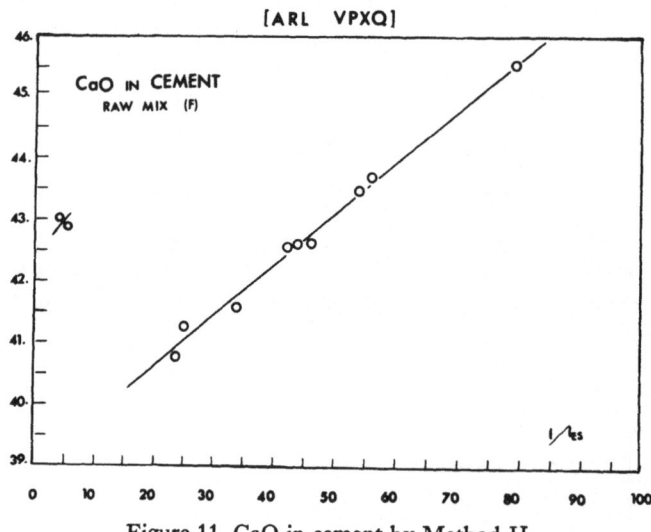

Figure 11. CaO in cement by Method II.

Table IV. Analytical Data for Raw Mix (Cement)

Element	Concentration range	Average deviation*		Precision		Concentration
		Raw	Fused	Raw	Fused	
CaO	38–48%	0.10%	0.11%	±0.02%	±0.048%	40.8%
SiO_2	8–18%	0.07%	0.10%	±0.11%	±0.18%	16.8%
Al_2O_3	1.5–5.5%	0.06%	0.08%	±0.05%	±0.15%	3.4%
Fe_2O_3	2.0–4.5%	0.08%	0.08%	±0.007%	±0.009%	3.4%

* Average of duplicate analyses: ARL—VPXQ, $2\frac{1}{2}$-min integration, 50 kv and 35 ma.

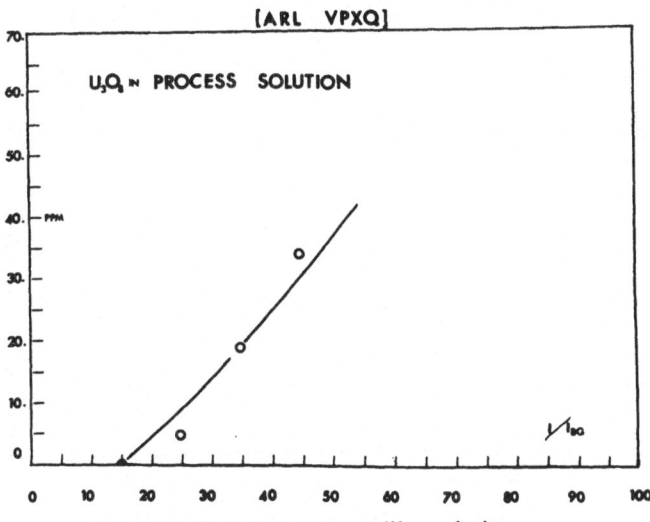

Figure 12. U_3O_8 in process tailing solutions.

U_3O_8 Process Solution. Tailing solutions of the uranium processing were analyzed for U_3O_8. In order to compensate for changes in solutions, the ratio of line/background was used to provide reliable data. The analytical curve for this analysis is shown in Figure 12 and provides excellent sensitivity and accuracy.

Phosphate Rock (Ore). A variety of phosphate ore samples from two sources were analyzed for SiO_2, P_2O_5, CaO, and Fe_2O_3. These samples were prepared by Method I using a boric acid binder. Typical analytical curves are shown in Figures 13 and 14. Calcium absorbs the iron radiation and requires that corrections be made to the iron data. A summary of the analytical data is shown in Table V.

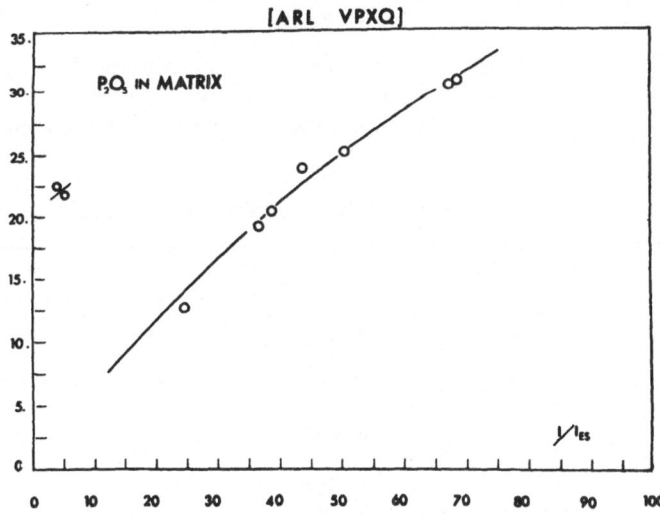

Figure 13. P_2O_5 in phosphate ore material.

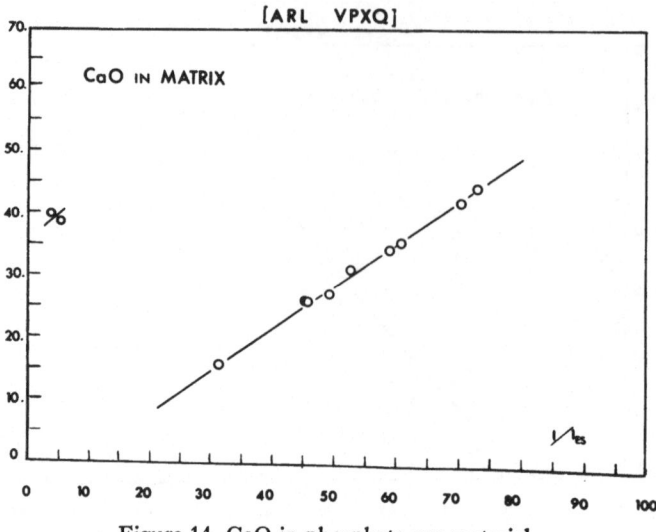

Figure 14. CaO in phosphate ore material.

Petroleum (Oil). The analysis of petroleum, one of our natural resources, has been included in our investigation to illustrate the versatility of X-ray fluorescence analyses. The analysis of most mineral materials involves the determination of major concentrations of elements or compounds; the analysis of lubricating oils involves the determination of trace quantities of added metals. Figures 15 and 16, showing the analytical curves for sulfur in oil using a ratio to external standard intensity and to scattered radiation intensity, respectively, illustrate an ideal application of the principle of scattered radiation. In this case, the scattered radiation used was the Compton scattered line of $W L_\alpha$. The summary of the analytical data for the oil samples is shown in Table VI.

SUMMARY

The results of a variety of investigations have been presented to demonstrate the performance and versatility of X-ray fluorescence methods. Because of the numerous problems and variety of materials associated with the nonmetallics, their analysis by X-ray

Table V. Analytical Data for Phosphate Rock

Element	Concentration range	Average deviation*	Precision	Concentration
SiO_2	10–45%	0.62%	±0.21%	24.5%
P_2O_5	10–35%	0.24%	±0.094%	24.7%
CaO	15–45%	0.36%	±0.066%	29.5%
Fe_2O_3	3–11%	0.095%	±0.005%	6.15%

* Average of duplicate analyses: ARL—VPXQ, 2-min integration, 50 kv and 35 ma.

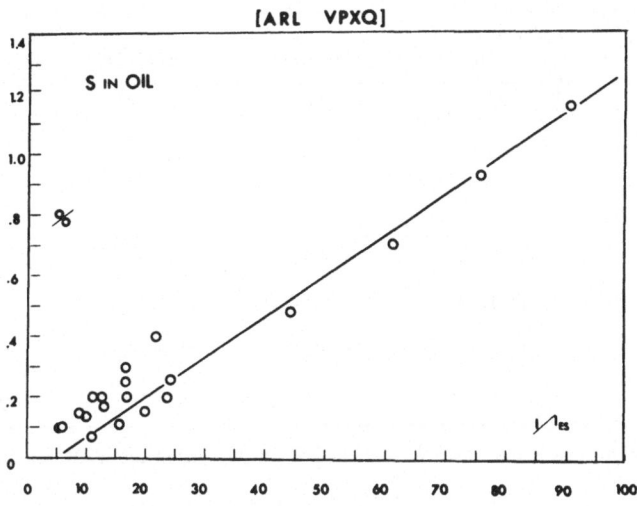

Figure 15. Sulphur in oil; concentration *vs.* I_s/I_{es}.

techniques has been considered extremely difficult and, in some cases, unreliable. However, our experiences and those of other authors have proven that the analysis of non-metallics can be readily accomplished by the proper selection of sampling, sample preparation, and analytical methods. Each type of material must be investigated to select the optimum conditions. X-ray analysis can then provide the analytical data more rapidly, with less human error, with more complete information, and at considerably less cost as compared to routine chemical analyses.

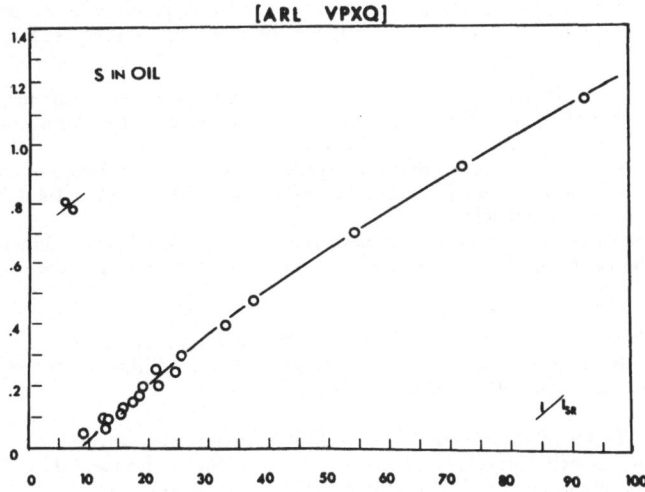

Figure 16. Sulfur in oil; concentration *vs.* $I_s/I_{s.r.}$ (s.r. at Compton scattered W L_x.)

Table VI. Analytical Data for Oils

Element	Concentration range	Average deviation*	Precision	Concentration
P	0.005–0.36%	0.008%	0.006%	0.10%
S	0.04–1.2%	0.012%	0.007%	0.25%
Zn	0.005–0.20%	0.002%	0.0003%	0.03%
Ba	0.1–0.7%	0.010%	0.005%	0.32%
Cl	0.2–1.2%	0.010%	0.007%	0.80%
Ca	0.08–0.60%	0.016%	0.003%	0.15%
Pb	0.2–1.2%	0.010%	0.007%	0.56%

* Average of duplicate analyses: ARL—VPXQ, 5-min integration, 50 kv and 35 ma.

REFERENCES

1. *ARL Spectrographer's News Letter* **2** (12), 1949.
2. F. Claisse, *Can. Dept. Mines Tech. Surveys Report*, PR No. 327, 1956.
3. G. Andermann, J. W. Kemp, and M. F. Hasler, International Colloquium on Spectroscopy, June 1961.
4. G. Andermann, *Anal. Chem.* **33**: 1689–1699, 1961.
5. I. Adler and J. Rose, International Colloquium on Spectroscopy, 1962.
6. S. Friedlander and A. Goldblatt, Pittsburgh Conference of Anal. Chem. and Applied Spectroscopy, 1959.
7. E. Davidson, A. W. Gilkerson, and H. Neuhaus, Pittsburgh Conference on Anal. Chem. and Applied Spectroscopy, 1960.
8. G. Andermann and J. W. Kemp, *Anal. Chem.* **30**: 1306, 1958.

DISCUSSION

A. K. Baird (Pomona College): I would like to ask about the grinding. First I would like to know the particle size you get, grinding time, and the total sample size you can grind. What is the function of the binder you use, and is this just for the grinding or is it used in the briquetting afterward?

H. T. Dryer: The Bleuler grinder has a capacity of 100 g, however loads under 50 g are preferred. For loads over 50 g, the grinding time must be increased to 4 or 5 min to compensate for compaction and the resulting loss in grinding action.

In regard to particle distribution, we have checked a variety of nonmetallics using 15-g sample weight and 1½-min grinding time. For most materials, about 95% will pass a 325-mesh screen and about 98% will pass a 200 screen.

The binder used serves two purposes: first, it serves as a lubricant, reducing contamination and facilitating the cleaning of the container; and second, it serves as a binder in the briquetting operation.

C. N. Schieltz (Colorado School of Mines): Did you ever check on the amount of impurity you pick up from the mill? In the grinding did you pick up any additional contamination? Did the lubricant you used make any difference in the contamination level?

H. T. Dryer: We have checked contamination from the container with and without the lubricant. The container is an iron–chrome alloy and contributes about 0.04% iron and 0.015% Cr, depending upon the hardness of the material being ground. Contamination is quite high if the grinding is performed without a lubricant in the Fe–Cr container. A tungsten carbide container is now available which should eliminate the contamination problem.

AUTHOR INDEX

Bold numbers refer to papers in this volume

A

Abrams, R., 210, 221
Adda, Y., 282, 283, 290
Adler, I., 449, 458
Aka, E. Z., 125, 135
Alexander, L. E., 18, 21, 24, 145, 147, 157, 193, 200, 216, 219, 221
Allen, J. D., 387, 388
Allison, S. K., 28, 41, 180, 184, 362, 364, 375
Amelinckx, S., 114, 120
Anantharaman, T. R., 102, 120
Anderko, K., 211, 221
Andermann, G., 387, 388, 423, 428, 449, 451, 458
Anderson, S., 140, 141
Andrews, C. L., 282, 290
Antonic, G., 315, 327
Arajs, Sigurds, **121–135**
Asp, E. T., 237, 241
Augustyniak, W., 165, 170
Austin, A. E., 283, 284, 290
Averbach, B. L., 35, 36, 42, 122, 135, 217, 221

B

Bacon, G. E., 158, 162
Baecklund, J., **328–338**
Baird, A. K., 361, **377–388**
Baker, R. M., 251, 261
Baldwin, W. M., Jr., 116, 120
Banister, A. J., 211, 221
Barkow, A. G., **210–222**
Barrett, C. S., 30, 34, 36, 41, 42, 74, 83, 101, 120, 217, 221
Bartram, S. F., 197, 200
Beck, P. A., 57, 61, 75, 83
Bennett, F. D., 211, 221
Berndt, A. F., **18–24**
Bernstein, F., **436–446**
Beycler, M., 282, 283, 290
Bharacha, R. M., 211, 221
Binns, R. A., 285, 290
Birks, L. S., 283, 289, 290, 321, 327
Black, I. A., 243, 248
Black, R. H., 320, 327
Blin, J., 35, 42
Blokhin, M. A., 364, 376
Böhm, B., 52, 61
Bollenrath, F., 98, 116, 120
Bolleter, W. T., 417, 421
Bolling, G. F., 100, 117, 120
Bolz, L. H., 76, 83, **242–249**
Bond, W. L., 149, 157, 166, 171

Borie, B. S., **177–184**
Bragg, W. L., 218, 221
Brauer, G., 53, 61
Braumann, F., 36, 42
Brentano, J. C. M., 217, 221
Bridgman, P. W., 165, 171
Brimhall, J. L., 100, 117, 120
Brooks, F. P., 243, 248
Brown, N., 98, 120
Brown, P. J., 58, 59, 61
Brown, R. N., 203, 209
Bryan, F. R., **339–344**
Buehler, W. J., 74, 75, 77, 83
Bueren, H. G. v., 137, 141, 160, 162
Bulatova, R. F., 242, 247, 248
Burton, R. L., 223, 230

C

Caley, J. L., **396–402**
Campbell, W. J., 250, 261, 314, 327, 444, 445
Capelli, R., 91, 94
Carpenter, R. D., 214, 221
Castaing, R., 37, 42, 276, 277, 283, 289, 290
Catterall, J. A., 93, 94
Cermak, J., 144, 157
Chace, W. G., 210, 221
Chessin, Henry, **121–135**
Chodos, A. A., 378, 388
Chollet, P., 282, 290
Claisse, F., 378, 387, 388, 437, 444, 445, 449, 458
Clement, J. R., 243, 249
Clusius, K., 202, 209, 243, 248
Cochran, W., 138, 141
Cohen, J. B., 102, 120
Cohen, M., 122, 135, 217, 221
Coles, B. R., 121, 123, 124, 130, 134, 135
Collen, B., 140, 141
Colodny, P. C., **202–209**
Colvin, R. V., 121, 134
Compton, A. H., 28, 41, 180, 184, 362, 364, 375
Conner, R. A., Jr., 91, 92, 94
Cork, J. M., 364, 365, 376
Cosslett, V. E., 276, 279, 280, 281, 282, 289, 290
Cullen, T. J., 423, 428
Cullity, B. D., 89, 214, 219, 221, 237, 241, 340, 343, 352, 359

D

Darwin, C. G., 218, 221
Das, B. N., 57, 61
Dauben, C. H., 180, 182, 184

Davey, A. R., 245, 249
Davidson, E., 449, 458
Davis, E. N., 403, 416
Decker, B. F., 237, 241
Deichert, R. W., **429–435**
Delavignette, P., 114, 120
de Smedt, J., 248, 249
de Wolff, P. M., **1–17**
Dhosi, J. M., **172–176**
Dobbs, E. R., 242, 248
Dolby, R., 279, 280, 281, 282, 290
Donachie, M. J., 98, 115, 116, 120
Dowling, P. H., 31, 42
Downey, J. W., 91, 92, 94
Dreyer, K. L., 117, 120
Druyvesteyn, M. J., 254, 261
Dryer, H. T., **447–458**
Duncumb, P., 276, 277, 280, 281, 287, 288, 289, 290
Dunne, J. A., **345–351**
Dunning, H. N., 403, 416
Duwez, P., 75, 83
Dwiggins, C. W., Jr., 403, 416
Dwight, A. E., 91, 92, 94

E
Eager, R. L., 437, 445
Eastman, W. B., **313–327**
Eatwell, A. J., 242, 248
Edwards, J. W., 250, 261
Eisenstein, A., 185, 190
Ekvall, R. A., 98, 120
Elbaum, C., 172, 176
Ellinger, F., 21, 24
Elliott, H. A., 137, 141
Ence, E., 75, 83
Engel, C. G., 378, 388
Engler, N., 223, 230
Erofeev, B. V., 203, 209
Eshelby, J. D., 126, 135, 139, 141
Evans, J. C., 146, 157
Ewald, P. P., 29, 41

F
Fairbairn, H. W., 383, 385, 388
Fankuchen, I., 301, 311
Faulring, G. M., 214, 221
Federighi, T., 40, 42
Ferro, R., 91, 94
Figgins, B. F., 242, 248
Finch, L. G., 98, 120
Fine, S., 279, 290, 364, 365, 376
Fish, B. R., 210, 218, 221
Flinn, P. A., 35, 36, 42, 122, 135
Flubacher, P., 243, 248
Fonda, G. R., 436, 445
Foote, F., 19, 24
Forsyth, W. J., 320, 327
Fournet, G., 30, 31, 42
Fraenkel, W., 26, 41
Frank, F. C., 46, 56, 61

Franklin, E. W., **250–261**
Franz, H., 125, 129, 135
Freda, A., 237, 241
Freeman, A. J., 140, 141, 180, 181, 184
Friedel, J., 126, 127, 135
Friedlander, S., 449, 458
Furnas, T. C., Jr., 149, 157, 165, 171, 223, 230, 366, 376

G
Garrod, R. I., 98, 120
Gatto, F., 40, 42
Gaylord, N. G., 417, 421
Gazzara, C. P., **172–176**
Geisler, A. H., 28, 29, 30, 34, 36, 37, 41, 42, 223, 230
Geller, R. F., 251, 261
Gerold, V., 37, 42
Gevers, R., 114, 120
Gilfrich, J. V., **74–84**
Gilkerson, A. W., 449, 458
Glocker, R., 31, 36, 42
Glusker, D. L., 18, 24
Goldblatt, A., 449, 458
Goldschmidt, G. H., 209
Goldschmidt, H. J., 52, 61
Goldschmidt, V. M., 123, 129, 135
Gordon, J., 183, 184
Gould, R. W., **62–73**
Graf, R., 31, 32, 33, 37, 38, 42
Grain, C., 250, 261
Greenough, G. B., 97, 98, 115, 120
Gregor, L. V., 183, 184
Grogan, J. D., 93, 94
Gross, S. T., 215, 221
Grosse, L., 282, 290
Grun, F., 231, 241
Gschneidner, K. A., Jr., 134
Guerrieri, F., 211, 221
Guinier, A., 26, 27, 28, 29, 30, 31, 33, 34, 35, 36, 37, 38, 41, 42, 276, 289
Gunn, E. L., **403–416**, 417, 421
Gupta, K. R., 57, 61
Guy, A., 30, 42, 282, 290

H
Haasen, P., 114, 120
Hägg, G., 173, 176
Handy, R. L., 197, 200
Hansen, D., 62, 73
Hansen, M., 75, 83, 211, 221
Hardy, H. K., 29, 30, 33, 36, 37, 39, 41, 42
Harker, D., 237, 241
Harvey, J., 211, 221
Hashin, Z., 134, 135
Hasler, M. F., 423, 428, 449, 458
Hauk, V., 98, 116, 120
Havighurst, R. J., 218, 221
Hayes, E. T., 62, 73
Heal, T. J., 29, 30, 33, 36, 37, 41, 42
Heffelfinger, C. J., 223, 230

Heinrich, K. F. J., **291–300**
Helion, J. C., 105, 120
Hendee, C. F., 31, 42, 279, 290, 364, 376
Henke, B. L., **361–376**, 378, 388
Henry, N. F. M., 159, 162
Henshaw, D. G., 248, 249
Herbstein, F. H., 177, 184
Hill, J. K., 28, 29, 30, 36, 37, 41
Hirabayashi, K., 211, 221
Hofmann, E., 16, 17
Holden, A. N., 225, 230
Holland, J. R., 223, 230
Honeycombe, R. W., 172, 176
Howe, S., 172, 176
Huang, K., 136, 141, 169, 171
Huggins, R. A., 100, 117, 120
Hume–Rothery, W., 56, 60, 61, 74, 75, 83, 121, 123, 124, 133, 134, 135
Hurst, D. G., 209

I
Iredale, A., 286, 287, 290
Ito, T., 1, 5, 6, 16, 17

J
Jacquet, P., 33, 42
Jagodzinski, H., 16, 17, 28, 41
James, R. W., 218, 221
Jette, E. R., 19, 24
Johns, T. F., 248, 249
Johnston, H. L., 250, 261
Jones, G. O., 242, 248
Jones, J. W., **223–230**
Jones, R. A., 403, 416
Jones, W. M., 183, 184

K
Kalakoutzky, N., 97, 119
Kang, C. C., 403, 416
Karioris, F. G., **210–222**
Kärlsson, N., 173, 176
Kartha, G., 138, 141
Kasper, J. S., 46, 48, 56, 61
Kauffman, J. W., 39, 42
Keel, E. W., 403, 416
Keesom, W. H., 248, 249
Kemp, J. W., 423, 428, 449, 451, 458
Ketelaar, J. A., 202, 203, 209
Kiefer, E. W., 403, 416
Kiley, W. R., 396, 397, 402, **429–435**
King, A., 114, 120
King, H. W., 132, 135, **142–157**
Kirianenko, A., 282, 283, 290
Kissinger, H. E., **164–171**
Klein, M. J., 100, 117, 120
Klemm, W., 52, 61
Klug, H. P., 18, 21, 24, 145, 147, 157, 193, 200, 216, 219, 221
Koehler, J. S., 39, 42, 165, 171
Kogan, V. S., 242, 247, 248

Kolb, K., 98, 116, 120
Köster, W., 31, 36, 42, 125, 129, 135
Kottwitz, D. A., 165
Kruis, A., 202, 209
Kuhlin, T. R., 31, 42
Kuhn, W., 231, 241
Kummer, E., 219, 221
Kuptsis, J. D., **389–395**
Kuylenstierna, U., 140, 141

L
Ladell, J., 150, 157
Lamb, F. W., 403, 416
Lambe, K. A. D., 183, 184
Lambert, M., 29, 41
Lamborn, R. E., **422–428**
Lambot, H., 36, 37, 38, 42
Lang, A. R., 144, 156
Lang, S. M., **250–261**
Lavrukhina, A. K., 211, 221
Lazarev, B. G., 242, 247, 248
Leipold, M. H., 251, 261
Levine, E. P., **158–163**
Levinson, D., 62, 73
Liebhafsky, H. A., 352, 359, 361, 375, 438, 445
Lindstrom, B., 282, 290
Lipson, H., 4, 5, 17, 159, 162
Lloyd, J. C., **389–395**
Long, E. A., 183, 184
Long, J. V. P., **276–290**, 299, 300
Lonsdale, K., 143, 156, 183, 184
Lublin, P., **185–190**
Lund, P. K., 372
Lutts, A., **25–42**

M
MacColl, R. S., 378, 387, 388
Macherauch, E., 98, 116, 120
Magneli, A., 140, 141
Mallery, J. H., 33, 42
Margolin, H., 75, 83
Mark, H. F., 417, 421
Martin, D. E., 215, 221
Massalski, T. B., 100, 117, 120, 132, 135
Mathews, H. I., 211, 221
Mathies, J., 372
Mauer, F. A., 76, 83, **242–249**
Maute, R. L., **417–421**
McCreery, G. L., 218, 221
McCune, R. A., **85–90**
McHargue, C. J., 100, 117, 120
McIntyre, D. B., **377–388**
McKeown, M., 165, 170
McLaren, A. C., 203, 209
McOmie, J. F. W., 211, 221
McQuillan, A. D., 140, 141
McQuillan, M. K., 140, 141
McReynolds, A. W., 165, 170
Meakin, J. D., 98, 120

Mehl, R. F., 30, 34, 36, 39, 41, 42
Mehta, S. M., 211, 221
Melford, D. A., 280, 281, 283, 286, 287, 290, 295, 300
Merica, P. D., 26, 41
Middleton, R. M., 172–176
Miller, A., 18, 24
Miller, D. C., 341, 343, 396, 402
Miller, D. S., 121–135
Miller, P. H., 126, 135
Mitius, A., 53, 61
Mitskevich, N. I., 203, 209
Mooser, E., 55, 61
Mooy, H. H., 248, 249
Morgan, P., 417, 421
Morningstar, D., 372
Muller, G. J., 202, 203, 209

N
Neerfield, H., 98, 120
Neighbors, J. R., 101, 106, 120
Nelson, J. B., 103, 120
Neuhaus, H., 449, 458
Nicholson, M. E., 132, 135
Niebylski, L. M., 403, 416
Nielsen, J. P., 75, 83
Nishida, D., 211, 221
Nixon, W. C., 287, 288, 290
Noggle, T. S., 165, 171
Norton, J. T., 57, 61, 98, 115, 116, 120
Novak, C., 2, 3, 6, 17
Nudelman, S. L., 18, 21, 24
Nutting, J., 114, 120

O
O'Brien, L. P., 268–275
O'Brien, W. L., 62, 73
Orcutt, D., 202–209
Ostapchencko, E. P., 185, 190
Ostwald, E., 98, 116, 120
Otte, H. M., 96–120

P
Panseri, C., 40, 42
Parr, J. G., 82, 83
Parrish, W., 31, 42, 85, 89, 133, 135, 142, 145, 146, 150, 156, 157
Parthe, E., 57, 61
Paterson, M. S., 98, 102, 103, 119, 120
Pauling, L., 123, 135
Pearson, W. B., 55, 61, 122, 127, 129, 132, 135
Peck, J. M., 121, 134
Peiser, H. S., 158–163, 243, 248
Perming, F. M., 254, 261
Pernot, B., 282, 283, 290
Pfeiffer, H. G., 438, 445
Philibert, J., 282, 290
Philip, T. V., 75, 83
Picklesimer, M. L., 64, 73
Piercey, D. C., 242, 248

Piesbergen, U., 243, 248
Pietrokowsky, P., 75, 83
Pike, E. R., 144, 145, 150, 154, 156, 157
Pleasance, R. J., 93, 94
Podchainova, V. N., 211, 221
Pollard, F. H., 211, 221
Poole, D. M., 75, 83, 291, 300
Post, B., 19, 24
Powers, W., 223, 230
Preston, G. D., 26, 30, 31, 41
Purdy, G. R., 82, 83

Q
Quinell, E. H., 243, 249
Qurashi, M. M., 140, 141

R
Rachinger, W. A., 100, 120
Rait, J. R., 159, 162
Rajan, N. S., 57, 61
Rau, R. C., 191–201
Ray, A. E., 58, 61
Raynor, G. V., 123, 132, 135
Reed, S. J. B., 283, 285, 290
Renfrew, A., 417, 421
Reuss, A., 98, 120
Rhines, F. N., 62–73
Rhines, F. W., 39, 42
Richard, N. A., 283, 284, 290
Richardson, W. H., 210, 221
Rickenbach, J. R., Jr., 352–360
Riley, D. P., 103, 120, 242, 248
Robertson, A. H., 62, 73
Röhner, F., 39, 42
Roof, R. B., Jr., 183, 184
Rooksby, H. P., 185, 190
Rosauer, E. A., 197, 200
Rose, H. J., 387, 388
Rose, J., 449, 458
Rosenblatt, D. B., 165, 170
Roth, E. L., 315, 327
Royster, G. W., Jr., 218, 221
Rudman, P. S., 35, 36, 42
Runge, G., 1, 5, 16, 17
Russell, B. R., 126, 135

S
Sack, R. A., 235, 240, 241
Saini, H., 202, 203, 209
Salmon, M. L., 301–312
Samson, C., 378, 387, 388, 437, 444, 445
Sandenaw, T. A., 183, 184
Sandstrom, A. E., 364, 375
Schael, G., 238, 241
Schanzer, W., 202, 209
Scherb, J., 31, 36, 42
Schleich, K., 243, 248
Schmalzried, H., 37, 42
Schmid, E., 26, 41
Schoening, F. R. L., 136–141

Schomaker, V., 18, 24
Schulz, L. G., 237, 241
Schulze, G. E. R., 54, 61
Scott, H., 26, 41
Scott, R. B., 243, 248
Seebold, R. E., 283, 290
Segmüller, A., 391, 395
Seiden, P. E., 389, 395
Seitz, F., 39, 42, 126, 135
Seng, R., 26, 41
Silcock, J. M., 30, 33, 42
Sivertsen, J. M., 132, 135
Skinner, B. J., 245, 249
Slater, J. C., 60, 61
Smith, B. L., 242, 248
Smith, C. S., 101, 106, 117, 120
Smith, G. D., **417–421**
Smith, J. F., 58, 61
Smith, S. L., 116, 120
Smithson, G. L., 437, 445
Smits, A., 202, 203, 209
Solomon, E., 403, 416
Sorenson, F. J., **422–428**
Sparks, C. J., **177–184**
Speiser, R., 250, 261
Sperandio, A., 243, 248
Spielberg, N., 376
Springer, G., 278, 289
Stadelmaier, H. H., 57, 61
Stammler, M., **202–209**
Stecura, S., 250, 261
Steigert, F. E., 33, 42
Stein, R. S., 231, 241
Steinert, H., 36, 42
Stewart, J. W., 243, 248
Stoffels, J. J., **210–222**
Stokes, A. R., 97, 119
Straumanis, M. E., 125, 135
Sturcken, E. F., 19, 24
Sutton, A. L., 133, 134, 135
Swann, P. R., 114, 120
Swanson, H. E., 8, 17, 125, 135

T
Tatge, E., 125, 135
Taylerson, C. O., 146, 157
Taylor, A., 100, 101, 105, 120
Taylor, J., 150, 157
Taylor, J. C., 134, 135
Taylor, J. L., 75, 83, 251, 261
Templeton, D. H., 180, 182, 184
Terada, J., 185, 190
Tetelman, A. S., 98, 99, 100, 120
Thatcher, J. W., 314, 327, 417, 421, 444, 445
Thewlis, J., 245, 249
Thomas, P. M., 291, 300
Thompson, L. T., 127, 132, 135
Thomson, J. R., **91–95**
Tonn, W., 62, 73
Torkildsen, R., **262–267**

Tucker, T. J., 220, 221
Tudbury, C. A., 251, 261
Tufts, G. F., 185, 186, 190
Turnbull, D., 39, 42

U
Ubaldini, I., 211, 221

V
Valentine, T. M., 183, 184
Van Cleave, A. B., 437, 445
Vand, V., 3, 17
van der Steinen, K. A., 39, 42
Van Nordstrand, R. A., 403, 416
Vassamillet, L. F., **142–157**
Vineyard, G. H., 134
Vogel, R., 62, 73
Voigt, W., 98, 120

W
Wagner, C. N. J., 98, 99, 100, 102, 105, 119, 120
Wagner, J. C., **339–344**
Walker, C. B., 34, 35, 42
Wallbaum, H. J., 53, 61, 62, 73
Waltenberg, R. G., 26, 41
Warburton-Brown, D., 251, 261
Warekois, E. P., 98, 106, 115, 120
Warren, B. E., 98, 99, 102, 103, 105, 106, 115, 118, 120, 165, 170, 178, 179, 182, 184
Warren, B. G., 122, 135
Waser, J., 18, 24
Wasserman, E., 26, 41
Wassermann, G., 33, 42
Watson, R. E., 180, 181, 184
Wechsler, M. S., 111, 117, 120
Weerts, J., 33, 42
Weigle, J., 202, 203, 209
Welch, D. O., **96–120**
Welday, E. E., **377–388**
Weltman, H. J., 204, 209
Wever, F., 1, 5, 17
White, E. W., 149, 157, 366, 376
Wilchinsky, Z. W., **231–241**
Wiley, R. C., 74, 75, 77, 83
Willard, M. L., 211, 221
Williams, J. M., 111, 117, 120
Willis, T. M., 183, 184
Wilman, H., 211, 221
Wilsdorf, H. G. F., 98, 120
Wilson, A. J. C., 97, 119, 142, 143, 144, 145, 146, 150, 156, 157
Winslow, E., 438, 445
Witt, F., **136–141**
Witte, H., 52, 56, 61
Wittig, W. J., 354, 359
Wolk, B., 185, 186, 190
Woltersdorf, G., 50, 61

Wood, G. C., 287, 290
Wood, W. A., 98, 100, 115, 116, 120
Wooster, W. A., 159, 162

Y
Yamaka, E., 185, 190
Youngkin, F. G., 75, 83

Z
Zachariasen, W. H., 21, 24
Zemany, P. D., 438, 445
Zerfoss, S., 211, 221
Ziegler, G., 31, 36, 42
Zingaro, P. W., 341, 343, 355, 359
Zintl, E., 50, 61
Zsoldos, L., 1, 3, 17

SUBJECT INDEX

A

Absolute intensity
  in spectrochemical analysis, 352–354
Absorption by oil
  in fluorescence, 403
Absorption coefficients
  for silica, 305
Accidental scattering
  in low-temperature aging, 28
Adler and Rose method
  in mining material analysis, 449
Aerosols
  in copper wire explosion, 210–222
Ag L excitation
  compared with Al K excitation, 366
  in silicate analysis, 385
Ag L radiation, 366
Ag–Pd system
  lattice parameters, 130
  solid solution, 121
Al–Ag
  low-temperature aging, 27–39
Al–20%Ag
  low-temperature aging, 29–34
Al–38%Ag, 35, 36
AlB₂-type structure, 54
Al–Cu
  low-temperature aging, 26–39
Al–Cu–Mg
Al–Mg–Cu
Al–Mg–Ge
Al–Mg–Si
Al–Mg–Zn
  preprecipitation stage, 36–40
Al–5.2%Cu
  diffusion coefficient at low temperature, 39
AlFe₃C-type structure
  in atom formation, 57
Alkalies
  in clay, 377
Alkaline earth oxide cathode
  X-ray diffraction study, 185–190
Al–Mg
  low-temperature aging, 40
Al–1.4%Mg₂Si, 37
Al₂O₃
  mining material analysis, 451, 452
Al–Zn
  low-temperature aging, 27–39
Al–25%Zn, 36
Aluminum
  fluorescence analysis, 361
  for carbon analysis, 317
  hot-dipped, 350
  in iron–aluminum alloys, 339–343

Aluminum (*continued*)
  in solid-propellant study, 423
  in three-rock standards, 373
  in Ziegler polyethylene, 417
  spectrographic analysis, 378
  with Ag L excitation, 385
Aluminum anode source, 366
Aluminum content
  by X-ray fluorescence, 340
  shown chemically, 340
Aluminum crystals
  cold-wash effect, 170
  under neutron irradiation, 164–170
Aluminum filter
  in suppressing emission, 306
Aluminum foil
  as flow counter window, 382
  in filter application, 301, 307–310
Aluminum foil epoxy
  in spectrographic analysis, 379
Aluminum powder
  in solid propellant, 422
Aluminum 8–quinolinolate
  contamination, 417
Aluminum radiation
  in helium path, 341
  in vacuum, 341
Aluminum single crystals
  produced for X-ray diffraction, 172–176
Aluminum stearate
  in calibration standards, 419
Ammonium bromide
  low-temperature study, 202
Ammonium chloride, 202
  diffraction pattern, 209
  low-temperature transition, 203, 208
Ammonium fluoroborate
  low-temperature transition, 204, 207, 208
Ammonium halides
  diffraction study, 202
Ammonium ion oscillation, 202
Ammonium ion (rotation)
  at liquid nitrogen temperature, 209
  at low temperatures, 202
Ammonium nitrate
  low-temperature transition, 203'
Ammonium paramolybdate crystal
  in mining materials control, 440
Ammonium perchlorate
  at low temperatures, 204, 208
  in solid propellant mix, 422
  particle size variation, 426
Ammonium phosphate
  low-temperature transition, 208

Ammonium salts
  with complex ions, 209
Amorphous substances
  compared with single crystals, 158
Amplifier gain
  in silicates analysis, 383
Amyl acetate
  in preparing beryllium oxide specimens, 197
Analytical line intensities
  in filter application, 305, 306
  reduced by aluminum filter, 306, 311
Analyzing crystal
  lithium fluoride, 302, 304
Anisotropic strain
  in polycrystalline sample, 97
Anodized structures
  photomicrographs, 65–69
Anthracite coal
  in carbon analysis, 320
Apparatus Standards Committee of A.C.A., 146
Application of Routine Methods of Crystallite
    Size Analysis, 191–201
Arc-melting
  in thorium and palladium investigation, 91
Argon
  in low-temperature test, 242, 245
  in resistance heating, 254, 256
Argon absorption, 370
Argon counter
  for chlorine through aluminum analyses, 371
Arsenic
  analytical lines, 302–306
Asphalt
  in petroleum analysis, 399
Asterism
  in Laue photographs, 158
Atomic diameter
  in measure of atom, 123
Atomic scattering factor
  for copper, 177
Austenite in steels
  with helium path, 89
Autocollimator
  in measuring backlash errors, 147
Axial divergence angles
  divergence error, 145
  for conventional diffractometers, 145
AZAR-type recorder
  in resistance heating, 259

B

Backreflection lines
  in oxide cathode study, 187
Backscattered electrons
  in electron beam scanning, 270
  in electron-probe analysis, 277–289
Backscattering
  with microprobe, 291–293
Bakelite powder
  in petroleum analysis, 398

Banbury mixer
  in determining contamination, 419
BaO
  solid solubility in SrO, 187
BaO–SrO solid solution
  in oxide cathode study, 189
Barium stearate
  soap film analyzer, 366
(BaSr)O system
  crystallite size, 185
Base metals
  exploded in argon, 210, 211
Bathophenanthroline
  to determine iron, 417
Bauxite
  analytical data, 453
Bell jar
  in work coil design, 254, 255
BeO
  for crystal measurement, 194–198
Beryllium filter
  for phosphorus measurement, 373
Beryllium oxide
  at varying temperatures, 193, 194
  in crystallite size analysis, 191–201
  preparation of specimens, 197
Bico Braun pulverizer
  in spectrograph operation, 314
Binary transition metals
  terminal solid solutions, 121
Binder
  in developing filter, 301
Bismuth
  with microprobe, 300
Blank matrix
  in filter application, 302
Bleuler mill
  mining sample analysis, 448, 449
Blood serum
  in fluorescence analysis, 372
Boltzmann factor
  in metals during heating, 39
Bond factor
  in element structure, 45, 59
Borax flux
  for Na analysis, 382
Bragg angle range
  for focusing monochromators, 268
Bragg angles
  from diffraction profiles, 143–145
  in determining sources of error, 143
  measuring orientation in polymers, 237
  study of cold-work filings, 97
  wire explosion test, 218
Bragg's law
  in solid solutions, 121
  in spectrochemical analysis, 352
  in study of large single crystals, 159
Bragg maxima
  with UO_2, 177, 184

Bragg peak
  observed intensity, 18
  with Cu₂O powder, 179
Bragg reflections
  intensity loss, 177
  using double scanning method, 150
  using wavelength value, 143
Brass (shim brass)
  in filter application, 302–308
Bridgman-grown crystal
  in neutron irradiation, 165
Bridgman technique (method)
  aluminum crystal production, 165
  modified, 165
Brillouin zones
  Cu₂O, 178
Bromine fluorescence
  experiment with oil, 405

C
CaCl₂
  comparison with CaGa₂, 54
CaCl-type structure, 209
CaF₂
  in rapid analysis, 314, 315
  in reduction cell electrolyte, 313
  -type structure, 55
CaGa₂
  comparison with CaAl₂, 54
Ca $K\alpha$
  in carbon analysis, 314
Calcium
  analysis of carbon materials, 313–317
  in petroleum analysis, 398
  measurement of wolframite, 444
  reduced background fluorescence, 363
  secondary fluorescence, 382
Calcium (in bath)
  in carbon analysis, 318, 319
Calibration curves
  in iron–aluminum study, 340
C $K\alpha$
  from graphite flakes, 280
CaPlug
  in filter application, 305
  in petroleum analysis, 396
Carbon
  in alloy analysis, 334
Carbon analysis
  in spectrograph operation, 315
Carbon materials analysis
  by X-ray spectrograph, 313–327
Carbonate powders
  in oxide cathode study, 189
Cargill's oil
  as immersion medium, 186
Castaing formula
  of fluorescence intensities, 284, 285
Castaing's thesis
  of electron probe analysis, 276–289

Cast iron
  analysis for carbon, 313, 316
Catalyst residues
  in polyolefins, 417
Cathodoluminescence
  with electron probe, 299, 300
Cathode-ray-tube screen
  in electron probe analysis, 278, 281
Cathodes—C–4, C–10
  oxide, study of crystal structure, 189
CaTiO₃ (perowskite structure), 57
Cauchy distribution curve
  in describing diffraction profile, 101, 105
Cement
  spectrographic analysis, 377
Cement (raw mix)
  analytical data, 454
Cementite, 1
Ceramics
  spectrographic analysis, 377
Ceramic susceptor
  in resistance heating, 260
Cerium
  spectrographic analysis, 389–395
Chart-scan technique
  in filter application, 303
Chart speeds
  in line-broadening work, 193
Chemical analysis
  compared with X-ray analysis, 317, 327
Chlorine
  concentrations, 417
  contamination, 418
  fluorescence analysis, 361
Chromium
  in alloy analysis, 328–336
  increase of lattice constants, 133
  in filter application, 302
  in spectrochemical analysis, 355, 358
  lattice expansions, 132
Chromium intensity
  in stainless steels, 334
Chromium $K\alpha$ radiation
  to raise intensity level, 85
Chromium radiation
  with Debye–Scherrer camera, 89
Chromium solutions
  lattice parameter comparison curves, 121
Chromium tube
  in iron–aluminum test, 339
Chromium X-rays
  with helium path, 89
Classical elasticity theory
  for solid solutions, 121
Classical precipitation theory
  in low-temperature aging, 26
Clay
  in spectrographic analysis, 377
Clement–Quinell equation
  in carbon resistance reading, 243

Close-packed structures
    in atom arrangement, 47
Coaxial line
    with multiturn primary coil, 257
Cobalt
    increase of lattice constants, 133, 134
    in spectrochemical analysis, 353
    with iron in sample, 310
    without filter, 311
Coherency strains
    in hardening effect, 27
CoK$_\alpha$ radiation
    with europium oxide filter, 311
    with scintillation detector, 85
Cold–aging, 25
Collimated white radiation
    of stationary crystal, 159
Collimation-crystal system, 363
Collimation
    in spectrographic analysis, 378
    parallel plate in measuring zinc coating, 348
    reduction, 363
Collimator
    entrance, 313
    exit, 313
    in on-stream analysis, 433
    in measuring zinc coatings, 349–351
    in rare earth analysis, 390
    measuring elements in oil, 404
    parallel blade in filter application, 304, 305
    with Philips instrument, 382
Compton effect
    in parasitic scattering, 28, 29
Compton scattered line
    in petroleum analysis, 456
Compton scattering
    in study of Cu$_2$O, 180
Computer program
    to measure crystallite size, 191
Computer techniques, 14
Concentration mapping
    for ratemeter output, 295
Conducting oxides
    in resistance heating, 251
Conellite
    in emission experiment, 444
Connection principle
    compared with space principle, 50
    in atom arrangement, 44–59
Constant Q
    in measuring zinc coating, 346
Coordination number
    in structure types, 44
Copolymers
    low-pressure analysis of, 421
Copper
    increase of lattice constants, 132
    in exploded wire test, 210
    in mining materials control, 440, 441
    in spectrochemical analysis, 353

Copper (*continued*)
    in work coils, 253, 257
    low-level background, 302, 305, 306
    silver-plated for Si and Al analysis, 380
Copper anode (silver-plated)
    for Ag L excitation, 370
Copper core mold
    for carbon analysis, 315, 317
Copper ores
    chalcopyrite, conellite, 444
Copper oxide powder
    in mining materials control, 441
Copper radiation
    in study of thorium and palladium, 91
    with GM diffractometer, 85
Copper–silicon alloy single crystal
    irradiation single effect, 170
Copper–silicon–manganese alloy
    effect of cold work, 96
Copper target X-ray tubes
    in crystallite size analysis, 195
Counter tube diffractometers
    using powder specimens, 142
Cr K_α
    with scintillation detector, 85
CrN$_2$, CrN$_5$
    in electron-probe analysis, 283
Crystal lattice
    with neutron irradiation, 164
Crystallite size analysis, 191–200
Crystallite size analysis curves, 191
Crystallographic direction R
    in reference direction Q, 231–240
Crystallographic planes
    in crystallite size analysis, 194
Crystal-monochromated radiation
    to determine austenite in steel, 85
Crystal monochromator
    in study of scattering, 30, 31
CuAl$_2$-type structure, 54, 55
Cu–Be
    reaction to aging temperatures, 33, 34
Cu$_2$O
    Debye–Waller factor, 177
Cu–Cu$_2$O–CuO
    from copper explosion in air, 210
Cu/Cu$_2$O ratio and Cu/CuO ratio
    in wire explosion test, 215
Cu K_α radiation, 216
    absorption coefficients, 214
Cu$_3$N, 211
CuO, Cu, Cu$_2$O, 214
CuO, 211
    in exploding wire test, 210
Cu$_2$O
    in exploded wire test, 210, 211
    in low-temperature range, 183
Cu$_2$O, CuO, Cu, 218, 219
CuO line, 218
Cu–Si–Mn alloy
    effect of cold-work, 96

Czochralski technique
  to grow aluminum crystals, 172

D
DBA (dichlorobenzoic acid)
  in solid propellant study, 423
Debye–Scherrer method
  deformation and heat treatment, 96
  in preprecipitation, 31
  in study of deformation, 101
  patterns, 21
  using powder specimens, 142–157
Debye temperature
  in oxygen–titanium study, 140
Debye–Waller factor
  for Cu_2O, 177–184
  in neutron irradiated crystal, 167
  oxygen in titanium, 136–140
Deformation faults
  in cold-work filings, 96–102
  residual elastic strains, 96
Degradation
  affecting tube output, 265
Detector slit
  in wire explosion test, 216
Detector voltage
  in silicates analysis, 383
Detector window materials
  in spectrographic analysis, 378
Diabase
  in standard preparation, 383
Dibromobenzene
  in absorption effect test, 404
Diffracted intensities
  quantitative measurement, 235
Diffraction analysis, 262
Diffraction angle
  instrumental broadening, 193
Diffraction errors
  arising from alignment, 147
  extrapolation procedures, 143–145
  geometry, instrumental, 145, 146
  physical effects, measurement, 143
Diffraction line-broadening
  for crystallite size analysis, 191
Diffraction pattern
  in crystallite measurement, 193
  with slow scanning speed, 216
Diffraction peak breadths
  related to crystal size, 193
Diffraction peaks
  in copper wire explosion, 216
Diffractometer
  data, 16
  focusing, modified, 15, 16
Diffractometer furnace
  in high-temperature methods, 250
Diffractometer geometry
  in developing heating systems, 253

Diffractometer method
  counter tube, 142
  in determining d spacings, 143
Diffractometer traces
  with air and helium X-ray paths, 88
Diffusion couples
  in electron probe, 282
Diffusion zones
  due to impurities, 283
Digital circuitry
  in on-stream analysis, 434
Digital timing clock
  in on-stream analysis, 433
Diode
  with oxide systems, 185, 186
Direction Q
  in orientation in polymers, 231–240
Direction R
  in orientation in polymers, 231–240
Discrete peaks, 18
Dispersion factor
  in diffraction error, 154
Displacement disorder
  revealing scattering, 28, 32
Distortion factor
  in titanium defects, 138
Divergence
  axial, 145
  horizontal, 145
Divergence slits
  to determine diffraction errors, 154
Domains
  coherently diffracting, 96
Domain size
  after deformation, heat treatment, 96, 97, 101
Double oxide system, 185
  plot of lattice constants, against composition
    for double oxide systems, 188
Double scanning method
  for determining diffraction error, 154
Dry nitrogen
  with oxide cathode, 185, 186
Duco cement
  as filter binder, 301
  in preparing beryllium oxide specimens, 197
Duncumb's apparatus
  in electron probe analysis, 289
D-values
  in low-temperature transitions, 207
Dysprosium
  spectrographic analysis, 389–395

E
EDDT crystal, 363
  in carbon analysis, 318
  in petroleum analysis, 398
  in spectrographic analysis, 382
EDTA determinations
  duplicate, in leakage check, 315
Elastic constants
  for titanium, 138

Elastic continuum
  theory, in lattice parameter size effect, 129
  with properties of titanium, 137
Elastic strain
  in bulk crystalline sample, 97
  residual, following deformation, 96
Electrochemical factor
  in element structure, 54, 59
Electrodes
  in resistance heating, 254
Electromagnetic transformer
  for inductive power, 252
Electron backscattering
  in electron-probe analysis, 277–289
Electron beam scanning system
  in cubic micron analysis, 268–272
Electronegative valance factor
  in lattice parameter changes, 122–132
Electronic density
  in a crystal, 27
Electronic discrimination
  in filter application, 308, 309
  with optic path, 377
Electronic line scans
  in microprobe, 292
Electronic time constant error
  elimination by scanning, 150
Electron microprobe, 276–300
Electron microscopy, 287
  to measure crystallite sizes, 193
Electron optical system
  for analyzing cubic microns, 268
  in electron probe analysis, 276, 279
Electron scattering
  in electron probe analysis, 287
Embrittlement
  in on-stream analysis, 432
Emery paper
  in polishing samples, 315
Emission gauges
  in process control, 436
Emission spectrography, 352–359
Entrance collimator, 313
Equiangular projection
  in preparing pole figures, 225
Equiarea projection
  in preparing pole figures, 225–229
Erbium
  spectrographic analysis, 389–395
Ethanol solution
  in rare earth analysis, 393
Ethocel
  in rare earth analysis, 389, 393
Ethyl acetate
  in preparing beryllium oxide specimens, 197
Ethylenediamine tetraacetic acid
  as titrating agent, 340
Europium
  spectrographic analysis, 389–395
Europium oxide (filter)
  in silicone fluid, 310

Eutectoid
  decomposition, reaction, structure as two-phase constituent, 65
Excitation potential
  in tube power supply, 265, 266
Exit collimator, 313
Extinction
  primary, secondary, in wire explosion test, 218, 219

F

Face-centered cubic lattice
  Cu_2O, UO_2, 178
Fankuchen's scheme
  in filter application, 303
Fe–Al alloys
  vacuum X-ray fluorescence, 343
Fe K_α radiation
  attenuated in zinc coating, 347
Fe K_β radiation
  with europium oxide filter, 311
Fe–Ge alloy
  Solid terminal solutions, 124
Fe–Mn system
  in measuring atomic volume, 134
Fe_2O_3
  in mining material analysis, 451, 452
  in solid propellant, 422
Fe–V system
  in lattice parameter relationships, 124
  lattice spacings, 126
  solid terminal solutions, 124
Filings
  cold-work, 96
Film techniques
  double oxide (BaSr)O system, 185
Filtered molybdenum radiation
  in oxygen–titanium study, 139
Filters
  to reduce unwanted spectra, 301–311
Flat crystal optics
  in sensing head channel, 433
Flow counter preamplification
  in spectrographic analysis, 383
Flow detector tube
  using PR gas in spectrochemical analysis, 353
Flow proportional counter
  in fluorescence analysis, 361–366, 370
  in spectrograph operation, 314
Fluid coke
  in carbon analysis, 320
Fluorescence analysis
  mining materials, 447–458
  silicates, 377
  sodium and magnesium, 361–375, 377–387
Fluorescence correction equation
  in electron-probe analysis, 283
Fluorescence
  of minerals, 285, 286

Fluorescent intensity
  in relation to particle size, 436
  in relation to thickness, 438
Fluorescent radiation
  as origin of parasitic scattering, 28
  elimination, 29
  in solid propellant study, 423
Fluorescent X-ray beam
  with filter, 311
Fluorine
  fluorescence analysis, 361
Fluorine analysis, 371, 372
Fluorite
  polished sample, 372
Flutter compensation
  in measuring zinc coating, 349
Flutter range
  in measuring zinc coating, 350
Foils
  in resistance heating, 251
Fourier analysis
  for deformation fault probability, 102
Frequency spectrum
  acoustic, optic in temperature calculation, 183
Fused silica vacuum furnace
  in heat treating binary alloys, 63
Fusion techniques
  in silicates analysis, 387

G
Galvanized steel
  measuring zinc coatings, 347
Gas
  composition, pressure in resistance heating,
    256
Gasoline
  in petroleum analysis, 401
Gaussian distribution
  in alloy analysis, 333
  in spectrochemical analysis, 354
Geiger–Muller tube
  in resistance heating, 259
Generator
  characteristics for maximum stability, 262
  furnace, 20-kW induction, 259
Germanium
  analytical lines, 302, 307
Glass
  in spectrographic analysis, 377
Glow discharge
  with induction heating, 256
Glusker and Miller theory
  of radial distribution analysis, 18
Gold
  analytical lines, 302
Goldschmidt radii, 123–129
Goldschmidt's rule
  in effect of temperature, 45
Goniometer scanning speeds
  in crystallite size analysis, 196

Granite powder
  in sodium analysis, 366
Granitic rock analyses, 377, 385
Graphite
  in resistance heating, 252
Grit belt
  in sample polishing, 315
Guinier–Preston zones
  in low-temperature aging of Al–Cu, 26
Guinier principle, 15
Guinier-type cameras, 15, 16
Guinier-type focusing camera
  with copper radiation, 91
Gypsum
  for Na and Mg analysis, 382
Gypsum crystal
  in fluorescence analysis, 363, 364, 366
  in silicates analysis, 387

H
H_3BO_3
  in solid propellant study, 423
Halogen
  in production of polyolefins, 417
Harmonics
  in generator study, 264, 267
Hastelloy
  in spectrochemical analysis, 354
Helical tungsten cathode
  in fluorescence analysis, 367
Helium
  in resistance heating, 254, 256
Helium flow
  in spectrograph operation, 314
Helium path
  for austenite and martensite, 89
  in carbon analysis, 318
  in iron-aluminum test, 341
  in petroleum analysis, 396
  to increase long-wavelength intensity, 85, 86
Helium path spectrometer
  in radiation detection, 85
Hematite, 445
Hexagonal titanium
  distortion by oxygen atoms, 136–141
High-temperature equipment
  in X-ray diffraction, 250
High-voltage Laue method
  with large single crystal, 158–159
HNO_3
  in carbon analysis, 320
Holmium
  spectrographic analysis, 389–395
Homogeneous isotropic metal
  in distortion study, 139
Horizontal divergence angle, 145
Human blood serum
  fluorescence analysis, 374
Hume–Rothery compounds
  in element structure, 49, 56
Hume–Rothery phases
  of bond formation, 55, 56

Hume–Rothery rule
  in element structure, 45
Hume–Rothery tendency, 45, 53, 54

I

Indexing methods, 1–16
Indium seal
  in spectrographic analysis, 379
Induction heating
  for X-ray diffraction, 250–256
  generators, 251
Infinite thickness
  of particles in mining materials control, 438
Images
  in electron-probe analysis, 278
Integrated intensity
  in orientation distribution, 235
Integrated radiation
  in propellant mix study, 423
Intensity data
  in steel analysis, 331
Intensity measurement
  of neutron-irradiated crystal, 166
Intergranular stress system model
  in study of residual strains, 98
Intermetallic compounds
  for structural applications, 74
Internal standard method
  direct comparison, 214
  in spectrochemical analysis, 352, 353
  single line, 214
Interplant checks
  on use of spectrograph, 327
Interstitial oxygen atoms
  with hexagonal titanium, 136
Iodide titanium
  in oxygen–titanium study, 139
Ionization
  in resistance heating, 254
  of gases in chamber, 252
Iridium
  in resistance heating, 251
Iron (see also Fe)
  analysis of carbon materials, 313, 317
  as hematite, as magnetite, 445
  atomic number 26, 377
  expansion of lattice, 133
  increase of lattice constants, 132
  in electron probe analysis, 284
  in filter application, 302
  in oil experiment, 404
  in spectrochemical analysis, 353
Iron absorption method
  in measuring zinc coating, 348
Iron contamination
  from process equipment, 417
Iron emission line
  measured by fluorescence, 345
Iron oxide
  in rare earth analysis, 390
Iron oxide blank samples, 310

Iron oxide determination
  by X-ray fluorescence, 422–428
Iron solid solutions
  lattice parameter–composition curves, 121
Iron stearate
  in calibration standard, 419
Ito's procedure, 2, 5, 6, 16
Irradiation
  of single aluminum crystal, 164
Isotropic strain
  of polycrystalline sample, 97

J

Johansson focusing
  in cold-work deformation, 101
Joule capacitator
  to explode wire, 210

K

KAP crystal
  in fluorescence analysis, 363, 364
Kasper's polyhedra, 46, 47
  in space filling, 47
K_β background
  from filter material, 305
K emission lines
  in electron probe analysis, 279
KNa_2, K_7Na, 21, 22, 43
K-phthalate
  for Na and Mg analysis, 382
Krypton
  in low-temperature test, 242, 245
Krypton-filled detector tube
  in spectrochemical analysis, 353
Krypton-filled Geiger counter
  to scan diffraction peaks, 102
Krypton-filled Geiger tube
  in detecting molybdenum radiation, 75
K-value
  in ammonium salts test, 207

L

L_α peak
  in filter application, 305
L_α tungsten lines, 302, 305
$LaNbTiO_6$, 9
Lattice constants, 242
  in study of oxide cathode, 186, 187
Lattice distortions
  of oxygen in titanium, 136–141
Lattice parameter
  changes after deformation, 96
  changes in cold-work deformation study, 105
  in study of cold-work deformation, 103
  of cubic crystal, 142, 145
  of oxide cathode, 185
  variation, 92
  with changes of composition, 121
Lattice parameter measurement
  of neutron-irradiated crystal, 166
  on zirconium filings, 64

Lattice spacings
  in solid solutions, 123
Lattice specimens
  in solid solutions, 121–135
Lattice strain
  computed by elasticity theory, 97
Laue–Bragg intensity
  static displacement of atoms, 136
Laue photographs
  of irradiated aluminum, 164
  of large single crystals, 158–163
  of neutron-irradiated aluminum, 164
  with neutron-irradiated crystal, 164
Laue spots
  caused by crystal imperfections, 160
Laue streaks, 36
Laue transmission
  in aging of an alloy, 37
  of aerosols, 218
Laue transmission method
  employing polychromatic radiation, 30
  in investigating single crystals, 34, 36
Layer faults
  as cause of line broadening, 98, 99
L_β tungsten lines
  in filter application, 302
Lead
  in petroleum analysis, 398, 401
L_γ tungsten lines
  in filter application, 302
LiF
  as analyzing crystal, 302–304
  in rare earth analysis, 390
  in silicates analysis, 387
  in spectrograph operation, 314
  with K_α lines, 321
Light elements analysis, 279
Light element range
  in spectrographic analysis, 378, 379
Light optical system
  in cubic micron analysis, 271
Limestone
  analytical data, 453
Limit of detectability
  in carbon analysis, 321
Linear absorption coefficient, 145
Line breadth measurements
  in oxide cathode study, 187
Linen phenolic
  in rare earth analysis, 389
Lipson's method, 4, 5, 16
Liquid gallium metal
  in preparing aluminum specimen, 173
Liquid helium cryostat
  in measuring lattice constants, 242–248
Logarithmic digital counter
  in on-stream analysis, 435
Long period
  in atomic measurement, 123, 124, 132
Lonsdale's suggestion
  in measuring PuO_2, 183

Lorentz factor
  in determining diffraction error, 143, 154
Lorentz polarization effect
  in oxygen–titanium study, 139
Lorentz polarization factor, 166
  across diffraction profile, 144, 154
Low-angle region
  in oxide cathode study, 186
Lucite plastic, 404

M
Macroscopic stress system
  in determining residual strain, 98
Macrostress
  in steel, 85
  with martensite, 89
Magnesium
  fluorescence analysis, 377–387
Magnesium analysis
  in rock sample, 370
Manganese
  analysis of cast iron, 313, 315, 316
  in analysis of cast iron, 316
  in experiment in oil, 404
  fluorescence analysis, 361
  size of atom, 133, 134
Manganese dioxide
  particle size intensity, 441
Magnetite
  in ore analysis, 445
Martensite
  with helium path, 89
Martensitic-like structure
  in TiNi examination, 76
Matrix absorption
  in relation to particle size, 437
Matrix effects
  in determining catalyst residues, 420
Medium resolution slits
  in determining axial divergence, 145
Mercury
  analytical lines, 302
Metal foils
  aluminum, 305, 306, 308
  used as filters, 301–311
$Mg_{17}Al_{12}$
  as crystallizing compound, 49
  in formation of alloy structure, 43
$MgCu_2$-type structure, 54, 58
MgO
  in analysis of variance, 385
  in mining material analysis, 452
  in soft excitation, 382, 383
Mg_2Si alloy
  as stable precipitate, 36
  in aging experiment, 37
MgX_2-type structure, 54
Mica windows
  to study crystal growth and size, 185

Microanalysis
  of organics, 273
Microbeam techniques
  to measure double oxide, 185
Microprobe analysis
  in examination of alloys, 62
  through electron beam scanning, 270
Mineralogical effects in control of mining
    materials
  in chemical combinations, 436, 437
  in X-ray emission analysis, 436–444
Mining materials
  X-ray control, 436
Mirror plane
  in orientation in polymers, 234
Molybdenum
  in absorption effect through oil, 404
  in alloy analysis, 328–336
  in resistance heating, 251
  in spectrochemical analysis, 355
Molybdenum crystals
  in mining materials control, 440, 441
Molybdenum determination
  for steel analysis, 328
Molybdenum wavelength
  with lithium fluoride crystal, 165
Monatomic cells
  displacement of atoms, 137
Motor change method
  in preparation of pole figures, 229
Multiplexing
  a readout mode, 293, 294
Multiturn primary coil
  in resistance heating, 257
Mylar windows
  for heavier element analysis, 370, 371
  in experiment with oil, 404
  in on-stream analysis, 429, 431, 432
  in petroleum analysis, 396, 399
  in sodium analysis, 366, 368

N
NaCl
  in silicates analysis, 383
$Na_2CO_3–H_3BO_3$
  fusion mix in carbon analysis, 320
NaI scintillation detector, 85
Na_2O
  analysis of variance, 385
  in drift analysis, 384
  in soft excitation, 382
NaTi-type structure, 57
Nb_2S_5
  in electron-probe analysis, 283
Neodymium
  spectrographic analysis, 389–395
Nd_4Cl
  in low-temperature transition, 208
Nelson–Riley function
  in cold-worked materials, 102

Nelson–Riley plot, 22, 103
Neon
  lattice constant, 248
  lattice constant at low temperature, 242
Neon-filled counter
  for chlorine through aluminum analysis, 371
Neutron diffraction
  of aluminum single crystal, 165
Neutron diffraction techniques
  with large crystals, 158
NH_4BF_4
  polymorphic transition, 202–209
NH_4Cl
  at low temperatures, 204
  polymorphic transition, 202–209
NH_4ClO_4
  in low-temperature transition, 208
  polymorphic transition, 202–209
$(NH_4)HPO_4$
  at low temperatures, 204
  polymorphic transition, 202–209
NH_4NO_3
  at low temperatures, 203
Nickel
  analysis of carbon materials, 313, 317
  in alloy analysis, 328–336
  in electron probe analysis, 284
  in filter application, 302–311
  in oil experiment, 404
  in spectrochemical analysis, 355, 358
  lattice expansions, 132
  parameter measurements exaggerated, 133
Nickel crucible lid
  in preparing samples, 314
Nickel filter, 301–311
  in study of oxide cathode, 186
Nickel solid solutions
  for calculating size effect, 131
  lattice parameter composition curves, 121
Ni–Ge system
  showing atomic volume of, 128–129
Nitinols, 74
Nitrides
  copper, in exploding wire test, 211
Nitrogen
  in resistance heating, 256
Nixon's arrangement
  in electron probe analysis, 287–289
Noble metals
  explosion in air, 210
  oxides, in exploding wire test, 211
Noggle method
  of crystal production, 165
Noggle-type crystals
  for neutron irradiation, 165
Nonmetallics, 456
  compared with metallics for analysis, 447
Nontriclinic crystals, 4
Novak's procedure, 3, 6, 8, 16
Nowotny phases of structure formation, 57

O

Oil
  for measuring elements, 403–416
  synthetic, 412
Oils
  analytical data, 458
On-stream X-ray analysis, 429–435
Open counter scan method, 165
Optical autocollimating system
  in study of diffractometers, 146
Orientation
  c-axis, q-axis, in measuring polymers, 238
Orienter
  in program control, 224
Orthographic projection
  in preparing pole figures, 225
Orthorhombic unit cell
  indexing Pu_3Ru, 18–23
Oxide cathode
  structure, alkaline earth, 185
Oxide-coated nickel cap
  with oxide systems, 185
Oxides
  conducting, 251
  in mixture as filter, 311
  in wire explosion test, 211, 215, 217
Oxide values
  in spectrographic analysis, 383
Oxygen
  in titanium crystal, 136–141

P

Palladium atoms, 91–94
Parameters
  orientation, 231–240
Parasitic scattering of X-rays, 28
Paterson's analysis
  of deformation faults, 103
$PbO–PbF_2$
  in rare earth analysis, 389
P-component system, 49
Pelletized samples
  in carbon analysis, 318
Pelletizing
  in on-stream X-ray analysis, 432
Perforated grids
  to increase signal on fluorine, 372
Peritectic reaction
  in Zr_2Fe microstructure, 72
Perowskite structure ($CaTiO_3$), 57
Petroleum
  analysis of mining material, 456
  influence of absorption, 403
  product analysis, 396–402
  refinery control, 396
Phenolic linen
  in rare earth analysis, 389
Phosphate rock
  analytical data, 456
Phosphorus
  fluorescence analysis, 361

Phosphorus (*continued*)
  in alloy analysis, 334
  in analysis of cast iron, 316
  in spectrograph analysis, 313, 315
  in three rock standards, 373
Photoluminescence
  with electron probe, 299
Photometric methods
  in carbon analysis, 318
Pike's analysis
  of axial divergence, 145
Pirani-type gauges, 7
  in resistance heating, 254
Plane orientation diagrams
  equiangular, equiarea, normal projections, 225
  preparation of pole figures, 226–228
Plane parallel electrodes
  in resistance heating, 254
Plant control
  analysis of carbon materials, 317
Platinum, 252
  analytical lines, 302
Plutonium
  low-temperature measurement, 183
Polarization factor
  in determining diffraction error, 154
Pole figures
  for polymers, 223–230
  in measuring orientation in polymers, 236–240
Polybutenes
  in low-temperature analysis, 421
Polycrystalline substances
  compared with single crystals, 158
Polyethylene
  in determining contamination, 420, 421
Polyganization
  as cause of Laue spots, 160
Polymer(s)
  films, for preparation of pole figures, 223–230
  in solid propellant analysis, 424–428
  measurement of orientation, 231–240
Polyolefins
  determining catalyst residues, 417–421
Polypropylene
  in determining contamination, 420, 421
  measuring orientation in polymers, 238
Polystyrene
  coating, 185
  in atomic scattering, 179
  vial, in rare earth analysis, 394
Post-irradiation etching
  of neutron-irradiated crystals, 167, 168
Potassium (see also K)
  analysis of carbon materials, 313, 315, 317
Potassium acid phthalate crystal
  in fluorescence analysis, 363, 372
Potassium bromide
  in experiment with oil, 404
Powder materials
  in filter application, 305
  in on-stream analysis, 433

Powder method
    of carbon analysis, 318
Powder pattern(s)
    from cubic crystals, 98
    of unknown structure, 18–24
Powder photographs
    to measure crystallite size, 191
Powder sample
    X-ray intensity and particle size, 436
Powder specimens
    in precision diffractometry, 142
Praseodymium
    spectrographic analysis, 389–395
Precipitate particles
    in decomposition process, 28
Preferential absorption
    in filter application, 310
    of Fe K_β radiation, 311
Preferred orientation(s)
    in quantitative analysis, 218
    value of pole figures, 223
Premix analysis
    in solid propellant study, 424
Preprecipitation (cold-aging)
    in alloys, 25
Primary standard
    in spectrochemical analysis, 354
Process control
    mining, cement industry, 445
    through X-ray analysis, 436
Propellant mixes
    composition, 426
Proportional counter
    xenon filled, for high-order interferences, 390
Protein matrix absorption, 373
Pulse amplitude discrimination
    in carbon analysis, 318, 321
    in spectrograph operation, 314
Pulse height analyzer
    in determining contamination, 418, 420
    in emission spectrography, 396
    oxide cathode study, 186
Pulse height discrimination
    measuring elements in oil, 404
    with soft X-ray excitation, 370
Pulse height electronics
    providing scaler readout, 275
Pulse height selection
    in spectrographic analysis, 378, 383
Pyrometer, 255

Q
Q as constant
    in measuring zinc coating, 346
Q-component system
    in formation of alloy structures, 43, 48, 49
Q compound, 43
    phase in Q component system, 49
Quantitative analysis
    of copper and oxide mixtures, 211
    of sodium and magnesium, 382

Quantitative calibration techniques
    in silicon analysis, 387
Quantitative diffraction analysis
    direct-comparison method, 219
Quantitative point analysis
    with microprobe, 291
Quantum intensity
    in electron-probe analysis, 279
Quartz crystal
    in petroleum analysis, 398
Quartz monochromator, 91
    for study of scattering, 30
Quartz monzonite
    in standard preparation, 383
Q-value(s), 1–14, 55
    for Pu_3Ru, 21
    indicating bond character, 53

R
R constant
    in measuring zinc coating, 346
Radial distribution analysis, 18, 19, 24
Radiation-type thermal measurement, 259, 260
Rare earths
    spectrographic analysis, 389–395
Ratemeter readouts
    with microprobe, 293
Ratio method
    in alloy analysis, 337
    of spectrochemical analysis, 352–354
Reactor-ambient temperature
    in aluminum crystal irradiation, 165
Reciprocal lattice, 5, 8, 16, 18
    in interpreting scattering, 29
    points, 32
Reciprocal space, 30, 31, 36
    intersecting nets, 2–11
Recorder chart speeds
    in crystallite size analysis, 196
Reduction cell electrolyte
    in spectrograph operation, 313
Reflection hemisphere, 225, 226
Reflections
    low, medium intensities, 207
Relative valency effect
    in lattice parameter changes, 122
Residual strains
    determined from diffraction lines, 96, 97
Resistance heating
    in X-ray diffraction, 250, 251
Rhenium
    analytical lines, 302
Rhodium
    in resistance heating, 251
Rhombohedral deformation
    in alloy structures, 44
Rock analysis, 375
Rotating anode tube, 31
Rotation axis
    in orientation in polymers, 233, 234

Runge's method, 1, 2, 5
R-values, 2–10

S
Salts
  in diffraction studies, 202, 204
Samarium
  spectrographic analysis, 389–395
Sample container system
  in filter application, 305
Sample excitation efficiency
  in generator study, 263
Sample spinner
  in filter application, 305
Sample surface
  in carbon analysis, 317
Sandenaw's value
  in low-temperature measurement, 183
Scanning speeds
  in line-broadening work, 193
Scanning technique
  in electron-probe analysis, 276
Scattered radiation
  in petroleum analysis, 456
  in solid propellant study, 423
Scattering
  due to air, 29
  factor for copper, 177–184
  parasitic, causes, 28
Scherrer equation
  in crystal size measurement, 193
Schiebold–Sauter camera, 31
Schmidt trigger
  in measuring zinc and aluminum coating, 350
Scintillation counter
  in oxide cathode study, 186
  used as detector in experiment with oil, 404
  detector, 165
Secondary standard
  in spectrochemical analysis, 354
Seeman–Bohlin powder method
  in preprecipitation, 31
Selenium
  analytical lines, 302
  in spectrochemical analysis, 353
Semiquantitative determinations
  of tungsten, 310
Sensing head
  in measuring zinc coating, 349–351
  in on-stream X-ray analysis, 433
Shear modulus
  in atom displacement, 137
Shim brass
  in filter application, 302–308
Siemens curved quartz crystal monochromator
  in cold-work deformation study, 101
Silica
  in mining material analysis, 451, 452
  in standard preparation, 383
  matrix, with nickel foil, 305

Silica (*continued*)
  sample, in filter application, 306, 310
  sample, with shim brass and nickel foil, 302
  with nickel filter, 306
  without filter, 302, 306
Silicates
  in fluorescence analysis, 377–387
Silicon
  fluorescence analysis, 361
  in alloy analysis, 334
  in analysis of cast iron, 316
  in carbon analysis, 318
  in diffraction errors, 152, 155
  in spectrograph analysis, 313, 315
  in three rock standards, 373
  with Ag L excitation, 387
Single crystals
  examined by high-voltage pictures, 158
Sinusoidal plot
  in lattice parameter determination, 153
Size effect theory
  in measuring atomic volume, 134
Size factor
  influencing structure, 51, 59
Slit systems
  in crystallite size analysis, 196, 198
Sodium (see also Na)
  fluorescence analysis, 361–375, 377–387
  analysis, in rock sample, 370
  spectrographic analysis, 378
Solid nonmetallics
  in mining material analysis, 450
Solid oxidizer
  in solid propellant, 422
Solid solutions
  atomic size effects, 121
  nickel, 131
  substitutional, 132
Solid state circuitry
  in on-stream analysis, 434, 435
Soller slits
  in copper wire explosion test, 216
  system, to determine axial divergence, 145–151
Space principle
  compared with connection principle, 50
  in changing character of compounds, 55
  of atom arrangement, 44, 45, 48, 52, 59
Spacing faults
  in cold-work deformation, 105
Spectra
  from synthetic sample, 302
  unwanted, 301
Spectral line breadth, 16
Spectrochemical analysis of alloys, 352–359
Spectrographic analysis
  of rare earths, 389–395
Spectrophotometric methods
  in carbon analysis, 322
SrO
  with BaO, 187

SrO–CaO solid solution
  in triple oxide system study, 189
Stacking fault(s)
  density, as determined from diffraction lines, 96, 97
  in copper–silicon–manganese alloy, 96
  in face-centered cubic metals and alloys, 100
  in study of deformation, 99
Standard photometric curve
  in carbon analysis, 322
Static lattice distortions
  oxygen in titanium, 136
Stationary counter
  in oxygen–titanium experiment, 139
Step-scanning
  in determining diffraction errors, 150
  in microprobe, 294, 295
  in preparation of pole figures, 229
Stereographic projection
  in preparing pole figures, 226
Strain anneal technique
  of growing aluminum crystals, 172, 176
Structure analysis, 1–16
Structure factor
  calculated for titanium, 140
  in distortion measurement, 137, 138
Substitutional disorder, 28
  revealing scattering, 32
Substitutional solid solutions
  in study of lattice parameter changes, 121
Sulfur
  fluorescence analysis, 361
  in alloy analysis, 334
  in analysis of cast iron, 313
  in carbon analysis, 318, 319
  in petroleum analysis, 398
  secondary fluorescence, 382
  spectrograph analysis, 313–317
Surface diffraction
  of large crystals, 158
Symmetry
  high crystallographic, 240
  in orientations distribution, 234
  principle, of atom arrangement, 44, 45, 52, 59
Synthetic standard
  in filter application, 302, 303, 307

T
Tantalum
  analytical lines, 302
  in resistance heating, 251
Tar
  in petroleum analysis, 399
Teflon window
  for fluorine analysis, 371, 372
Temperature factor
  in element structure, 45, 59
  Goldschmidt's rule, 45
Tensile deformation
  in study of cold-work, 96

Terbium
  spectrographic analysis, 389–395
Terminal solid solutions
  of binary transition metals, 121
Tetrachloro–bis–phenol A
  in calibration standard, 419
Tetragonal deformation
  in alloy structures, 44
Tetragonal system
  of polymer crystallization, 238
Tetragonal unit cell, 18, 19, 23
Thermal distortion
  in resistance heating, 251
Thermal expansion
  in resistance heating, 251
Thermal gradients
  leading to elastic stresses, 97
Thermal measurement
  at high temperatures, 251
  radiation-type, 259, 260
Thermocouple
  in ammonium chloride study, 204
  sensing, in powder sample preparation, 259
Thomas–Fermi atomic scattering factor for uranium, 182
Thorium atoms, sites, 92, 94
Thorium oxide
  with microprobe, 300
$ThPd_3$, $ThPd_4$, 91–94
Thulium
  spectrographic analysis, 389–395
Ti K radiation, 366
Tin
  in spectrochemical analysis, 353
Ti_2Ni, $TiNi_3$, 75
Ti–Ni system
  in diffraction studies, 74
TiO_2
  in mining material analysis, 451
  with scanning electron probe, 297
Titanium
  distortion by oxygen atoms, 136
  in alloy analysis, 328–330
  increase of lattice constants, 132, 134
  in experiment with oil, 404
  in spectrochemical analysis, 353–359
  in Ziegler polyethylene, 417
Titanium-chromotropic acid
  in determining catalyst residues, 417
Titanium dioxide
  in calibration standard, 419
Titanium lattice
  with interstitial oxygen, 137
Titration
  in carbon analysis, 322
Triclinic crystals, 5
Triple oxide, 189
Triple oxide systems
  in oxide cathode study, 189
Tungsten
  in high-current operation, 370

Tungsten (*continued*)
  in resistance heating, 251
Tungsten line intensities
  in filter application, 305–310
Tungsten target excitation
  with nickel disc, 405
Tungsten-target tube
  in iron–aluminum study, 339
  with lithium fluoride crystal, 302
Twin fault density
  in cold-work strain, 102

U

UCu₅
  in electron-probe analysis, 282
UO₂
  Bragg maxima, 177–184
U₃O₈
  in mining material analysis, 452
Uniaxial tension
  in loading of specimen, 102
UPd₃, 93
Uranium
  diffuse scattering test, 182
  dioxide, in test of atomic scattering, 182
  in flotation concentrates, 437
  in mining material analysis, 455
  wire, in explosion test, 218
USBS white cast iron standards in spectrograph analysis, 315

V

V₄Al₂₃ structure, 58
Vacuum mode of operations
  in mining material analysis, 450
Vacuum spectrograph
  in fluorescence analysis, 361–363
  with pulse height analyzer, 418
Valence electron factor, 132
Valency effect
  in the noble metals, 132
Vanadium
  analysis of carbon materials, 313, 317
  dissolved as a naphthenate, 404
  increase of lattice constants, 132
Vanadium atoms
  in Kasper polyhedra form, 58
Vand's method, 3, 5
Vegard's rule
  in evaluating isotopes, 247
  on atomic size, 123
Volatiles
  determination of, 286

W

Warren's investigation
  of atomic scattering, 182
Warren's TDS
  of Cu₂O and UO₂, 178
Warren's theory
  for close-packed cubic powders, 178

Water cooling
  in resistance heating, 253
  in spectrographic analysis, 379
Wet-chemical results
  in alloy analysis, 354, 358
Wilson's analysis
  of diffractometer error, 148
WO₃
  in silica matrix, 309
Work coil
  in resistance heating, 252, 253

X

Xenon
  in low-temperature test, 242, 245
X-ray analysis, quantitative, 211, 214
X-ray analyzing crystals
  in diffraction and emission, 436
X-ray fluorescence
  absorption effects, 403
  as affected by grain size, 437
  in analyzing mining materials, 447–458
  in determination of aluminum, chlorine, iron, titanium (in polyolefins), 417–421
  with petroleum products, 396, 401, 402
X-ray quantum counter
  in measuring orientation in polymers, 236
X-ray scattering
  to evaluate atomic size, 122
X-ray spectrography
  application of filters, 301–311
  for silicates analysis, 387
X-ray spectrograph
  automatic multielement, 396, 401, 402
X-ray spectrographic techniques
  for analysis of raw materials, 429
X-ray transmission
  in dry air and helium, 85
X-ray tube output
  in generator study, 262
X-ray tube target materials
  in spectrographic analysis, 378
XRD-5 X-ray spectrograph, 353

Y

YFe garnet powders
  rare earth analysis, 389–395
Young's modulus
  in determining fracture strength, 103, 106
Ytterbium
  spectrographic analysis, 389–395
YXₙ groupings
  in structure arrangement, 46, 47

Z

Zero tungsten concentrations
  in filter application, 310
Zinc
  area density, 346
  as naphthenate, 404

Zinc (*continued*)
   coating on steel, 345–351
   in filter application, 308
   in mill feeds, concentrates, middlings and
      tailings, 429–435
   with microprobe, 300
Zinc emission line
   measured by fluorescence, 345
Zinc emission method
   of measuring zinc coating, 347
Zinc intensity ratio
   in measuring zinc coating, 348
Zinc ore benefication plant
   using on-stream analyzer, 429
Zintl line
   in examination of crystal structure, 26, 44,
      45, 48
Zintl phases
   of bond formation, electrochemical factor,
      structure types, 54–56
Zircaloy
   in diffraction study, 62
Zirconium
   in compounding alloys, 62
   with microprobe, 300

Zirconium block
   in on-stream analysis, 431–435
Zirconium-gettered argon atmosphere
   in study of thorium and palladium, 91
Zirconium intensity
   in on-stream analysis, 431
Zirconium system
   in study of selected alloys, 62, 64
Zirconium radiation absorption
   in monitoring pulp density, 435
$ZnCu_2Al$
   as crystallizing compound, 49
Zn K_α area scan
   with microprobe, 300
Zone detection method, 2, 5
Zone-matrix interface, 27
Zr–Fe alloy powders, 70
Zr–Fe phase diagram
   proposed revision, 62
$ZrFe_2$, Zr_2Fe_3, 65, 69
Zr_4Fe, Zr_2Fe
   in study of selected alloys, 62–69
Zsoldos' method, 3, 5
Zytel holder
   in carbon analysis, 320